Wind Turbine Aerodynamics

Wind Turbine Aerodynamics

Special Issue Editor

Wen Zhong Shen

MDPI • Basel • Beijing • Wuhan • Barcelona • Belgrade

Special Issue Editor
Wen Zhong Shen
Technical University of Denmark
Denmark

Editorial Office
MDPI
St. Alban-Anlage 66
4052 Basel, Switzerland

This is a reprint of articles from the Special Issue published online in the open access journal *Applied Sciences* (ISSN 2076-3417) from 2018 to 2019 (available at: https://www.mdpi.com/journal/applsci/special_issues/wind_turbine_aerodynamics)

For citation purposes, cite each article independently as indicated on the article page online and as indicated below:

LastName, A.A.; LastName, B.B.; LastName, C.C. Article Title. *Journal Name* **Year**, *Article Number*, Page Range.

ISBN 978-3-03921-524-9 (Pbk)
ISBN 978-3-03921-525-6 (PDF)

Cover image courtesy of Wen Zhong Shen.

Contents

About the Special Issue Editor

Wen Zhong Shen (Dr) is Professor in Computational Aero-Acoustics and Wind Energy at the Department of Wind Energy, Technical University of Denmark (DTU) since 2014. He obtained his bachelor's degree at Wuhan University, China, in 1988, and master's and Ph.D. degrees in fluid mechanics at Paris-Sud University, France, in 1989 and 1993, respectively. His research areas cover the fields of wind turbine aerodynamics, aero-acoustics, computational fluid dynamics, wind turbine airfoil/rotor design, and wind farm optimization.

Editorial

Special Issue on Wind Turbine Aerodynamics

Wen Zhong Shen

Department of Wind Energy, Technical University of Denmark, Nils Koppels Alle, Building 403, 2800 Lyngby, Denmark; wzsh@dtu.dk

Received: 20 April 2019; Accepted: 22 April 2019; Published: 26 April 2019

1. Introduction

In order to reach the goal of 100% renewable energy in energy systems, wind energy, as a pioneer of renewable energy, is developing very quickly all over the world. To reduce the levelized cost of energy (LCOE), the size of a single wind turbine has been increased to 12 MW nowadays and it will increase further in the near future. Big wind turbines and their associated wind farms have many advantages but also challenges in aerodynamics, aero-elasticity and aeroacoustics. The typical effects are mainly related to the increase in Reynolds number and blade flexibility. This Special Issue collects some important works addressing the aerodynamic challenges appearing in such development. Aerodynamics of wind turbines is a classic concept and is the key for wind energy development as all other parts rely on the accuracy of its aerodynamic models. There are numerous books and articles dealing with wind turbine aerodynamic problems and models. As a good example, the wind energy handbook by Burton et al. [1] gives an overview of wind turbine aerodynamics and its related problems. There are also several special issues on wind turbine aerodynamics. This author edited a special issue on aerodynamics of offshore wind energy systems and wakes [2] in 2014, which collected state-of-the-art research articles on the development of offshore wind energy.

2. Current Status in Wind Turbine Aerodynamics

In the context introduced above, this special issue collects the latest research articles on various topics related to wind turbine aerodynamics, which includes atmospheric turbulent flow, wind turbine flow modeling, wind turbine design, wind turbine control, wind farm flow in complex terrain, wind turbine noise modeling, vertical axis wind turbine, and offshore wind energy. A summary of the collected papers is given below in the order mentioned above.

There are 4 papers dealing with atmospheric turbulent flow. Olivares-Espinosa et al. [3] addressed an assessment of turbulence modeling in the wake of an actuator disc with a decaying turbulence inflow and developed a method to replicate wake measurements obtained in a decaying homogeneous turbulence inflow produced by a wind tunnel. Ehrich et al. [4] compared the blade element momentum theory, actuator line method, and computational fluid dynamics in the case of a wind turbine under turbulence inflow by checking its sectional and integral forces in terms of mean, standard deviation, power spectral density and fatigue loads. As wind flow measurements are often carried out at low heights, Xu et al. [5] evaluated the power-law wind-speed extrapolation method for flows over different terrain types and a new wind-speed extrapolation method based on atmospheric stability classification methods was developed. Schaffarczyk and Jeromin [6] analyzed the measured high-frequency atmospheric turbulence and its impact on the boundary layer of wind turbine blades, and found that the stability state in the atmospheric boundary does not seem to depend on simple properties, but on higher statistical properties, such as shape factors.

There are 4 papers dealing with wind turbine simulation modeling. Xu et al. [7] developed a simplified free vortex wake model of wind turbines using vortex sheet for near wake and using ring wake for far wake. Zhu and Wang [8] studied dynamic stall phenomenon and rotational augmentation under pitch oscillation and oscillating freestream on wind turbine airfoil and blade using unsteady Reynolds averaged Navier–Stokes equations. Zhong et al. [9] analyzed the wind turbine stall prediction using Reynolds averaged Navier–Stokes simulations by turbulence coefficient calibration and found that the primary reason for the inaccuracy of rotor simulations is not rotational effects, but a turbulence-related modeling problem. Li et al. [10] conducted a study on the aerodynamic performance of a wind turbine airfoil DU 91-W2-250 under dynamic stall and found that an increased-reduced frequency leads to a lower aerodynamic efficiency during the upstroke process of pitching motions.

There are also 4 papers dealing with wind turbine design. Wu et al. [11] presented a framework to optimize the design of large-scale wind turbines by using different design objectives, such as levelized cost of energy, net present value, internal rate of return, and discounted payback time, and found that the blade obtained with economic optimization objectives has a much large relative thickness and smaller chord distributions than obtained with high aerodynamic performance design. Yang et al. [12] investigated the design of wind turbine in low wind speed areas by considering both blade length and hub height. Vorspel et al. [13] developed an optimization tool for rotor blades using bend-twist-coupling which allows the computation of gradients based on flow field at low cost. Cao et al. [14] developed a computational fluid dynamics/actuator disc-based wind turbine rotor optimization tool by considering wake effects and letting different designs for upstream and downstream turbines.

Concerning wind turbine control, Sanati et al. [15] investigated condition monitoring of wind turbine blades using both active and passive thermography in order to find the failure of wind turbine blades. Different image processing algorithms were used to increase the thermal contrasts of subsurface defects in thermal images obtained by active thermography, while a method of step phase and amplitude thermography was used for passive thermography.

There are 2 papers dealing with wind farm flow and layout optimization in complex terrain. Sessarego et al. [16] simulated wind turbine/farm flows in complex terrain using both Reynolds averaged Navier–Stokes with an actuator disc model and large eddy simulation with actuator line model and results were compared with detailed field measurements from two met-masts and SCADA (supervisory control and data acquisition) data. Feng et al. [17] developed an optimization framework for wind farm layout optimization in complex terrain, which employs a CFD (Computational Fluid Dynamics) wind resource assessment tool, an engineering wake model adapted in complex terrain and an advanced wind farm layout optimization algorithm, and it was found that the framework can provide a better layout than the original layout.

On wind turbine noise modeling, Sun et al. [18] developed an efficient numerical method for wind turbine noise modeling under multi-wake conditions. This model can predict simultaneously wind turbine flow and wake, wind turbine noise source and wind turbine noise propagation under different atmospheric conditions.

There are 2 papers dealing with the development of vertical axis wind turbines. Li et al. [19] designed a Lanzhou University of Technology (LUT) airfoil for straight-bladed vertical axis wind turbines with an objective of maximizing its aerodynamic performance. Zhao et al. [20] introduced a variable pitch approach for improving the performance of straight-bladed vertical axis wind turbine at rated tip-speed-ratio, and it was found that the new variable pitch approach can achieve an 18.9% growth of the peak power coefficient of the vertical axis wind turbine at the optimum tip-speed-ratio.

Offshore wind energy is a growing topic and this special issue collected 3 papers on this topic. Zhang and Kim [21] developed a fully coupled computational fluid dynamics method to analyze a semi-submersible floating offshore wind turbine under wind-wave excitation conditions and validated the model by inputting gross system parameters supported in the offshore code comparison, collaboration, continued with correlations project. Ke et al. [22] analyzed aerodynamic force and mechanical performance of a large wind turbine during typhoons using a weather research forecasting and computational fluid dynamics nesting method and found that typhoons increased the aerodynamic force and structure responses and decrease the overall buckling stability and ultimate bearing capacity of a 5 MW wind turbine. Guo et al. [23] studied the aerodynamic and motion performance of a 5 MW

H-type floating vertical axis wind turbine and it was found that the H-type floating VAWT (vertical axis wind turbine) has a better motion performance and the mean values of surge, heave and pitch of the H-type floating VAWT are smaller than those of the Φ-type floating VAWT.

3. Future Research Need

Although this special issue has been closed, more in-depth research in wind turbine aerodynamics is expected as the goal of wind energy research is to help the technological development of new environmental friendly and cost-effective wind energy systems.

Funding: This research received no external funding.

Acknowledgments: This special issue would not be possible without the contributions of various talented authors, hardworking and professional reviewers, and the dedicated editorial team of Applied Sciences. Congratulations to all the authors. I would like to take this opportunity to record my sincere gratefulness to all the reviewers. Finally, I place on record my gratitude to the editorial team of Applied Sciences, and special thanks to Nicole Lian, Assistant Managing Editor from MDPI Branch Office, Beijing.

Conflicts of Interest: The author declares no conflict of interest.

References

1. Burton, T.; Jenkins, N.; Sharpe, D.; Bossanyi, E. *Wind Energy Handbook*, 2nd ed.; John Wiley & Sons, Ltd.: Hoboken, NJ, USA, 2011.
2. Shen, W.Z.; Sørensen, J.N. Special Issue on Aerodynamics of Offshore Wind Energy Systems and Wakes. *Renew. Energy* **2014**, *70*, 1–2. [CrossRef]
3. Olivares-Espinosa, H.; Breton, S.P.; Nilsson, K.; Masson, C.; Dufresne, L.; Ivanell, S. Assessment of Turbulence Modelling in the Wake of an Actuator Disk with a Decaying Turbulence Inflow. *Appl. Sci.* **2018**, *8*, 1530. [CrossRef]
4. Ehrich, S.; Schwarz, C.M.; Rahimi, H.; Stoevesandt, B.; Peinke, J. Comparison of the Blade Element Momentum Theory with Computational Fluid Dynamics for Wind Turbine Simulations in Turbulent Inflow. *Appl. Sci.* **2018**, *8*, 2513. [CrossRef]
5. Xu, C.; Hao, C.; Li, L.; Han, X.; Xue, X.; Sun, M.; Shen, W. Evaluation of the Power-Law Wind-Speed Extrapolation Method with Atmospheric Stability Classification Methods for Flows over Different Terrain Types. *Appl. Sci.* **2018**, *8*, 1429. [CrossRef]
6. Schaffarczyk, A.P.; Jeromin, A. Measurements of High-Frequency Atmospheric Turbulence and Its Impact on the Boundary Layer of Wind Turbine Blades. *Appl. Sci.* **2018**, *8*, 1417. [CrossRef]
7. Xu, B.; Wang, T.; Yuan, Y.; Zhao, Z.; Liu, H. A Simplified Free Vortex Wake Model of Wind Turbines for Axial Steady Conditions. *Appl. Sci.* **2018**, *8*, 866. [CrossRef]
8. Zhu, C.; Wang, T. Comparative Study of Dynamic Stall under Pitch Oscillation and Oscillating Freestream on Wind Turbine Airfoil and Blade. *Appl. Sci.* **2018**, *8*, 1242. [CrossRef]
9. Zhong, W.; Tang, H.; Wang, T.; Zhu, C. Accurate RANS Simulation of Wind Turbine Stall by Turbulence Coefficient Calibration. *Appl. Sci.* **2018**, *8*, 1444. [CrossRef]
10. Li, S.; Zhang, L.; Yang, K.; Xu, J.; Li, X. Aerodynamic Performance of Wind Turbine Airfoil DU 91-W2-250 under Dynamic Stall. *Appl. Sci.* **2018**, *8*, 1111. [CrossRef]
11. Wu, J.; Wang, T.; Wang, L.; Zhao, N. Impact of Economic Indicators on the Integrated Design of Wind Turbine Systems. *Appl. Sci.* **2018**, *8*, 1668. [CrossRef]
12. Yang, H.; Chen, J.; Pang, X. Wind Turbine Optimization for Minimum Cost of Energy in Low Wind Speed Areas Considering Blade Length and Hub Height. *Appl. Sci.* **2018**, *8*, 1202. [CrossRef]
13. Vorspel, L.; Stoevesandt, B.; Peinke, J. Optimize Rotating Wind Energy Rotor Blades Using the Adjoint Approach. *Appl. Sci.* **2018**, *8*, 1112. [CrossRef]
14. Cao, J.; Zhu, W.; Shen, W.; Sørensen, J.N.; Wang, T. Development of a CFD-Based Wind Turbine Rotor Optimization Tool in Considering Wake Effects. *Appl. Sci.* **2018**, *8*, 1056. [CrossRef]
15. Sanati, H.; Wood, D.; Sun, Q. Condition Monitoring of Wind Turbine Blades Using Active and Passive Thermography. *Appl. Sci.* **2018**, *8*, 2004. [CrossRef]

16. Sessarego, M.; Shen, W.Z.; Van der Laan, M.P.; Hansen, K.S.; Zhu, W.J. CFD Simulations of Flows in a Wind Farm in Complex Terrain and Comparisons to Measurements. *Appl. Sci.* **2018**, *8*, 788. [CrossRef]
17. Feng, J.; Shen, W.Z.; Li, Y. An Optimization Framework for Wind Farm Design in Complex Terrain. *Appl. Sci.* **2018**, *8*, 2053. [CrossRef]
18. Sun, Z.; Zhu, W.J.; Shen, W.Z.; Barlas, E.; Sørensen, J.N.; Cao, J.; Yang, H. Development of an Efficient Numerical Method for Wind Turbine Flow, Sound Generation, and Propagation under Multi-Wake Conditions. *Appl. Sci.* **2019**, *9*, 100. [CrossRef]
19. Li, S.; Li, Y.; Yang, C.; Zhang, X.; Wang, Q.; Li, D.; Zhong, W.; Wang, T. Design and Testing of a LUT Airfoil for Straight-Bladed Vertical Axis Wind Turbines. *Appl. Sci.* **2018**, *8*, 2266. [CrossRef]
20. Zhao, Z.; Wang, R.; Shen, W.; Wang, T.; Xu, B.; Zheng, Y.; Qian, S. Variable Pitch Approach for Performance Improving of Straight-Bladed VAWT at Rated Tip Speed Ratio. *Appl. Sci.* **2018**, *8*, 957. [CrossRef]
21. Zhang, Y.; Kim, B. A Fully Coupled Computational Fluid Dynamics Method for Analysis of Semi-Submersible Floating Offshore Wind Turbines Under Wind-Wave Excitation Conditions Based on OC5 Data. *Appl. Sci.* **2018**, *8*, 2314. [CrossRef]
22. Ke, S.; Yu, W.; Cao, J.; Wang, T. Aerodynamic Force and Comprehensive Mechanical Performance of a Large Wind Turbine during a Typhoon Based on WRF/CFD Nesting. *Appl. Sci.* **2018**, *8*, 1982. [CrossRef]
23. Guo, Y.; Liu, L.; Gao, X.; Xu, X. Aerodynamics and Motion Performance of the H-Type Floating Vertical Axis Wind Turbine. *Appl. Sci.* **2018**, *8*, 262. [CrossRef]

Article

Assessment of Turbulence Modelling in the Wake of an Actuator Disk with a Decaying Turbulence Inflow

Hugo Olivares-Espinosa [1,2,*], **Simon-Philippe Breton** [2,3], **Karl Nilsson** [2], **Christian Masson** [1], **Louis Dufresne** [1] **and Stefan Ivanell** [2]

[1] Department of Mechanical Engineering, École de Technologie Supérieure, 1100 Notre-Dame Ouest, Montréal, QC H3C 1K3, Canada; Christian.Masson@etsmtl.ca (C.M.); Louis.Dufresne@etsmtl.ca (L.D.)
[2] Wind Energy Section, Department of Earth Sciences, Uppsala University Campus Gotland, Cramérgatan 3, 62165 Visby, Sweden; Simon-Philippe.Breton@geo.uu.se (S.-P.B.); karl.nilsson@geo.uu.se (K.N.); stefan.ivanell@geo.uu.se (S.I.)
[3] Nergica, 70 rue Bolduc, Gaspé, QC G4X 1G2, Canada
[*] Correspondence: hugo.olivares@geo.uu.se

Received: 21 June 2018; Accepted: 27 August 2018; Published: 1 September 2018

Abstract: The characteristics of the turbulence field in the wake produced by a wind turbine model are studied. To this aim, a methodology is developed and applied to replicate wake measurements obtained in a decaying homogeneous turbulence inflow produced by a wind tunnel. In this method, a synthetic turbulence field is generated to be employed as an inflow of Large-Eddy Simulations performed to model the flow development of the decaying turbulence as well as the wake flow behind an actuator disk. The implementation is carried out on the OpenFOAM platform, resembling a well-documented procedure used for wake flow simulations. The proposed methodology is validated by comparing with experimental results, for two levels of turbulence at inflow and disks with two different porosities. It is found that mean velocities and turbulent kinetic energy behind the disk are well estimated. The development of turbulence lengthscales behind the disk resembles what is observed in the free flow, predicting the ambient turbulence lengthscales to dominate across the wake, with little effect of shear from the wake envelope. However, observations of the power spectra confirm that shear yields a boost to the turbulence energy within the wake noticeable only in the low turbulence case. The results obtained show that the present implementation can successfully be used in the modelling and analysis of turbulence in wake flows.

Keywords: wind turbine wakes; turbulence; actuator disk; LES; wind tunnel; OpenFOAM

---

1. Introduction

Studies of turbulence in the wake of wind turbines are often made with the aim of analyzing the influence of the flow and rotor models in the reproduction of wake characteristics. Investigations with various rotor models in the Atmospheric Boundary Layer (ABL) have been made either with the goal of improving the production efficiency of a cluster of turbines (e.g., [1–4]) or aimed at comparing the characteristics of wakes with respect to measurements of real or downscaled turbines (e.g., [5–8]). When numerical simulations of large wind parks are made, computational limitations oblige employing simpler rotor models that, while numerically less expensive, are requested to produce a minimum level of detail in wake features that yields an acceptable reproduction of the interaction of the ensuing wakes and the downstream wind turbines, as well as other wakes. Amongst the simplest formulations of the rotor model is the Actuator Disk (AD) [9,10] which reproduces the effect of the rotor in the incoming flow by means of a permeable surface in the shape of a disk that acts as a momentum sink. In its most basic conception, the AD constitutes a one-dimensional force opposite to the flow, perpendicular to the rotor plane, with no wake rotation or airfoil properties considered. Different studies [8,11,12] have shown the capability of the AD technique to approximate the characteristics of a real turbine wake within its far region. An experimental investigation of the wakes produced by porous disks has been performed by [13,14], where the disks are made of metallic meshes representing different solidities. In recent work, similar research has been made of wakes produced by a decaying turbulence inflow [15,16]. Unlike the ABL, no shear-produced turbulence occurs, greatly simplifying the modelling of the flow physics and permitting to isolate the features of the wake from those deriving from the inflow (e.g., the vertical momentum flux in the ABL). With data collected in a related measurement campaign, the authors of [17] employed Reynolds-Averaged Navier–Stokes (RANS) models to reproduce measurements in the wakes produced by the porous disks. Although good results are obtained, the experimental study represents an opportunity to perform comparisons with numerical simulations that allow a greater detail in the reproduction of the turbulence characteristics. Therefore, the reproduction of these experiments employing Large-Eddy Simulations (LES) seems appealing.

To reproduce the inflow properties measured experimentally, first it is necessary to model the flow of decaying turbulence produced in a wind tunnel. In this regard, different works have been dedicated to investigate various methodologies to produce adequate inlet conditions [18–20]. Amongst the different techniques, the method developed by Mann [21–23] to create a synthetic turbulence field has often been used to create inflow conditions for simulations in ABL [5,6,24,25] as well as in homogeneous turbulence [26–28]. In these works, it has been shown that the Mann method is capable of producing realistic turbulence fields, resembling the second order statistics of real turbulence [5,29]. This algorithm permits creating synthetic fields of homogeneous turbulence by prescribing two parameters, the turbulence lengthscale and (albeit indirectly) the turbulence intensity. An anisotropy factor is also used in the algorithm to create boundary layer fields. The transition and evolution of turbulence characteristics when synthetic fields are introduced in LES domains, especially of integral lengthscales, has been previously studied by [27] for homogeneous turbulence and recently by [24,30] (using the turbulence generation method of [20]) in sheared flows.

The objective of this work is the study of turbulence characteristics in the wake flow of a wind turbine model. To fulfill this purpose, a methodology is developed to replicate: (1) the turbulence characteristics of a homogeneous wind tunnel flow and (2) the wake field arising from the introduction of porous disks representing the wind turbine. This procedure is employed to reproduce the inflow and wake characteristics measured in the experimental campaign carried out by G. Espana and S. Aubrun [17,31], so the comparison with wind tunnel data serves as method of validation. The study of diverse features of the turbulence field both in the free decaying flow and in the wake is presented next to such comparison. The computational platform employed is OpenFOAM® v.2.1 [32,33] (the OpenFOAM Foundation Ltd., London, UK) an open-source code amply used in flow simulations, chosen for its availability and access to apply ad-hoc modifications to existing solvers and utilities. While the synthetic turbulence is created with the Mann method, the flow is reproduced using LES computations. The methodology implemented follows—in part—a procedure developed on the CFD software EllipSys3D [34–36] (Department of Mechanical Engineering and RisøCampus, DTU, Lyngby, Denmark) to simulate wake flows with turbulent inflows [5]. It has been shown that this method provides good results to introduce ABL as well as homogeneous turbulence conditions [5,6,24–28]. By using this approach, we expect to assess: (a) how well the main turbulence characteristics can be reproduced (e.g., turbulence intensity and integral lengthscale) in the inflow and in the wake with LES, (b) how the main turbulence characteristics in the wind tunnel change due to the presence of the wind turbine model and (c) how the LES modelling change along the wake compared to the undisturbed flow (resolved/modelled velocity fluctuations). It should be noted that these questions are formulated in the context of a *limited mesh resolution*, which makes it more relevant for the wind energy field since it is often desired to minimize the computational requirements while successfully reproducing the requested flow features, which in this case consists mainly of the integral lengthscale.

The work presented here is organized as follows: a brief theoretical background is provided about the homogenenous isotropic turbulence and its experimental approximation, the decaying homogeneous turbulence, next to a description of the measurement campaign and data to be reproduced. Later, the methodology is described in detail, with special attention to the generation of turbulence inflow and the reproduction of the adequate characteristics in the absence of disk. This also comprises the description of the methodology employed to introduce the synthetic turbulence in the LES domain. Next, the results are presented and discussed for the different quantities computed in the turbulence field of the wakes. A final section presents a summary and the conclusions of the work.

2. Homogeneous Turbulence

For a given flow, the instantaneous velocity vector is referred to as u_i whose components in the streamwise, vertical and spanwise directions (x, y, z) are $u_i = (u, v, w)$. The Reynolds decomposition in the streamwise direction is defined as $u = \langle U \rangle + u'$, where "$\langle \cdot \rangle$" denotes time-average (capital letters are also used to emphasize averaged magnitudes). The characteristic size of the largest eddies is identified as the distance L required to nullify the correlation function $\mathcal{R}_{ij}(\mathbf{r}, t)$. With this assumption, the integral lengthscale

$$L_{ij}^{(d)} = \int_0^\infty \mathcal{R}_{ij}(\mathbf{e}_d r, t)\, dr \tag{1}$$

in the direction d (set by the unitary vector $\mathbf{e}$) is defined [37]. From all scales defined by this expression, those most commonly used are the longitudinal integral lengthscale $L_1 = L_{11}^{(1)}$ as well as the transversal one $L_2 = L_{22}^{(1)}$. Similarly, the longitudinal Taylor lengthscale (or micro-scale) $\lambda_1 = \lambda_{11}^{(1)}$ is defined by the osculating parabola to the correlation function $\mathcal{R}_{ij}(\mathbf{r}, t)$. If isotropy is assumed (or at least between the 1 and 2 directions) the equivalences $L_{22}^{(1)} = L_{11}^{(2)}$ and $\lambda_{22}^{(1)} = \lambda_{11}^{(2)}$ are also valid.

In the absence of shear, the Taylor hypothesis of frozen turbulence can be adopted more comfortably. This is, it is assumed that the turbulence field does not change as it is convected by the mean wind at $\langle U \rangle$, which yields the equivalence between the spatial and temporal correlations. In this way, correlations can be made from the time series of each velocity component. In particular, the autocorrelation will provide the integral time scales $\mathcal{T}_{11}$ and $\mathcal{T}_{22}$ from where the integral lengthscales can be computed by means of $L_1 = \langle U \rangle \mathcal{T}_{11}$ and $L_2 = \langle U \rangle \mathcal{T}_{22}$. Likewise, the longitudinal Taylor lengthscale can be calculated from the expression

$$\frac{1}{\lambda_1^2} = \frac{\langle U \rangle^{-2}}{2 \left\langle u_1'^2 \right\rangle} \left\langle \left(\frac{\partial u_1'}{\partial t} \right)^2 \right\rangle \tag{2}$$

as seen in [38] (note that Equation (2) differs from the one presented in that report by a factor of $\sqrt{2}$). Considering the turbulent kinetic energy of a field $k = \frac{1}{2}\left(\langle u_1'^2 \rangle + \langle v_2'^2 \rangle + \langle w_3'^2 \rangle \right)$ and the assumption of isotropy, λ_2 is related to the amount of dissipation of k by

$$\varepsilon = \frac{15\nu \left\langle u_1'^2 \right\rangle}{\lambda_2^2} = \frac{30\nu \left\langle u_1'^2 \right\rangle}{\lambda_1^2}\ . \tag{3}$$

When grid turbulence is used to approximate the theoretical case of decaying isotropic turbulence, the characteristics observed at different positions downstream from the grid correspond to the time evolution of isotropic turbulence with zero mean velocity. Thus, a decay during the interval Δt is approximated by that occurring within Δx in a wind tunnel. In this manner, it has been observed that the decay of k follows the expression

$$\frac{k}{\langle U \rangle^2} = c_A \left(\frac{x - x_0}{M} \right)^{-n}, \tag{4}$$

where M is the turbulence grid size, c_A (also written as $1/A$) is a fitting parameter, n is the decay exponent and x_0 a virtual origin. Equation (4) is commonly employed to track the streamwise turbulence intensity decay, replacing k for $\left\langle u_1'^2 \right\rangle$. While Ref. [37] mentions $1.1 \leq n \leq 1.3$ and $c_A \simeq 1/30$, Ref. [39] reports to have observed $n = 1.25$ with $x_0 = 0$, whereas Ref. [40] mentions $1.15 \leq n \leq 1.45$, remarking that c_A varies greatly depending on the geometry of the grid and Re_λ. Ref. [37] indicates that the integral lengthscale evolves downstream according to

$$L_2 \simeq c_{B_1} M \left[\frac{x - x_0}{M} \right]^{n_1} \tag{5}$$

with $c_{B_1} = 0.06$ and $n_1 = 0.35$.

3. Experimental Setup and Measurement Campaigns

The measurements used in this work come from experiments performed in the Eiffel-type wind tunnel of the Prisme laboratory of the University of Orléans. The test section has a width and a height of 0.5 m and a length of 2 m. Two different grids were used to generate turbulence at the entrance of the test section, resulting in two different turbulence intensities. At a distance of $x = 0.5$ m from that grid, the reported values of turbulence intensity and integral lengthscale (measured at the centreline) were TI = 3% and $L_1 = 0.01$ m as well as TI = 12% and $L_1 = 0.03$ m. These two inflow cases are henceforward identified as Ti3 and Ti12, respectively. The streamwise position where these values are reported is referred to as the *target position* x_D.

Later, disks made of a metallic mesh were located at x_D to simulate the effect of the AD model (a porous surface) on the flow. Two disks were used, each with a diameter of $D = 0.1$ m but made with a different wire to produce different induction factors. The thrust coefficient C_T of each disk is calculated following the procedure presented by [13] and revisited by [17], based on the measurement of the velocity deficit in the wake. In total, six experimental cases are considered for this work as shown in Table 1. Complete details about the experimental setup, the measurement techniques as well as the characteristics of the flows generated by this wind tunnel can be found in [15,31].

Table 1. Reference parameters of flow and disks used in the experiments.

TI [%]	L_1 [m]	Case
		No-disk
3	0.01	$C_T = 0.42$
		$C_T = 0.62$
		No-disk
12	0.03	$C_T = 0.45$
		$C_T = 0.71$

The data used in this work was obtained using two different techniques. Firstly, with the aim of obtaining time-series of the flow velocity, a Hot-Wire Anemometry (HWA) probe was located along vertical lines at $x = 3D$, $4D$ and $6D$ from the disk centre (the origin of the reference system $x, y, z = 0, 0, 0$ is set there). The probe moved along each vertical line between $0 \leq y/D \leq 1.5$, registering data at different steps (data recorded every $0.1D$ is employed). A scheme of the measuring locations with respect to the experimental arrangement is shown in Figure 1. At each probe position, data was acquired with a sampling frequency of $f_{acq} = 2$ kHz during about 1 min. A low-pass filter was also used, with a cut-off frequency fixed at $f_c = 1$ kHz. The reference velocity during these measurements was $U_\infty = 3$ m/s. Of the measurements made with this technique, only the database corresponding to the cases of Ti12 is used in the comparisons shown here since the sampling rate was assessed to be too low for the turbulence scales of the Ti3 case. Due to this, HWA measurements from [16] are used to complement the experimental data for the comparison of the Ti3 case. These were made using the same experimental setup as the other HWA measurements, with TI $\simeq$ 3% also at the target position but with $U_\infty = 20$ m/s and in consequence a higher Reynolds number.

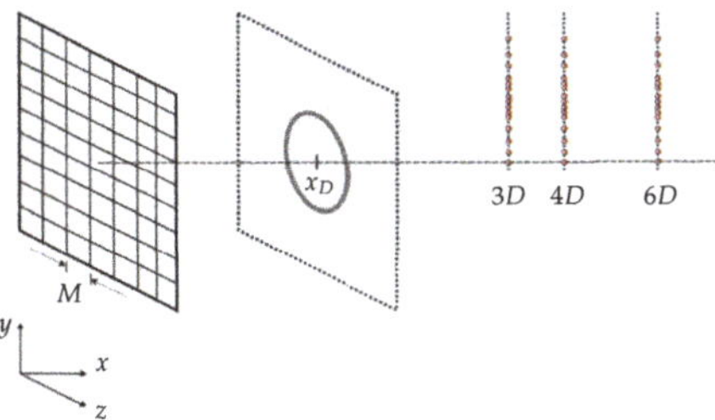

Figure 1. Representation of the measurement positions of the hot-wire. The turbulence is generated by a grid of spacing M. The reported values of TI are measured at $x = 0.5$ m from such grid, where the ADs are subsequently located. This position is referred to as x_D. Time-series of the velocity are recorded at various positions along vertical lines at 3D, 4D and 6D.

Secondly, a Laser-Doppler-Anemometer (LDA) was used to simultaneously measure two components of velocity (u, v) with the main purpose of obtaining time-averaged information of the wake. Measurements behind the disks were made along the vertical directions at $x = 2D$, 4D, 6D, 8D and 10D from the disk centre. The recording positions were aligned in the vertical direction, performed in steps (generally) of $0.1D$ between $-1.5 \leq y/D \leq 1.5$. For the Ti3 cases, the positions in the vertical direction where data is available vary slightly, but most of them are made in steps of $0.1D$ between $-1.0 \leq y/D \leq 1.0$. Measurements were made using a non-uniform sampling frequency, with an average of 1 KHz during 90 s. The reference velocity was $U_\infty = 10$ m/s for the cases Ti3 and $U_\infty = 6$ m/s for Ti12. As it was shown by [41] and later work, various estimations in grid-generated turbulence can be considered Reynolds independent (but not for observations such as the scaling region of the spectrum, as shown by [42]. Therefore, non-dimensional results of mean velocities and root-mean-square (RMS) statistics obtained with the LDA technique will be used despite the differences in reference velocity, set in the simulations to $U_\infty = 3$ m/s. LDA data are used to compare with the obtained results of velocity deficit and k, while the velocity time-series from HWA are used to compute the other quantities examined in this work.

4. Model Description

4.1. Numerical Model

In order to reproduce the measured characteristics of the wake field, an LES model is employed. This allows for resolving the large (energy-containing) motions, whereas the effects of the smaller eddies are modelled. To achieve this, the Navier–Stokes equations are decomposed into a filtered (or resolved) component and a residual (or subgrid scale, SGS) component. The classic Smagorinsky model [43] is used for the parametrization of the residual scales. There, the subgrid viscosity ν_{SGS} is assumed to be proportional to the Smagorinsky coefficient C_s and the local cell length $\Delta = (\Delta_x \Delta_y \Delta_z)^{1/3}$ and the resolved strain-rate $\overline{S}_{ij}$ as $\nu_{\text{SGS}} = (C_s \Delta)^2 \sqrt{2\overline{S}_{ij}\overline{S}_{ij}}$. The value of C_S is set to 0.168, which comes from adjustments made to reproduce decaying homogeneous turbulence [44]. This model is chosen for its ubiquity and because the absence of its best known disadvantage, associated to bounded flows: overdissipation close to walls [40,45]. The interpolation scheme for the convective term consists of a dynamic blend where, depending on the velocity flux and the magnitude of the velocity gradients at the cell faces, an amount of up to 20% upwind is used in combination with the linear interpolation. This scheme is called *filteredLinear* within OpenFOAM v2.1 . In this way, the upwind part is employed only in regions of steep velocity gradients while the flow maintains the second-order accuracy of the linear scheme elsewhere. The PISO algorithm is employed for the solution of the pressure–velocity equations.

4.2. Computational Domain and Grid Resolution

The dimensions of the computational domain are set to imitate those of the measuring region in the wind tunnel. The domain and grid sizes of the LES computations as well as of the synthetic velocity field, identified as *turbulence box* (use as inflow) are listed in Table 2. As in the experiments, x_D corresponds to the origin of the coordinate system for the computations, at the centre of the crosswise y–z plane and at $5D$ from the inlet. The reason to imitate the dimensions of the experiment, in particular in the crosswise directions, is to reproduce the potential effects of blockage on the wake development. A small blockage of 1.3% in average has been reported for measurements in this wind tunnel [17].

The grid size is determined by the optimum number of cells per L, or in fact L_1. Unlike the ABL, where L_1 is typically two to three times the diameter of the rotor, the turbulence grids used in the wind tunnel produce turbulence with an eddy size approximately ten to three times smaller—at x_D—than the diameter of the AD. Evidently, this imposes a strict demand for the cell resolution, particularly for the turbulence box as the turbulence scale there ($L_{1,B}$) should be even smaller to account for its increase along the flow direction once turbulence is introduced in the LES domain.

On account of these limitations, the determination of the cell resolution of the turbulence boxes is based on what is physically realizable, due to the constraints represented by the total number of cells. For this, it should be considered that a box with twice the length of the desired lateral dimensions must be generated, as suggested by [22] since the simulated velocity field is periodic in all directions, so it is recommended to create a turbulence box with cross-flow dimensions twice as big as the desired size and only use one quarter of the simulated field. The box length is determined by the recycling period of the box into the computational domain (having assumed the equivalence between the streamwise direction of the box and the time for its convection in the domain), which is wished to be kept to a minimum. It is chosen to create boxes with length equivalent to at least two longitudinal flow-times, abbreviated as LFT (1 LFT is defined as $L_x / \langle U_\infty \rangle$). Considering these arguments, two turbulence boxes were created using the Mann implementation for each TI case. The parameters of these boxes are listed in Table 2. Note that the dimensions of the boxes are set to multiples of 2^n ($n \in \mathbb{N}^+$) due to the Fourier techniques used in the generation algorithm. As the grid size limit of the OpenFOAM installation in the cluster where computations were submitted has been found to be $\sim180 \times 10^6$ cells (perhaps due to the floating-point precision used to store the grid data) it is easy to see that a larger mesh than the one used for the Ti3 case (e.g., $2048 \times 512 \times 512$) would have surpassed this ceiling. The turbulence generator has been implemented outside the OpenFOAM framework so when the turbulence is originally generated, with twice as many points in the lateral directions, the cell number is not restricted by this limit. Since the mesh of the Ti12 case is coarser, it was possible to increase the length of the domain, allowing for a smaller recycling rate of the turbulence box.

In each TI case, the mesh of the computational domain is set according to the resolution used for the corresponding turbulence box, so that the cell sizes in the domain and box are equal. Unlike the turbulence box, the grid used in the LES is not completely uniform across the domain. Instead, only a central region of $20D \times 3.6D \times 3.6D$ of uniform (cubic) cells is defined. This region is needed to assure a consistent filtering for the SGS scales, as implicit filtering is used in the LES. In addition, the uniformly distributed cells should comprise all the positions of measurement, which includes those made up to $y = 1.5D$ from the centreline. Outside the uniform grid region, the cells are stretched towards the lateral boundaries with an aspect ratio $\Delta z_{max} / \Delta z_{min} = \Delta y_{max} / \Delta y_{min} = 4$, where $\Delta z_{min} = \Delta y_{min} = \Delta x = \Delta$ is the cell side length in the uniform region. This central region has approximately the same cell size as in the corresponding turbulence boxes. The slight differences arise from the purpose of accommodating an integer number of cells along the diameter of the AD. This way, the central region in each case consists of uniform cells with a side length Δ of 0.002 m (Ti3) and 0.004 m (Ti12). Hence, taking L_1 at x_D as a reference, the cell resolution of the integral scales L_1/Δ corresponds to 5 and 7.5 cells, respectively.

Table 2. Main parameters of the computational domains of LES and synthetic field (turbulence box). Dimensions of computational domains are given as $L_x \times L_y \times L_z$ with grids containing $N_x \times N_y \times N_z$ cells. Synthetic field domains are given as $L_{B,x} \times L_{B,y} \times L_{B,z}$ containing $N_{B,x} \times N_{B,y} \times N_{B,z}$ cells. Lengths measured in $D = 0.1$ m.

	LES Domain Size	$20D \times 5D \times 5D$
	Layout	Uniform region $20D \times 3.6D \times 3.6D$
Case Ti3	LES domain grid	$1000 \times 208 \times 208$ cells
	Turbulence box	$40D \times 5D \times 5D$
	Box grid	$2048 \times 256 \times 256$ cells
Case Ti12	LES domain grid	$500 \times 104 \times 104$ cells
	Turbulence box	$80D \times 5D \times 5D$
	Box grid	$2048 \times 128 \times 128$ cells

4.3. Generation of Turbulent Inflow, Introduction into the Computational Domain and Boundary Conditions

In the homogeneous case, the use of the Mann method requires adjusting two parameters to produce the turbulence with the demanded characteristics: the lengthscale L and TI. The latter is normally controlled by means of varying the coefficients $\alpha\varepsilon^{2/3}$ (α is the Kolmogorov constant) of the von Kármán energy spectrum [22] until the desired TI is achieved. As it is not straightforward to give an exact relationship between ε and the generated TI, the procedure suggested by [46] is used. Instead of changing ε, a scaling factor $SF = \sqrt{\sigma^2_{target}/\sigma^2}$, is used, where σ^2_{target} is the desired average variance of the turbulence and σ^2 is the variance of the turbulence field in each direction. In this way, the desired TI can be obtained by multiplying SF by each velocity component of the turbulence box. It is expected that when the HIT field is convected at a uniform velocity, the TI will decay monotonically in the streamwise direction. To estimate the turbulence intensity value that the box (TI_B) should have in order to attain the desired TI at the given position, empirical relations obtained from fits to experimental results can be used [37,39,40]. However, such relations depend on fitting parameters that are reported to vary within a certain margin. This fact and the expected numerical dissipation cause that TI_B cannot be estimated beforehand with precision. Furthermore, the averaged value of TI at the position where turbulence is introduced in the computational domain does not correspond to TI_B (due to the interpolations between turbulence planes and the adjustment of the synthetic velocity field to the LES conditions, to be discussed later on). In consequence, some testing was necessary to find the right value. Likewise, empirical relations for the development of L_1 [37] are not sufficient to predict the value of $L_{1,B}$, so tests were necessary to determine its magnitude. The values found for TI_B and $L_{1,B}$ are presented in the results section.

Boundary conditions are set to replicate the conditions in the wind tunnel. Thus, slip conditions are used for the lateral boundaries, whereas the outlet is set to zero gradient. For the time derivative, a second-order backward scheme is employed. When the disks are introduced, these are located at x_D, this is $x = 0.5$ m from the inlet, at the centre of the y–z plane. Assuming the Taylor hypothesis of equivalent spatial and temporal correlations, the streamwise axis of the turbulence box is assumed equivalent to time. To introduce the synthetic turbulence into the computational domain, cross-sectional planes of the turbulence box are taken for every available longitudinal position and their velocity values are mapped onto the inlet of the computational domain. The procedure is illustrated in Figure 2, where the planes extracted from the synthetic turbulence field are separated by Δx_B. As the crosswise locations of the cell centres of the synthetic turbulence do not exactly correspond to those of the computational domain, linear interpolations are used to evaluate the velocity values at the required positions. Different Courant–Friedrichs–Lewy (CFL) conditions are used in each case. For the Ti3, CFL ≈ 0.8 while for the Ti12, CFL ≈ 0.5. These are the maximum CFL values over the whole computational domain, which are attained next to the inlet, where the velocity fluctuation is the highest. In comparison, their domain-averaged values are ≈ 0.3 and ≈ 0.1, respectively. The time-steps used in the computations are $\Delta t = 2 \times 10^{-4}$ s and $\Delta t = 1.2 \times 10^{-4}$ s for each case. Linear interpolations are used in the streamwise direction (i.e., between planes of the turbulence box) to compute the required velocity values at the given time.

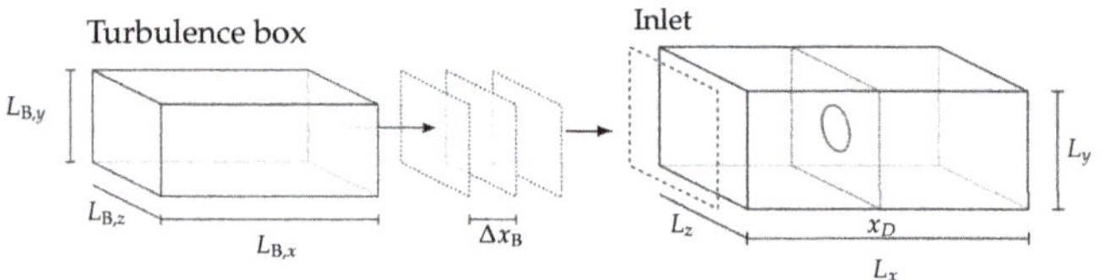

Figure 2. Introduction of synthetic turbulence field into the LES.

Simulations are allowed to run initially for 5 LFT to allow the stabilization of the flow. After this, measurements are made during a time equivalent to 20 LFT, which is equal to approximately 13.33 s in real time. Since the turbulence boxes defined in Table 2 are only enough to supply an inflow during 2 LFT (Ti3) and 4 LFT (Ti12), the boxes are recycled for the duration of the computations. Velocity data are sampled at every time-step, resulting in a higher sampling frequency than the one used in the experiments, although it is made during a shorter period (13.33 s compared to $\sim$60 s). The length of the computations is chosen as to maximize the bandwidth of the data employed for the calculations of spectra and correlations and to avoid fringe patterns in the average fields.

4.4. Estimation of Integral Lengthscales

The integral lengthscales are calculated from the autocorrelation curves of u and v in the longitudinal direction in the synthetic turbulence or from their time-series in the LES. In this way, L_1 and L_2 are obtained by making use of Equation (1). The method used to compute L_i consists of approximating the autocorrelation curve by a sum of six decaying exponentials. A similar procedure has been also used by [16,31], based on a technique first suggested by [47]. This technique is used as it avoids the uncertainty of determining the crossing of the oscillating function $\mathcal{R}$ around zero as well as approximating better the expected asymptotic behaviour of a theoretical autocorrelation sampled to infinity, $\lim_{x \to \infty} \mathcal{R}_{ii}(x, t) = 0$, yielded by the exponentials. While this method provides L_i for the synthetic turbulence, in the LES, the autocorrelations provide a time-scale that can be translated into a spatial one only under the assumption of the Taylor hypothesis (here, the integral time-scale obtained from the autocorrelations is multiplied by the average streamwise velocity at the point where the data are registered, which can be slightly different from U_∞).

4.5. Actuator Disk Model

In line with the experiments, the rotor of a horizontal-axis wind turbine is modelled in the computations as an actuator disk [9], where the effect of the blades on the wind flow is reproduced by forces distributed over a disk. As the actual geometry of the blades is not reproduced, the load of the turbine is distributed over the area swept by the rotor. The simplest conception of the model is employed here, where it is assumed that the forces over the AD point only in the axial direction, opposite to the flow and are distributed uniformly over the disk. If U_∞ is the inflow velocity, the force is calculated as

$$F_x = -\frac{1}{2} \rho U_\infty^2 C_T A, \tag{6}$$

where A is the area of the disk. The diameter of the disk is $D = 0.1$ m and C_T is equal to the values shown in Table 1. Since the introduction of the forces represents an abrupt discontinuity in the flow field, large velocity gradients occur in the vicinity of the AD and spatial oscillations (wiggles) on the velocity field may appear due to pressure-velocity decoupling inherent to collocated grids. To avoid this effect, the forces that comprise the AD are distributed in the axisymmetric direction. This is done by taking the convolution of the forces with a Gaussian distribution

$$g(x) = \frac{1}{\sigma\sqrt{2\pi}} \exp\left(-\frac{x^2}{2\sigma^2}\right). \tag{7}$$

In this manner, the value of the standard deviation σ (i.e., the distribution width) will define the thickness of the disk. The force distribution is defined between the limits $[-3\sigma, 3\sigma]$ so that it contains 99.7% of magnitude of the forces computed for the original—one cell thick—disk. A value of $\sigma = 2\Delta x$ is used, yielding a disk thickness equal to $12\Delta x$. Therefore, the absolute magnitude of the thickness changes according to the cell length.

4.6. About RANS Results

In a study by [17], RANS computations were performed to reproduce the same LDA measurements used in our study. In their work, a RANS turbulence model, identified as "Sumner and Masson", based on modifications to the k-ε model of [48] is proposed. While the latter model attempts to correct the well known overestimation of turbulent stresses [49] by introducing a dissipative term proportional to the turbulence production in the ε-equation, Sumner and Masson pursue the same objective by neglecting some terms of turbulence production also in the vicinity of the disk (the cylindrical volume centred at the AD, extending $\pm 0.25D$ in the axial direction), obtaining a good comparison for the velocity deficit and k along the wake of the disks. The results obtained with this model are also included in the comparisons as they serve as a reference element of the capabilities of an industry standard to reproduce the evolution of turbulence features in the wake. Note that since the simulations of [17] were made for only half the wake, their results (velocity deficit, k and ε) are shown mirrored in the vertical direction.

5. Results and Discussion

5.1. Turbulence Decay and Integral Lengthscale Development without Disk

The first step of the investigation consists of the calibration of the parameters of the synthetic turbulence. This is finding $\mathrm{TI_B}$ and $L_{1,\mathrm{B}}$ so that when a turbulence box is introduced in the computational domain, the desired target values are attained after a distance of $5D$, at x_D. The parameters of the synthetic turbulence for all boxes are shown in Table 3. These were computed longitudinally and averaged over the whole volume. It is immediately noticed that high TI values were necessary to reproduce the evolution of the turbulence intensities reported by the experiments. Consequently, the approach followed could be seen as rather crude, on account of the Taylor approximation. However, the results reproduce, for the most part, the longitudinal evolution of turbulence predicted also by the empirical relations found in the literature. This can be seen in Figure 3, which shows the free (no disk) homogeneous turbulence decay in each TI case obtained with LES. There, every value represents the average of the TIs computed from time-series stored in nine probes distributed in a crosswise plane, in turn located at every longitudinal position indicated by the marks in the curves. LES results are compared to the values measured in the wind tunnel and to the least-squares fit with Equation (4). As mentioned in Section 3, experimental values from [16] are used in the Ti3 case. To complement HWA data (which in the Ti12 case has only 3 points), LDA measurements are used, obtained in the experiment with the low thrust disks but outside the wake envelope ($y = \pm 1D$). It can be seen that for both Ti3 and Ti12 the decay predicted by the LES follows fairly well the experimental values. The subframes in Figure 3 represent the same data but plotted in a log-log scale, so the power law decay of the TI (of slope

$-n$ in Equation (4) can be better appreciated. This permits seeing that, after some distance, the decay rate is approximately equal for both TIs. Furthermore, it is also seen that the LDA (measurements outer wake) in the Ti3 case deviates from a constant decay rate and therefore from the LES. Conversely, the LDA data in the Ti12 compares very well with the LES prediction.

Table 3. Characteristics of the different boxes of synthetic turbulence used as inflow for the LES.

	TI_B [%]	$L_{1,B} \times 10^{-3}$ [m]	$k/\frac{1}{2}U_\infty^2$ [-]
Case Ti3	35.0	5.82	0.37
Case Ti12	60.2	15.3	1.08

Equation (4) is an empirical relation first proposed by [41] to describe the decay in homogeneous turbulence seen in a wind tunnel. The applicability of this relation has been proven in a wide range of Re flows in later work [39,42,50,51]. In most of the results reported in the literature, a fit is produced setting the virtual origin x_0 to zero in the equation, which neglects the agreement close to the grid or the place where turbulence originates as the stations where measurements or calculations are reported are generally far from such region. However, as the complete evolution of TI is monitored, a better fit is obtained by setting x_0 to a position different from where the turbulence is introduced (in particular, to an upstream location). The fit of Equation (4) for the curves shown in Figure 3 yields the results shown in Table 4. If the fit is made using $x_0 = 0$, the parameters are closer to those reported in the literature (see Section 2), although the curve would display a much higher TI at the inlet than the one given with $x_0 \neq 0$, this is, $\sim$60% for Ti3 and $\sim$100% for Ti12. The mesh spacings used for the fits are $M = 0.0225$ m for Ti3 and $M = 0.20$ m for Ti12. The problem of setting x_0 has been discussed by [52], which show that when x_0 is properly determined, its value and the exponent of the power law decay n is Re-independent, while A is indeed a function of initial conditions (including Re), so the turbulence decay takes a universal self-similar behaviour.

Table 4. Parameters of the fit of TI decay of LES to Equation (4). x_0 denotes the virtual origin of the curve with respect to the inlet.

	$x_0 = 0$		$x_0 \neq 0$		
	$1/A$	n	$1/A$	n	x_0 [m]
Case Ti3	24.11	1.281	9.85	1.519	−0.021
Case Ti12	28.49	1.15	11.43	1.661	−0.0845

Table 3 also shows the magnitude of the $L_{1,B}$ in the synthetic turbulence. According to these results, it is seen that the requirement of two cells per L_1 (assuming the applicability of the minimum condition to represent a wavelength [53]) is sufficiently fulfilled. In effect, for the case Ti3, $L_1/\Delta \simeq 3$ is obtained. The resolution of eddies is somewhat improved in the case of Ti12, where $L_1/\Delta \simeq 4$. These resolutions that seem a priori rather coarse are the result of a series of compromises that have been previously explained. In [30] a similar resolution was used for the synthetic inflow in LES computations of the wake of a rectangular channel, obtaining a good comparison with experimental results related to the flow structures. The results of the turbulence decay in the absence of the disk show that these values are enough to supply an integral lengthscale to procure the desired magnitudes $L_1 = 0.01$ m (Ti3) and $L_1 = 0.03$ m (Ti12) at x_D. Figure 4 shows the development of the longitudinal and transversal integral lengthscales, L_1 and L_2, along the flow. These are computed using velocity time-series (recorded at the same positions for TI above) and employing the method described in Section 4.4. The measurements from [16] are also used for comparison in the Ti3 case. There, the comparison of the measurements with the LES shows a small overestimation of L_1, although it should be considered that the difference is increased in the figure as the experimental value is below the target $L_1 = 0.01$ m. A fit of Equation (5) is also made for L_2 obtained with LES. In the Ti3 case, the least-squares fit method applied yields

$B_1 = 0.089$, $n_1 = 0.392$ and an origin set at $x_0 = -0.188$ m (upstream) from the inlet. Therefore, according to reference values provided along Equation (5), the LES slightly overestimates L_2 for Ti3. In the Ti12 case, the comparison with measurements is made with the only three points available so it is less decisive, although they suggest a higher growth rate. The fit of Equation (5) to the LES predicted L_2 yields $B_1 = 0.064$ and $n_1 = 0.254$ with $x_0 = -0.068$ m, which, compared to the values in the literature, also indicates a slight overestimation by the computations. For the purposes of this work and considering the mesh restrictions, the development of the integral lengthscales in the flow predicted by the LES is deemed satisfactory.

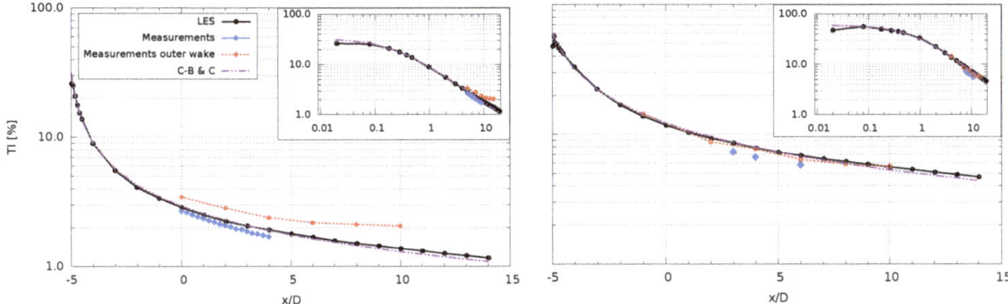

Figure 3. Comparison of the TI decay for the Ti3 (**left**) and Ti12 (**right**) cases without disk. C-B & C corresponds to Equation (5) as found in [41]. The inserted frames contain the same data but with a logarithmic scale also in both axes and denoting the distance from the inlet in x/D units.

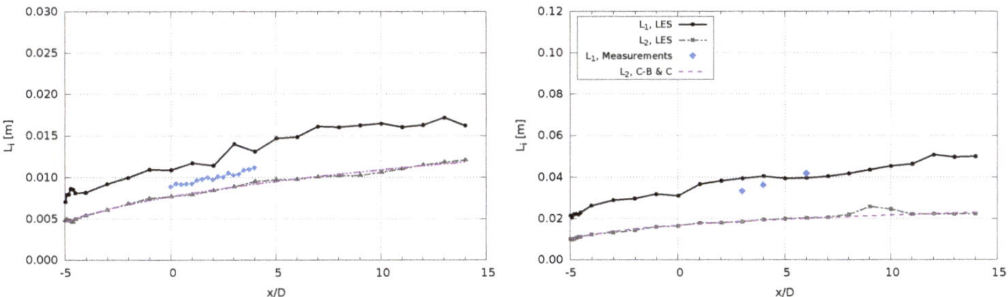

Figure 4. Longitudinal evolution of L_1 and L_2 for the Ti3 (**left**) and Ti12 (**right**) cases without disk.

In both TI and L_i results, we can observe small oscillations for the first positions next to the inlet, which are likely the result of the adjustment of the synthetic velocity field to the LES. Specifically, the incompressibility conditions imposed by the solver as the original formulation of the Mann algorithm does not produce divergence free fields, as pointed out by [29].

5.2. Velocity Deficit

The results for the wake simulations obtained with the AD are now shown. The first comparison is made from the results of the streamwise velocity deficit along the vertical direction at different longitudinal positions, normalized by the freestream velocity at $y = 1.5D$. For these and other quantities extracted across the wake field shown in Sections 5.2–5.5, the values are sampled at every cell centre along a line in the y-direction (at the mid-plane $z = 0$) at different x/D positions. These quantities correspond to time averages made during 20 LFT (13.33 s). No further spatial averaging is made. Note that LES results at $x = 3D$ have been added to the available x-positions of the LDA data as this is a position where values computed from HWA are later shown.

Figure 5 shows the results for the high and low solidity disks under the inflow Ti3. The agreement to the experimental results is very good, with the larger difference observed around the shear layer (i.e., the wake envelope at $y \simeq \pm 0.5D$) from the disk edges, especially for the disk with higher thrust.

In the case of the Ti12 inflow, Figure 6 shows a minor reduction in the agreement of the LES with the measurements. The predictions commence to differ when moving further into the far wake, around the centreline and shear layer regions. Remarkably, the results of RANS are almost identical to those from LES in both TI cases.

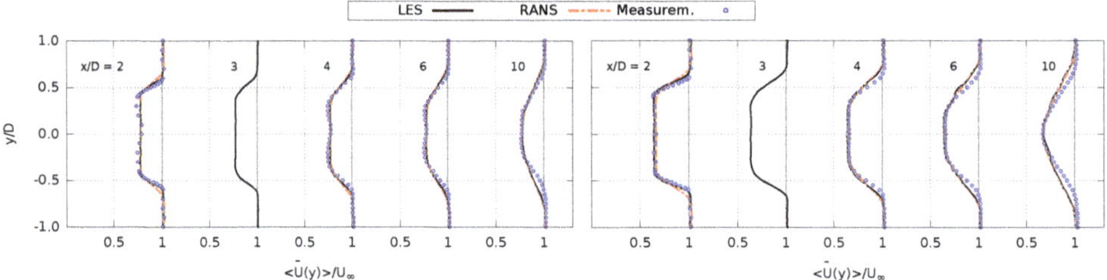

Figure 5. Vertical profiles of velocity deficit behind the disk $C_T = 0.42$ (**left**) and $C_T = 0.62$ (**right**), Ti3 case. RANS results in all figures are by Sumner et al. [17].

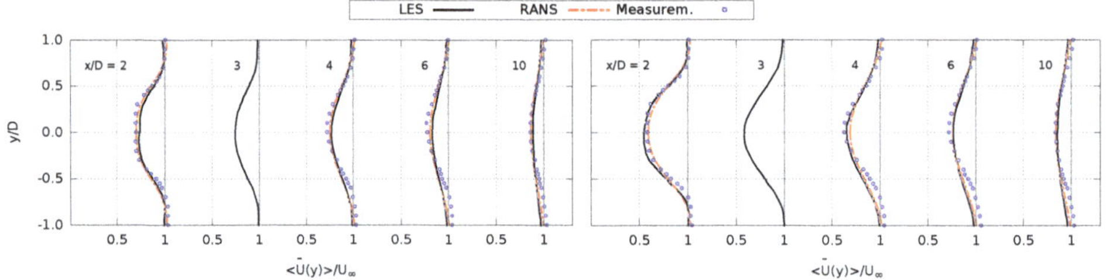

Figure 6. Vertical profiles of velocity deficit behind the disk $C_T = 0.45$ (**left**) and $C_T = 0.71$ (**right**), Ti12 case.

5.3. Turbulence Kinetic Energy in the Wake

It is expected that the wake created by the disks augments the turbulence level with respect to the ambient value. It is now investigated how the computations of the added turbulence compare to the experimental results within the wake. Figure 7 shows the profiles of k (this is, $k_{tot} = k_{\mathrm{SGS}} + k_{res}$ for the LES) at different downstream positions along the wake, when the inflow of the case Ti3 is used. There, it is observed that the LES results match quite well the measured (LDA data) turbulence levels behind both disks. However, it is noticed that, except for the nearest position to the disk, the LES predicts a higher diffusion of shear turbulence in the crosswise direction, an effect that is increased with the disk thrust.

For the disks in the Ti12 case in Figure 8, the LES compare mostly well with the experimental data, although a small overestimation of k can be seen just behind the disk ($x = 2D$). It is also observed that the shear layer originating at the edges of the disk is mixing faster with the ambient turbulence compared to the Ti3 inflow. Indeed, the effect of shear prevails deeper into the wake in the LES with the highest thrust disk, whereas it is mixed faster into the ambient turbulence when the thrust is lower. Results from the RANS computations are discussed in the next section.

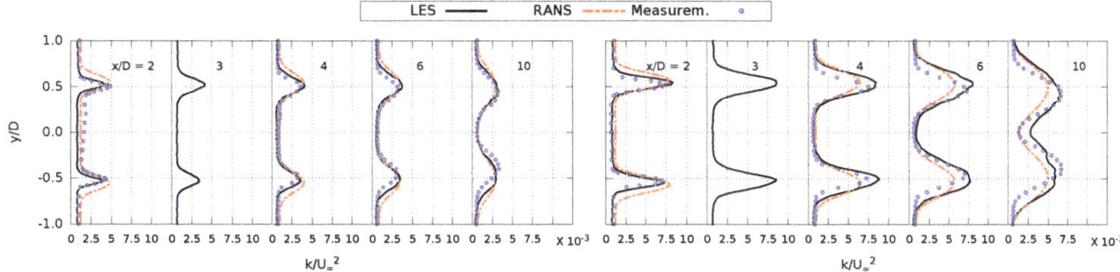

Figure 7. Vertical profiles of k behind the disk $C_T = 0.42$ (**left**) and $C_T = 0.62$ (**right**), Ti3 case.

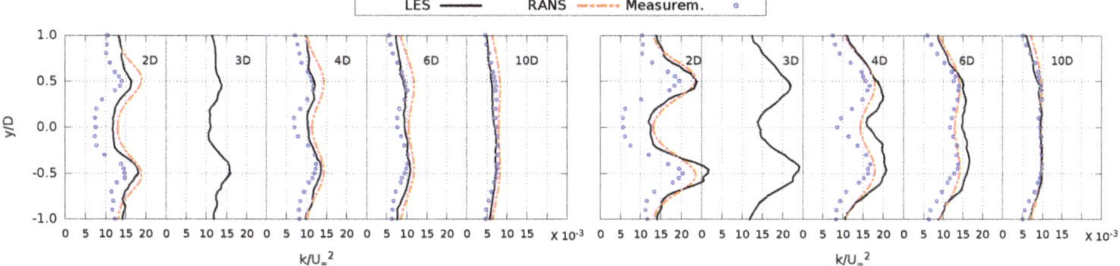

Figure 8. Vertical profiles of k behind the disk $C_T = 0.45$ (**left**) and $C_T = 0.71$ (**right**), Ti12 case.

It is also noticed that some inhomogeneities appear along the curves of LES, very noticeable in the simulations with the Ti12 inflow. These seem to indicate a footprint of the turbulence structures of the inflow turbulence. Although this feature could provide evidence of the need of performing averages in the azimuthal direction or creating synthetic turbulence that would cover longer simulation periods, it is thought that the results shown in the figures are sufficient for the purpose of these comparisons.

5.4. Turbulence Dissipation in the Wake

In the LES computations, the dissipation shown corresponds to $\varepsilon_{tot} = \varepsilon_{res} + \varepsilon_{SGS}$. The dissipation is calculated from the HWA data using Equations (2) and (3), which was only available in the Ti12 case. Therefore, results of Ti3 are shown only for completeness in Figure 9, where it can be seen that, as before, differences between LES and RANS are small, as the curves differ only at $x = 2D$ where RANS predicts a higher dissipation within the shear layer.

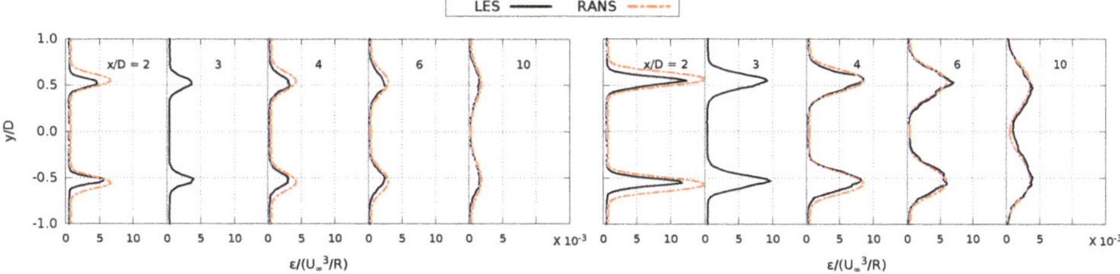

Figure 9. Vertical profiles of ε behind the disk $C_T = 0.42$ (**left**) and $C_T = 0.62$ (**right**), Ti3 case.

For the results with the Ti12 inflow in Figure 10, it is seen that, for the disk $C_T = 0.45$, the LES predictions compare well with measured values. For the disk $C_T = 0.71$, the measurements reveal a large increase of dissipation within the shear layer, compared to the data of computations with the lower thrust AD. Furthermore, at least within the three longitudinal positions available, measured dissipation in the shear layer is more or less maintained. The computations with $C_T = 0.71$ also display a somewhat stronger mixing of turbulence from $x = 4D$, where dissipation becomes more uniform and less predominant in the shear layer.

The RANS computations with the modified $k - \varepsilon$ model of Sumner and Masson have been previously shown capable of reproducing the turbulence level in the wake [17]. In the present comparison, it is seen that for the Ti3 inflow the agreement is very good for the disk $C_T = 0.42$ while it falls somewhat behind in the far wake of $C_T = 0.62$. However, in both cases, the agreement in the computed dissipation of RANS and LES is very good except for $x = 2D$. Interestingly, it is the vicinity of the disk where the k-ε is often corrected by adding dissipative terms to the ε equation to overcome the miscalculated turbulence stresses [49]. The results with the Ti12 show the opposite picture with regard to the estimation of k, as the agreement with measurements becomes better only for farther distances from the disk. For the closest position, the turbulence level is overestimated (as it is in the LES) despite the drop of the turbulence production terms near the disk ($x = 2D$ is outside this region). Dissipation seems overestimated in the case of $C_T = 0.45$ when comparing to the measurements. This is less certain for the higher thrust disk, where at $x = 4D$ the peak value of dissipation seems equal to the measured one, but much smaller in the case of $x = 6D$. Notably, ε from RANS is always higher than any LES in the wakes of the Ti12 inflow. Previous work [49] has shown that, in the ABL, the k-ε model overestimates the dissipation around the disk when comparing with LES. This has been observed to occur even upstream of the disk, where ε has been seen to increase unlike computations of LES, where this value does not grow until $0.5D$ downstream from the rotor.

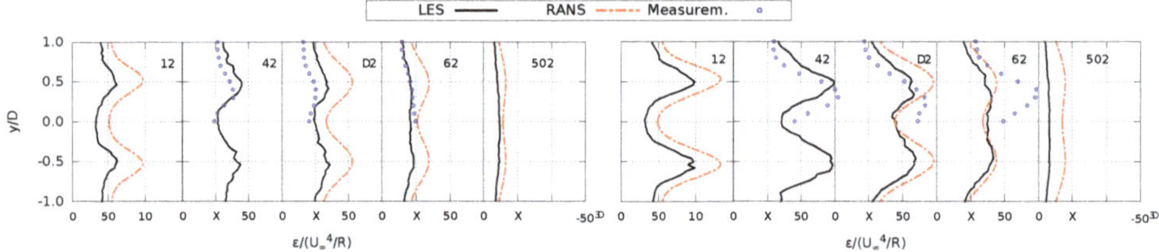

Figure 10. Vertical profiles of ε behind the disk $C_T = 0.45$ (**left**) and $C_T = 0.71$ (**right**), Ti12 case. The scale for the curves at $x = 2D$ has been doubled to accommodate the larger values.

It should be remarked that Sumner and Masson [17] showed that results with various turbulence closures (standard k-ε, RNG, El Kasmi and Masson model [48] and their own model) compare, in essence, equally well to the measurements, with no apparent advantage of their proposed correction to the k-ε model (although ε yielded by the different closures was not compared). The fact that all models compare well to measurements appears to contradict the otherwise inadequate results obtained in simulations of wakes in the ABL flow cited in that work. There, it is also argued that this is due to the relative decrease of the modelled turbulent viscosity ν_t in the reproduction of wind-tunnel wakes with homogeneous inflow with respect to its proportion in the modelling of atmospheric flow. In those conditions, previous work by [49] has successfully proved the advantages of LES to estimate the velocity deficit and turbulence levels in the wake.

5.5. LES Modelling in the Wake

The previous results for k and ε indicate that the LES running in OpenFOAM is able to predict with relative accuracy not only the velocity deficit in the wake, but also the level of turbulence and its dissipation in the case where the TI in the inflow is low ($\sim$3%). For the high TI inflow ($\sim$12%), the prediction becomes more imprecise, according to the comparison with the experimental data. In the absence of disks, it is observed that the fraction of the turbulence kinetic energy that is resolved by the LES with respect to the total k_{res}/k_{tot} occurs for the most part in the resolved scales, at around 90%

in both TI cases throughout the domain and only somewhat smaller close to the inlet [54]. With the introduction of the disks, this does not appreciably change except for the shear layer nearby the AD at around $x = 1D$ in the Ti3 case, where the resolved part is slightly reduced [54]. In the Ti12 case, this difference is not noticeable and the value of k_{res}/k_{tot} remains throughout.

Figure 11 shows the ratio of subgrid dissipation with respect to the total value $\varepsilon_{SGS}/\varepsilon_{tot}$ along the wake for the Ti3 case. Note that, as the positions of comparison are no longer restricted to those of the available experimental data, profiles are shown at different longitudinal positions from other figures. In Figure 11, it is noticed that the LES has an appreciable increment in subgrid dissipation within the shear layer. Furthermore, this increase persists longitudinally even as far as when the wake appears to reach a state of transversally-uniform dissipation, i.e. at $x = 12D$ with disk $C_T = 0.62$. This is consistent with the hypothesis that turbulence is created at smaller lengthscales than the ambient turbulence at the disk edge. Due to the limited resolution of turbulence lengthscales in the Ti3 flow (missing in the synthetic flow as well), the increase in subgrid dissipation is produced at scales that seem absent in the incoming flow. These results also show that the wake envelope becomes the main carrier of dissipation. The subgrid dissipation part is also larger with higher thrust, yet by a small margin. It is observed that, in the absence of disks, most of the dissipation comes from the resolved fluctuations, in a proportion very similar to that shown outside the wake, at $y \pm 1.0D$ [54]. It is seen in Figure 12 that when the inflow turbulence raises (which comprises better resolved lengthscales), the increment of subgrid dissipation in the region of the wake envelope is greatly reduced. As a result, the modelling ratio seen outside the wake is essentially conserved. From these results, it can be deduced that the LES modelling across the wake is largely determined by the ambient turbulence.

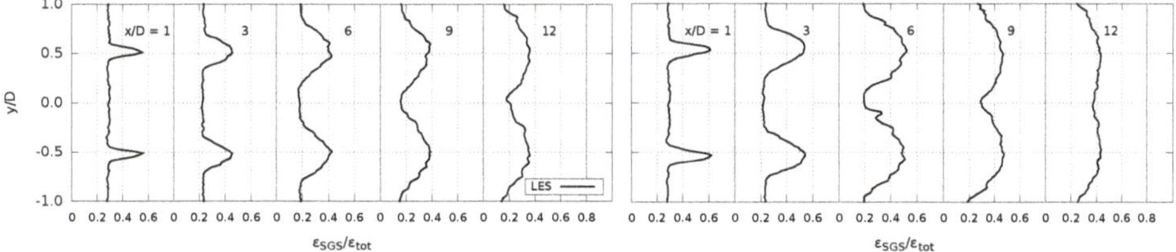

Figure 11. Vertical profiles of $\varepsilon_{SGS}/\varepsilon_{tot}$ behind the disk $C_T = 0.42$ (**left**) and $C_T = 0.62$ (**right**), Ti3 case.

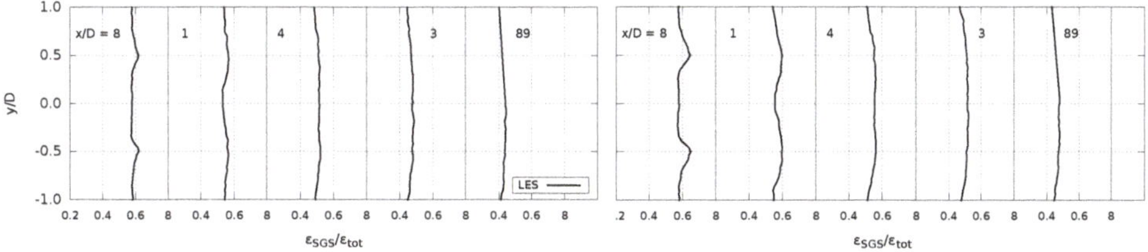

Figure 12. Vertical profiles of $\varepsilon_{SGS}/\varepsilon_{tot}$ behind the disk $C_T = 0.45$ (**left**) and $C_T = 0.71$ (**right**), Ti12 case.

5.6. Integral Lengthscale in the Wake

The changes in L_1 caused by the presence of the AD and the wake are now investigated. The computation of L_1 is performed as described in Section 4.4, which involves the assumption of the Taylor hypothesis to transform the computed time-scales into lengthscales. Evidently, this supposition becomes more difficult to accept when shear is present in the flow. However, previous work has reported satisfactory results in wake studies that support the continuing applicability of the hypothesis. For instance, Ref. [16] has compared the lateral distribution of L_1 behind the wake produced by a porous disk (in a setting similar to the experiments used in this work) computed from

HWA with the one obtained from PIV. They did not find a difference in the results obtained from either technique, despite the fact that HWA uses the local mean velocity to calculate the lengthscale, compared to the direct spatial measurement offered by PIV. Making the same assumption, the evolution of the integral lengthscale behind the AD computations is studied.

Figure 13 displays the values of L_1 computed from the LES in each code with the Ti3 inflow, from $y = 0$ to $y = 1.5$, at $3D$, $4D$ and $6D$, which correspond to the positions where HWA data for the Ti12 inflow is available. It is first noticed that there is not a clear influence of the shear layer in the size of the turbulence scales. However, for a region about $0.5 \leq y/D \leq 1.0$, next to the the shear layer, larger lengthscales can be discerned amongst the variations in the profile. Indeed, the maximum values of L_1 at each $x-$position are at close $y = 0.5D$ in the wake of the disk $C_T = 0.42$. This is consistent with the previous results with regard to the location of the shear layer along the wake (e.g. k and ε). Conversely, for the other disk the maxima of L_1 would suggest a wake that expands to about $y = 0.75D$ at $x = 6D$, which is larger than what the previous computations indicate.

Results for the Ti12 inflow are shown in Figure 14. Notably, the computed values from the experimental time-series do not reveal a variation of the lengthscale values at the shear layer. In fact, there is no evident change in L_1 within the wake. This trait is similarly observed in the LES results. The only variations in computations are observed at the upper part of the curves or, in the case of the disk $C_T = 0.45$, towards the bottom part where L_1 is larger (but this effect is reduced further downstream).

Previous experimental work by [16] showed that in the wake of a porous disk with a solidity of 45%, L_1 is approximately 1.5 times larger within the shear layer with respect to the values within the wake or outside the envelope. However, these measurements were obtained using an inflow with very low turbulence (TI $< 0.4\%$), which clearly sets a different scenario in comparison to this study. Precisely, the absence of a variation of L_1 in the shear layer can be explained considering the previous results, which point at a dominance of the ambient turbulence characteristics over the wake in the case of the inflow Ti12. Although the turbulence production is visibly higher when the disk thrust is larger (Figure 8), its effect does not appear to have an impact on the turbulence lengthscales. Similarly, the use of a lower turbulence inflow (Ti3) does not seem to decidedly increase the magnitude of the lengthscales in the area of turbulence production, or at least not in the computations performed for this work. In this regard, the fact that the characteristic lengthscales of the Ti12 inflow are better resolved by the mesh and the LES compared to the Ti3 cases can be a factor to consider. This is, if resolution is not adequate within the shear layer, it is to be expected that a sizeable part of the turbulence being produced would fall into the modelled part instead of being resolved, therefore affecting the magnitude of the computed scales. This has been studied in Section 5.5, where it is shown that the LES modelling does not vary within the wake with respect to the external flow aside from very close to the disk ($x = 1D$) in both codes. Nevertheless, it has been seen that despite the limited resolution, the LES computations have been able to reproduce other principal features along the wake, such as the turbulence levels.

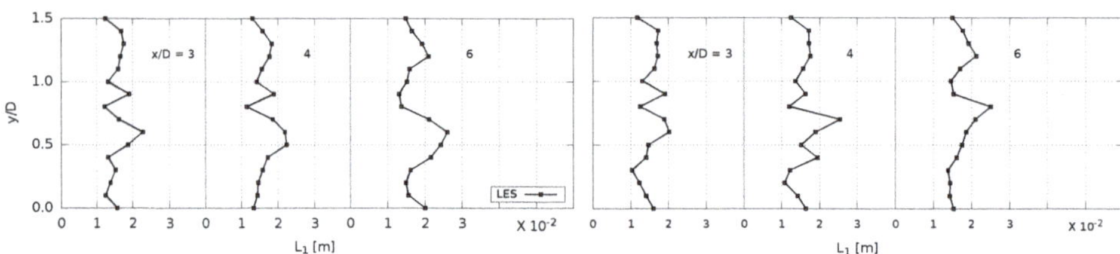

Figure 13. Vertical profiles of L_1 behind the AD with inflow Ti3, disks: $C_T = 0.42$ (**left**) and $C_T = 0.62$ (**right**).

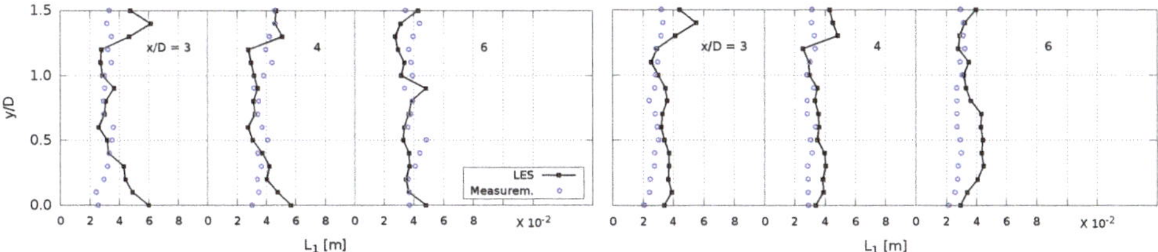

Figure 14. Vertical profiles of L_1 behind the AD with inflow Ti12, disk $C_T = 0.45$ (**left**) and $C_T = 0.71$ (**right**).

5.7. Spectra behind Disks

To study the redistribution of turbulence energy along the wake, the spectra obtained at different longitudinal positions for every disk are compared with the spectrum of the free decaying turbulence. Power Spectral Density (PSD) curves are calculated from only one measuring position at centreline, so to reduce the noise in the spectral curves, the time-series of each register are divided into eight non-overlapping blocks with an equal number of samples. Then, the PSD of all blocks are averaged to produce the curve at each longitudinal position. However, noise remains along the curves that make the comparison very difficult, so a smoothing procedure is needed to be performed. To this aim, an exponential moving average is used to filter the spectra computed at each longitudinal position (a rational transfer function is employed, see [55]). Hence, the spectra shown in the following figures have been processed with this technique, with the sole exception of that obtained from measurements without a disk, which was spatially averaged with results obtained at the other eight locations distributed crosswisely. As the spectra are calculated from data at a fixed location (sampled in time), the Taylor hypothesis is applied to transform the frequency spectra into a wavenumber spectra using $\kappa_1 = 2\pi f / \langle U \rangle$, where f is f_{acq} for measurements or $f = 1/\Delta t$ for the LES. In this way, it is possible to also compare with the PSD from the synthetic turbulence, which is calculated as the volume average of the spectra computed in the longitudinal direction.

The results for the inflow Ti3 are shown in Figure 15. In the results without the AD, a constant decay of energy is observed as the flow moves downstream. The spectra from the synthetic box serve to mark the extension of the resolved wavenumbers ($\kappa_{max} = \pi/\Delta = 1571$ m^{-1}) since the spatial resolution in the box is the same as in the LES. Note that the abrupt drop in the turbulence energy spectra is attributed to a combination of numerical diffusion and the limited spatial resolution [5]. In case of the disk $C_T = 0.42$, a gain in turbulence energy is seen immediately behind the disk, as the curves at $-1D$ and $1D$ are almost identical. A small decay of energy is observed at $4D$ and, from there, an increase in turbulence energy around the highest levels (lowest κ). For the disk $C_T = 0.62$, the effects are accentuated, and the curves at $4D$ are the only ones displaying a decay and yet only around the inertial range. The energy of the next two longitudinal positions, $10D$ and $14D$, increases for all wavenumbers, which represents an increment of about one order of magnitude at the lowest wavenumber, with respect to the levels displayed by the decaying turbulence without disks. Notably, the spectra of the last two positions seemingly exhibit an inertial range, characterized by the slope of $-5/3$ in the decay rate.

The results for the Ti12 inflow are shown in Figure 16. In this case, the spectra computed from experimental results are also included. The spectra obtained from measurements with disks extend to larger wavenumbers than in the cases without disk, due to the use of a different frequency in the low-pass filter. For the case without disks, the energy at the lowest wavenumbers obtained from the LES proves to decay less as it has been shown before. However, they display a steady decay that adjusts well to the characteristic slope of the intertial range, also discernable in the experimental results. These observations are analogous for the results with the disk $C_T = 0.45$. In contrast to the Ti3 inflow where energy is seen to increase beyond $x = 4D$ for the disk with the same porosity, here a reduction in the contribution of shear towards the increase of energy along the wake is observed. Although the overall levels of turbulence energy in the wake are higher than in the free decaying turbulence, they maintain more or less the same relative decay from one to another. This behaviour is similar in the case of the disk $C_T = 0.71$. There, only the curve at $4D$ shows an increase in energy compared to the previous disk (also matching fairly well the experimental results in the inertial range).

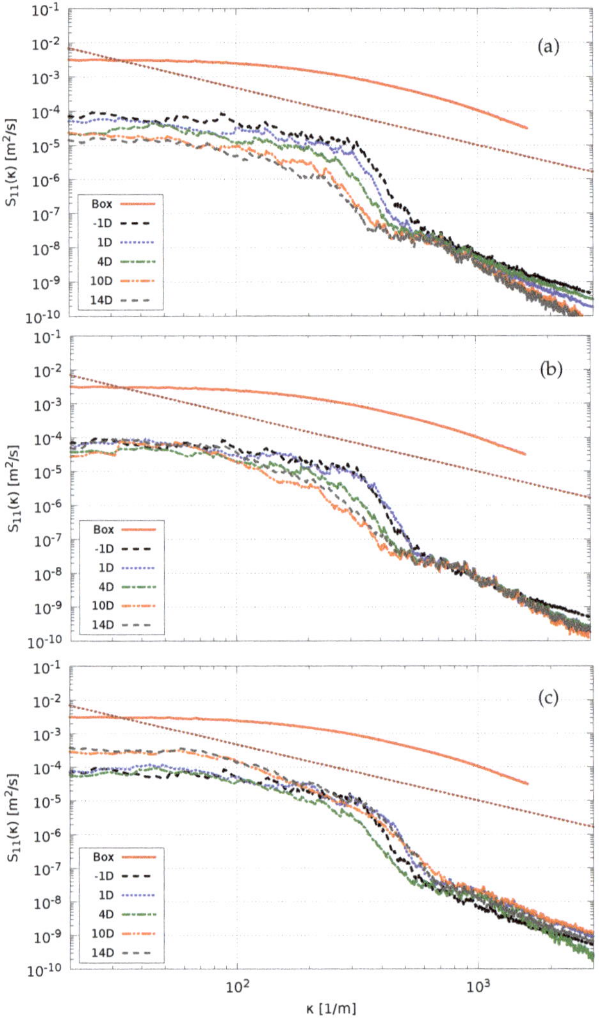

Figure 15. Longitudinal evolution of spectra at centreline using the Ti3 inflow. The results for the free decaying turbulence (without AD) are shown in the top row (**a**), results with disk $C_T = 0.42$ are shown in the middle row (**b**) and results with disk $C_T = 0.62$ are shown in the bottom row (**c**). The straight dotted line marks the $-5/3$ slope that characterizes the inertial range.

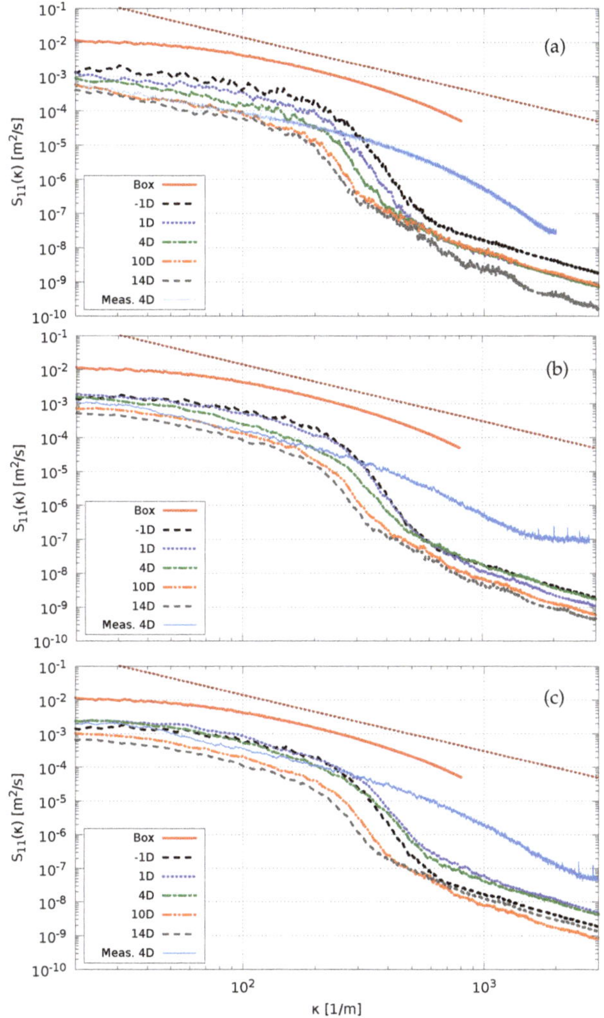

Figure 16. Longitudinal evolution of spectra at centreline using the Ti12 inflow. The results for the free decaying turbulence (without AD) are shown in the top row (**a**), results with disk $C_T = 0.45$ are shown in the middle row (**b**) and results with disk $C_T = 0.71$ are shown in the bottom row (**c**). Spectra computed from measurements are included only for the position $x = 4D$. The straight dotted line marks the $-5/3$ slope that characterizes the inertial range.

5.8. Vorticity Contours

Lastly, to complement all previous results, Figure 17 shows a comparison between the the contours of vorticity obtained in the Ti3 and Ti12 cases and using the highest porosity disk, $C_T = 0.62$ and $C_T = 0.71$, respectively. The images are taken at the x–y plane, at $z = 0$ and correspond to the vorticity field computed at the last time step of the LES runs. Make note that black bars are used to represent the disk position but do not portray the complete longitudinal region where the forces modelling the AD act, distributed using Equation (7). This figure permits visualizing the dominant effect of the turbulence structures from the inflow of the Ti12 case, which, unlike the Ti3 inflow, prevail along the wake. With the Ti3 inflow, there is a clear shear region arising from the edges of the disk with distinct turbulence structures. Conversely, with the use of the Ti12 inflow, the shear region is substantially less noticeable in the higher ambient turbulence. Indeed, the vorticity contours from the inflow appear to dominate in the vicinity of the AD. This is consistent with the comparison of k in Figures 5 and 6, where its increase within the shear layer is less prominent with the increase of ambient TI. In Figure 17, it can also be seen that, for the Ti12 inflow, structures that appear to arise from within the wake become more apparent than those outside from approximately $x \simeq 6D$.

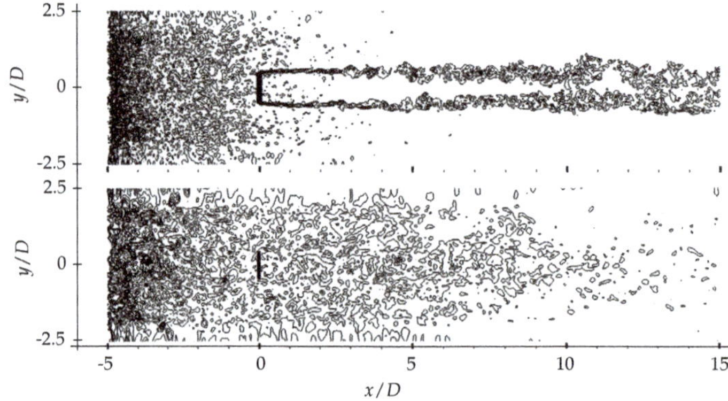

Figure 17. Contours of the vorticity field obtained with the Ti3 inflow and disk $C_T = 0.62$ (**top**) and the Ti12 inflow and disk $C_T = 0.71$ (**bottom**).

6. Summary and Conclusions

This work has been dedicated to study the flow characteristics as well as the modelling of turbulence in wakes produced by a porous disk. For this, disks of two porosities have been used as well as two instances of decaying and homogeneous turbulence inflows, identified by streamwise turbulence intensities (TI) of approximately 3% and 12% and corresponding longitudinal integral lengthscales (L_1) of 0.01 m and 0.03 m. To perform this study, a methodology has been developed and implemented in OpenFOAM that uses Large-Eddy Simulations (LES) to reproduce the main turbulence features of a turbulent flow generated in a wind tunnel. The Actuator Disk (AD) technique was used to model the porous disk and replicate the turbulence properties measured in the wake flow. The employed methodology is based on a procedure previously shown to provide good results in the simulation of wind turbine wakes in homogeneous turbulence and atmospheric flows. A comparison with prior work made with RANS was made wherever possible.

To reproduce the grid-generated turbulence in the wind tunnel, a synthetic field of homogeneous isotropic turbulence was introduced at the inlet of an LES. An analysis of the decaying turbulence simulated by the LES demonstrated that the simulations reproduce the TI decay and the evolution of L_1 predicted by the parametrizations predominant in the literature, in addition to replicating fairly well the experimental values at the measured locations. The wake field results obtained with the introduction of the AD showed a good agreement with the wind tunnel experiments for the velocity deficit, turbulence kinetic energy k and its dissipation ε. It was also found that L_1 evolves, for the most part, as in the decaying homogeneous turbulence. Remarkably, there was no apparent variation of L_1 within the shear layer compared to the center of the wake or outside the wake envelope.

A study of how the LES performs in the wake was also carried out, with the main interest of observing how the modelling ratio (the contribution of the resolved or SGS parts to the total value) is affected by the regions of high shear behind the AD in comparison with the outflow and the center of the wake. It was found that while the modelling ratio of k in the wake is largely maintained with respect to the outside flow, the subgrid part of ε increases within the shear layer, but only when the low TI inflow is used. From the results, it is seen that modelling in the freestream flow prevails in the wake just as the level of inflow turbulence increases. On the other hand, the wake results compare well with the velocity and k obtained with RANS. Likewise, LES and RANS results of ε match also in the low TI case, whereas the values yielded by RANS seem overestimated—in comparison with both measurements and LES—in the high TI case, particularly in regions of strong shear.

Spectra computed at different axial positions in the wake reveal that shear causes a noticeable boost in the energy content of turbulence, but only in the low TI case. This causes that for the two furthermost positions ($x = 10D$ and $14D$), the energy levels are higher or at least as energetic as in the upstream region near the disks. Moreover, the turbulence at those positions shows a clear inertial range that was absent in the decaying turbulence at low TI. Conversely, for the high TI inflow, it is seen that, despite the fact that turbulence energy levels rise in the wake with respect to the decaying homogeneous flow, the relative decay is maintained from one position to the other.

Considering the above, the central conclusions of this investigation are:

- The evolution of L_1 was not noticeably different in the wake in comparison to what was observed in the decaying homogeneous turbulence.
- Turbulence scales in the wake appear to be dominated by the inflow characteristics and this effect increases with the level of TI in the inflow.
- It is seen that the resolved/subgrid modelling ratio in the freestream flow prevails in the wake in relation to the increasing level of ambient turbulence.

The comparison of the LES with experimental results has permitted validating the methodology presented in this work as a suitable approach to investigate the turbulence features in the wake flow of wind turbine models. With this method, it can be possible to study the spatial and transient development of wake characteristics and assess the influence of, for example, a variety of inflow turbulence conditions or wind turbine models of varying sophistication. These aspects are considered for future work. In the same manner, an analogous study can be performed for an array of wind turbine models with the additional interest of observing the influence of downstream turbines in the wake turbulence field, for different layouts. Clearly, a CFD tool like the one presented in this work bears the advantage of changing the simulation setup without needing to perform new experiments each time, while also removing the spatial or temporal limitations associated with the measuring techniques.

Author Contributions: Conceptualization, H.O.-E. and S.-P.B.; Methodology, H.O.-E. and S.-P.B.; Software, H.O.-E., S.-P.B. and K.N.; Validation, H.O.-E.; Formal Analysis, H.O.-E.; Investigation, H.O.-E.; Writing—Original Draft Preparation, H.O.-E. with contributions of all authors to its review and editing; Supervision, C.M. and L.D.; Project Administration, C.M., L.D. and S.I.; Funding Acquisition, H.O.-E., C.M., L.D. and S.I.

Funding: We wish to thank the financial support provided by the Canadian Research Chair on the Nordic Environment Aerodynamics of Wind Turbines and the Natural Sciences and Engineering Research Council (NSERC) of Canada, as well as the Consejo Nacional de Ciencia y Tecnología (CONACYT) of Mexico for their financial support received under their scholarship program for postgraduate studies (number 213254). The program Vindforsk from the Swedish Energy Agency is also acknowledged for providing research funds for this work.

Acknowledgments: We thank Sandrine Aubrun and Yann-Aël Muller of Université d'Orléans for facilitating the wind tunnel measurements used in this work. Likewise, we thank Jonathon Sumner from Dawson College (Montreal, Canada) for the RANS data. The computations were performed on resources provided by the Calcul Québec-Compute Canada consortium. Complementary calculations were performed on resources provided by the Swedish National Infrastructure for Computing (SNIC) at the National Supercomputer Centre (NSC).

Conflicts of Interest: The authors declare no conflict of interest.

References

1.	Crespo, A.; Hernandez, J.; Frandsen, S. Survey of modelling methods for wind turbine wakes and wind farms. *Wind Energy* **1999**, *2*, 1–24. [CrossRef]
2.	Calaf, M.; Meneveau, C.; Meyers, J. Large eddy simulation study of fully developed wind-turbine array boundary layers. *Phys. Fluids* **2010**, *22*, 015110. [CrossRef]

3. Churchfield, M.J.; Lee, S.; Moriarty, P.J.; Martinez, L.A.; Leonardi, S.; Vijayakumar, G.; Brasseur, J.G. A large-eddy simulation of wind-plant aerodynamics. In Proceedings of the 50th AIAA Aerospace Sciences Meeting Including the New Horizons Forum and Aerospace Exposition, Nashville, TN, USA, 9–12 January 2012; p. 537.

4. Nilsson, K.; Ivanell, S.; Hansen, K.S.; Mikkelsen, R.; Sørensen, J.N.; Breton, S.P.; Henningson, D. Large-eddy simulations of the Lillgrund wind farm. *Wind Energy* **2015**, *18*, 449–467. [CrossRef]

5. Troldborg, N. Actuator Line Modeling of Wind Turbine Wakes. Ph.D. Thesis, Technical University of Denmark, Lyngby, Denmark, 2008.

6. Ivanell, S.A. Numerical Computations of Wind Turbine Wakes. Ph.D. Thesis, Royal Institute of Technology, Stockholm, Sweden, 2009.

7. Chamorro, L.P.; Porté-Agel, F. A wind-tunnel investigation of wind-turbine wakes: Bundary-layer turbulence effects. *Bound.-Layer Meteorol.* **2009**, *132*, 129–149. [CrossRef]

8. Porté-Agel, F.; Wu, Y.T.; Lu, H.; Conzemius, R.J. Large-eddy simulation of atmospheric boundary layer flow through wind turbines and wind farms. *J. Wind Eng. Ind. Aerodyn.* **2011**, *99*, 154–168. [CrossRef]

9. Sørensen, J.N.; Myken, A. Unsteady actuator disc model for horizontal axis wind turbines. *J. Wind Eng. Ind. Aerodyn.* **1992**, *39*, 139–149. [CrossRef]

10. Ammara, I.; Leclerc, C.; Masson, C. A viscous three-dimensional differential/actuator-disk method for the aerodynamic analysis of wind farms. *J. Sol. Energy Eng.* **2002**, *124*, 345–356. [CrossRef]

11. Jiménez, A.; Crespo, A.; Migoya, E.; Garcia, J. Large-eddy simulation of spectral coherence in a wind turbine wake. *Environ. Res. Lett.* **2008**, *3*, 015004. [CrossRef]

12. Aubrun, S.; Loyer, S.; Hancock, P.; Hayden, P. Wind turbine wake properties: Comparison between a non-rotating simplified wind turbine model and a rotating model. *J. Wind Eng. Ind. Aerodyn.* **2013**, *120*, 1–8. [CrossRef]

13. Aubrun, S.; Devinant, P.; Espana, G. Physical modelling of the far wake from wind turbines. Application to wind turbine interactions. In Proceedings of the European Wind Energy Conference, Milan, Italy, 7–10 May 2007; pp. 7–10.

14. Espana, G.; Aubrun, S.; Loyer, S.; Devinant, P. Spatial study of the wake meandering using modelled wind turbines in a wind tunnel. *Wind Energy* **2011**, *14*, 923–937. [CrossRef]

15. Espana, G.; Aubrun, S.; Loyer, S.; Devinant, P. Wind tunnel study of the wake meandering downstream of a modelled wind turbine as an effect of large scale turbulent eddies. *J. Wind Eng. Ind. Aerodyn.* **2012**, *101*, 24–33. [CrossRef]

16. Thacker, A.; Loyer, S.; Aubrun, S. Comparison of turbulence length scales assessed with three measurement systems in increasingly complex turbulent flows. *Exp. Therm. Fluid Sci.* **2010**, *34*, 638–645. [CrossRef]

17. Sumner, J.; Espana, G.; Masson, C.; Aubrun, S. Evaluation of RANS/actuator disk modelling of wind turbine wake flow using wind tunnel measurements. *Int. J. Eng. Syst. Model. Simul.* **2013**, *5*, 147–158. [CrossRef]

18. Tabor, G.R.; Baba-Ahmadi, M. Inlet conditions for large eddy simulation: a review. *Comput. Fluids* **2010**, *39*, 553–567. [CrossRef]

19. Lund, T.S.; Wu, X.; Squires, K.D. Generation of turbulent inflow data for spatially-developing boundary layer simulations. *J. Comput. Phys.* **1998**, *140*, 233–258. [CrossRef]

20. Klein, M.; Sadiki, A.; Janicka, J. A digital filter based generation of inflow data for spatially developing direct numerical or large eddy simulations. *J. Comput. Phys.* **2003**, *186*, 652–665. [CrossRef]

21. Mann, J. The spatial structure of neutral atmospheric surface-layer turbulence. *J. Fluid Mech.* **1994**, *273*, 141–168. [CrossRef]

22. Mann, J. Wind field simulation. *Probab. Eng. Mech.* **1998**, *13*, 269–282. [CrossRef]

23. Peña, A.; Hasager, C.; Lange, J.; Anger, J.; Badger, M.; Bingöl, F.; Bischoff, O.; Cariou, J.P.; Dunne, F.; Emeis, S.; et al. *Remote Sensing for Wind Energy*; Technical Report; DTU Wind Energy: Roskilde, Denmark, 2013.

24. Keck, R.E.; Mikkelsen, R.; Troldborg, N.; de Maré, M.; Hansen, K.S. Synthetic atmospheric turbulence and wind shear in large eddy simulations of wind turbine wakes. *Wind Energy* **2014**, *17*, 1247–1267. [CrossRef]

25. Nilsson, K. Numerical Computations of Wind Turbine Wakes and Wake Interaction. Ph.D. Thesis, Royal Institute of Technology, Stockholm, Sweden, 2015.

26. Bechmann, A. Large-Eddy Simulation of Atmospheric Flow over Complex. Ph.D. Thesis, Risø Technical University of Denmark, Roskilde, Denmark, 2006.

27. Gilling, L.; Sørensen, N.N. Imposing resolved turbulence in CFD simulations. *Wind Energy* **2011**, *14*, 661–676. [CrossRef]

28. Troldborg, N.; Zahle, F.; Réthoré, P.E.; Sørensen, N.N. Comparison of wind turbine wake properties in non-sheared inflow predicted by different computational fluid dynamics rotor models. *Wind Energy* **2015**, *18*, 1239–1250. [CrossRef]

29. Gilling, L. *TuGen: Synthetic Turbulence Generator, Manual and User's Guide*; Technical Report DCE-76; Department of Civil Engineering, Aalborg University: Aalborg, Denmark, 2009.

30. Nilsen, K.M.; Kong, B.; Fox, R.O.; Hill, J.C.; Olsen, M.G. Effect of inlet conditions on the accuracy of large eddy simulations of a turbulent rectangular wake. *Chem. Eng. J.* **2014**, *250*, 175–189. [CrossRef]

31. Espana, G. Étude expérimentale du sillage lointain des éoliennes à axe horizontal au moyen d'une modélisation simplifiée en couche limite atmosphérique. Ph.D. Thesis, Université d'Orléans, Orléans, France, 2009. (In French)

32. Weller, H.G.; Tabor, G.; Jasak, H.; Fureby, C. A tensorial approach to computational continuum mechanics using object-oriented techniques. *Comput. Phys.* **1998**, *12*, 620–631. [CrossRef]

33. The OpenFOAM Foundation. OpenFOAM: The Open Source CFD Toolbox; User Guide. 2016. Available online: https://cfd.direct/openfoam/user-guide (accessed on 29 August 2018).

34. Michelsen, J. *Basis3D—A Platform for Development of Multiblock PDE Solvers*; Technical Report AFM 92-05; Technical University of Denmark: Lyngby, Denmark, 1992.

35. Michelsen, J.A. *Block Structured Multigrid Solution of 2D and 3D Elliptic PDE's*; Technical Report AFM 94-06; Technical University of Denmark: Lyngby, Denmark, 1994.

36. Sørensen, N.N. General Purpose Flow Solver Applied to Flow over Hills. Ph.D. Thesis, Risø Technical University of Denmark, Lyngby, Denmark, 1995.

37. Bailly, C.; Comte-Bellot, G. *Turbulence*; CNRS éditions: Paris, France, 2003. (In French)

38. Jiménez, J. (Ed.) *A Selection of Test Cases for the Validation of Large-Eddy Simulations of Turbulent Flows*; Technical Report AGARD Advisory Report No. 345; Working Group 21 of the Fluid Dynamics Panel, North Atlantic Treaty Organization: Neuilly-sur-Seine, France, 1997.

39. Kang, H.S.; Chester, S.; Meneveau, C. Decaying turbulence in an active-grid-generated flow and comparisons with large-eddy simulation. *J. Fluid Mech.* **2003**, *480*, 129–160. [CrossRef]

40. Pope, S.B. *Turbulent Flows*; Cambridge Univ Press: Cambridge, UK, 2000.

41. Comte-Bellot, G.; Corrsin, S. The use of a contraction to improve the isotropy of grid-generated turbulence. *J. Fluid Mech.* **1966**, *25*, 657–682. [CrossRef]

42. Mydlarski, L.; Warhaft, Z. On the onset of high-Reynolds-number grid-generated wind tunnel turbulence. *J. Fluid Mech.* **1996**, *320*, 331–368. [CrossRef]

43. Smagorinsky, J. General circulation experiments with the primitive equations: I. The basic experiment*. *Mon. Weather Rev.* **1963**, *91*, 99–164. [CrossRef]

44. Muller, Y.A. Étude du méandrement du sillage éolien lointain dans différentes conditions de rugosité. Ph.D. Thesis, Université d'Orléans, Orléans, France, 2014. (In French)

45. Porté-Agel, F.; Meneveau, C.; Parlange, M.B. A scale-dependent dynamic model for large-eddy simulation: Application to a neutral atmospheric boundary layer. *J. Fluid Mech.* **2000**, *415*, 261–284. [CrossRef]

46. Larsen, T.J. *Turbulence for the IEA Annex 30 OC4 Project*; Technical Report I-3206; Risø Technical University of Denmark: Roskilde, Denmark, 2013.

47. Kaimal, J.C.; Finnigan, J.J. *Atmospheric Boundary Layer Flows: Their Structure and Measurement*; Oxford University Press: Oxford, UK, 1994.

48. El Kasmi, A.; Masson, C. An extended k–ε model for turbulent flow through horizontal-axis wind turbines. *J. Wind Eng. Ind. Aerodyn.* **2008**, *96*, 103–122. [CrossRef]

49. Réthoré, P.E.M. Wind Turbine Wake in Atmospheric Turbulence. Ph.D. Thesis, Technical University of Denmark, Risø National Laboratory for Sustainable Energy Risø National laboratoriet for Bæredygtig Energi, Lyngby, Denmark, 2009.

50. Comte-Bellot, G.; Corrsin, S. Simple Eulerian time correlation of full-and narrow-band velocity signals in grid-generated, 'isotropic'turbulence. *J. Fluid Mech.* **1971**, *48*, 273–337. [CrossRef]

51. Mydlarski, L.; Warhaft, Z. Passive scalar statistics in high-Péclet-number grid turbulence. *J. Fluid Mech.* **1998**, *358*, 135–175. [CrossRef]

52. Mohamed, M.S.; Larue, J.C. The decay power law in grid-generated turbulence. *J. Fluid Mech.* **1990**, *219*, 195–214. [CrossRef]
53. Fletcher, C. *Computational Techniques for Fluid Dynamics 1*; Springer Science & Business Media: Berlin, Germany, 1991; Volume 1.
54. Olivares-Espinosa, H. Turbulence Modelling in Wind Turbine Wakes. Ph.D. Thesis, École de Technologie Supérieure, Université du Québec, Montreal, QC, Canada, 2017.
55. Oppenheim, A.V.; Schafer, R.W.; Buck, J.R. *Discrete-Time Signal Processing*; Prentice Hall: Upper Saddle River, NJ, USA, 1999.

 *applied sciences*

Article

Comparison of the Blade Element Momentum Theory with Computational Fluid Dynamics for Wind Turbine Simulations in Turbulent Inflow

Sebastian Ehrich [1,*], Carl Michael Schwarz [1], Hamid Rahimi [1,2], Bernhard Stoevesandt [2] and Joachim Peinke [1,2]

[1]   ForWind, University of Oldenburg, Institute of Physics, Küpkersweg 70, 26129 Oldenburg, Germany; carl.michael.schwarz@uni-oldenburg.de (C.M.S.); hamid.rahimi@uni-oldenburg.de (H.R.); joachim.peinke@uni-oldenburg.de (J.P.)

[2]   Fraunhofer IWES, Küpkersweg 70, 26129 Oldenburg, Germany; bernhard.stoevesandt@iwes.fraunhofer.de

[*]   Correspondence: sebastian.ehrich@uni-oldenburg.de; Tel.: +49-441-798-5057

Received: 30 September 2018; Accepted: 2 December 2018; Published: 6 December 2018

Abstract: In this work three different numerical methods are used to simulate a multi-megawatt class class wind turbine under turbulent inflow conditions. These methods are a blade resolved Computational Fluid Dynamics (CFD) simulation, an actuator line based CFD simulation and a Blade Element Momentum (BEM) approach with wind fields extracted from an empty CFD domain. For all three methods sectional and integral forces are investigated in terms of mean, standard deviation, power spectral density and fatigue loads. It is shown that the average axial and tangential forces are very similar in the mid span, but differ a lot near the root and tip, which is connected with smaller values for thrust and torque. The standard deviations in the sectional forces due to the turbulent wind fields are much higher almost everywhere for BEM than for the other two methods which leads to higher standard deviations in integral forces. The difference in the power spectral densities of sectional forces of all three methods depends highly on the radial position. However, the integral densities are in good agreement in the low frequency range for all methods. It is shown that the differences in the standard deviation between BEM and the CFD methods mainly stem from this part of the spectrum. Strong deviations are observed from 1.5 Hz onward. The fatigue loads of torque for the CFD based methods differ by only 0.4%, but BEM leads to a difference of up to 16%. For the thrust the BEM simulation results deviate by even 29% and the actuator line by 7% from the blade resolved case. An indication for a linear relation between standard deviation and fatigue loads for sectional as well as integral quantities is found.

Keywords: wind turbine; aerodynamics; turbulent inflow; Computational Fluid Dynamics; blade element momentum theory; actuator line method; Fatigue Loads

---

1. Introduction

One of the most common tools for wind turbine design and load calculation is the Blade Element Momentum method (BEM). Although BEM based tools are known to be very efficient, they are based on many simplifications. Among others, some worth mentioning are stationary environment, 2D airfoil characteristics and inviscid fluid behavior [1]. While preserving its efficiency, over the years a lot of correction models have been developed to improve the accuracy, e.g., by using tip loss or dynamic stall models. A comprehensive overview of those models can be found in the work of Schepers [1]. However, especially for turbulent inflow scenarios, the reliability of such estimations is still unknown. This is mainly due to the relatively coarse discretization of wind fields and rotor blades, the incorrect dynamic aerodynamic response to complex inflow, the unresolved flow field but also the assumption of inviscid and stationary flow. Nevertheless, wind turbines are operating under atmospheric wind conditions and the dynamics of the forces that are acting on the blades are still unknown. Most of the effects were measured in the wind tunnel with uniform or very simplified changing inflow conditions. However, turbulent inflow scenarios are relevant to the design and certification process, e.g., to the design load cases 1.1–1.3 of the IEC standard [2]. The research conducted in this field is limited with recent studies done by Madsen et al. [3] and in the AVATAR project [4]. In both cases BEM is compared to higher fidelity codes. In the work by Madsen et al. a 2.5MW turbine was simulated with BEM and Blade Resolved (BR) Computational Fluid Dynamics (CFD) with a turbulent inflow based on the model by Mann [5]. The wind fields for BEM were extracted at the rotor position in the empty CFD domain. Additionally, measurement data of this turbine and BEM results for the same data were compared. In both cases spectra of sectional forces were investigated. In the AVATAR project similar quantities were in the focus of the study, but one Mann box was generated and used for the CFD as well as the BEM simulation. In CFD the natural decaying process of the wind fields leads to a decreased turbulence intensity and different turbulent structures at the turbine compared to BEM. Therefore the simulation results of BEM and CFD are not comparable. Also global quantities like thrust and torque as well as their corresponding fatigue loads were not investigated. Although those works are giving a valuable insight to BEM modeling uncertainties when combined with turbulent inflow, there is still a lack of comparable studies and a best practice guide for the comparison of CFD and BEM simulations is still missing.

The present work aims to compare BEM based simulations for turbulent inflow conditions with two higher fidelity CFD methods, namely the BR simulation and the Actuator Line (AL) model [6]. New in this work is the combination of the investigation of frequency resolved integral forces, the analysis of fatigue loads and a bridge between BR and BEM calculations, namely the AL method. The goal of this study is to shed light into the difficulties and expectations of the different methods proposed and information for future comparisons of CFD and BEM based simulations shall be provided.

This paper is organized as following. The basic setup of the turbine, the turbulent wind fields, the CFD and BEM simulations, and the fatigue load estimation are addressed in Section 2. In Section 3 key aerodynamic quantities of the turbine such as power, thrust and sectional forces and the corresponding equivalent fatigue loads as well as spectral properties are investigated. Finally a conclusion including future work is presented in Section 4.

2. Methodology

In this section the basic methodology used in this work concerning the wind turbine, the wind fields and the numerical setup as well as a short introduction to the fatigue load estimation procedure is presented.

2.1. The Wind Turbine Model

The numerical studies are performed on the NREL-5MW reference wind turbine with a rotor diameter of 126 m [7]. This three bladed horizontal axis wind turbine with a blade length of 61.5 m was designed by the National Renewable Energy Laboratory. It is close to the current average rotor size of newly installed wind turbines which accounts to 100–120 m. More details on the reference turbine can be found in Table 1.

Table 1. Overview of NREL-5MW turbine specification.

Parameter	Value
Number of rotor blades	3
Rotor diameter	126 m
Rated aerodynamic power	5.3 MW
Rated wind speed	11.4 m/s

Table 1. *Cont.*

Parameter	Value
Rated rotational speed	12.1 rpm
Blade cone angle	2.5°
Shaft tilt angle	5.0°
Blade length	61.5 m
Blade mass	17,740 kg

2.2. The Wind Fields

It is assumed that the specific inflow turbulence model does not change the general trend of the outcome of the study presented in this paper as long as some main properties of the wind fields are fulfilled like spatial and temporal correlations, Gaussianity and stationarity. However, the model used here should also be as simple as possible with a very low memory consumption which is for spectral methods usually very high if the time and space domains are large. Therefore velocity fluctuations are represented by a Gaussian process realized at the inflow patch of the CFD domain. In particular, coupled stochastic Ornstein-Uhlenbeck processes of the form

$$\frac{\mathrm{d}u_i^{(k)}(t)}{\mathrm{d}t} = -\gamma \left(u_i^{(k)}(t) - \bar{u}^{(k)} \right) + \sqrt{D} \sum_{j}^{N} H_{ij} \Gamma_j^{(k)}(t) \tag{1}$$

with an exponentially decaying spatial and temporal correlation function for all wind components are used without taking shear into account. In this stochastic differential equation each subscript $i, j = 1 \ldots N$ denotes a specific cell on the inlet patch where the turbulence is generated and each superscript $k = x, y, z$ corresponds to the longitudinal, lateral or vertical velocity component. $\gamma = 0.97 \text{ s}^{-1}$ is a damping factor, $\bar{u}^{(y,z)} = 0 \text{ ms}^{-1}$ and $\bar{u}^{(x)}$ the mean velocities in x, y and z direction, $D = 0.62 \text{ m}^2\text{s}^{-3}$ a diffusion constant, H the lower Cholesky-decomposed correlation matrix and Γ a Gaussian, delta-correlated random noise vector with variance 2.

Fields with a mean wind speed of $\bar{u}^{(x)} = 11.4 \text{ ms}^{-1}$ corresponding to the rated operating condition of the wind turbine and a Turbulence Intensity (TI) of 20% are fed into the CFD domain. This relatively high TI had to be chosen, because of the natural decaying process leading to much smaller turbulence intensities. namely 11%, in the vicinity of the turbine in the absence of shear. To make the comparison between different codes in a similar manner, the pitch and rotor speed are fixed and the effect of structural deformations, tilt and cone as well as the tower and nacelle are neglected.

2.3. Numerical Methods

In the following the BR, AL and BEM method are described in more detail:

- BR: The blade resolved simulations are performed with the CFD software OpenFOAM V4.1 [8], which is an open-source package with a collection of modifiable libraries written in C++. The simulation domain is discretized using the bladeblockMesher [9] and windTurbine mesher tools [10] with a combination of structured and unstructured grids consisting of a total of 36M cells. To exclude influences of the mesh on the solution, a mesh study has been done in our previous work [11]. The grid of the complete simulation domain and the rotating part are shown in Figure 1 and a sectional view on the mesh near the blade in Figure 2. The cell length at the inlet, which is three rotor diameters upstream of the rotor, is 1m. This mesh is uniformly extruded up to the near blade mesh where a refinement takes place and the smallest cell length of approximately 0.2 m is reached. In order to limit the cell aspect ratio in radial direction, adaptive wall functions are being applied. The solution for the flow field is achieved by solving the incompressible Delayed Detached Eddy Simulations equations. It is assumed that the flow is fully turbulent, i.e., the laminar-turbulent transition in the boundary layer is not considered, because the inflow itself is turbulent and therefore the laminar boundary layer is not expected to have a strong influence on the blade. The closure problem is treated by use of the turbulence model proposed by Spalart et al. [12] with the proposed standard parameters. In order to solve the pressure-velocity coupling, the PIMPLE algorithm, is used, which is a combination of the loop structures of SIMPLE [13] and PISO [14]. Integration in time was performed using the implicit second order backward scheme together with the second order linear upwind scheme for discretizing the convection terms. For these simulations 360 cores were used for approximately 20 days resulting in 600 s simulation time with a time step of 0.01 s. This corresponds to 121 revolutions of the rotor.

- AL: The rotor is simplified by an AL model [6], where all rotor blades are reduced to rotating lines with a radial distribution of body forces at 60 positions along the lines. A Gaussian filtering kernel with the dimensionless smoothing parameter ε equal to 4 was used to obtain well fitting sectional forces and power output to the BR case. Based on the angle of attack at the given positions on the line, body forces are gained from lookup tables containing polars of two dimensional airfoils. A Prandtl tip [15] and root loss correction, as well as a 3D correction proposed by Du and Selig [16] are applied to the airfoil data. Except for a coarser resolution in the vicinity of the rotor with an average cell length of 1m the domain is exactly the same as for the BR case leading to a domain size of 22M cells. The flow field is simulated by Large Eddy Simulations (LES) with a Smagorinsky subgrid-scale model [17]. These simulations have been conducted on 360 cores for 25 h real time reaching 600 s simulation time with a time step of 0.01 s.

- BEM: This method uses wind fields recorded in the CFD domain on an equidistant 31×31 grid with a grid cell size of 4.5 m in both directions. In this case the fields are obtained at the position of the rotor from the same mesh as for the AL method but in an empty CFD domain. This seems to be a reasonable choice, because under this condition the TI at the rotor will be comparable for the CFD and BEM simulations. The BEM based tool FAST v8 [18] in combination with AeroDyn v15 [19] with the Beddoes Leishman type unsteady aerodynamics model [19,20] and an equilibrium wake model is used in this work. The BEM model relies on the very same airfoil data and correction models as the AL, but with only 17 blade sections. Because of the simplicity of the BEM approach the simulation with the same time stepping and total time as for the higher fidelity approaches only took two minutes on one core.

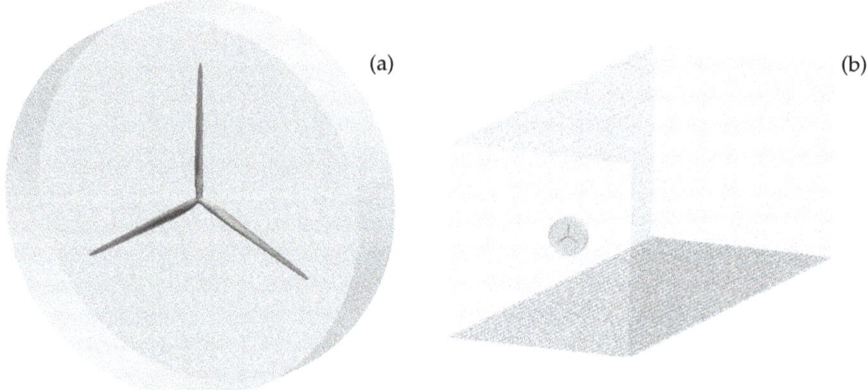

Figure 1. Grids of (**a**) rotating part and (**b**) full domain of Blade Resolved (BR) simulation.

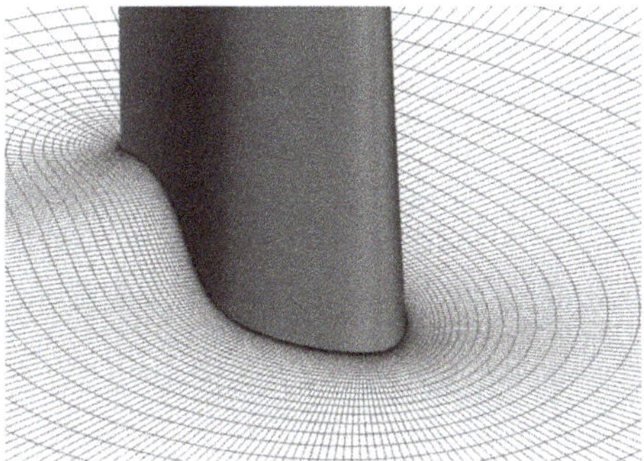

Figure 2. View onto the grid at the pressure side of the rotor blade.

2.4. Fatigue Load Estimation

One of the most important quantities relevant for wind turbine designs are the fatigue loads on different parts of the turbine estimated by damage equivalent loads. They are also known as Equivalent Fatigue Loads (EFL), which will play a role in the load analysis of the three methods investigated here. EFL can be defined as

$$EFL = \left(\frac{\sum n_i r_i^m}{T} \right)^{\frac{1}{m}},$$ (2)

where r_i are the specific load ranges, n_i the corresponding amount of load ranges, T the reference number of ranges (here the dimensionless time in seconds) and $m = 8$ the Wöhler exponent. More information on the estimation of the EFL and the underlying rainflow algorithm can be found in [21,22].

3. Results

In this section the results of this study are presented. The focus lies on the wake structure, average and standard deviation of sectional and integral forces, as well as fatigue loads and Power Spectral Densities (PSD).

3.1. Wake Dynamics

Snapshots of the velocity magnitude for the BR and AL cases are shown in Figure 3. Despite the differences in the methods both simulations show similar fields in terms of the velocity magnitude upstream of the turbine. This leads to the conclusion that the induction zone of the turbine is also similar for both simulations. For a more detailed view of the similarities and differences of the two compared CFD methods the wake structure is analyzed by the time series of the main stream component at 0.5, 1 and 2 diameters at the centerline downstream of the nacelle in Figure 4. It can be seen that the wind velocity is in good agreement in most time intervals. As a quantitative measure of similarity of the wake structure, the coefficients for the correlation between both simulations are calculated by

$$\rho(x) = \frac{\langle (u_{AL}(x,t) - \bar{u}_{AL}(x)) \, (u_{BR}(x,t) - \bar{u}_{BR}(x)) \rangle_t}{\sigma_{AL}(x)\sigma_{BR}(x)}$$ (3)

where we used the time averaging operation $\langle . \rangle_t$, the standard deviations σ_{AL} and σ_{BR} and the local mean values $\bar{u}_{AL}$ and $\bar{u}_{BR}$ of the time series of AL and BR simulations. The correlation is 0.54, 0.41 and 0.51 for 0.5, 1 and 2 diameters downstream, respectively. This shows that the wind field dynamics are rather similar and not uncorrelated at different distance from the rotor for both simulation methods. However, the BR simulation is resolving the tip and root vortex in a more physical way which results in smaller structures in the near wake.

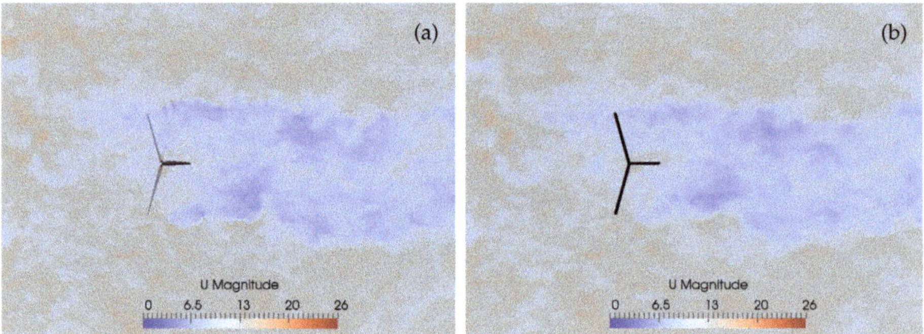

Figure 3. Snapshots of the velocity magnitude for (**a**) BR and (**b**) Actuator Line (AL) simulations.

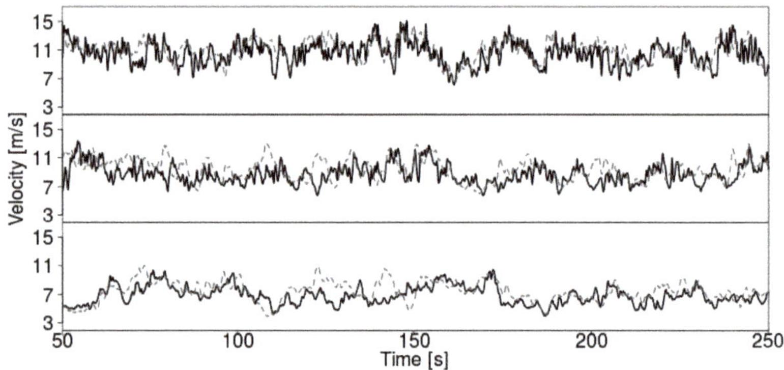

Figure 4. Wind velocity time series of BR (black, solid) and AL (grey, dashed) simulation for 0.5, 1 and 2 diameters (from top to bottom) at the centerline downstream of the rotor nacelle.

3.2. Sectional Forces

All three simulation methods (BR, AL and BEM) are compared based on averaged sectional quantities, namely axial and tangential forces, the corresponding standard deviations and the EFLs. The sectional axial and tangential forces on the blade do not reveal big differences for the average between all cases in the mid-span as shown in Figure 5. However at the root (up to 30% of span) and the very tip (from 80% of span onwards) some differences can be observed. BEM and AL rely on similar root and tip correction models and the very same airfoil data which explains their good agreement. In comparison to the more realistic BR case, where 3D effects are modeled physically, the forces at the root and tip section are far off supporting the well known issues with 3D effects for BEM and AL. The more developed vortices in the BR simulation induce much higher loads in the root and tip region than for the other methods.

The standard deviations for the tangential and axial force distributions, shown in more detail in Figure 6, are very different for all three methods. With the BR case as the reference the standard deviation of sectional forces tend to be lower at the root of the blade for the BEM and AL methods whereas they are larger in the mid-span. Close to the tip at approximately 95% of the blade the AL method gives much smaller values while BEM is in the range of the BR simulation. The corresponding errors are estimated by following three steps. In the first step the time series for the specific forces at each radial position are separated into five distinct sub time series. In the second step the standard deviation for each sub time series is calculated. Finally the standard deviation of the set of the standard deviations from the second step is computed which provides the error for the whole time series.

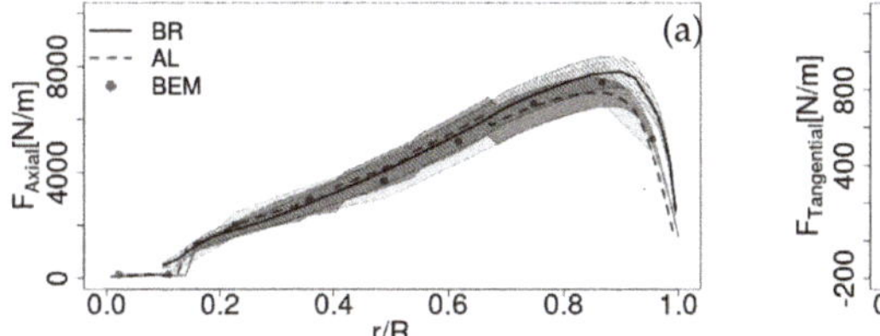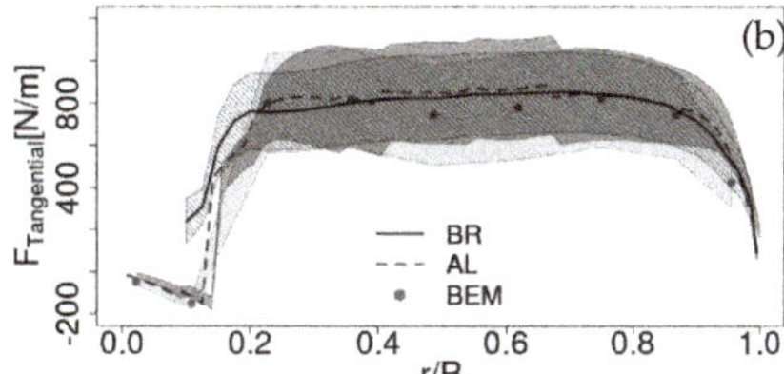

Figure 5. Radial distributions of (**a**) axial and (**b**) tangential forces per unit span. The shaded areas illustrate one standard deviation of the fluctuations.

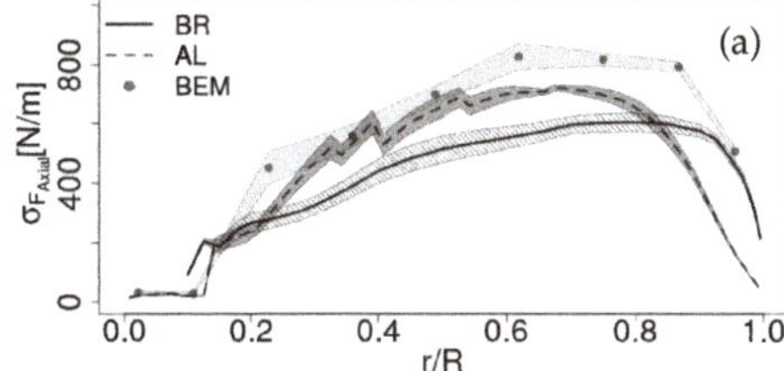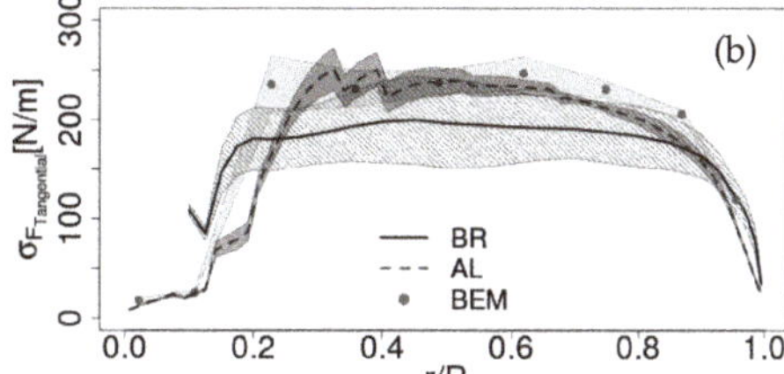

Figure 6. Radial distribution of standard deviation of (**a**) axial and (**b**) tangential forces per unit span. The shaded areas correspond to errors.

A more detailed insight in the standard deviation is gained by looking at the PSDs of axial and tangential forces at 22%, 50%, 75% and 95% radial position, shown in Figures 7 and 8.

All blade sections have in common that the spectra of the highly resolved BR case drop fastest between 2 and 10 Hz to a constant noise level. This is due to the expected bigger structures generated by the blade geometry but also due to the DDES model which is treated as RANS in the boundary layer acting as a filter.

For the AL method the energy drop happens at much higher frequencies because of the missing geometry and the LES model, which generally filters at smaller scales. The noisy behavior at high frequencies is part of every simulation of this kind and is due to numerical instabilities resulting from the convolution of the forces given at the airfoil sections with a Gaussian kernel [23].

The BEM spectra show filtering effects at higher frequencies corresponding to a multiplication with a *sinc*-like-function but with different frequencies for different radial positions. The origin of this effect could not be found and has to be investigated further. Due to the missing geometry and the lack of unsteady wake effects BEM is also expected to make worse predictions than the higher fidelity methods in the high frequency range. Therefore this high frequency part has to be taken with care like for the other two methods with their aforementioned problems of smoothing for the AL and the RANS treatment for the BR case and none of them can be considered trustworthy in this range.

In the low frequency range, for axial as well as tangential forces it can be observed that in comparison to the other two methods BEM has a much higher energy content for the inboard section. This is responsible for the high standard deviation at this section.

At radial position of 50% and 75% of the blades all the spectra are very similar in the low frequency range but it is still obvious that BEM has a higher energy content up to the 1P frequency, especially for the axial forces. Furthermore all three methods are able to capture the 1P frequency which is based on the passing of the blades through long time correlated structures.

At the outboard section at 95% the energy content of the BEM results in low frequencies is in the same range as for the BR case, but the AL method shows significant smaller values and the 1P frequency is not captured anymore, especially for the axial forces, which is due to a smeared out tip vortex.

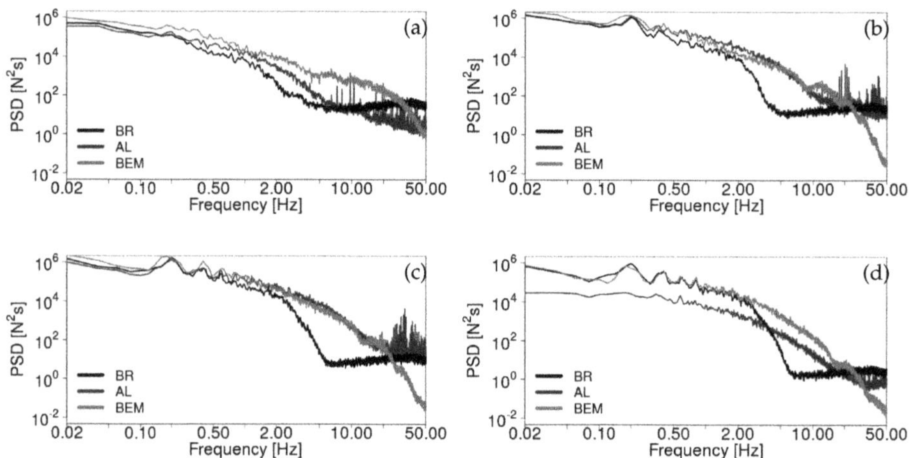

Figure 7. Power Spectral Density (PSD) of axial forces at (**a**) 22.5%; (**b**) 50%; (**c**) 75% and (**d**) 95% radial position.

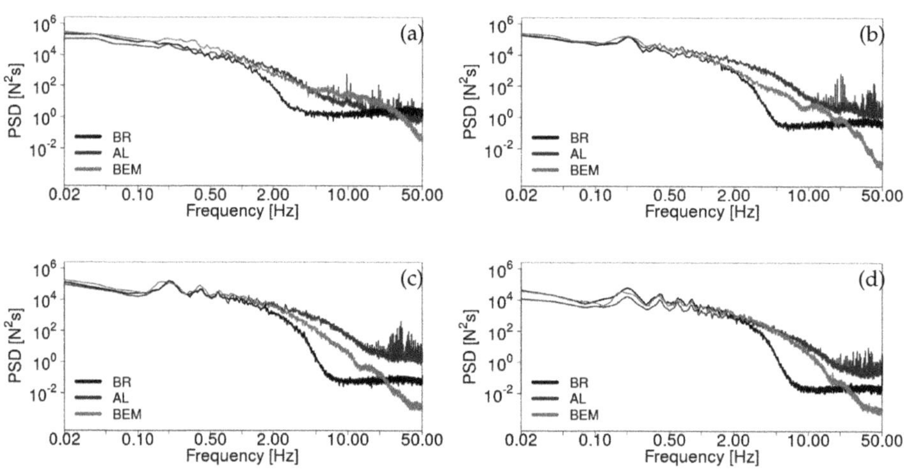

Figure 8. PSD of tangential forces at (**a**) 22.5%; (**b**) 50%; (**c**) 75% and (**d**) 95% radial position.

To illustrate the different flow structure sizes in the different simulation methods, the two point correlation of the sectional forces along the blade are measured. The argument for this approach is that the correlation of sectional forces is similar to the correlation of the wind velocity close to the blade surface, which is strongly connected to the integral scales of wind. In Figure 9 the correlation at each radial position with the forces at a reference position, here 22.5% radial location, is shown. This specific position is in principle arbitrary, but if the focus lies on the measurement of big structures, this point should not be too close to the tip or root where smaller flow structures dominate. However, even if the tendency for the correlation functions of the three different models is the same, the larger correlation values for the BR case along the whole blade indicate that the structure sizes are also bigger than for the AL and BEM cases.

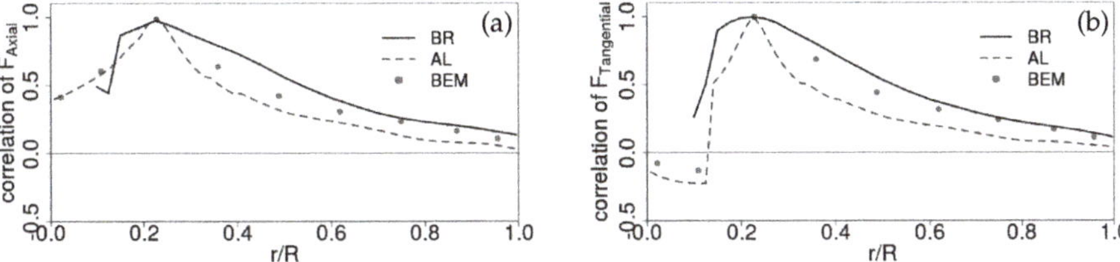

Figure 9. Correlation of (**a**) axial and (**b**) tangential forces at different radial positions with 22.5% position.

To give a more detailed description of the effect of forces on the rotor blades, fatigue loads at each section have been investigated. As observed in Figure 10, the EFL distribution over the whole blade is similar to the distribution of standard deviation in Figure 6. The BEM method gives higher EFL for every section for axial forces and almost everywhere higher values for tangential forces. All other points mentioned for sectional standard deviations also hold for sectional EFL. The errors are calculated with the same method as for Figure 6.

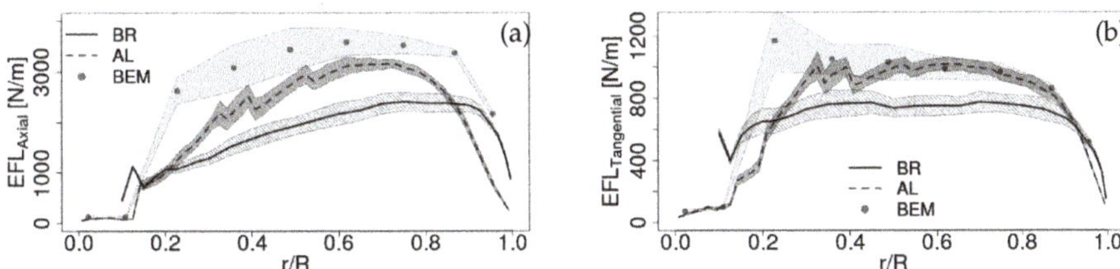

Figure 10. Radial distribution of (**a**) axial and (**b**) tangential equivalent fatigue loads per unit span. The shaded areas correspond to the errors.

The sectional EFL are linearly dependent on the sectional standard deviation for all simulation methods as seen in Figure 11. This raises the point that the dynamics quantified by two different properties of the load time series, namely standard deviation and EFL, are similar for all three methods. With the Wöhler Exponent $m = 8$ and the wind fields used here, this linear relation follows approximately $EFL = 4\sigma$ for axial and tangential forces.

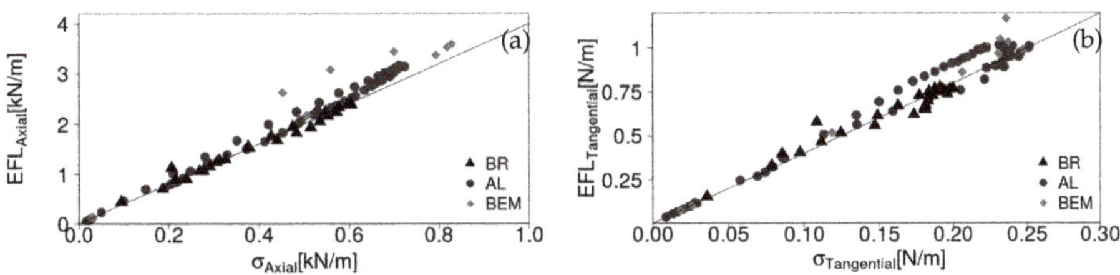

Figure 11. Equivalent Fatigue Loads (EFL) dependence on standard deviation for (**a**) axial and (**b**) tangential forces for all three methods. The diagonal line is governed by $EFL = 4\sigma$.

3.3. Integral Forces

In the following, integral forces resulting from sectional forces from Section 3.2 are discussed. Time series of the aerodynamic thrust and torque are shown in Figure 12. As can be observed, the thrust as well as the torque evolution is very similar for BR and AL. With the definition of the correlation coefficient

$$\rho = \frac{\langle (F_1(t) - \bar{F}_1)(F_2(t) - \bar{F}_2) \rangle_t}{\sigma_1 \sigma_2} \tag{4}$$

where F_1 and F_2 are the considered forces and σ_1 and σ_2 the corresponding standard deviations, the correlation is calculated as 0.90 and 0.93 for thrust and torque respectively, even if the distribution of sectional forces and the wake dynamics are more different (see Figures 4–6). A comparison between BR and BEM reveals a much smaller correlation with values of 0.08 and 0.09 for thrust and torque. This supports the argument that the wind dynamics are handled in a very different way for BEM and CFD simulations for the current setup.

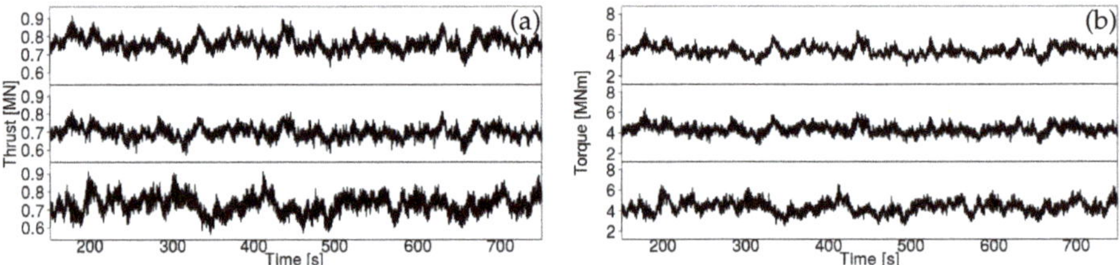

Figure 12. Time series for (**a**) thrust and (**b**) torque for Blade Resolved (BR), Actuator Line (AL) and Blade Element Momentum (BEM) methods (from top to bottom).

In Table 2 the means, standard deviations and EFLs of torque and thrust for all three methods are illustrated with their corresponding errors. Those errors are obtained by separating the corresponding time series' into five distinct equally sized ensembles and calculating the ensemble standard deviation. For comparison reasons the ratios of the calculated quantities for AL and BR as well as BEM and BR are shown, because the BR case is assumed to be the highest fidelity approach. For aerodynamic thrust and torque AL and BEM show lower values than the BR case with a minimum ratio of 91.7%. which can also be estimated from the sectional forces in Figure 5.

Table 2. Thrust and torque comparison by mean, standard deviation σ and EFL between BR, AL and BEM.

		BR	**AL**	**BEM**	**AL/BR [%]**	**BEM/BR [%]**
	mean	761 ± 7.3	698 ± 4.0	729 ± 9.6	91.7	95.8
Thrust [kN]	σ	37.9 ± 4.3	33.9 ± 2.6	49.7 ± 6.5	89.4	131.1
	EFL	147.6 ± 9.7	136.7 ± 4.2	190.1 ± 8.0	92.6	128.8
	mean	4.43 ± 0.10	4.32 ± 0.08	4.35 ± 0.13	97.5	98.2
Torque [MNm]	σ	0.495 ± 0.044	0.452 ± 0.038	0.599 ± 0.061	91.3	121.0
	EFL	1.90 ± 0.15	1.90 ± 0.13	2.29 ± 0.12	99.6	116.0

Analog to the relation between mean sectional forces and mean integral forces, the standard deviations of tangential and axial forces in Figure 6 correspond to a standard deviation in the thrust and torque. The standard deviation for the AL case is for both quantities approximately 10% smaller than for the BR simulation, because of the missing variance in the root and tip region. Especially for the torque the tip region plays a significant role. In contrast to that, for the BEM simulation an approximately 20 to 30% increase in the standard deviation is obtained which is also very clear from the much higher sectional standard deviation except at the root. A reason for the high standard deviation could be the equilibrium wake model used here, where the wake dynamics are not captured in a proper way. The relation between EFL and the standard deviation of integral forces is shown in Figure 13 where we additionally added the diagonal line governed by $EFL = 4\sigma$ like for sectional forces in Figure 11. Sectional as well as integral forces seem to follow the same linear trend.

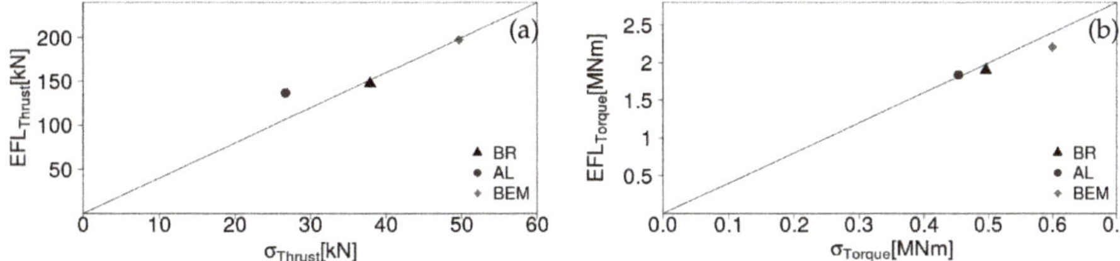

Figure 13. EFL dependence on standard deviation for (**a**) thrust and (**b**) torque for all three methods. The diagonal line is governed by $EFL = 4\sigma$.

Further analysis of the EFL by the underlying range histograms in Figure 14 shows that for the thrust and torque force all three methods have in general a similar trend. But especially in the moderate ranges for the thrust and for large ranges in both quantities BEM has more counts than AL and BR. Due to their large weight the most important part for the EFLs are the high ranges. Those high ranges must not be neglected, even if they have very few counts, because very important quantities like the difference of global maxima and minima and the standard deviation contribute to them. Also the similarity of the inflow conditions justifies the comparison of the loads. The contribution of all ranges to the EFLs in Equation (2), namely $n \cdot r^m$, is shown in Figure 15, where the big contribution of the highest ranges to fatigue loads is very clear.

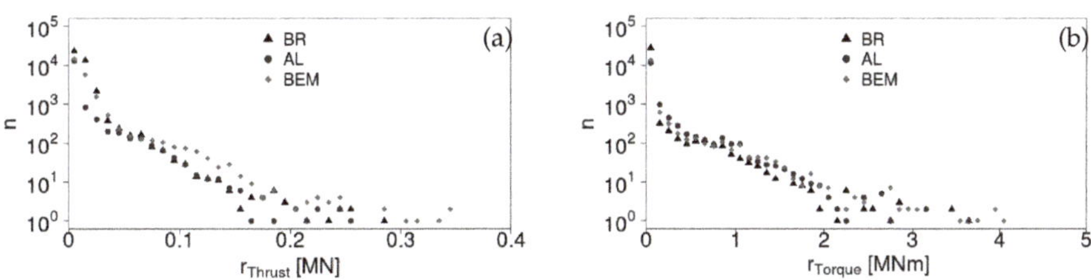

Figure 14. Range histograms for (**a**) thrust and (**b**) torque.

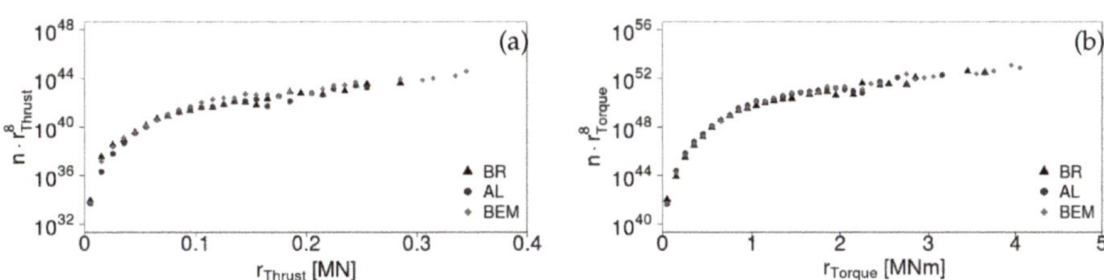

Figure 15. Contribution of ranges to EFL for (**a**) thrust and (**b**) torque.

For an analysis of the dynamics of torque and thrust, PSDs are investigated and shown in Figure 16. All three methods show a very good agreement in the shape of spectra for low frequencies in region I up to 1.5 Hz including the 3P frequency and the first harmonic. For higher frequencies up to approximateley 4 Hz, here region II, BEM and AL are very similar but the BR method drops very quickly and saturates in a white noise like behaviour for higher frequencies in region III. The reason for this is the big structures generating blade geometry and the RANS model working in the boundary layer of the blade while the AL does not have a geometry and it uses LES everywhere and therefore does not smooth out the wind field in the vicinity of the blade as has been discussed already for the sectional forces. It again has to be amplified that the BR case has the highest resolution in the blade region and therefore has the most reliable results. The difference between BEM and AL in the high frequency range in region III could be according to the missing damping, filtering and dynamic wake effects in the BEM case. The AL method uses a Gaussian filter to smooth out the forces along the blade and the wake is much more complex than in the static wake model of BEM. It has to be amplified, that especially for the torque, a transition from BEM to AL can not be made just by filtering the signals, because filtering corresponds to a drop in the PSD, which is not observed in the mid range.

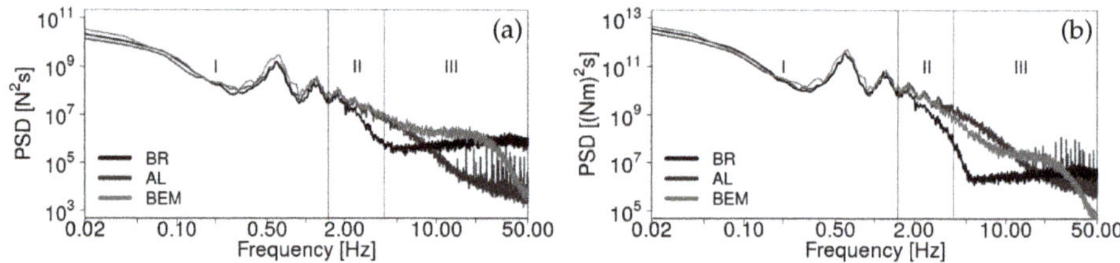

Figure 16. PSD of (a) thrust and (b) torque.

For a more detailed analysis, the cumulative frequency dependent variance normalized by the total variance, i.e.,

$$\hat{\sigma}^2(f) = \frac{\int_0^f PSD(f')\mathrm{d}f'}{\sigma^2} \tag{5}$$

has been investigated in Figure 17. This is equivalent to the energy content of the spectrum at frequencies smaller than f, normalized by the total energy contained in the spectrum. The energy content for the different regions can be determined as $\hat{\sigma}^2_{\text{region}} = \hat{\sigma}^2(f_{\text{max}}) - \hat{\sigma}^2(f_{\text{min}})$ with f_{max} the upper frequency limit and f_{min} the lower frequency limit of the region. For all three method the conclusion can be drawn that by far the most energy is contained in region I, in particular for the thrust: BR 95.4%, AL 93.3% and BEM 94.9%. Region II also contains a small amount of energy with 2.3% for BR, 5.8% for AL and 3.2% for BEM. The torque is distributed similarly with 97.7% for BR, 91.7% for AL and 91.1% for BEM in region I and 2.2%, 7.1% and 3.5% in region II. The variance contained in region III is neglegible for both quantities. Based on the high percentage of variance contained in region I it can be concluded that the big difference in the total variance between BEM and the other two methods mainly stems from this frequency range. Therefore the extraction of wind fields at the position of the rotor plane has to be seen as a first step for this comparison and it has to be investigated if it is more appropriate to extract the fields downstream of the rotor position or to artificially damp the wind input in BEM on large scales.

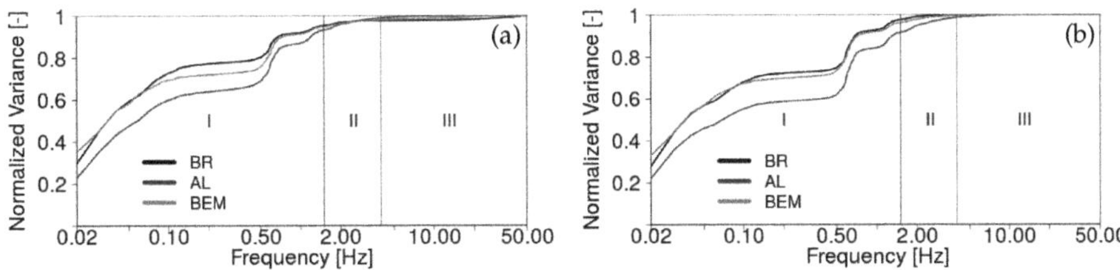

Figure 17. Normalized cumulative frequency dependent variance of (**a**) thrust and (**b**) torque.

4. Discussion

From this work it can be concluded that the AL, BR and the BEM method with extracted wind fields at the rotor plane from an empty domain are very similar with respect to averaged tangential and axial forces, except close to the root and the tip. However, a comparison of BEM with the CFD methods with respect to the standard deviations of sectional and integral forces shows very different results. For the integral quantities thrust and torque, BEM tends to overestimate the standard deviation compared to the other two methods. It was found out that the standard deviation differences between all methods mainly come from the different amplitudes in the first part of the spectrum and not from the filtering effects on larger frequencies. This leads to the conclusion that, if similar results for the standard deviation for BEM and CFD should be achieved, either the wind fields have to be extracted downwind of the rotor position from the empty domain simulation or a correction of wind fields for the BEM case has to be considered. A good starting point could be to substitute the equilibrium wake model by a dynamic wake model which could damp the oscillations and with it the standard deviation. This and extracting the wind fields in an empty domain downwind from the position where the turbine is normally located will be part of the future work. Another contribution to the differences, which also should not be neglected, comes from the fact that the turbulent wind field decays across the CFD domain due to the missing energy injection. In an empty domain, the turbulence intensity is higher upstream and lower downstream of the position, where the turbine would be located. This decay is not considered within the BEM simulation and could be a reason for the differences.

An indication for a linear dependence between standard deviation and EFL could be found for sectional as well as integral forces which seem to follow the same trend. This shows that the dynamics characterized by the EFL and by the standard deviation are similar for all three methods. However, this dependence is assumed to be existent only if the wind fields are comparable with respect to temporal and spatial correlations to assure similar turbulent structures which play a big role for EFL estimations. Out of scope of this paper, an advantage of this linear relationship could be that the computational expensive rainflow count algorithm is not needed if one is interested in equivalent fatigue loads. A simple calculation of the standard deviation could give this information as well. Additionally the equivalent fatigue loads could also be estimated if the same simulation is redone with a different standard deviation for the wind velocities.

Even if no experiments for comparison were performed, the BR simulation with its high resolution close to the blades and the full geometry consideration can be expected to be the most physical and therefore it should be seen as a benchmark case. However, a further part of the future work should be a fair comparison with experiments where the focus should be set on the statistical equivalence of the wind fields at least across the full rotor size in the CFD domain, in the BEM simulation and in the experiments.

Author Contributions: Conceptualization S.E., C.M.S. and H.R.; Software, S.E.; Methodology, S.E., C.M.S. and H.R.; Formal Analysis, S.E.; Writing-Original Draft Preparation, S.E. and H.R.; Writing-Review & Editing, J.P.; Supervision, J.P.; Project Administration, B.S.

Funding: This research received no external funding.

Acknowledgments: The simulations were performed at the HPC Cluster EDDY, located at the University of Oldenburg (Germany) and funded by the Federal Ministry for Economic Affairs and Energy (Bundesministeriums für Wirtschaft und Energie) under grant number 0324005.

Conflicts of Interest: The authors declare no conflict of interest.

Abbreviations

The following abbreviations are used in this manuscript:

AL	Actuator Line
BEM	Blade Element Momentum
BR	Blade Resolved
CFD	Computational Fluid Dynamics
EFL	Equivalent Fatigue Loads
PSD	Power Spectral Density
TI	Turbulence Intensity

References

1. Schepers, J.G. Engineering Models in Wind Energy Aerodynamics. Ph.D. Thesis, Delft University of Technology, Delft, The Netherlands, 2012. [CrossRef]
2. International Electrotechnical Commission. *WIND TURBINES—PART 1: DESIGN REQUIREMENTS;* Standard IEC 61400-1; International Electrotechnical Commission: Geneva, Switzerland, 2005.
3. Madsen, H.A.; Sørensen, N.N.; Bak, C.; Troldborg, N.; Pirrung, G. Measured aerodynamic forces on a full scale 2MW turbine in comparison with EllipSys3D and HAWC2 simulations. *J. Phys. Conf. Ser.* **2018**, *1037.* [CrossRef]
4. Kim, Y.; Lutz, T.; Jost, E.; Gomez-Iradi, S.; Muñoz, A.; Méndez, B.; Lampropoulos, N.; Sørensen, N.; Madsen, H.; van der Laan, P.; et al. AVATAR Deliverable D2.5: Effects of Inflow Turbulence on Large Wind Turbines. AVATAR Deliverable D2.5, Technical Report. Available online: http://www.eera-avatar.eu/fileadmin/avatar/user/avatar_D2p5_revised_20161231.pdf (accessed on 4 December 2018).
5. Mann, J. Wind field simulation. *Probabilistic Eng. Mech.* **1998**, *13*, 269–282. [CrossRef]
6. Sørensen, J.N.; Shen, W.Z. Numerical Modeling of Wind Turbine Wakes. *J. Fluids Eng.* **2002**. *124*, 393–399. [CrossRef]
7. Jonkmann, J.; Butterfield, S.; Musial, W.; Scott, G. *Definition of a 5-MW Reference Wind Turbine for Offshore System Development;* Technical Report TP-500-38060; National Renewable Energy Laboratory: Golden, CO, USA, 2009.
8. OpenCFD. *OpenFOAM—The Open Source CFD Toolbox—User's Guide;* 1.4 ed.; OpenCFD Ltd.: Reading, UK, 2007.
9. Rahimi, H.; Daniele, E.; Stoevesandt, B.; Peinke, J. Development and application of a grid generation tool for aerodynamic simulations of wind turbines. *Wind Eng.* **2016**, *40*, 148–172. [CrossRef]
10. Rahimi, H. Validation and Improvement of Numerical Methods for Wind Turbine Aerodynamics. Ph.D. Thesis, University of Oldenburg: Oldenburg, Germany, 2018.
11. Rahimi, H.; Garcia, A.M.; Stoevesandt, B.; Peinke, J.; Schepers, J.G. An engineering model for wind turbines under yawed conditions derived from high fidelity models. *Wind Energy* **2018**, *21*, 618–633. [CrossRef]
12. Spalart, P.R.; Deck, S.; Shur, M.L.; Squires, K.D.; Strelets, M.K.; Travin, A. A New Version of Detached-eddy Simulation, Resistant to Ambiguous Grid Densities. *Theor. Comput. Fluid Dyn.* **2006**, *20*, 181. [CrossRef]
13. Patankar, S.; Spalding, D. A calculation procedure for heat, mass and momentum transfer in three-dimensional parabolic flows. *Int. J. Heat Mass Transfer* **1972**, *15*, 1787–1806. [CrossRef]
14. Issa, R. Solution of the implicitly discretised fluid flow equations by operator-splitting. *J. Comput. Phys.* **1986**, *62*, 40–65. [CrossRef]
15. Prandtl, L. *Ludwig Prandtl Gesammelte Abhandlungen: Zur Angewandten Mechanik, Hydro- und Aerodynamik;* Springer: Berlin, Germany, 2013.
16. Du, Z.; Selig, M. A 3-D stall-delay model for horizontal axis wind turbine performance prediction. In Proceedings of the 1998 ASME Wind Energy Symposium, Reno, NV, USA, 12–15 January 1998; p. 21.
17. Smagorinsky, J. General Circulation Experiments with the Primitve Equations. *Mon. Weather Rev.* **1963**, *91*, 99–164. [CrossRef]

18. Jonkman, B.; Jonkman, J. *FAST v8.16.00a-bjj*; Technical Report; NREL: Golden, CO, USA, 2016.
19. Jonkman, J.; Hayman, G.; Jonkman, B.; Damian, R. *AeroDyn v15 User's Guide and Theory Manual (Draft Version)*; Technical Report; NREL: Golden, CO, USA, 2016.
20. Leishman, J.G.; Beddoes, T.S. A Semi-Empirical Model for Dynamic Stall. *J. Am. Helicopter Soc.* **1989**, *34*, 3–17. [CrossRef]
21. Madsen, P.H. *Recommended Practices for Wind Turbine Testing and Evaluation. 3. Fatigue Loads*; Riso National Laboratory: Roskilde, Denmark, 1990.
22. Rychlik, I. A new definition of the rainflow cycle counting method. *Int. J. Fatigue* **1987**, *9*, 119–121. [CrossRef]
23. Martínez-Tossas, L.A.; Churchfield, M.J.; Leonardi, S. Large eddy simulations of the flow past wind turbines: Actuator line and disk modeling. *Wind Energy* **2015**, *18*, 1047–1060. [CrossRef]

 *applied sciences*

Article

Evaluation of the Power-Law Wind-Speed Extrapolation Method with Atmospheric Stability Classification Methods for Flows over Different Terrain Types

Chang Xu [1], Chenyan Hao [1], Linmin Li [1,*], Xingxing Han [1], Feifei Xue [1], Mingwei Sun [2] and Wenzhong Shen [3]

[1] College of Energy and Electrical Engineering, Hohai University, Nanjing 211100, China; zhuifengxu@hhu.edu.cn (C.X.); HAOcy@hhu.edu.cn (C.H.); hantone@hhu.edu.cn (X.H.); xuefeifeihhu@163.com (F.X.)
[2] College of Naval Coast Defence Arm, Naval Aeronautical University, Yantai 264001, China; sunmingwei1993@163.com
[3] Department of Wind Energy, Fluid Mechanics Section, Technical University of Denmark, Nils Koppels Allé, Building 403, 2800 Kgs. Lyngby, Denmark; wzsh@dtu.dk
*   Correspondence: lilinmin@hhu.edu.cn

Received: 11 June 2018; Accepted: 18 August 2018; Published: 22 August 2018

Abstract: The atmospheric stability and ground topography play an important role in shaping wind-speed profiles. However, the commonly used power-law wind-speed extrapolation method is usually applied, ignoring atmospheric stability effects. In the present work, a new power-law wind-speed extrapolation method based on atmospheric stability classification is proposed and evaluated for flows over different types of terrain. The method uses the wind shear exponent estimated in different stability conditions rather than its average value in all stability conditions. Four stability classification methods, namely the Richardson Gradient (RG) method, the Wind Direction Standard Deviation (WDSD) method, the Wind Speed Ratio (WSR) method and the Monin–Obukhov (MO) method are applied in the wind speed extrapolation method for three different types of terrain. Tapplicability is analyzed by comparing the errors between the measured data and the calculated results at the hub height. It is indicated that the WSR classification method is effective for all the terrains investigated while the WDSD method is more suitable in plain areas. Moreover, the RG and MO methods perform better in complex terrains than the other methods, if two-level temperature data are available.

Keywords: wind speed extrapolation; atmospheric stability; wind shear; wind resource assessment

1. Introduction

In the feasibility study and microsite selection stage of wind farms, using the wind measurement data is a key step to evaluate the wind resources at the hub height. At present, wind farms are being developed towards low wind speed and high hub height. Since the height of existing wind mast towers is often lower than the hub height which results in a lack of wind data at high hub heights, it is important to develop a reliable method to evaluate the accurate wind resource for wind farm development. The motivation of wind speed extrapolation is to characterize the wind shear and the importance of wind shear characterization for wind speed calculation was mentioned in a number of studies [1–3].

Wind turbines typically operate at altitudes below 200 m, while the height of the convective boundary layer is on the order of 1000 m [4]. The wind shear within 100–200 m is a function of wind speed, atmospheric stability, surface roughness, and height spacing [5,6]. At present, the power law and logarithmic law are commonly used to extrapolate the wind speed. The logarithmic law can represent the vertical wind-speed profiles fairly well under neutral conditions [7] and was extended by using the Monin–Obukhov similarity theory [8] under stable or unstable conditions to include thermal stratification effects [9]. According to Optis et al. [10], the logarithmic wind-speed profile based on the Monin–Obukhov similarity theory was found inaccurate under strong stable conditions due to the shallow surface layer. For unstable conditions, Lackner et al. [11] found that the power law is robust to give a realistic wind profile than the logarithmic law.

In the near-surface layer, the wind speed affected by surface friction and atmospheric stability, varies significantly with the height [12]. Holtslag [13] analyzed the 213-m-high Cabauw meteorological tower data and found that the measured wind profile was consistent with the Monin–Obukhov similarity theory (surface layer scaling) even for the height above 100 m. Gualtieri [14] investigated the relationship between the wind shear coefficient and atmospheric stability based on the dataset recorded from the met mast of Cabauw (Netherlands), and found that the Panofsky and Dutton [15] model appeared particularly skillful during the diurnal unstable hours and during the nighttime under moderately stable conditions. Mohan and Siddiqui [16] summarized various methods for atmospheric stability classification. In the radiation method, the intensity of radiation is determined based on the amount of observed cloud which is difficult to obtain, and in the M-O length method, the heat flux data are required. In addition to these two methods, there are four methods based on the gradient Richardson number, horizontal wind standard deviation, vertical temperature gradient and wind speed ratio to classify the atmospheric stability. The parameters used in these methods can be calculated using measured wind speed, wind direction, and temperature data. Wharton and Lundquist [17] analyzed the wind farm operating data and found that, when the atmosphere was stable and the wind speed was in the range of 5~8.5 m/s, the output power was obviously greater than that in unstable conditions with strong convection, and the average difference was close to 15%. Gryning et al. [18] applied the similarity theory to establish a new model for wind-speed profile by considering the effects of atmospheric stability, and the model was applied to the height of the entire atmospheric boundary layer in a flat terrain. Đurišić and Mikulović [19] proposed a model of vertical wind-speed extrapolation for improving the wind resource assessment using the WAsP (Wind Atlas Analysis and Application Program) software [20], and found that if the estimation is carried out with the fixed wind shear exponent, the overestimated wind speed in the day time (unstable atmosphere) and underestimated wind speed in the night time (stable atmosphere) are usually obtained. Gualtieri and Secci [21] tested the power law extrapolation model over a flat rough terrain in the Apulia region (Southern Italy), and investigated the effect of atmospheric stability and surface roughness on wind speed. They found that the empirical JM Weibull distribution extrapolating model was proved to be preferable. Different extrapolating models were compared and their advantages and disadvantages were investigated. However, there are few studies on the effect of atmospheric stability on wind shear and wind speed extrapolation in different terrain types. At the same time, there are few discussions about the results of wind speed extrapolation above different ground topographies.

The present work aims to evaluate the wind speed extrapolation method using the power law and the atmospheric stability classification. For different terrain types, the characteristics of wind shear exponent are discussed, and four atmospheric stability classification methods are employed and investigated. The model is validated against the measured data, and the suitability of different atmospheric stability classification methods for flows over different types of terrain is indicated.

2. Model Description

2.1. Power Law

It is usually necessary to calculate the wind shear exponent by the wind speeds at two different levels. In the neutral condition, the wind-speed profile can be calculated using the empirical formula:

$$u(z) = \frac{u^*}{\kappa} \ln \frac{z}{z_0} \tag{1}$$

where κ is the Von Karman constant equal to 0.4, u^* is the friction velocity calculated as $u^* = (\tau/\rho)^{1/2}$, τ is the wall friction, ρ is the density, z_0 is the roughness length, and z is the height.

The power law (PL) method is widely used for estimating the wind speed at a wind generator hub height [22], which is defined as

$$u_2 = u_1(z_2/z_1)^{\alpha} \tag{2}$$

where u_1 is the wind speed at the height z_1, u_2 is the wind speed at the height z_2, and α is the wind shear exponent.

From Equation (2), α can be calculated by u_1 and u_2:

$$\alpha = \ln(u_2/u_1)/\ln(z_2/z_1) \tag{3}$$

2.2. Atmospheric Stability Classification

The atmospheric thermal stability is suppressed or enhanced by the vertical temperature difference. The two atmospheric stability classification methods commonly used are those from Pasquill [23] and IAEA (International Atomic Energy Agency) [24]. The Pasquill method (abbreviated to P·S) proposed in the Chinese standards includes six categories: highly unstable, moderately unstable, slightly unstable, neutral, moderately stable and extremely stable. They are denoted as A, B, C, D, E, and F.

Taking the turbulence and thermal factors into account, there are three methods for classifying the atmospheric stability, namely the Monin-Obukhov method, Bulk Richardson number method, and Richardson Gradient method. Among them, theBulk Richardson number method does not have a uniform atmospheric stability classification standard, so it is not used in the present work. When the measured wind data only contains the wind direction and speed data, the wind direction standard deviation method and wind speed ratio method can be used to classify the atmospheric stability.

2.2.1. Richardson Gradient (RG) Method

The Richardson number, Ri, synthesizes the effects of thermodynamic and kinetic factors caused by the turbulence, reflecting more turbulent information [25]. Therefore, the RG method can distinguish the atmospheric stability more accurately. The Ri in the surface layer can be expressed as [26]

$$Ri = \frac{gz}{T} \frac{\Delta T}{\Delta u^2} \ln \frac{z_2}{z_1} \tag{4}$$

where ΔT is the difference of temperatures at z_2 and z_1. Δu is the wind speed difference. T is the average atmospheric temperature, g is the acceleration of gravity, and z is the average geometric height calculated as $z = \sqrt{z_1 z_2}$. Table 1 shows the classification standard of Ri in different terrains [27].

Atmospheric stability depends on the net heat flux to the ground, which is equal to the sum of incident radiation, radiation emitted, latent heat, and sensible heat exchange between the atmosphere and underlying surface. When the radiation incident on the ground is dominant, the air parcels at the lower part rise, leading to a vertical air motion, and making the atmosphere unstable. Therefore, the thermal effect aggravates the air movement and prevents the wind speed from changing dramatically in the vertical direction. In this case, the Richardson number is negative. When the ground cools down, the temperature increases with increasing the height, which weakens the vertical air movement. The situation is recognized as a stable state and a positive Richardson number is obtained. The neutral stability corresponds to the case where the thermal effect is not significant. This situation occurs when the cloud is dense, and the Richardson number is zero.

Table 1. Classification of stability based on *Ri* in different terrain conditions.

Stability Conditions	Mountain	Plain
A	$Ri < -100$	$Ri < -2.51$
B	$-100 \leq Ri < -1$	$-2.51 \leq Ri < -1.07$
C	$-1 \leq Ri < -0.01$	$-1.07 \leq Ri < -0.275$
D	$-0.01 \leq Ri < 0.01$	$-0.275 \leq Ri < 0.089$
E	$0.01 \leq Ri < 10$	$0.089 \leq Ri < 0.128$
F	$10 \leq Ri$	$0.128 \leq Ri$

2.2.2. Wind Direction Standard Deviation (WDSD) Method

The wind direction pulsation is a direct indicator of atmospheric turbulence. The magnitude of the wind direction pulsation angle is directly related to the diffusion parameter, so it can be used as an indicator to classify the atmospheric stability. However, the measurement of pulsation angle is easily influenced by the local influence of sampling location and instrument performance, which makes the method unrepresentative. So this method is more suitable for flat terrain. Table 2 shows the classification criteria recommended by Sedefian [28] based on the horizontal wind direction standard deviation, σ_θ. When σ_θ is used, the classification standard should be corrected according to the actual surface roughness. The classification standard is recommended by the United States Environmental Protection Agency (1980), and the actual area is corrected with roughness as $(z_0/0.15)^{0.2}$ (the corrected value is obtained by multiplying the values in Table 2).

Table 2. Relationship between σ_θ and the atmospheric stability.

Parameter	Atmospheric Stability					
	A	**B**	**C**	**D**	**E**	**F**
$\sigma_\theta/°$	$\sigma_\theta \geq 22.5$	$22.5 > \sigma_\theta \geq 17.5$	$17.5 > \sigma_\theta \geq 12.5$	$12.5 > \sigma_\theta \geq 9.5$	$9.5 > \sigma_\theta \geq 3.8$	$3.8 > \sigma_\theta$

2.2.3. Wind Speed Ratio (WSR) Method

The wind speed ratio U_R is defined as the ratio of the wind speeds at two different heights:

$$U_R = \frac{u_2}{u_1} \tag{5}$$

According to the rule of Pasquill stability classification, Chen [29] divided it into six categories according to the ratio of the wind speeds at two different heights (Table 3).

Table 3. Relationship between U_R and the atmospheric stability.

Atmospheric Stability	A	B	C
Range	$U_R < 1.0032$	$1.0032 \leq U_R < 1.0052$	$1.0052 \leq U_R < 1.0101$
Atmospheric Stability	**D**	**E**	**F**
Range	$1.0101 \leq U_R < 1.5717$	$1.5717 \leq U_R < 2.1963$	$U_R \geq 2.1963$

2.2.4. Monin–Obukhov (MO) Method

In the Monin–Obukhov theory, the atmospheric stability is described by the Obukhov length L, which is determined directly from sonic anemometer measurements of friction velocity and heat flux:

$$L = -\frac{(u^*)^3}{\kappa \frac{g}{\bar{T}} \overline{w'T'_S}} \tag{6}$$

where $\overline{w'T'_S}$ is the covariance of temperature and vertical wind speed fluctuations and g is the gravitational acceleration. When only the temperature gradient is available, L can be obtained by [30].

$$\begin{cases} L = \frac{z}{Ri} & (Ri < 0) \\ L = \frac{z(1-5Ri)}{Ri} & (Ri > 0) \end{cases} \tag{7}$$

According to the relationship between L and Ri, Table 4 shows the classification of stability based on L for flows above different terrains [27].

Table 4. Classification of stability based on L in different terrains.

Stability Conditions	Mountain	Plain
A	$L > -0.032$	$L > -0.316$
B	$-3.162 < L \leq -0.032$	$-3.162 < L \leq -0.316$
C	$-316.228 < L \leq -31.623$	$-63.246 < L \leq -3.162$
D	$L \leq -316.228, L > 158.114$	$L \leq -63.246, L > 158.114$
E	$-154.952 < L \leq 158.114$	$-154.952 < L \leq 158.114$
F	$L \leq -154.952$	$L \leq -154.952$

2.3. Wind Speed Extrapolation Method Based on Atmospheric Stability

The power law does not fully consider the effect of atmospheric stability on wind shear. According to the characteristics of atmospheric stability, the wind speed extrapolation (WSE) method based on the atmospheric stability is proposed. The measured dataset by wind towers is named as dataset Q, including time, wind speed, wind direction and wind direction standard deviation. The low wind speed data less than the cut-in wind speed in the dataset Q is first removed to obtain the filtered dataset W. Then the dataset W is classified into different stability conditions using the RG, MO, WSR, and WDSD methods. After that, Equation (3) is used to calculate the average wind shear exponent using the dataset W under each atmospheric stability condition. Finally, using the wind speeds at the lower heights in dataset Q, the higher level wind speed is predicted by Equation (2).

Four kinds of atmospheric stability classification methods, RG, WDSD, WSR and MO are adopted to develop four new methods, namely WSE-RG, WSE-WDSD, WSE-WSR and WSE-MO, for the wind speed extrapolation.

3. Cases and Measurements

3.1. Case Definition

Three cases are chosen to test all the extrapolation methods as shown in Table 5. The terrains include mountain, plateau and plain, and the surface types include wasteland, shrubbery and farmland.

Table 5. Wind tower site information.

Number	Site Name	Terrain	Surface	Elevation (m)
1	SJB	Plateau	Wasteland	1678
2	GFC	Mountain	Shrubbery	713
3	HNH	Plain	Farmland	66

Figure 1 shows the topographic maps of the three cases. The wind tower site SJB is located at an edge of a plateau (Figure 1a,d). The elevation of SJB in the plateau is 1678 m, and the surface type is wasteland. There is a hillside in the southwest of SJB, and flat grassland in the northeast. The GFC wind tower sits on the mountaintop, which has an altitude of 713 m (Figure 1b,e). The surface type of GFC is shrubbery. Due to the special weather conditions in the area, underlying shrub is evergreen throughout the year. The HNH wind tower is located at the elevation of 66 m, where the terrain is flat (Figure 1c,f). The land surface type is farmland and there are a large number of villages in the area.

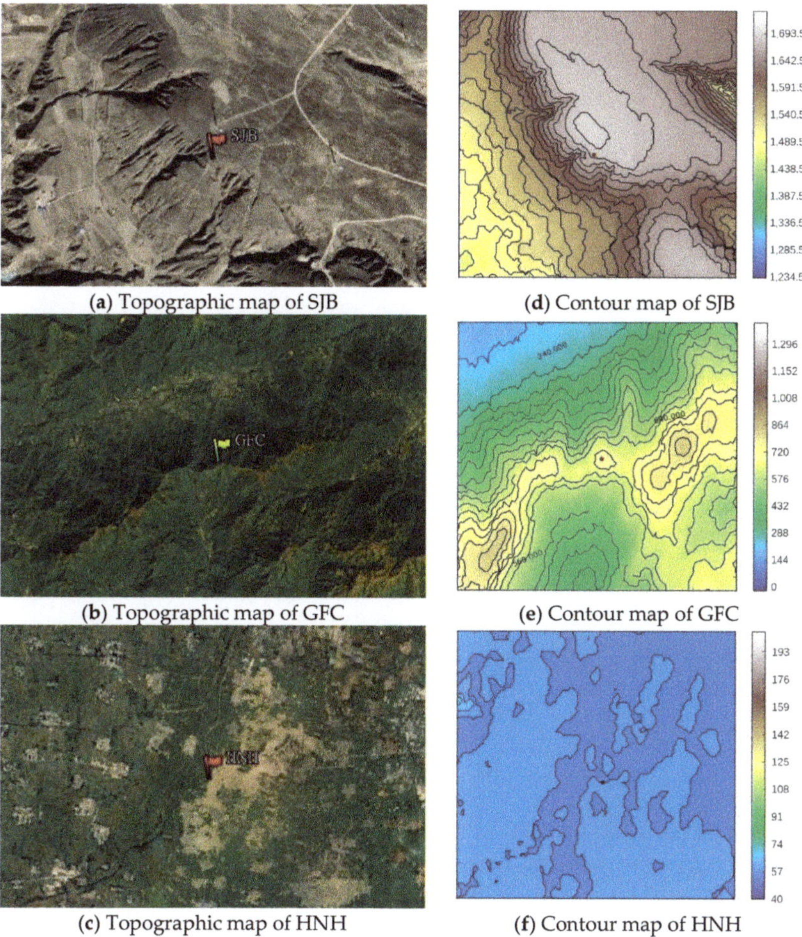

(**a**) Topographic map of SJB   (**d**) Contour map of SJB

(**b**) Topographic map of GFC   (**e**) Contour map of GFC

(**c**) Topographic map of HNH   (**f**) Contour map of HNH

Figure 1. Topographic maps of three cases: (**a**) SJB; (**b**) GFC and (**c**) HNH; Contour maps of the three cases: (**d**) SJB; (**e**) GFC and (**f**) HNH.

3.2. Measurements

The wind resource in the three cases is usually measured at a site using a wind tower, equipped with wind speed and wind direction sensors. The locations of the wind towers are shown in Figure 1. The datasets of the three cases are obtained as follows:

(1)　SJB: The measured data includes wind speed, wind direction and temperature data. The cup anemometers are placed at 30, 50 and 70 m, and the wind vines are placed at 30 and 70 m. The temperature observations are also used to obtain the temperature data at 30 and 70 m.
(2)　GFC: The cup anemometers are at 30, 50 and 80 m, and the wind direction is obtained with wind vines placed at 30 and 80 m. Unfortunately, there is no temperature sensor installed on the tower.
(3)　HNH: Wind speeds are measured by the cup anemometers placed at 10, 60 and 100 m. Wind vanes are placed at 10 and 100 m. The temperature data is also unavailable in the HNH area.

The datasets at upper heights are used to verify the reliability of the wind speed extrapolation method. The measurement parameters of each case including height, time interval and data collection period are presented in Table 6.

Table 6. Summary of measurement parameters.

Number	Site Name	$h1$ (m)	$h2$ (m)	$h3$ (m)	Time Interval (min)	Data Collection Period
1	SJB	30	50	70	10	27 August 2015~26 August 2016
2	GFC	30	50	80	10	1 August 2012~31 July 2013
3	HNH	10	60	100	10	15 September 2014~16 September 2015

4. Results and Discussion

4.1. Overall Meteorological Characteristics

In Table 7, the annual mean meteorological parameters observed for the three cases are provided where $U1$ and $U2$ are the average wind speeds at the heights of $h1$ and $h2$, $U3$ is the average wind speed at the hub height of $h3$, $T1$ and $T2$ are the mean temperatures at the heights of $h1$ and $h2$. Besides, α_{h1-h2} and α_{h2-h3} are the mean wind shear exponents calculated with velocities at $h1$ and $h2$, and $h2$ and $h3$ respectively.

Table 7. Wind data information.

Site Name	$U1$ (m/s)	$U2$ (m/s)	$U3$ (m/s)	$T1$ (°C)	$T2$ (°C)	α_{h1-h2}
SJB	5.34	5.46	5.71	8.27	8.06	0.0435
GFC	5.07	5.10	5.17	-	-	0.0115
HNH	2.73	4.61	5.44	14.30	-	0.2924

The results indicate the difference of the wind speed distribution between the plain and mountain areas. In the HNH case, where the terrain is flat, the inconspicuous mixing of atmospheric turbulence and the weak exchange of momentum between the vertical layers result in a vertical gradient of wind speed and a large wind shear exponent. Because of the acceleration effect on the wind speed on the mountaintop, the wind shear exponents in GFC and SJB are smaller than the one in HNH.

For the SJB plateau complex terrain, at 30 m (Figure 2a) and 70 m (Figure 2b), the south direction is the most frequent wind direction with the occurrence percentages of 15.81% and 14.50%, respectively. The changes of wind speed and direction in this area are limited, because at 30 m, the wind direction is less affected by the terrain, and the height difference between the two measurement levels is only 40 m. At the same time, under the complex terrain condition, the mixing of atmospheric turbulence, the momentum exchange in the vertical direction are strong, resulting in a small wind shear exponent.

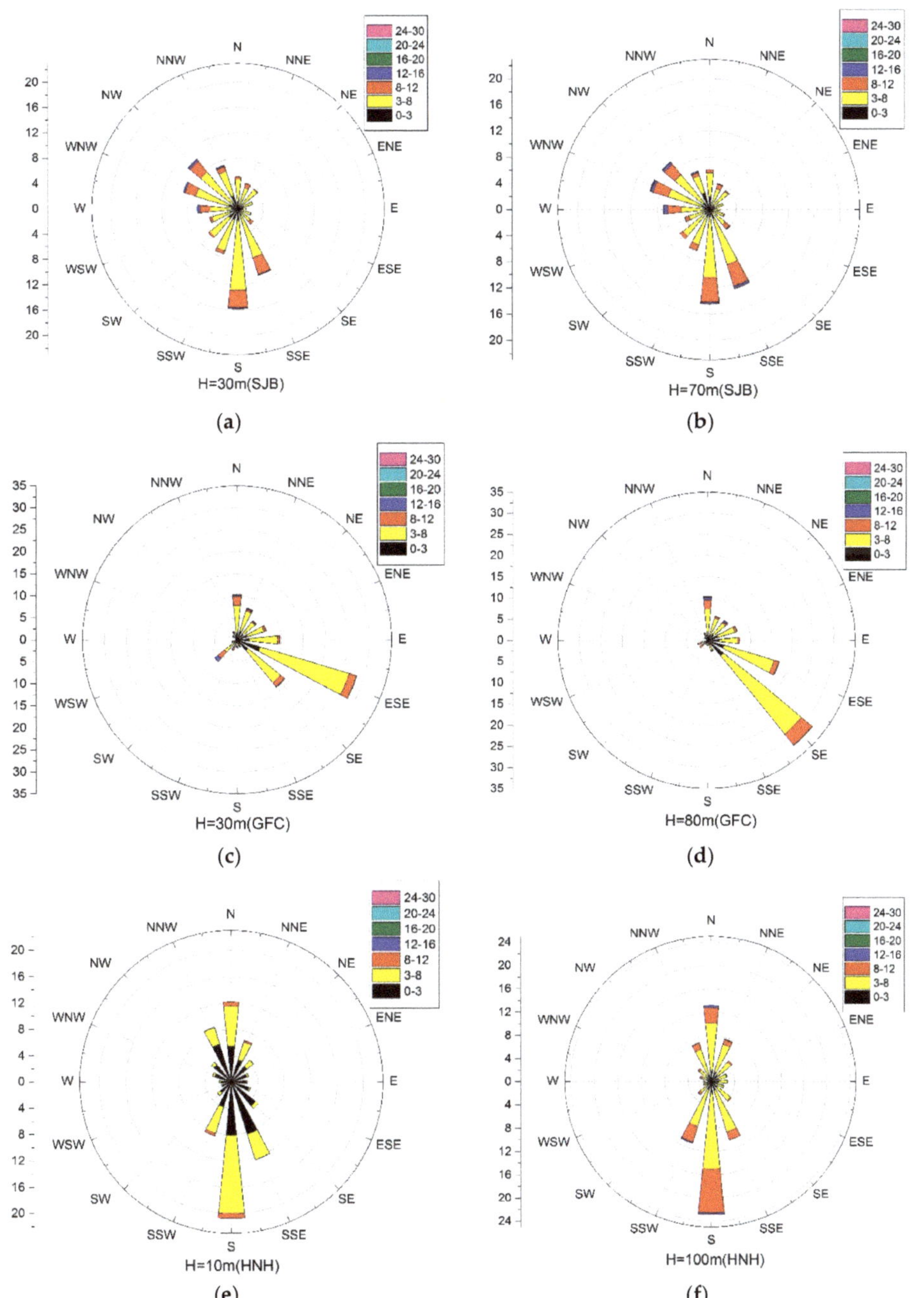

Figure 2. Wind rose measured at the three sites: (**a**) 30 m at SJB; (**b**) 70 m at SJB; (**c**) 30 m at GFC; (**d**) 80 m at GFC; (**e**) 10 m at HNH; (**f**) 100 m at HNH.

The complex topography in the GFC site strongly affects the wind behaviors at both 30 and 80 m (Figure 2c,d). At 30 m (Figure 2c), the most frequent direction is ESE (28.37%) while at 80 m

(Figure 2d), the SE direction is the most frequent direction (32.17%), and ESE (17.63%) is the secondary predominant direction. At the same time, the overall wind speed interval is dominated by a low wind speed interval of 3~8 m/s. This is because the complex mountain terrain has a great impact on the wind direction. At the same time, the topographic factors also cause strong mixing in vertical, which results in smaller values of wind shear exponent, so there is no significant change in the wind speed.

For the plain area HNH, at 10 m (Figure 2e), the S direction occurs most frequently (20.75%), and at 80 m (Figure 2f), it is also the S direction (22.90%). The wind direction at 10 m and 100 m are similar, but the wind speed at 100 m is significantly higher. This is related to the flat terrain condition of this area. The atmosphere is inadequately mixed in the vertical direction in this terrain condition, which leads to a larger wind shear exponent and causes a significant change in the vertical wind speed.

4.2. Wind Shear Characteristics of Different Terrains

Figure 3 shows the wind shear characteristics as the diurnal and monthly changes. As expected, the monthly variations of wind shear exponent at SJB and HNH are relatively high in Winter and low in Summer. The main reason is that, in Summer, the temperature is high, the mixing of near-surface air is sufficient, and the wind shear exponent is small. In Winter, the temperature is low and the air mixing is weak, usually resulting in a large wind shear exponent. However, the wind shear exponent at GFC is low in Winter and high in Summer, which deviates from the trend at SJB and HNH. This is because the area has more rains in Spring and Summer, making the air more humid and the frequency of low-level jets higher. The terrain has a dynamic lifting effect on the mountainous airflow and is prone to a strong wind shear. Furthermore, the daily change of wind shear exponent is closely related to the ambient temperature as seen in Figure 3b. The main reason for the diurnal variation of wind shear exponent is the cyclic variation of temperature. The wind shear exponent in the daytime is lower than that of the nighttime. This result is coincident with the results in the previous studies [31–33].

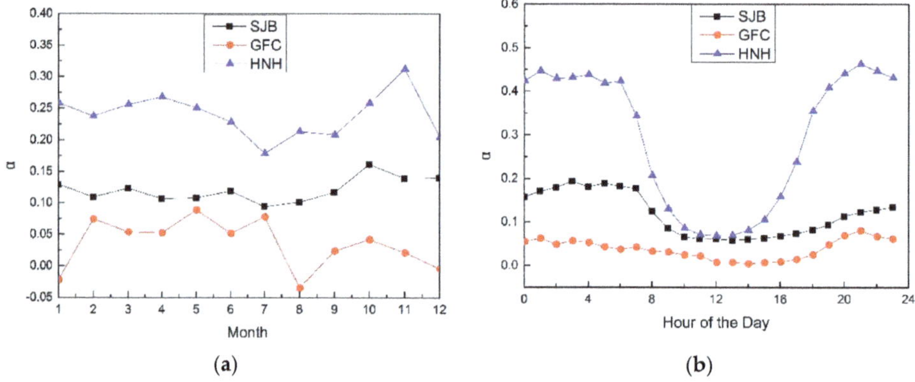

Figure 3. Monthly and diurnal variations of $h1$–$h2$ wind shear exponents in the three cases: (**a**) monthly variation; (**b**) diurnal variation.

Figure 4 shows the monthly and daily variations of atmospheric stability at SJB. Comparing the atmospheric stability variation with the variation of wind shear exponent, it is found that the wind shear exponent is lower when the atmosphere is more unstable.

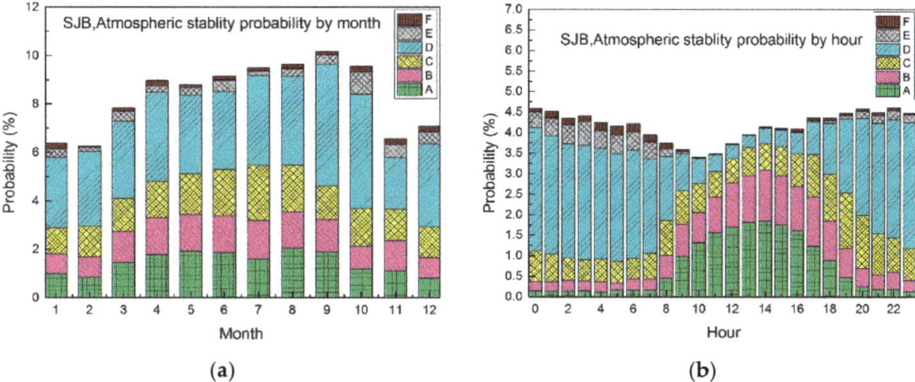

Figure 4. Probability of atmospheric stability classification at SJB: (**a**) monthly variation; (**b**) daily variation.

4.3. High Level Wind Speed Extrapolation and Validation

In the SJB area, starting from the 30 m and 50 m 10-min observations, the $\alpha_{30\text{--}50}$ is calculated and then used to estimate the wind resource at a higher level. In particular, the 10-min $\alpha_{30\text{--}50}$ is calculated using the filtered dataset W by implementing the methods of PL, WSE-RG, WSE-WDSD, WSE-WSR and WSE-MO. All the methods are adopted to calculate the wind shear exponents for the three areas and the results are listed in Table 8. At the same time, because the two-level temperature data at GFC and HNH is unavailable and the WSE-RG and WSE-MO methods cannot be used, the wind shear exponents of the two areas are calculated using the PL, WSE-WDSD, and WSE-WSR methods. It is shown that the wind shear exponent is larger under stable conditions and smaller under unstable conditions, especially in HNH. It is found that, using the WSE-WSR method, the variation of wind shear exponent with the atmospheric stability is more obvious than that using the WSE-WDSD method.

Using the results of wind shear exponent in Table 8, the wind speeds at the hub height for the three cases are also calculated. The wind speed mean relative error (MRE), root-mean-square error (RMSE) and mean wind speed at the hub height are shown in Table 8. For SJB, the MREs between the measurements and predicted wind speeds obtained by the PL, WSE-WSR, WSE-WDSD, WSE-RG and WSE-MO methods have the following sequence: MRE(PL) > MRE(WSE-WSR) > MRE(WSE-WDSD) > MRE(WSE-RG) > MRE(WSE-MO). For GFC, MRE(PL) > MRE(WSE-WSR) > MRE(WSE-WDSD). Besides, for HNH, it is MRE(PL) > MRE(WSE-WSR) > MRE(WSE-WDSD). The improvements of MRE range from 1.58 to 0.22%, 0.33 to 0.02%, 7.38 to 3.17% for the three cases using the new wind speed extrapolation methods.

It is proposed that the new WSE method based on the atmospheric stability better reflects the true changes of wind speed in two dimensions: height and time. It takes the effect of atmospheric stability on the wind profile into account and makes separate calculations for the wind resources at different conditions of atmospheric stability, which can reflect the mixing of atmosphere in the vertical direction. On this basis, the accuracy of atmospheric stability classification will directly affect the accuracy of wind speed estimation. When the two-level temperature data sets are available, it can be seen from Table 9 that the WSE-RG and WSE-MO methods can better estimate the wind speed for the SJB plateau area. For the GFC mountainous area without temperature measurements, both the WSE-WDSD and WSE-WSR methods can better estimate the wind speeds and the WSE-WDSD method is more accurate than the WSE-WSR method. For the HNH plain area, where the underlying surface is farmland, both the WSE-WDSD method and WSE-WSR method can better estimate the wind speeds than the PL method.

Table 8. Wind shear exponents in different areas.

Area	Method	Wind Shear Exponent under Different Stability Conditions					
		A	**B**	**C**	**D**	**E**	**F**
SJB	PL				0.0679		
	WSE-RG	0.1861	0.0558	0.0676	0.3825	0.0809	0.0981
	WSE-MO	0.2875	0.0474	0.0789	0.2753	0.1156	0.0963
	WSE-WDSD	0.0648	0.0774	0.0931	0.0608	0.0779	0.0523
	WSE-WSR	0.0863	-	0.0637	0.1309	0.4704	-
GFC	PL				-0.0023		
	WSE-WDSD	0.0066	0.0224	0.0307	0.0030	0.0239	-0.0175
	WSE-WSR	-0.1111	0.0081	0.0147	0.1500	0.9619	-
HNH	PL				0.1790		
	WSE-WDSD	0.1466	0.1455	0.1373	0.1740	0.2587	0.3703
	WSE-WSR	-0.0079	-	-	0.1227	0.3278	0.4807
Area	**Method**	**Wind Shear Exponent under Different Stability Conditions**					
		A	**B**	**C**	**D**	**E**	**F**
SJB	PL				0.0679		
	WSE-RG	0.1861	0.0558	0.0676	0.3825	0.0809	0.0981
	WSE-MO	0.2875	0.0474	0.0789	0.2753	0.1156	0.0963
	WSE-WDSD	0.0648	0.0774	0.0931	0.0608	0.0779	0.0523
	WSE-WSR	0.0863	-	0.0637	0.1309	0.4704	-
GFC	PL				-0.0023		
	WSE-WDSD	0.0066	0.0224	0.0307	0.0030	0.0239	-0.0175
	WSE-WSR	-0.1111	0.0081	0.0147	0.1500	0.9619	-
HNH	PL				0.1790		
	WSE-WDSD	0.1466	0.1455	0.1373	0.1740	0.2587	0.3703
	WSE-WSR	-0.0079	-	-	0.1227	0.3278	0.4807

Table 9. Mean wind speed relative error (MRE), root-mean-square error (RMSE) of wind speed and mean speed at the hub height in the three cases.

Area	Methods	MRE	RMSE (m/s)	Calculated Mean Speed (m/s)	Measured Mean Speed (m/s)
SJB	PL	-1.58%	0.378	6.6909	
	WSE-WDSD	-1.58%	0.3781	6.6911	
	WSE-MO	0.22%	0.3337	6.8229	6.8210
	WSE-RG	0.63%	0.3231	6.8081	
	WSE-WSR	-1.52%	0.3311	6.7242	
GFC	PL	-0.33%	0.543	6.2069	
	WSE-WDSD	-0.02%	0.5276	6.2386	6.2904
	WSE-WSR	-0.03%	0.536	6.2346	
HNH	PL	-7.38%	0.8732	7.2567	
	WSE-WDSD	-3.17%	0.7931	7.2681	7.5972
	WSE-WSR	-3.26%	0.7035	7.2815	

Figure 5 shows MRE and RMSE between the measured and calculated wind speeds in different atmospheric stability conditions. For SJB, MREs and RMSEs of WSE-WDSD under different atmospheric stability conditions are close to each other and are similar to the results of PL. It can be concluded that the WDSD method cannot be applied to the complex terrain of plateau to classify the atmospheric stability. In contrast, combining the results of MREs and RMSEs, it is found that the accuracy of the WSE-RG method is improved in the cases of A, B and C, and the WSE-WSR method is suitable for the cases of C, D and E, while the WSE-MO method improves the calculation accuracy in all the conditions. For GFC, both MREs and RMSEs of the WSE-WSR method are large when the atmospheric stability condition is E. Through the comparison of RMSEs, it is seen that the WSE-WDSD method performs well in almost all the conditions. For HNH, the absolute values of MREs of the WSE-WDSD and WSE-WSR methods are smaller than those obtained by the PL method. The accuracy of both the WSE-WDSD and WSE-WSR methods has been greatly improved. It is shown that these two atmospheric stability classification methods are suitable for flat terrain. And the WSE-WSR method is more effective than the WSE-WDSD method especially in unstable conditions.

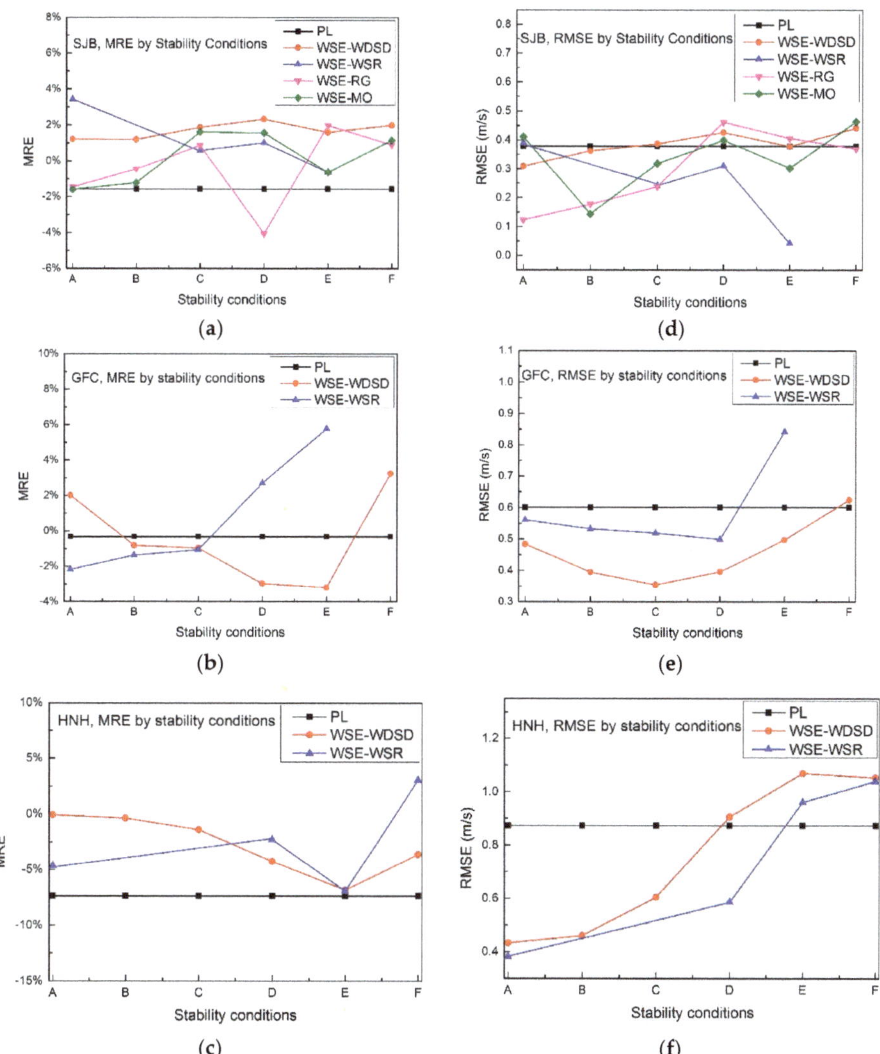

Figure 5. (**a–c**) MREs and (**d–f**) RMSEs between the measured and calculated values under different atmospheric stabilities conditions in the three different areas.

The monthly and daily variations of RMSE of the three cases using different wind speed extrapolation methods are shown in Figure 6. For SJB, the results of WSE-WDSD are very close to the traditional PL method, indicating that the WSE-WDSD method is inaccurate in the complex terrain plateau. In contrast, the results of the WSE-RG method are obviously better than those obtained by the other methods. At the same time, the results of the WSE-WSR and WSE-MO methods are similar especially for daily variations. From the daily variation, it is found that the WSE-WSR and WSE-MO methods are superior to the WSE-RG method in the night, but worse in the daytime. For the GFC and HNH, both the WSE-WDSD and WSE-WSR methods can effectively reduce RMSE, among which the WSE-WSR method is more effective. By analyzing RMSEs of the WSE-WDSD method in the HNH area where the surface is farmland, the WSE-WDSD method performs better in Winter and Spring. This is because that there are no crops in these periods, the roughness length is small, and the WDSD method is more suitable for that case.

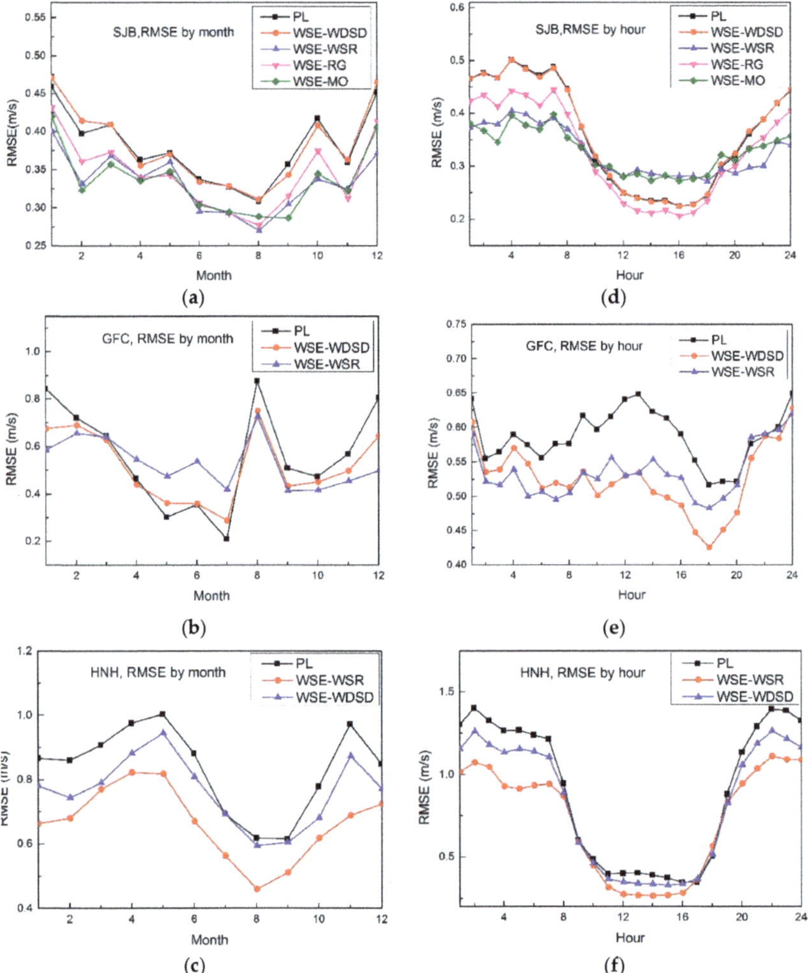

Figure 6. (**a–c**) Monthly and (**d–f**) daily variations of RMSE using different methods in the three different cases.

As a result, the WSE-WSR method is confirmed to be suitable for all the above mentioned terrain types. The WSE-WDSD method is more prominent under flat terrain and mountaintop. Meanwhile, both the WSE-RG and WSE-MO methods are good choices when the measurement data has the two-level temperatures.

5. Conclusions

A new wind shear extrapolation method based on theatmospheric stability was proposed in order to calculate the wind speed at the hub height and compared with the traditional PL method. Particularly, four methods for the classification of atmospheric stability were incorporated into the WSE method. Calculations were performed for flows in three different areas, namely SJB, GFC and HNH, to verify the suitability of the proposed methods. Conclusions can be drawn as follows:

1. For SJB, the plateau where the surface is wasteland, the WSE-RG, WSE-WSR and WSE-MO methods can well calculate the wind speed at the hub height. When two-level temperature data is available, the WSE-RG and WSE-MO methods are more effective, of which MRE of WSE-MO is 0.22% (MRE of PL is −1.58%) and RMSE of WSE-RG is 0.3231 m/s (RMSE of PL is 0.3780 m/s). When there are not enough temperature data, the WSE-WSR method is most effective, of which MRE is −1.52% and RMSE is 0.3311 m/s.
2. For GFC, the mountain where the surface is shrubbery, the WSE-WDSD and WSE-WSR methods perform well and the WSE-WDSD method is most effective, of which MRE is −0.02% (MRE of PL is 0.33%) and RMSE is 0.5276 m/s (RMSE of PL is 0.5430 m/s).
3. For HNH, the plain where the surface is farmland, the WSE-WDSD and WSE-WSR methods are also suitable. MREs of the WSE-WDSD and WSE-WSR methods are −3.17% and −3.26%, respectively (MRE of PL is −7.38%) and RMSEs of the WSE-WDSD and WSE-WSR methods are 0.7931 m/s and 0.7035 m/s, respectively (RMSE of PL is 0.6005 m/s). The WSE-WSR method is recommended when the atmosphere is unstable in most of the time.
4. The new WSE model proposed in the present work has advantages over the traditional PL method. Besides, the WSE-WDSD method for extrapolating the wind speed at the hub height is more effective in plain terrain. WSE-WSR is suitable in complex terrain. Besides, the WSE-RG and WSE-MO methods have more advantages when Ri and L can be calculated.

Author Contributions: Methodology, C.X.; Writing-Original Draft Preparation, C.H. and L.L.; Writing-Review and Editing, X.H., F.X., M.S. and W.S.

Funding: The Jiangsu Science Foundation for Youths (Grant No. BK20180505), the China Postdoctoral Science Foundation (No. 2018M630502), the Fundamental Research Funds for the Central Universities (No. 2018B01614) and the International Cooperation of Science and Technology Special Project (No. 2014DFG62530).

Acknowledgments: Authors are grateful to the support by the Jiangsu Science Foundation for Youths (Grant No. BK20180505), the China Postdoctoral Science Foundation (No. 2018M630502), the Fundamental Research Funds for the Central Universities (No. 2018B01614) and the International Cooperation of Science and Technology Special Project (No. 2014DFG62530).

Conflicts of Interest: The authors declare no conflict of interest.

List of Symbols

PL	Power law
WSE	Wind speed extrapolation method
WSE-RG	Wind speed extrapolation method based on the Richardson gradient method
WSE-WSR	Wind speed extrapolation method based on the wind speed ratio method
WSE-WDSD	Wind speed extrapolation method based on the wind direction standard deviation method
WSE-MO	Wind speed extrapolation method based on the Monin–Obukhov method
MRE	Mean relative error
RMSE	Root-mean-square error, m/s

κ	Von Karman's constant
u^*	Friction velocity, m/s
z	Height, m
z_0	Roughness length, m
u	Wind speed, m/s
α	Wind shear exponent
A~F	Classification of atmospheric stability: highly unstable, moderately unstable, slightly unstable, neutral, moderately stable and extremely stable
Ri	Gradient Richard number
ΔT	Temperature difference between two levels of height of z_1 and z_2, °C
Δu	Wind speed difference between two levels of height of z_1 and z_2, m/s
T	Atmospheric average absolute temperature, °C
σ_θ	Horizontal wind direction standard deviation, °
U_R	Wind speed ratio
L	Obukhov length
$\overline{w'T'_S}$	Covariance of temperature and vertical wind speed fluctuations at the surface
g	Gravitational acceleration
Q	Measured wind tower data set
W	Filtered dataset
$h1$	Height of a low level, m
$h2$	Height of a medium level, m
$h3$	Height of a high level, m
$U1$	Mean wind speed at the height of $h1$, m/s
$U2$	Mean wind speed at the height of $h2$, m/s
$U3$	Mean wind speed at the height of $h3$, m/s
$T1$	Mean temperature at the height of the low level, °C
$T2$	Mean temperature at the height of the high level, °C

References

1. Motta, M.; Barthelmie, R.J.; Vølund, P. The influence of non-logarithmic wind speed profiles on potential power output at Danish offshore sites. *Wind Energy* **2005**, *8*, 219–236. [CrossRef]
2. En, Z.; Altunkaynak, A.; Erdik, T. Wind Velocity Vertical Extrapolation by Extended Power Law. *Adv. Meteorol.* **2012**, *2012*, 885–901.
3. Rehman, S.; Al-Abbadi, N.M. Wind shear coefficients and their effect on energy production. *Energy Convers. Manag.* **2005**, *46*, 2578–2591. [CrossRef]
4. Huang, W.Y.; Shen, X.Y.; Wang, W.G.; Huang, W. Comparison of the Thermal and Dynamic Structural Characteristics in Boundary Layer with Different Boundary Layer Parameterizations. *Chin. J. Geophys.* **2015**, *57*, 543–562.
5. Justus, C.G. *Winds and Wind System Performance*; Franklin Institute Press: Philadelphia, PA, USA, 1978.
6. Irwin, J.S. A theoretical variation of the wind profile power-law exponent as a function of surface roughness and stability. *Atmos. Environ.* **1979**, *13*, 191–194. [CrossRef]
7. Troen, I.; Petersen, E.L. European Wind Atlas. Available online: http://orbit.dtu.dk/files/112135732/European_Wind_Atlas.pdf (accessed on 20 August 2018).
8. Monin, A.S.; Obukhov, A.M. Basic regularity in turbulent mixing in the surface layer of the atmosphere. *Trans. Geophys. Inst. Acad. Sci. USSR* **1954**, *24*, 163–187.
9. Jensen, N.O.; Petersen, E.L.; Troen, I. *Extrapolation of Mean Wind Statistics with Special Regard to Wind Energy Applications*; World Meteorological Organization: Geneva, Switzerland, 1984.
10. Optis, M.; Monahan, A.; Bosveld, F.C. Moving Beyond Monin-Obukhov Similarity Theory in Modelling Wind Speed Pro les in the Stable Lower Atmospheric Boundary Layer. *Bound. Layer Meteorol.* **2014**, *153*, 497–514. [CrossRef]
11. Lackner, M.A.; Rogers, A.L.; Manwell, J.F.; Mcgowan, J.G. A new method for improved hub height mean wind speed estimates using short-term hub height data. *Renew. Energy* **2010**, *35*, 2340–2347. [CrossRef]

12. Li, P.; Feng, C.; Han, X. Effect Analysis on the Wind Shear Exponent for Wind Speed Calculation of Wind Farms. *Electr. Power Sci. Eng.* **2012**, *28*, 7–12.

13. Holtslag, A.A.M. Estimates of diabatic wind speed profiles from near-surface weather observations. *Bound. Layer Meteorol.* **1984**, *29*, 225–250. [CrossRef]

14. Gualtieri, G. Atmospheric stability varying wind shear coefficients to improve wind resource extrapolation: A temporal analysis. *Renew. Energy* **2016**, *87*, 376–390. [CrossRef]

15. Panofsky, H.A.; Dutton, J.A. *Atmospheric Turbulence: Models and Methods for Engineering Applications*; Prentice-Hall: Upper Saddle River, NJ, USA, 1983.

16. Mohan, M.; Siddiqui, T.A. Analysis of various schemes for the estimation of atmospheric stability classification. *Atmos. Environ.* **1998**, *32*, 3775–3781. [CrossRef]

17. Wharton, S.; Lundquist, J.K. Atmospheric stability affects wind turbine power collection. *Environ. Res. Lett.* **2012**, *7*, 17–35. [CrossRef]

18. Gryning, S.E.; Batchvarova, E.; Brümmer, B.; Jørgensen, H.; Larsen, S. On the extension of the wind profile over homogeneous terrain beyond the surface layer. *Bound. Layer Meteorol.* **2007**, *124*, 251–268. [CrossRef]

19. Đurišić, Ž.; Mikulović, J. A model for vertical wind speed data extrapolation for improving wind resource assessment using WAsP. *Renew. Energy* **2012**, *41*, 407–411. [CrossRef]

20. Wind Resource Assessment, Siting & Energy Yield Calculations. Available online: http://www.wasp.dk/wasp (accessed on 30 July 2018).

21. Gualtieri, G.; Secci, S. Extrapolating wind speed time series vs. Weibull distribution to assess wind resource to the turbine hub height: A case study on coastal location in Southern Italy. *Renew. Energy* **2014**, *62*, 164–176. [CrossRef]

22. Hellmann, G. *Über die Bewegung der Luft in den untersten Schichten der Atmosphäre*; Kgl. Akademie der Wissenschaften: Copenhagen, Denmark, 1919. (In Germany)

23. Pasquill, F. *Atmospheric Diffusion*, 2nd ed.; Ellis Horwood: Chichester, UK, 1974.

24. Agency, I.A.E. *Atmospheric Dispersion in Nuclear Power Plant Siting: A Safety Guide*; International Atomic Energy Agency: Vienna, Austria, 1980.

25. Balsley, B.B.; Svensson, G.; Tjernström, M. On the Scale-dependence of the Gradient Richardson Number in the Residual Layer. *Bound. Layer Meteorol.* **2008**, *127*, 57–72. [CrossRef]

26. Ma, Y. The basic physical characteristics of the atmosphere near the ground over the Qinghai-Xizang Plateau. *Acta Meteorol. Sin.* **1987**, *2*, 188–200.

27. Deng, Y.; Fan, S.J. Research on the surface layer's stability classifying schemes over coastal region by Monin-Obukhov length. In Proceedings of the National Conference on Atmospheric Environment, Nanning, China, 25 October 2003; pp. 136–141.

28. Sedefian, L.; Bennett, E. A comparison of turbulence classification schemes. *Atmos. Environ.* **1980**, *14*, 741–750. [CrossRef]

29. Chen, P. A comparative study on several methods of stability classification. *Acta Sci. Circumstantiae* **1983**, *3*, 77–84.

30. Businger, J.A. Flux profile relationships in the atmospheric surface layer. *J. Atmos. Sci.* **1971**, *28*, 181–189. [CrossRef]

31. Rehman, S.; Al-Abbadi, N.M. Wind shear coefficient, turbulence intensity and wind power potential assessment for Dhulom, Saudi Arabia. *Renew. Energy* **2008**, *33*, 2653–2660. [CrossRef]

32. Greene, S. Analysis of vertical wind shear in the Southern Great Plains and potential impacts on estimation of wind energy production. *Int. J. Energy Issues* **2009**, *32*, 191–211. [CrossRef]

33. Fox, N.I. A tall tower study of Missouri winds. *Renew. Energy* **2011**, *36*, 330–337. [CrossRef]

Article

Measurements of High-Frequency Atmospheric Turbulence and Its Impact on the Boundary Layer of Wind Turbine Blades

Alois Peter Schaffarczyk * and Andreas Jeromin

Mechanical Engineering Department, Kiel University of Applied Sciences, D-24149 Kiel, Germany; andreas.jeromin@fh-kiel.de
* Correspondence: Alois.Schaffarczyk@FH-Kiel.de; Tel.: +49-431-210-2600

Received: 12 June 2018; Accepted: 8 August 2018; Published: 21 August 2018

Abstract: To gain insight into the differences between onshore and offshore atmospheric turbulence, pressure fluctuations were measured for offshore wind under different environmental conditions. A durable piezo-electric sensor was used to sample turbulent pressure data at 50 kHz. Offshore measurements were performed at a height of 100 m on Germany's FINO3 offshore platform in the German Bight together with additional meteorological data provided by Deutscher Wetterdienst (DWD). The statistical evaluation revealed that the stability state in the atmospheric boundary does not seem to depend on simple properties like the Reynolds number, wind speed, wind direction, or turbulence level. Therefore, we used higher statistical properties (described by so-called shape factors) to relate them to the stability state. Data was classified to be either within an unstable, neutral, or stable stratification. We found that, in case of stable stratification, the shape factor was mostly close to zero, indicating that a thermally stable environment produces closer-to Gaussian distributions. Non-Gaussian distributions were found in unstable and neutral boundary layer states, and an occurrence probability was estimated. Possible impacts on the laminar-turbulent transition on the blade are discussed with the application of so-called laminar airfoils on wind turbine blades.

Keywords: turbulence; super-statistics; piezo-electric flow sensor; ABL stability; laminar-turbulent transition

---

1. Introduction

The use of wind energy has been very successful during the last decades [1], reaching a nearly stable annual investment corresponding to 50 GW of rated power word-wide. This was in connection and in parallel to an impressive development in Wind Turbine Aerodynamics [2] and even a special branch of *wind energy meteorology* has been established [3].

Due to the increased number of annual new installations, site assessment for wind farms has become more and more important and sophisticated, even under offshore conditions, with application to tailored turbine design as well. For a selection of important references on properties of the atmospheric boundary layer on- and off-shore see, for example, reference [4–7]. In most cases, turbulence has been treated in the context of loads, and frequency ranges higher than a few Hertz have generally been considered to be of no importance. However, this high frequency turbulence plays a significant role in laminar-to-turbulent transition inside the boundary layer of blades for airplanes [8,9] and wind turbines [10,11]. Due to the much higher drag of turbulent parts, this may give rise lower wind turbine efficiency (in terms of c_P, the number $0 \leq c_P \leq 0.596$ measures the fraction of extracted power from the wind) as desired, and the proper choice of airfoils is crucial.

In this paper, these high-frequency turbulent statistics were studied with respect to higher order statistical moments in some detail using the method of super-statistics [12]. The shape factor from reference [13]—which gives estimates of the extent to which turbulent fluctuations diverge from Gaussian behavior—was of special interest.

Unlike earlier investigations, the non-Gaussianity of turbulent pressure fluctuations could not be linked to one of the more popular atmospheric parameters like the Reynolds number based on Taylor's micro-scale [14], wind speed, or others. Therefore, we propose characterizing our findings according to the stability of the atmospheric boundary layer in terms of Richardson's number. This paper is organized as follows: Firstly, the measurement setup is described together with the locations at which they were performed. Then, we describe the procedure of data analysis. After that, we present and discuss our findings and finally, draw some conclusions.

Parts of the material have been presented earlier in unpublished proceedings of ICOWES2013 [15].

2. Measurements

Our first set of high-resolution measurements (during the years 2009 until 2011) was performed using piezoelectric pressure sensors from PCB Piezotronics (Figure 1) that were connected to an imc Meßsystem GmbH CS-1208 data logger. The diameter of the sensing element was 15 mm. The minimal pressure resolution was 0.13 Pa, and the possible temporal resolution ranged from 2.5 Hz to 80,000 Hz.

Reference [16] discusses how the **length**, which is usually more than 100 times larger than the diameter of a hot-wire, influences the spatially resolution.

In our measurements, the pressure data was sampled at 50 kHz with a total duration of 100 s. The wind speed and temperature were sampled with a 1 Hz temporal resolution.

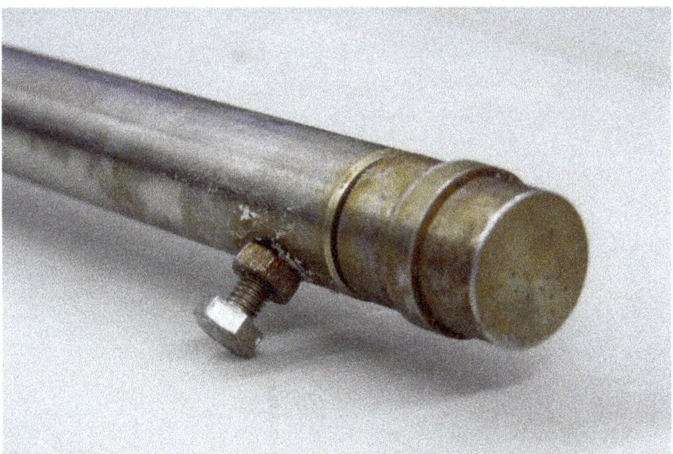

Figure 1. Pressure sensor after six months of offshore service (diameter = 12 mm).

The piezoelectric pressure sensor was calibrated against a hot wire anemometer in a wind tunnel at the University of Oldenburg. The turbulent wind speed from the hot wire anemometer and the variation in turbulent pressure showed the same statistical properties up to 3 kHz (see [17]).

Onshore measurements were conducted at the Kaiser–Wilhelm–Koog test site of Germanischer Lloyd/Garrad Hassan (see Figure 2). A lattice tower of 60 m height provided booms to mount the pressure sensor on and other measurement equipment. The pressure sensor was mounted at a height of 55 m. The wind speed and wind direction were recorded at a height of 55 m by calibrated cup anemometers and wind vanes, respectively. The temperature was recorded at height of 53 m by a resistor-type thermometer.

Figure 2. Locations of onshore (Kaiser–Wilhelm–Koog) and offshore (FINO3 platform) test sites, 80 km west of the island of Sylt. ©FEZ FH Kiel GmbH, Graphics: Bastian Barton.

FEZ Kiel's platform FINO3 was the location for the offshore measurements. About 80 km west of the island of Sylt (see Figure 2) the platform was constructed close to (at that time not) operational wind farms like DanTysk and Sandbank 24. The tower is a lattice tower type with booms of sufficient length for undisturbed measurements. Two pressure sensors (Figure 3) were mounted at a height of about 100 m above the mean sea level for parallel operation. The data acquisition equipment for meteorological signals was similar to that used onshore. The wind speed and wind direction were measured at a height of 100 m, and the temperature was measured at a height of 95 m above the mean sea level.

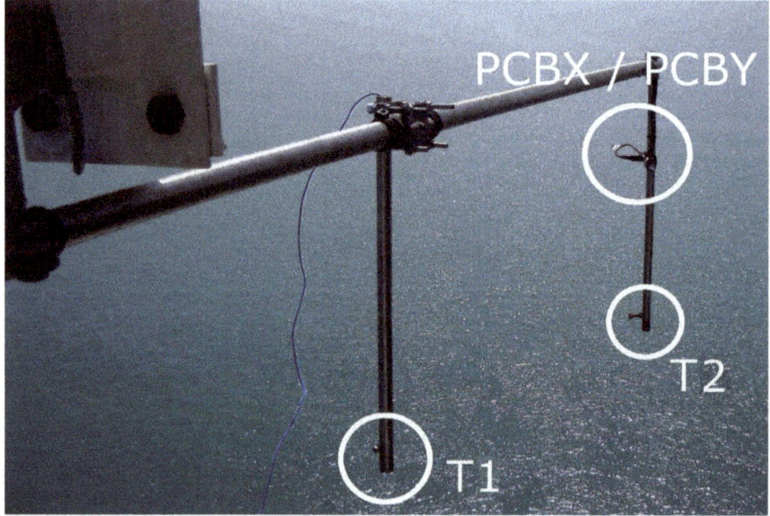

Figure 3. Installation of two parallel pressure sensors (T1 and T2) at the FINO3 platform.

3. Analysis

Our recorded data sets were put into wind speed classes starting at 6 m/s, 12 m/s, and 16 m/s. Measurements were triggered manually if wind conditions were regarded as suitable. The bin size for each class was generally ± 2 m/s, and for some cases (offshore class 10 m/s, offshore class 12 m/s), it was less (lower limit: -1 m/s). For each measurement, the statistical characteristics were analyzed with respect to the power-spectral density, incremental distributions, shape factors, structure functions, auto-correlations and Taylor's micro-scale. (It may be interesting to note that Taylor's micro-scale may have an complementary interpretation to the *Markov–Einstein coherence length* [18]. To use this one seems to be much more physical than the somewhat vague interpretation of the average turbulent vortex size.) We now explain our methods in more detail.

The increments of measured quantities were defined as

$$\Delta u = u(t + \Delta t) - u(t) \qquad \Leftrightarrow \qquad \Delta p = p(t + \Delta t) - p(t), \tag{1}$$

where u is the velocity, p is the pressure, t is the time, and Δt is a fixed time increment. The increments also eliminated the mean value of the time series and thus, stochastic fluctuations remained for a specific time scale (Δt). These increments were the basic statistical quantities used for more sophisticated analyses like structure functions or the so-called shape parameter. Reference [13] suggested that the distribution of increments may be described as a superposition of two distributions, one of them being lognormal. The shape factor then may be regarded as a measure of the level of intermittency. In accordance with Beck's approach from reference [12], the shape parameter can be calculated by the 2nd and 4th order moments of the distribution:

$$s_u^2 = \ln\left(\frac{1}{3}\frac{\langle \Delta u^4 \rangle}{\langle \Delta u^2 \rangle^2}\right) \qquad \Rightarrow \qquad s_p^2 = \ln\left(\frac{1}{3}\frac{\langle \Delta p^4 \rangle}{\langle \Delta p^2 \rangle^2}\right), \tag{2}$$

where s_u^2 is the shape factor for the velocity and s_p^2 is the shape factor for the pressure. The brackets $\langle x \rangle$ define the mean value of a quantity (x). Further properties are given in reference [12]. A value of $s^2 \approx 0$ indicates a normal distribution. It is worth noting that another important parameter for the deviation from Gaussian behavior, skewness, which is related to third-order moments, was found to be small [17].

It has well-known, at least since the work of reference [19], that turbulent velocities and pressures obey different scaling laws ($\sim k^{-5/3}$ and $\sim k^{-7/5}$, respectively), although they are seemingly related by $p \sim v^2$ according to Bernoulli's law. A direct comparison of a hot-wire with piezo-electric pressure sensor data showed comparable power density spectra up to 4 kHz (see [17], unpublished), however. Theoretical as well as experimental investigations of reference [20] are still incomplete, so that there is no clear answer so far about how s_u^2 and s_p^2 are related to each other.

When heat transfer occurs, such as in the atmospheric boundary layer if the sea is warmer than the air (autumn), thermal stratification becomes more important. Simple static stability of the boundary layer may be used to characterize levels of turbulence. As was shown by reference [21], the turbulence intensity differs remarkably for stable and unstable flow when used to describe fatigue loads of wind turbines.

A simple characterization of a stratified boundary layer exposed to heat transfer was introduced by reference [4] as being stable, neutral, or unstable. Different calculation methods have been used to distinguish these states ([4]) depending on what is known in terms of input. In our case, a Richardson's number approach was used, in accordance with common practice in wind energy meteorology [22–24]. This was defined as

$$Ri = \frac{g}{T}\frac{\partial\theta/\partial z}{(\partial v/\partial z)^2}, \tag{3}$$

where g is the acceleration of gravity, T is the mean absolute temperature, z is the height normal to the surface, and $\partial v / \partial z$ is the mean velocity the gradient. The z-direction was assumed to be vertical. The potential temperature (θ) is defined as

$$\theta = T \left(\frac{p_0}{p} \right)^{R/c_p},$$ (4)

where p is the pressure, and $p_0 = 1000$ hPa is the reference pressure, $R = 289 \frac{J}{kg\,K}$ is the specific gas constant, and $c_p = 1005 \frac{J}{kg\,K}$ is the specific heat capacity of air. It has to be noted that we checked for the influence of humidity and found that the influence on density as well as specific humidity [5] was below 1%. We therefore neglected all moisture corrections to Richardson's Number (Ri).

In Equation (3), the quantity $\frac{g}{T} \partial \theta / \partial z$ describes the forces introduced by heat transfer in the boundary layer. The term $(\partial v / \partial z)^2$ represents the momentum forces in the boundary layer. If we consider two points in the boundary layer at heights z_1 and z_2 where $z_1 < z_2$ and $\Delta z = z_2 - z_1$, Ri from (3) can be rewritten with the differences between the two locations as

$$Ri = \frac{g}{T} \frac{\Delta\theta/\Delta z}{(\Delta v/\Delta z)^2} = \frac{g}{T} \frac{(\theta_2 - \theta_1)/\Delta z}{((v_2 - v_1)/\Delta z)^2}.$$ (5)

We always assumed the presence of positive velocity gradients so only the convective part of Equation (5) remained. Three typical situations were then distinguished as follows:

$\theta_2 > \theta_1$ The surface is colder than the fluid and the gradient becomes $\partial\theta/\partial z > 0$ and therefore, $Ri > 0$. Heat is transported by conduction only, and a convection flow does not occur. In this case, the stratification is strong, and turbulence gets damped. The boundary condition is stable for $Ri > 0$.

$\theta_2 = \theta_1$ The temperature gradient is zero and therefore, $Ri = 0$. There is no temperature gradient and therefore, no conduction nor convection. This condition is called neutral.

$\theta_2 < \theta_1$ The surface is warmer than the fluid and the gradient becomes $\partial\theta/\partial z < 0$ and therefore, $Ri < 0$. Heat is transported by conduction and by convection from the surface to the fluid. The convection results in a vertical, upward component of the flow that interacts with the horizontal velocity component. This leads to the production of turbulence in the boundary layer and therefore, is called unstable.

From our measurements we obtained the necessary values for p, T_{air} at a height of 100 m , wind speed v_{air}, and wind direction as well as the sensor signal of the piezoelectric microphone. By assuming the air was a perfect gas, the density ρ was calculated by the perfect gas law.

For the derivatives $\partial\theta/\partial z$ and $\partial v/\partial z$, the temperature and wind speed at a second height needed to be known. Unfortunately, these data were not available during most of our measurements. Therefore, these gradients had to be approximated in a different way.

The ground temperature (T_{gnd}) was estimated using data from the Deutscher Wetterdienst (DWD). For the onshore measurements, the ground temperature at a nearby location was available at hourly samples. The temperature of the sea for the offshore measurements was interpolated from measurement stations on the island of Helgoland and at List on the island of Sylt.

The sign of Ri mostly depends on the temperature gradient, so the velocity gradient can be regarded as of lower importance. Therefore, as a rough estimate, $v_{gnd} \approx 0$ was used to calculate the velocity gradient. This mostly affected the absolute value of Ri but not the heat transfer conditions in the atmospheric boundary layer.

With this procedure, estimates for Ri were possible. The identification of stable or unstable states was simple. However, the neutral state required a value of exactly $Ri = 0$ which was difficult to attain precisely within our set of approximations. To indicate a nearly neutral state, a bandwidth of $|Ri| < 0.02$ was used instead.

4. Results

The computation of the shape parameter was straightforward, and the results are shown in Figure 4. The scale for the shape factor is shown on the left. Colors indicate locations, and symbols represent velocity classes. Time series with $s_p^2 < 2$ were grayed out, and will not be considered further in the following analysis. For purposes of comparison, shape factors from a particular velocity measurement ([25], $\bar{U}$ = 7.6 m/s, u' = 1.36 m/s) were included (black line) as well.

The general behavior of both shape parameter functions for velocity and pressure was similar, whereas absolute values differed significantly. This behavior was also observed in reference [26]. Surprisingly, most of the pressure measurements resulted in s_p^2 close to zero.

We noted differences between the measurements of velocity and pressure concerning sampling rates and length of the time series.

In Figure 4, offshore curves are distinguished with numbers in brackets. They were all recorded on the same day with a short delay at the following times:

(1)   19 October 2010, 8:44 a.m.
(2)   19 October 2010, 8:54 a.m.
(3)   19 October 2010, 8:55 a.m.
(4)   19 October 2010, 8:57 a.m.

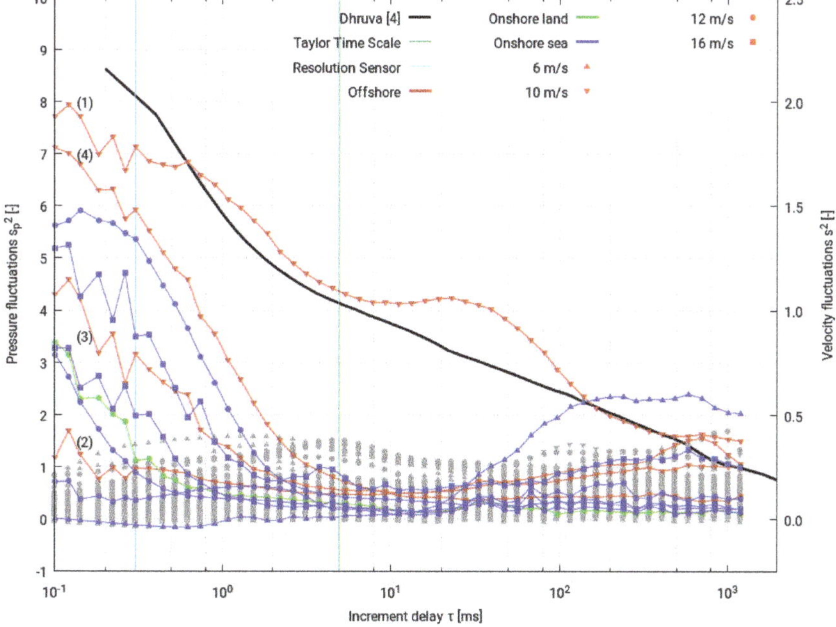

Figure 4. Shape parameters for all collected measurements. Colors represent locations: offshore, onshore with wind coming from land or sea side. The symbols represent the velocity classes for the wind speed. Measurements with $max(s_p^2) < 2$ are shown in grey. For reasons of comparison, Taylor's and Kolmogorov's time scales are given in 5 msec and 0.05 msec, respectively. The sensor's resolution goes down to approximately 0.3 msec only.

Pressure time series were recorded for a period of 100 s. For the first data set (1) s_p^2 was high. However, then, the pressure data resulted in $s_p^2 \approx 0$ until the values increased eight minutes later at

(2). The time series (2), (3), and (4) were consecutive, so it can be said that the shape parameter for the turbulent pressure fluctuations did not remain constant in time for this interval. When all time series from 8:44 a.m. to 8:57 a.m.were concentrated into one and evaluated according to a common s_p^2, a curve similar to the curve for velocity fluctuations was obtained. This has already been presented in reference [14].

We now try to relate these different statistical behaviors to the atmospheric boundary layer state. With our estimation of the Richardson number, a stability state was determined. Selected data from all our 119 measured data sets are shown in Tables 1 and 2. Measurements with $s_p^2 > 0$ are marked with a star in the last column.

As can be seen from Table 1, high s_p^2 ($s_p^2 > 0$) behavior corresponded to unstable to neutral Atmospheric Boundary Layer (ABL) conditions: $-0.13 \leq Ri \leq 0.01$. However, a significant correlation with the Richardson number was not found.

Table 1. Summary of selected Atmospheric Boundary Layer (ABL) stability cases: wind speeds (v_W) and turbulence intensities (Ti) from cup anemometers, Ri numbers and estimated ABL states.

Date & Time	Location	v_W (m/s)	Ti (%)	Ri (-)	Boundary Layer State	
21 October 2010, 7:56 a.m.	Offshore	6.6	8.2	-0.54	Unstable	
28 April 2008, 9:35 a.m.	Onshore	6.1	11.9	-0.15	Unstable	
19 October 2010, 8:57 a.m.	Offshore	10.8	2.2	-0.13	Unstable	*
19 October 2010, 8:44 a.m.	Offshore	10.2	3.1	-0.12	Unstable	*
19 August 2010, 8:13 a.m.	Offshore	12.7	2.8	-0.06	Unstable	
25 March 2008, 2:55 p.m.	Onshore	11.3	10.8	-0.03	Unstable	*
29 March 2008, 11:44 a.m.	Onshore	15.3	10.6	-0.01	Neutral	
30 March 2008, 6:04 p.m.	Onshore	15.7	5.6	0.00	Neutral	*
28 April 2008, 2:20 p.m.	Onshore	5.8	2.1	0.01	Neutral	*
1 May 2008, 2:55 a.m.	Onshore	6.0	6.0	0.10	Stable	
12 April 2008, 7:24 p.m.	Onshore	5.2	11.0	0.28	Stable	

4.1. Time Development of a Sample Time Series

Therefore, especially for the measurement on 19 October 2010, a possible transient behavior of the shape factor was investigated in more detail. The time development of measured data is presented in Table 2.

Table 2. Development of Time Series on October 19. Potential temperatures (θ), wind speeds (v_W) and turbulence intensities (Ti) from cup anemometers, Ri numbers, and estimated ABL states.

Time	θ_{air} (K)	θ_{gnd} (K)	v_W (m/s)	Ti (%)	Ri (-)	Boundary Layer State	$s_p^2 \neq 0$
8:44 a.m.	283.6	286.8	10.2	3.1	-0.12	Unstable	*
8:45 a.m.	283.5	286.8	10.2	3.9	-0.13	Unstable	
8:47 a.m.	283.4	286.8	10.0	7.5	-0.14	Unstable	
8:49 a.m.	283.4	286.8	10.1	7.3	-0.13	Unstable	
8:50 a.m.	283.5	286.8	9.3	6.7	-0.16	Unstable	
8:52 a.m.	283.5	286.8	11.2	6.0	-0.11	Unstable	
8:54 a.m.	283.5	286.8	10.8	7.7	-0.11	Unstable	*
8:55 a.m.	283.2	286.8	11.2	4.2	-0.12	Unstable	*
8:57 a.m.	283.1	286.8	10.8	2.2	-0.13	Unstable	*

The measurements started at 8:44 a.m. when s_p^2 was high and the turbulence intensity from cup anemometers was low. In the following 7 min, the time series were found to have $s_p^2 \approx 0$, with the mean velocity and potential temperature of the air remaining constant. Only the turbulence intensity increased from about 4 to about 8%.

At 8:54 a.m., the shape factor began to rise (see also curve (2) of Figure 4). Promptly, the turbulence intensity dropped more than 5.5 percentage points, and also, the potential temperature of the air decreased by 0.4 K with the rising shape factor.

Our interpretation is summarized as follows: the boundary layer, corresponding to time interval from 8:45 a.m. to 8:49 a.m. was stratified, and turbulent pressure fluctuations were distributed in a Gaussian way. At 8:50 a.m., a cluster of warm air rose up from the warmer sea surface to higher (and colder) regions ($\theta_{air} \approx$ consant, v_W fluctuating). This may have led to a stronger vertical shear, thereby disturbing the (Gaussian) turbulent structures.

4.2. Occurrence Probabilities

As was seen in our evaluations of boundary layer states and shape factors, the occurrence of high values for s_p^2 (a highly non-Gaussian behavior) stems from a non-linear, dynamic process with chaotic phases, for which the term *intermittency* was introduced [27]. The occurrence probabilities for all of our 119 measurements are listed in Table 3.

In the first row, the number of states is listed for all measurements. We see much more unstable (84) than neutral (23) or stable (12) conditions. The subsequent two rows list the numbers for offshore and onshore locations with their states. In the fourth row, the tallies for high shape factor events are listed for all measurements. The probability for the occurrence of a high-s_p^2 event at a specific boundary layer state is shown in the next line. The same was done for both locations, offshore and onshore.

To correctly interpret the data, one has to know the circumstances of the measurements. The onshore measurements took place in spring 2008 between March and May. During this time of year, the ground begins to warm up slowly while the air heats up much faster, yielding stable conditions. Project specific constraints limited the offshore measurement campaign to a period in late summer and autumn 2010. At this time of year, the sea is still heated up from the summer and warms the air near the surface, while the air temperature subsides, and many unstable conditions can be observed.

The focus was laid on the onshore measurements for the relation between the boundary layer state and the occurrence of high shape factors in the turbulent pressure fluctuations. With a joint probability of about 16.7%, a high-s_p^2 was found for unstable and neutral states (7 out of 42, see Table 3). However, in stable conditions, a high-s_p^2 was not observed. So, we propose that turbulent fluctuations in the pressure may deviate from a normal distribution much more frequently under unstable or neutral boundary conditions than under stable conditions.

Table 3. Correlation of the occurrence probability of a high shape factor to boundary layer state for all 119 measurements.

	Boundary Layer State			
	Unstable	Neutral	Stable	Total
All	84	23	12	119
Offshore	65	0	0	65
Onshore	19	23	12	54
All $s_p^2 \neq 0$	7	4	0	11
Probability (%)	8.3	17.4	0.0	9.2
Offshore $s_p^2 \neq 0$	4	0	0	4
Probability (%)	6.2	-/-	-/-	-/-
Onshore $s_p^2 \neq 0$	3	4	0	7
Probability (%)	15.8	17.4	0.0	-/-

4.3. Confidence Considerations

To check the confidence of our above assumption, possible sources for errors were investigated. Individual measurements at onshore locations were recorded at different hours and days, so they can be assumed to be independently and randomly sampled. The probability of occurrence of a high s_p^2 event under combined unstable or neutral conditions was estimated to be $\Phi_0 = 0.167$. The statistical test was based on a binomial distribution with the states *event occurred* and *no event occurred*.

To check the occurrence probability of an event in the onshore stable boundary layer states, for the null-hypothesis, an occurrence probability was assumed to be the same as for unstable/neutral states, and finding no event was just bad luck. The alternative hypothesis postulates a much lower probability of $\Phi_1 = 0.062$ for high-s_p^2 events under stable conditions (see Table 4).

Table 4. Error estimation for the occurrence of non-Gaussian turbulence for onshore stable conditions.

Hypotheses:	
H_0	Occurrence probability is $\Phi_0 = 0.167$ in stable conditions
H_1	Occurrence probability is $\Phi_1 < \Phi_0$ in stable conditions
	Error Type I
Number of samples	12
Occurrence probability	$\Phi_0 = 0.167$
Expected occurrences	$E_0 = 2.00$
Variance	$V_0 = 1.67$
Acceptance region for H_0	$\{1 \ldots 12\}$
Rejection region for H_0	$\{0\}$
Probability of error	11.2%

The number of expected events for H_0 in onshore stable conditions is given in Table 4, and $E_0 = 2$ and H_0 were accepted if at least one event was found. However, no event of a non-Gaussian distribution was found under onshore stable conditions, and H_0 was rejected. The type I error rate was about 11.2%. This is also the level of significance (α). To improve the statistical significance, the number of samples in stable conditions had to be raised up to 17 for $\alpha < 5\%$ (or 26 for $\alpha < 1\%$), but has not been done so far.

Since the offshore measurements were often consecutive within the same conditions, and no observations in neutral or stable conditions were acquired, the statistical confidence could not be estimated for the offshore turbulence. However, the calculated occurrence probability gave a good approximation to allow the quality of the type II error for the onshore unstable and neutral boundary layers to be checked. The setup for this test is presented in Table 5 and the errors were 11.4% for type I (H_0 was rejected despite it being true) and 14.9% for type II (H_0 was accepted when it was really false).

Table 5. Error estimation for the occurrence of non-Gaussian turbulence for the onshore **unstable/neutral** conditions.

Hypotheses:		
H_0	Occurrence probability is $\Phi_0 = 0.062$ in unstable/neutral conditions	
H_1	Occurrence probability is $\Phi_1 > \Phi_0$ in unstable/neutral conditions	
	Error Type I	**Error Type II**
Number of samples		42
Occurrence probability	$\Phi_0 = 0.062$	$\Phi_1 = 0.167$
Expected occurrences	$E_0 = 2.58$	$E_1 = 7.00$
Variance	$V_0 = 2.43$	$V_1 = 5.83$
Acceptance region for H_0		$\{0 \ldots 4\}$
Rejection region for H_0		$\{5 \ldots 42\}$
Probability of error	11.4%	14.9%

These error estimates were acceptable with the small number of measurements and the occurrence probability onshore was reasonable.

5. Impact on Boundary Layer Transition on a Wind Turbine Blade

In the preceding sections, we have described that small-scale turbulence below a Taylor's length-scale of less than 50 mm are likely to show non-Gaussian behavior. We now briefly describe how this may be related to turbulent-laminar transitions on wind turbine blades. It has to be noted that more detailed investigations have been performed in the upper atmosphere for small aircraft [8,9]. Corresponding outdoor experiments on wind turbines were reported in reference [10,28]. It can be concluded that despite a high integral turbulence intensity of more that 10%, the energy content suitable for the receptivity of TS-waves is even smaller than in a wind-tunnel environment. More evidence for this **low-frequency cut-off** for aerodynamic important turbulence can be given by a simple argument to estimate an **upper** frequency $f^\star$ for a boundary layer (BL) responding to an oscillating outer flow. Stokes [29], pp 191 ff, showed that this outer flow (with ω) is damped out by viscosity according to $U = u_0 \cdot e^{-ky} \cdot cos(\omega t - ky)$ with $k = \sqrt{\omega/2\nu}$, ν being the kinematic viscosity and U being the main flow in the x-direction, with y being perpendicular to that. Now, if we introduce a *Stokes boundary layer thickness* by $\delta_S = 2\pi/k$, we get $f^\star = 4\pi\nu/\delta^2 \approx 200\ Hz$. Here, we have set δ_S to an approximate value of 1 mm, a typical value for laminar boundary layers on airfoils at Reynolds numbers of several million. Therefore, this high-frequency, non-Gaussian regime may play an important role in triggering the TS-type of (blade) boundary layer instability on the pathway towards fully developed turbulence.

6. Conclusions

High-frequency (above 100 Hz) resolved measurements of pressure fluctuations for different average wind velocities under onshore and offshore conditions were investigated with respect to higher order statistical properties and were related to atmospheric boundary layer stability. It was found that up to a frequency of 4 kHz, pressure and velocity fluctuations obey the same power spectrum, and further, out of 119 time series in total, high shape factors were shown to occur with a total probability of about 10%, independent of the Taylor-length based Reynolds number. Sixty-four percent of all high shape factor events were recorded during unstable thermal stratification, and the level of occurrence seemed to be almost doubled when compared to onshore cases. When a continuous sequence of time series was investigated, strong variations of the shape factor were sometimes observed even during short periods of about half an hour.

We argue for a separation of time scales to distinguish between instationary and turbulent flow at about 100 Hz, because it is clear that a severe reduction of total turbulence intensity takes place which makes the usage of laminar airfoils (NACA 63-215, for example) meaningful.

Author Contributions: A.P.S. and A.J. conceived and designed the experiments; A.J. performed the experiments; A.P.S. and A.J. analyzed the data; A.P.S. wrote the paper.

Funding: We thank the State of Schleswig-Holstein, Ministry of Science, Economics and Transportation for funding this project under Contract No.122-09-023.

Acknowledgments: We would like to thank the Germanischer Lloyd/Garrad Hassan (formerly WINDTEST) for their support during the measurement periods, onshore as well as offshore. We thank the Department of Physics at the University of Oldenbourg for their beneficial assessments and testing of our pressure sensors. Our thanks go also to the Deutscher Wetterdienst (DWD) for providing additional local temperature data. Discussion with S. Emeis, Institute of Meteorology and Climate Research, Karlsruhe Institute of Technology, Germany, is gratefully acknowledged.

Conflicts of Interest: The authors declare no conflict of interest.

References

1. Schaffarczyk, A. (Ed.) *Understanding Wind Power Technology*; Wiley: Chichester, UK, 2014.
2. Schaffarczyk, A.P. *Introduction to Wind Turbine Aerodynamics*; Springer: Berlin, Germany, 2014.

3.   Emeis, S. *Wind Energy Meteorology*; Springer: Heidelberg, Germany, 2013.

4.   Obukhov, A.M. Turbulence in an atmosphere with a non-uniform temperature. *Bound.-Lay. Metereol.* **1971**, *2*, 7–29. [CrossRef]

5.   Kaimal, J.; Finnigan, J. *Atmospheric Boundary Layer Flows: Their Structure and Measurements*; Oxford University Press: Oxford, UK, 1994.

6.   Mahrt, L.; Vickers, D.; Howell, J.; Høstrup, J.; Wilzak, J.; Edson, J.; Hare, J. Sea surfuace drag coeficiecient in the Risøe Air Sea Experiment. *J. Geophys. Res.* **1996**, *14*, 327–335.

7.   Grachev, A.; Leo, L.; Fernando, H.; Fairall, C.; Creegan, E.; Blomquist, B.W.; Christman, A.; Hocut, C. Air-sea/land interaction in the costal zone. *Bound.-Lay. Meteorol.* **2018**, *167*, 181–210.

8.   Reeh, A.D.; Weissmüller, M.; Tropea, C. Free-Flight Investigation of Transition under Turbulent Conditions on a Laminar Wing Glove. In Proceedings of the 51st AIAA Aerospace Sciences Meeting, Grapevine, TX, USA, 7–10 January 2013.

9.   Reeh, A.D.; Tropea, C. Behaviour of a natural laminar flow airfoil in flight through atmospheric turbulence. *J. Fluid Mech.* **2015**, *767*, 394–429. [CrossRef]

10. Schaffarczyk, A.; Schwab, D.; Breuer, M. Experimental detection of Laminar-Turbulent Transition on a rotating wind turbine blade in the free atmosphere. *Wind Energy* **2016**, *19*. [CrossRef]

11. Schwab, D. Aerodynamische Grenzschichtuntersuchungen an Einem Windturbinenblatt im Freifeld. Ph.D. Thesis, Helmut-Schmidt-Universität, Hamburg, Germany, 2018.

12. Beck, C. Superstatistics in hydrodynamic turbulence. *Physica D* **2004**, *193*, 195–207. [CrossRef]

13. Castaing, B.; Gagne, Y.; Hopfinger, E. Velocity probability density functions of high Reynolds number turbulence. *Physica D* **1990**, *64*, 177–200. [CrossRef]

14. Jeromin, A.; Schaffarczyk, A.P. Advanced statistical analysis of high-frequency turbulent pressure fluctuations for on- and off-shore wind. In Proceedings of the EUROMECH Colloquium 528: Wind Energy and the Impact of Turbulence on the Conversion Process, Oldenburg, Germany, 22–24 February 2012.

15. Jeromin, A.; Schaffarczyk, A.P. Relating high-frequency offshore turbulence statistics to boundary layer stability. In Proceedings of the 2013 International Conference on Aerodynamics of Offshore Wind Energy Systems and Wakes (ICOWES2013), Frankfurt, Germany, 19–21 November 2013; Shen, W.Z., Ed.; 2013; pp. 162–172.

16. Segalini, A.; Ramis, O.; Schlatter, P.; Henrik Alfredsson, P.; Rüedi, J.D.; Alessandro, T. A method to estimate turbulenFphysicace intensity and transvers Taylor microscale in turbulent flows from spatially averaged hot-wire data. *Exp. Fluids* **2011**, *51*, 693–700. [CrossRef]

17. Jeromin, A.; Schaffarczyk, A.P. *Statistische Auswertungen Turbulenter Druckfluktuationen auf der Off-Shore Messplattform FINO3*; Technical Report Unpublished Interal Report No. 78; University of Applied Sciences Kiel: Kiel, Germany, 2012.

18. Lück, S.; Renner, C.; Peinke, J.; Friedrich, R. The Markov-Einstein coherence length—An new meaning for the Taylor length in turbulence. *Phys. Lett.* **2006**, *359*, 335–338. [CrossRef]

19. Batchelor, G. Pressure fluctuations in isotropic turbulence. *Proc. Camb. Philos. Soc.* **1951**, *47*, 359–374. [CrossRef]

20. Xu, H.; Ouellette, T.; Vincenzi, D.; Bodenschatz, B. Acceleration correlations and pressure structure functions in high-reynolds number turbulence. *Phys. Rev. Lett.* **2007**, *99*, 204501. [CrossRef] [PubMed]

21. Sathe, M.; Mann, J.; Barlas, T.; Bierbooms, W.; van Bussel, G. Influence of atmospheric stability on wind turbine loads. *Wind Energy* **2013**, *16*, 1013–1032. [CrossRef]

22. Coelingh, J.; van Wijk, A.; Holtslag, A. Analysis of wind speed observations over the North Sea. *J. Wind Eng. Ind. Aerodyn.* **1996**, *61*, 51–69. [CrossRef]

23. Oost, W.; Jacobs, C.; van Oort, C. Stability effects on heat and moisture fluexes at sea. *Bound.-Lay. Meteorol.* **2000**, *95*, 271–302. [CrossRef]

24. Emeis, S. Upper limit for wind shear instable stratified conditions expressed in terms of a bulk Richardson number. *Meteorol. Z.* **2017**, *16*, 421–430. [CrossRef]

25. Dhruva, B. An Experimental Study of High Reynolds Number Turbulence in the Atmosphere. Ph.D. Thesis, Yale University, New Haven, CT, USA, 2000.

26. Jeromin, A.; Schaffarczyk, A.; Puczylowski, J.; Peinke, J.; Hoelling, M. Highy resolved measurements of atmospheric turbulence with the new 2D-atmospheric Laser Cantilever Anemometer. *J. Phys. Conf. Ser.* **2014**, *555*, doi:10.1088/1742-6596/555/1/012054. [CrossRef]

27. Lohse, D.; Grossmann, S. Intermittency in turbulence. *Physica A* **1993**, *194*, 519–531. [CrossRef]

28. Madsen, H.; Fuglsang, P.; Romblad, J.; Enevoldsen, P.; Laursen, J.; Jensen, L.; Bak, C.; Paulsen, U.S.; Gaunna, M.; Sorensen, N.N.; et al. The DAN-AERO MW experiments. *AIAA* **2010**, 645, doi:10.2514/6.2010-645. [CrossRef]
29. Batchelor, G. *An Introduction to Fluid Dynamics*; Cambridge University Press: Cambridge, UK, 1967.

Article

A Simplified Free Vortex Wake Model of Wind Turbines for Axial Steady Conditions

Bofeng Xu [1,*], Tongguang Wang [2], Yue Yuan [1], Zhenzhou Zhao [1] and Haoming Liu [1]

[1]   College of Energy and Electrical Engineering, Hohai University, Nanjing 211100, China; yyuan@hhu.edu.cn (Y.Y.); zhaozhzh_2008@hhu.edu.cn (Z.Z.); liuhaom@hhu.edu.cn (H.L.)
[2]   Jiangsu Key Laboratory of Hi-Tech Research for Wind Turbine Design, Nanjing University of Aeronautics and Astronautics, Nanjing 210016, China; tgwang@nuaa.edu.cn
*   Correspondence: bfxu1985@hhu.edu.cn

Received: 18 April 2018; Accepted: 22 May 2018; Published: 25 May 2018

Abstract: A simplified free vortex wake (FVW) model called the vortex sheet and ring wake (VSRW) model was developed to rapidly calculate the aerodynamic performance of wind turbines under axial steady conditions. The wake in the simplified FVW model is comprised of the vortex sheets in the near wake and the vortex rings, which are used to replace the helical tip vortex filament in the far wake. The position of the vortex ring is obtained by the motion equation of its control point. Analytical formulas of the velocity induced by the vortex ring were introduced to reduce the computational time of the induced velocity calculation. In order to take into account both accuracy and calculation time of the VSRW model, the length of the near wake was cut off at a 120° wake age angle. The simplified FVW model was used to calculate the aerodynamic load of the blade and the wake flow characteristic. The results were compared with measurement results and the results from the full vortex sheet wake model and the tip vortex wake model. The computational speed of the simplified FVW model is at least an order of magnitude faster than other two conventional models. The error of the low-speed shaft torque obtained from the simplified FVW model is no more than 10% relative to the experiment at most of wind speeds. The normal and tangential force coefficients obtained from the three models agree well with each other and with the measurement results at the low wind speed. The comparison indicates that the simplified FVW model predicts the aerodynamic load accurately and greatly reduces the computational time. The axial induction factor field in the near wake agrees well with the other two FVW models and the radial expansion deformation of the wake can be captured.

Keywords: wind turbine; simplified free vortex wake; vortex ring; aerodynamics; axial steady condition

---

1. Introduction

Over the past four decades, free vortex wake (FVW) methods have emerged as robust and versatile tools for modeling the aerodynamic performance of wind turbine rotors. Unlike the blade element momentum (BEM) theory [1,2] where annular average induction is found, the FVW method can determine vortical induction directly at the blade elements from the effect of the modeled wake and the method is more efficient than computational fluid dynamic (CFD) methods. However, the computational cost of the FVW method is much higher than that of the BEM method. Perhaps this is the reason that the FVW method is not commonly used for predicting the aerodynamic loads of wind turbine rotors in the wind energy community.

FVW methods are based on a discrete representation of the rotor vorticity field and a Lagrangian representation of the governing equations for the wake elements. In essence, the wake elements are allowed to convert and deform under the action of the local velocity field to force-free locations.

Clearly, the ability to predict the aerodynamic performance of wind turbines is strongly dependent on the ability to predict the highly complex wake geometry. Free wake analyses are fundamentally better suited to the complex flowfields generated by wind turbines and avoid the difficulty of prescribing a wake geometry but, in so doing, introduce more computational costs. Gohard [3] presented the first full vortex sheet wake (FVSW) model for horizontal axis wind turbines, in which the unconstrained wake was permitted to move freely with the local velocities existing in the wake. The initial wake geometry of the FVSW model is shown in Figure 1a. FVSW methods for the analysis of wind turbine aerodynamics were also used by Arsuffi et al. [4], Garrel [5], Sant et al. [6] and Sebastian et al. [7]. These methods are computationally expensive, making them somewhat impractical as a design tool. To remedy this problem, Rosen et al. [8] divided the wake into two regions, i.e., near and far wakes. The calculations associated with the far wake are speeded up by some approximations. The most popular approximation is the tip vortex wake (TVW) model [9–12] as shown in Figure 1b, in which, the far wake extends beyond the near wake and is comprised of a single helical tip vortex filament, which is appropriate based on the physics of the flow. The strength of the tip vortex is determined by assuming that the sum of the blade bound vorticity outside of the maximum is trailed into the tip vortex. The release point of the tip vortex is usually the tip of the blade. More significant simplifications to the FVW models for wind turbines are attributed to Miller [13], Afjeh et al. [14], and Yu et al. [15]. The vortex wake system was simplified by Miller [13] and Afjeh et al. [14] into three parts: a number of straight semi-infinite vortex lines, a series of vortex rings, and a semi-infinite vortex cylinder. The method has been used for comprehensive rotorcraft and wind turbine analyses. Yu et al. [15] developed a free wake model that combines a vortex ring model with a semi-infinite cylindrical vortex tube, in which the thrust coefficient of the rotor must be known and is used to calculate the strength of the new vortex ring produced in a time step. Except for the approximations of the wake, parallelization techniques have been used to address the computational expense problem of the FVW model. Farrugia et al. [16] modified the FVW code Wake Induced Dynamics Simulator (WInDS) [7] to enable parallel processing. Later, Elgammi et al. [17] used the modified model to investigate the cycle-to-cycle variations in the aerodynamic blade loads experienced by yawed wind turbine rotors operating at a constant speed with a fixed yaw angle. Turkal et al. [18] used a Graphics Processing Unit (GPU), to exploit the computational parallelism involved in the free-wake methodology.

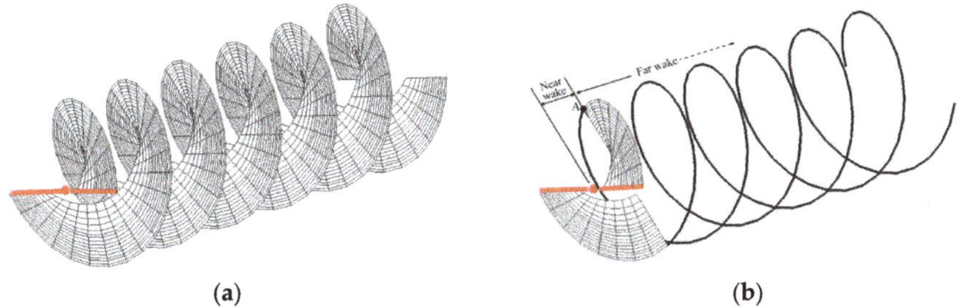

(a) (b)

Figure 1. Schematic of the initial wakes of different FVW models. (**a**) FVSW model. (**b**) TVW model.

In FVW methods, the number of discrete elements per vortex filament can be very large, making the tracking process memory-intensive and computationally demanding, although still considerably less demanding than when using CFD methods. The FVSW model has a computation time of about tens of minutes at one steady condition, and the TVW model has a computation time of about several minutes. Although the TVW model requires much less computational time than the FVSW model because it has fewer wake nodes in the far wake, the computational cost makes it somewhat impractical as an iterative optimization design tool. The main objective of this study is to reduce the number of wake nodes that need to be computed and achieve lower computation time in the axial steady condition. The velocity in the plane of rotation of the blade induced by the near wake is remarkable, so the near wake is still modeled by the vortex sheets in order to reflect the actual flow in this study. The effect of the far wake on the plane of rotation is relatively smaller due to the increase in distance, so the far wake is simplified into a series of vortex rings through learning from the studies of Miller [13] and Yu et al. [15]. Therefore, the new simplified FVW model is referred to as the vortex sheet and ring wake (VSRW) model.

In Section 2, the development of the VSRW model is described. Section 3 describes the analysis of the effect of the length of the near wake. The results are presented in Section 4 and include the rate of convergence of the wake iteration, the wake geometry, the induction factor in the wake, and the low-speed shaft torque (LSST). The conclusions are drawn in Section 5.

2. Simplified Free Vortex Wake

The development of the simplified FVW model will be presented in this section. The blade model is first briefly introduced and then the representation of the wake model is provided, including the near wake model and the far wake model. For completeness, the calculation procedure of the simplified FVW model is detailed.

2.1. Blade Model

The FVW model assumes that the flow field is incompressible and potential. Figure 2 shows the rotor body frame coordinates with the z-axis pointing downstream. The blade is modeled by the Weissinger-L model [19], which is a good compromise between the lifting line and the lifting surface models, as a series of straight constant-strength vortex segments lying along the blade quarter chord line. The control points are located at the 3/4-chord line at the center of each vortex segment. The wake vortices extend downstream from the 1/4-chord and form a series of horseshoe filaments. The trailing and shed vortices are modeled by the trailing and shed straight-line vortex filaments as shown in Figure 2.

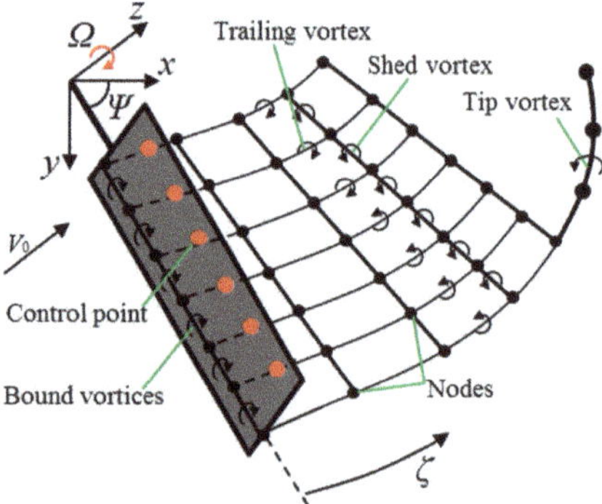

Figure 2. Schematic of the blade model and the vortex wake model.

The blade root section corresponds to the boundary of the first blade element. The remaining boundary distribution along each blade is achieved using the following "arc-cosine" relationship:

$$(\bar{r}_b)_i = \frac{(r_b)_i}{R} = \frac{2}{\pi}\arccos\left(1 - \frac{i-1}{N_E}\right) \tag{1}$$

where N_E is the number of blade elements and i is the element boundary number ($i = 1, \ldots, N_E + 1$). Consequently, there are N_E element control points and ($N_E + 1$) boundary points.

The strength, Γ_b, of each blade element is evaluated by the application of the Kutta-Joukowski theorem on the basis of airfoil data. The bound vorticity for the ith blade element is given by:

$$(\Gamma_b)_i = \frac{1}{2} W_i C_l c_i \tag{2}$$

where W_i is the resultant velocity at the ith control point and c_i is the chord of the ith blade element. To take into account the three-dimensional rotational effect, the 2D airfoil data is modified by the Du-Selig stall-delay model [20].

2.2. Near Wake Model

The wake that extends downstream from the blade is divided into two parts: the near wake and the far wake (Figure 1b). The near wake is modeled by the vortex sheets, which consist of the trailing and shed vortex filaments. The circulation of each blade element is assumed to be constant. However, the adjacent segments may have unequal loadings and, therefore, have different circulation strengths. The circulation strength of each trailing vortex is equal to the difference between the bound vortex strengths of the two adjacent segments. The circulation strength of each shed vortex is equal to 0 in the axial steady condition. The vortex filaments are allowed to freely distort under the influence of the local velocity field. The convection of these vortex filaments can be described by the Helmholtz equation. For a blade with fixed coordinates, the governing equation of the vortex filaments can be written as the partial differential form:

$$\frac{\partial r_v(\psi,\zeta)}{\partial \psi} + \frac{\partial r_v(\psi,\zeta)}{\partial \zeta} = \frac{1}{\Omega}[V_0 + V_{ind}(\psi,\zeta)] \tag{3}$$

where r is the position vector of the vortex collocation point; V_{ind} equals the mean value of the induced velocities at the surrounding four grid points [9].

To solve the convection equation of the vortex filaments, numerical solutions have been investigated over the last decades, including relaxation methods [21] and time-marching methods [9,11]. The time-marching methods can potentially provide the best level of approximation to the rotor wake problem with the fewest application restrictions. However, these methods have been found to be rather susceptible to numerical instabilities [22]. On the other hand, the relaxation methods improve the ability to control the non-physical wake instabilities by explicit enforcement of the wake periodicity. The relaxation methods can only be used for steady-state problems.

In this study, the axial steady conditions are predicted. It is sufficient to apply a steady relaxation scheme to solve the convection equation of the vortex filaments in the near wake. A five-point central difference approximation is used for both the temporal and spatial derivatives. The detailed information can be found in Reference [21]. A pseudo-implicit technique [23,24] has been introduced to improve the stability of the free-vortex iteration. The effective range of the azimuthal discretization is usually between $\Delta\psi = 5°$ and $\Delta\psi = 20°$ [22,25] and $\Delta\psi = 10°$ was selected for use in this study. The spatial step $\Delta\zeta$ uses the same value as the azimuthal step.

The velocity in the plane of rotation of the blade induced by the near wake is remarkable, so the length of the near wake is significant for the accuracy of the proposed model. The determination of the length of the near wake will be discussed further in Section 3.

2.3. Far Wake Model

In the TVW model, the far wake is modeled by the tip vortex model. In this study, vortex rings are used to replace the helical tip vortex filament as shown in Figure 3. Point A is the release point of the tip vortex and point C is the origin point of the 2nd circular tip vortex filament. The intersection point B of the vortex ring and the replaced helical tip vortex filament is located at the middle of every circular helical filament. The center of the ring is on the rotor's rotation axis. The point 1 opposite the point B on the ring is the control point of the vortex ring. The wake length in this paper is 4R. The number of the vortex rings is determined by the wake length:

$$N_C = \text{INT}\left(\frac{4R}{V_0} \cdot \frac{\Omega}{2\pi}\right) + 1 \tag{4}$$

where the equation INT is an integer-valued function; therefore, we need to add 1 to consider the missing part.

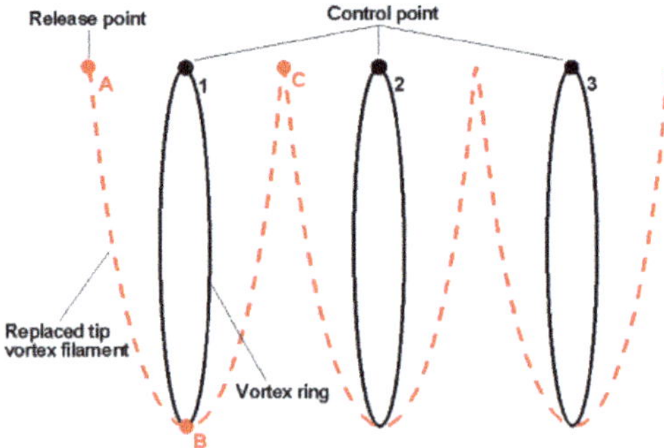

Figure 3. Schematic of the far wake.

As shown in Figure 3, points 1, 2, and 3 are the control points of the 1st, 2nd, and 3rd vortex rings respectively. In the axial steady condition, the vortex rings in the far wake are stationary when the wake geometry is convergent. It is assumed that point A will arrive at the position of point 1 when it moves under the influence of the velocity of point A for one half of the rotational period. Therefore, one period will be experienced between the two control points of the adjacent vortex rings. The position of the nth control point can be represented by its axial position $Z_{cr(n)}$ and its radial position $R_{cr(n)}$. The axial position can be obtained by:

$$
\begin{aligned}
Z_{cr(1)} &= Z_A + \sum_{k=1}^{N_T/2} \frac{\Delta\psi}{\Omega}\left(V_0 + V_{ind,Z}^A\right) \\
Z_{cr(n)} &= Z_{cr(n-1)} + \sum_{k=1}^{N_T} \frac{\Delta\psi}{\Omega}\left(V_0 + V_{ind,Z}^{n-1}\right), \quad 2 \leq n \leq N_C
\end{aligned}
\tag{5}
$$

The radial position can be obtained by:

$$
\begin{aligned}
R_{cr(1)} &= R_A + \sum_{k=1}^{N_T/2} \frac{\Delta\psi}{\Omega} V_{ind,R}^A \\
R_{cr(n)} &= R_{cr(n-1)} + \sum_{k=1}^{N_T} \frac{\Delta\psi}{\Omega} V_{ind,R}^{n-1}, \quad 2 \leq n \leq N_C
\end{aligned}
\tag{6}
$$

2.4. Velocity Induced by the Near Wake and the Far Wake

2.4.1. Velocity Induced by the Near Wake

The vortex filaments of the near wake comprise a series of straight-line vortex elements. The velocity induced by the vortex filaments equals the sum of the velocities induced by the straight-line vortex elements at the control nodes of the vortex elements, which are calculated using the Biot–Savart law as:

$$V_{ind} = \frac{\Gamma}{4\pi h}(\cos\theta_A - \cos\theta_B)\frac{r_A \times r_B}{|r_A \times r_B|} \tag{7}$$

where h, θ_A, θ_B, r_A, and r_B are as shown in Figure 4. However, if a collocation point is positioned very close to the vortex-line segment, then ($h \to 0$) will result in a very high induced velocity. In addition, the self-induced velocity ($h = 0$) exhibits a logarithmic singularity. These two phenomena cause convergence problems. To avoid these numerical problems, some vortex core models based on the Lamb-Oseen vortex model [26], Ramasamy and Leishman vortex model [27], Vatistas vortex model [28] and β-Vatistas vortex model [29,30] have been applied in a FVW model. In our recent work, we compared the effects of the β-Vatistas model and Lamb-Oseen model in the FVW model [29] and further studied the application [30] of the β-Vatistas model. In this study, a vortex core model based on the Lamb-Oseen vortex model, which is the most widely adopted and simpler model, was used. To account for the effect of viscous diffusion, the vortex core radius growth is used in the Biot-Savart law and is modified by using an empirical viscous growth model [31]. The stretching effect of the vortex filaments is taken into account by the application of a model developed by Ananthan and Leishman [32].

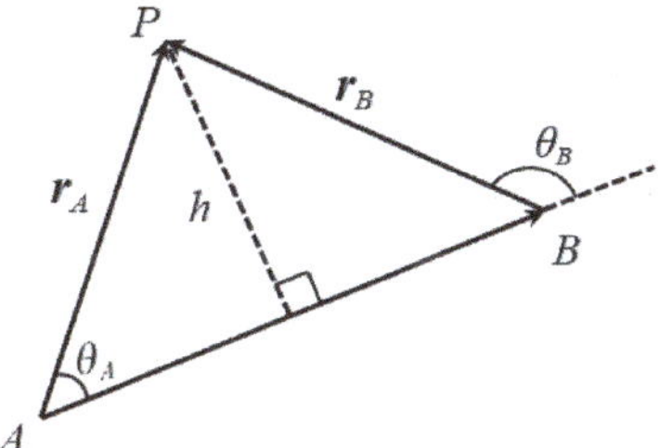

Figure 4. Schematic of the straight-line vortex elements.

2.4.2. Velocity Induced by the Far Wake

The analytical formulas of the velocity induced by a vortex ring are given by Yoon and Heister [33]. Figure 5 shows the Cartesian coordinate system used in this study for the induced velocity calculation. The axial and radial velocities at an arbitrary point P induced by the nth vortex ring are given by:

$$v_{ind,z} = \frac{\Gamma_n}{2\pi a}\left[K(m) - \frac{b^2 + R_p^2 - R_{cr(n)}^2}{a^2 - 4R_p R_{cr(n)}} E(m) \right] \tag{8}$$

$$v_{ind,r} = -\frac{b\Gamma_n}{2\pi R_p a}\left[K(m) - \frac{b^2 + R_p^2 + R_{cr(n)}^2}{a^2 - 4R_p R_{cr(n)}} E(m) \right] \tag{9}$$

where

$$a = \sqrt{\left(Z_p - Z_{cr(n)} \right)^2 + \left(R_p + R_{cr(n)} \right)^2} \tag{10}$$

$$b = Z_p - Z_{cr(n)} \tag{11}$$

$K(m)$ and $E(m)$ are the complete elliptic integrals of the first and second kind, where m is given by:

$$m = \frac{4R_p R_{cr(n)}}{a^2} \tag{12}$$

A fast method [34] is used for evaluating the first and second integrals. When point P is on the vortex ring, the induced velocity is calculated by Equation (7) to avoid the self-induced numerical problem.

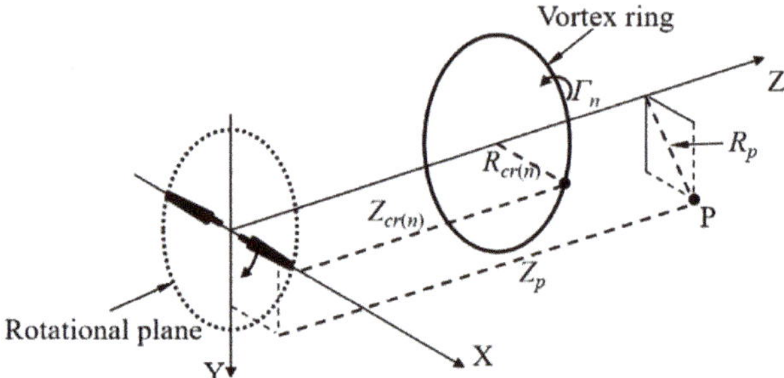

Figure 5. Schematic of the coordinate system for the induced velocity calculation of the vortex ring.

2.5. Calculation Procedure

The calculation process of the proposed simplified FVW model is shown in Figure 6. The following provides detailed explanations for some steps in the flowchart.

- In step 2, the initial wake geometry of the near wake consists of a set of regular helixes. The initial wake geometry of the far wake is calculated by Equations (5) and (6) and the initial induced velocities in the two equations equal 0.
- In step 5 and step 7, the velocity at the node and the control point induced by the vortices is calculated using the methods in Section 2.4.
- In step 6, the five-point central difference approximation is used to solve the convection equation of the vortex filaments in the near wake and Equations (5) and (6) are used to obtain the shape and location of the vortex rings in the far wake.
- In step 8, the root mean square (RMS) change between the new wake geometry and the old wake geometry of the two iteration steps is calculated. If the RMS change is less than a prescribed tolerance of 10^{-4}, convergence is achieved. Otherwise, return to step 3.

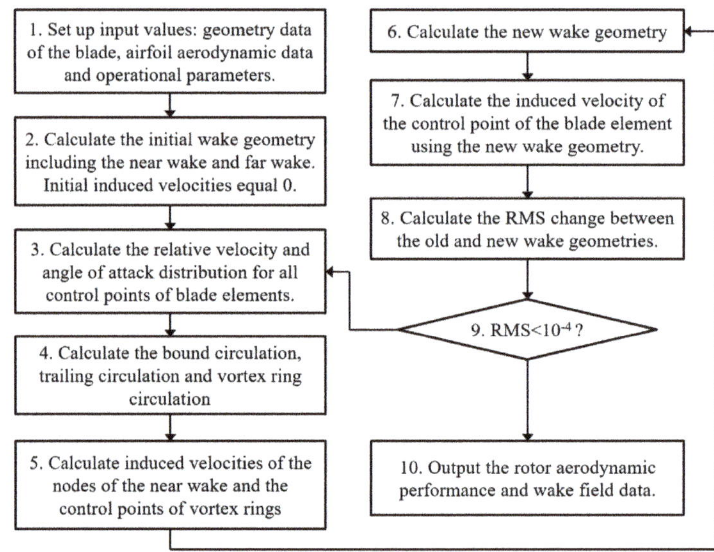

Figure 6. Flowchart of the simplified FVW model.

3. Length of Near Wake

To validate the VSRW model, the National Renewable Energy Laboratory (NREL) Phase VI wind turbine is used in this paper as a test case. The NREL Phase VI is a stall-regulated turbine. This turbine was designed by the NREL. The experiments were performed in the National Aeronautics and Space Administration (NASA) Ames wind tunnel (24.4 × 36.6 m) [35], which is considered a benchmark for the evaluation of wind turbine aerodynamic methods.

In the TVW model, the near wake is truncated after a short azimuthal distance, which is typically an azimuth angle of about 60° [36] in the helicopter field. Beyond this point, the far wake extends and is comprised of a single tip vortex filament. In the proposed VSRW model, one should ensure that the vortex sheet extends sufficiently far downstream. The aerodynamic performance of the rotor should be independent of the length of the vortex sheet. The length of the vortex sheet (or length of the near wake) can be represented by the wake age angle of the near wake. The spatial step equals 10° as mentioned above.

The LSST has been calculated used different lengths of the near wake ranging from 10° to 360°. Figure 7 shows the LSST along the length of the near wake at different wind speeds. As the length of the near wake increases from 10°, the LSST increases at the beginning and reaches the first maximum value at a certain near wake length. When the length of the near wake is longer than this certain length, the LSST is nearly constant at this maximum value. This certain length of the near wake is defined as the dividing length in this study. It is evident that the dividing length of the near wake is about 120° at wind speeds of 13 m/s, 20 m/s and 25 m/s, about 150° at wind speeds of 7 m/s and 10 m/s and about 140° at a wind speed of 15 m/s. At wind speeds of 7 m/s, 10 m/s and 15 m/s, the changes of the LSST between the near wake length of 120° and the dividing length are about 0.73%, 0.08% and 0.04% of the first maximum value. This illustrates that when the length of the near wake is about 120°, the LSSTs achieve nearly constant values at all wind speeds.

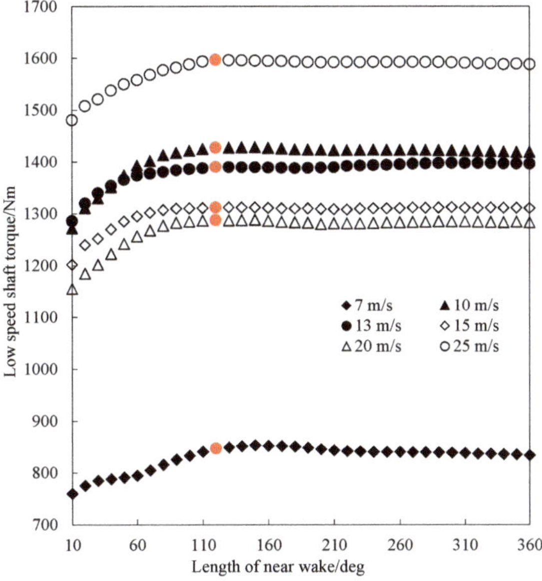

Figure 7. LSST along the length of the near wake at different wind speeds (The red points are the LSST values with a near wake age angle of 120°).

The calculation time of the VSRW model along the length of the near wake was also examined. The central processing unit (CPU) of the computer is an Intel Core i5-4200U and the memory is 8 GB. Figure 8 shows the wake iteration time along the length of the near wake at 7 m/s and 15 m/s. It is apparent that the number of discrete nodes used to approximate the near wake increases as the length of the near wake increases. However, the computational cost for the induced-velocity calculation changes as the square of the number of discrete nodes [25].

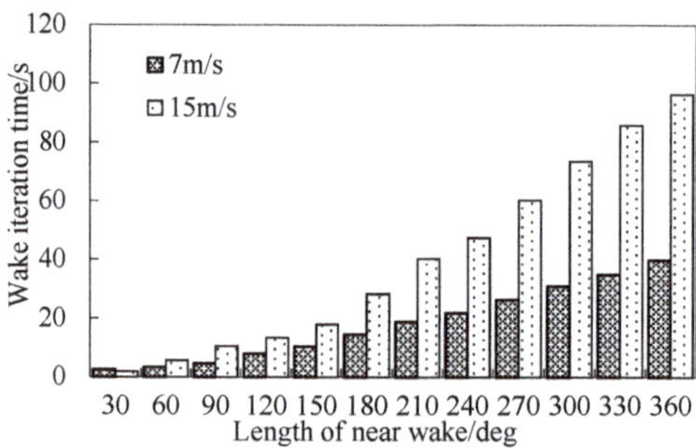

Figure 8. Wake iteration time along the length of the near wake at 7 m/s and 15 m/s.

The vortex sheet of the near wake is cut off at a 120° wake age angle in the VSRW model after considering the accuracy and calculation time. The red points in Figure 7 are the LSST values with a near wake age angle of 120° at every wind speed. It is observed that all LSST values reach the stationary values when the length of the near wake is 120°. It should be noted that this cut-off wake age angle is just an empirical angle rather than a real roll-up angle. In the following section, the aerodynamic analysis is conducted using the VSRW model with the near wake age angle of 120°.

4. Description of TVW and FVSW Models

To compare the calculation time and the accuracy of the proposed VSRW model with other models, the existing and mature models TVW and FVSW are used in this study. The blade models of the TVW and FVSW models are the same as the blade model of the VSRW model. As shown in Figure 2, the wake of the TVW model consists of vortex sheets in the near wake and tip vortex filaments in the far wake; the wake of the FVSW model consists of only vortex sheets in the near wake and far wake. The wake lengths of the two models are 4R and the near wake length of the TVW model was also set at 120°. The five-point central difference approximation [21] and the pseudo-implicit technique [23,24] are used to solve the convection equation of the vortex filaments in all the wakes of the TVW and FVSW models. The prescribed tolerances of the RMS change are all set as 10^{-4}.

5. Results and Discussions

5.1. Wake Iteration

In this study, we focus on decreasing the calculation time of the vortex model, which is approximately equal to the computational cost for the wake iteration. Figure 9 shows the wake iteration time of the VSRW model, TVW model, and FVSW model at the wind speeds of 7 m/s, 15 m/s, and 25 m/s. The wake iteration of the VSRW model requires 7.94 s, 13.37 s, and 20.24 s at the three wind speeds respectively; the computation speed is far faster than the two other models. The wake iteration time of the VSRW model at 7 m/s is two orders of magnitude less than for the FVSW model and the other times are one order of magnitude less. The VSRW model can generate results much faster than the two conventional methods mainly for two following reasons.

(1) In the VSRW model, the position of the vortex ring is determined by its control point, so we just
     have to calculate the induced velocity and the position of the control point of the vortex ring
     in the far wake. However, in the conventional methods, induced velocities and positions of all
     nodes of the vortex filaments in the far wake need to be calculated.

(2) The analytical method described in Section 2.4.2 is used to calculate the velocity induced by the
     far wake in the VSRW model. In the conventional methods, the velocity induced by the far wake
     is the sum of the velocities induced by the straight-line vortex elements, which are calculated
     using the Biot–Savart law.

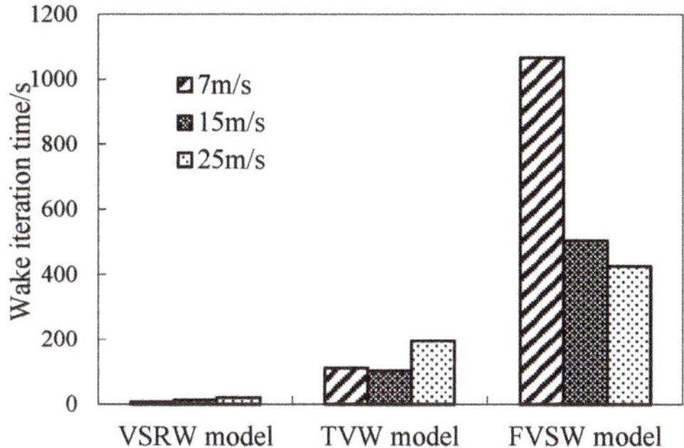

Figure 9. Wake iteration time of the VSRW model, TVW model, and FVSW model at 7 m/s, 15 m/s,
and 25 m/s.

Figure 10 shows the convergence histories of the RMS of the error in the wake geometries at wind
speeds of 7 m/s, 15 m/s, and 25 m/s using the three wake models. The convergence iteration numbers
of the VSRW and TVW models are higher when the wind speed is higher. However, the convergence
iteration number of the FVSW model does not change much with increasing wind speed. The wake
iteration time depends on the convergence iteration number and the computational time of each step.
In the TVW and FVSW models, more resources are needed to calculate the induced velocity field and
the positions of the wake nodes; this results in a longer wake iteration time as shown in Figure 9. This is
apparent in the FVSW model because of the dense vortex sheets, especially at a wind speed of 7 m/s.
The introduction of the vortex ring technology in the far wake of the proposed VSRW model saves a
lot of calculation time for determining the induced velocity field of the vortex rings and the positions
of the vortex ring control points; as a result, the computational time for each step is greatly reduced.
Therefore, the wake iteration time is still very short at a high wind speed although the convergence
iteration number increases markedly.

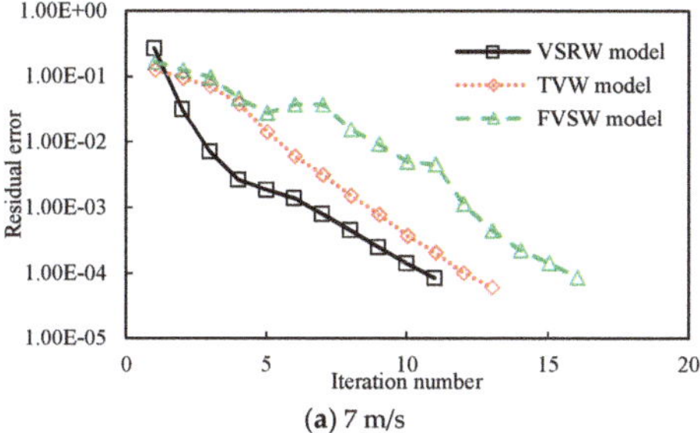

Figure 10. *Cont.*

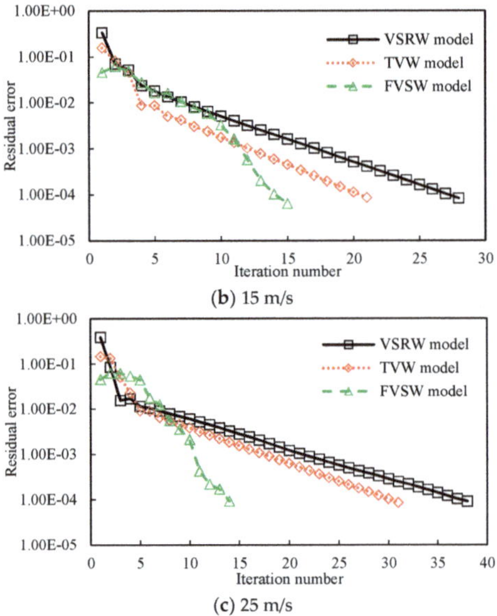

(b) 15 m/s

(c) 25 m/s

Figure 10. Convergence history of the RMS of the error in the wake geometry.

5.2. Low Speed Shaft Torque

The computational time of the proposed VSRW model can be greatly reduced and in the following section, we analyze its accuracy. The aerodynamic load of the LSST is predicted by the three different wake models. Figure 11 shows the aerodynamic estimates from the models compared with the values measured directly at the shaft [35]. The trends of the calculated results are consistent with the measurement results. Above 15 m/s, the estimated values begin to differ from each other and from the value measured at the shaft. The reason is that most of the blade is undergoing a stall flow. Although the 2D airfoil data are modified by the stall-delay model, the accurate prediction of the aerodynamic load in rotational and stall conditions is still challenging. This is also reflected in the studies by Breton et al. [37] who used a prescribed vortex wake technique and by Sant et al. [6] who used a free vortex wake technique. All the results from the VSRW model are slightly higher than those from the TVW and FVSW models. Except for 20 m/s, the results from the VSRW model could have errors no more than 10% relative to the experiment data. At 20 m/s, the error relative to the experiment data is about 16%. However, it is gratifying that the simplification of the far wake in the VSRW model did not significantly affect the accuracy of the aerodynamic load prediction.

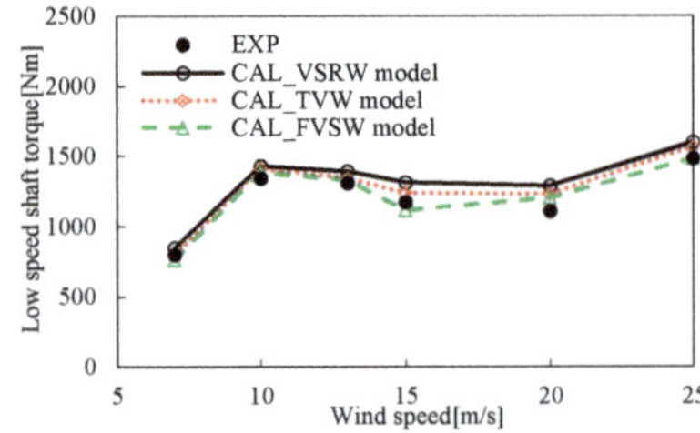

Figure 11. Measured and calculated results of LSST of NREL Phase VI wind turbine for wind speeds ranging from 7 m/s to 25 m/s.

5.3. Radial Distribution of Blade Airloads

The distributions of the normal and tangential force coefficients (C_n and C_t) at different wind speeds are computed by the proposed VSRW model, the TVW model, and the FVSW model. Figures 12 and 13 show the normal force coefficients and the tangential force coefficients respectively at 7 m/s, 15 m/s, and 25 m/s from the models and comparisons with the measurement results [35]. It is evident that the trends of the distributed airloads along the blade span are consistent with the measurement results. At 7 m/s, C_n and C_t derived by the three models agree well with the measurement results; C_n and C_t predicted by the VSRW model could have errors no more than 5% and 9% respectively relative to the experiment data. At high wind speeds of 15 m/s and 25 m/s, the estimated values, especially C_t, begin to differ from each other and from the measurement results because most of the blade is undergoing a stall flow. The comparison results exhibit similarities to the calculation of the LSST and illustrate that the proposed VSRW model predicts the radial distribution of the blade airloads as well as the TVW and FVSW models.

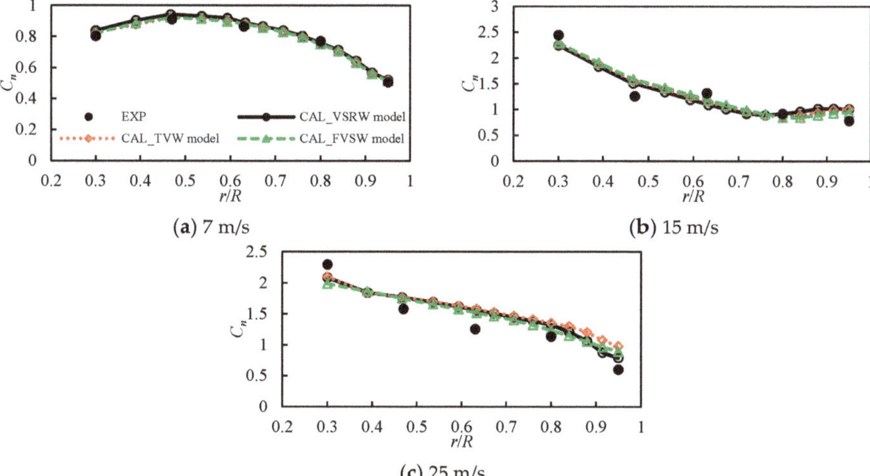

Figure 12. Comparison of the distributions of the computed normal force coefficients at (**a**) 7 m/s, (**b**) 15 m/s, and (**c**) 25 m/s along the blade radial positions for NREL Phase VI in axial steady conditions.

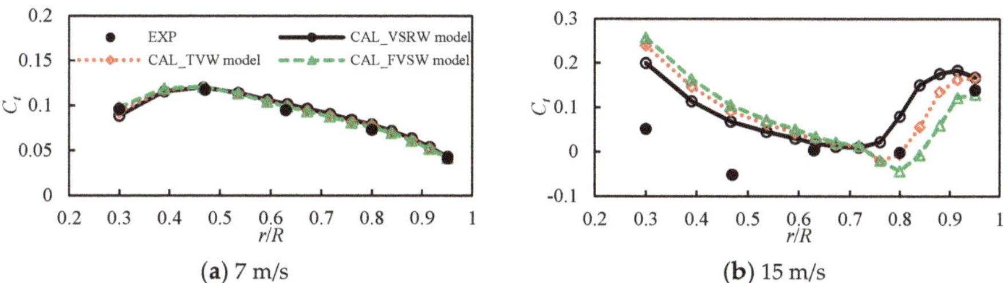

Figure 13. *Cont.*

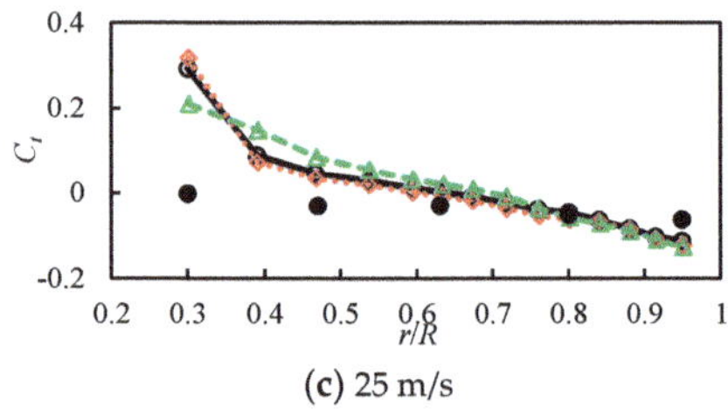

(**c**) 25 m/s

Figure 13. Comparison of the distributions of the computed tangential force coefficients at (**a**) 7 m/s, (**b**) 15 m/s, and (**c**) 25 m/s along the blade radial positions for NREL Phase VI in axial steady conditions.

5.4. Wake Geometry

Figure 14 shows the wake geometries at 10 m/s computed using the three different wake models. The azimuthal angle of the wake geometry is 0°. The wake deformation computed by using the FVSW model differs from that shown in Figure 2a. The wake deformation of the near wake of the VSRW and TVW models is also shown. The number of discrete wake nodes is far higher for the FVSW model than for the VSRW and TVW models; therefore, more computational time is required. It should be noted that the vortex rings from the two blades overlap at the same axial position. Therefore, there are actually six vortex rings in Figure 14a.

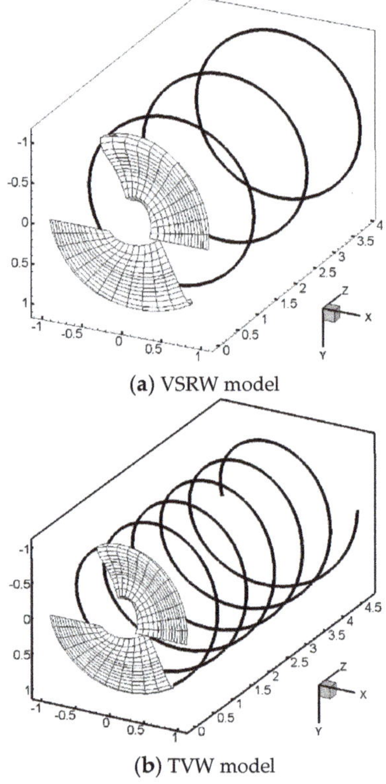

(**a**) VSRW model

(**b**) TVW model

Figure 14. *Cont.*

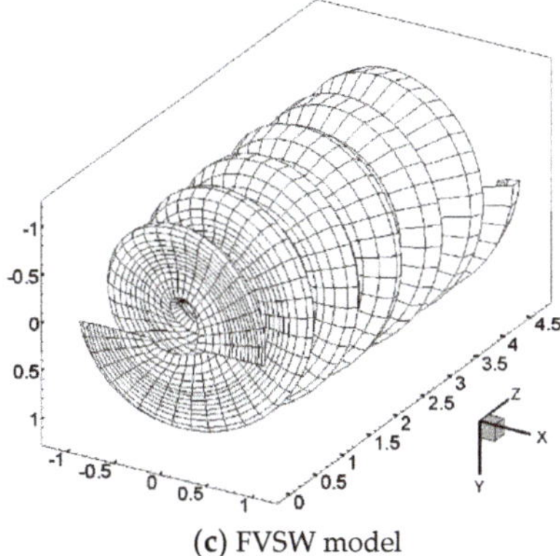

(**c**) FVSW model

Figure 14. Wake geometries (front view) at four times the radius distance behind the rotor at 10 m/s calculated using the (**a**) VSRW model, (**b**) TVW model, and (**c**) FVSW model.

5.5. Induction Factor in the Wake

Figure 15 shows the contours of the axial induction factor in the XOZ plane at the blade azimuthal angle of $0°$. In the front of the plane $Z/R = 0.5$ (on the left of the red dashed line in Figure 14), the distributions of the axial induction factor are similar for the three wake models. On the right of the red dashed line, the patterns of the contours of the axial induction factor are similar for the TVW and FVSW models but a marked difference is observed for the VSRW model. This is because the vortex rings are used to replace the helical tip vortex filament in the far wake. The radial expansion deformation of the wake can be captured by the VSRW model, as well as the other two models. The two overlapping vortex rings at the same axial position are the reason that the maximum axial induction factor around the vortex ring is twice as high as the maximum value in the TVW and FVSW models. On the motion trail of the tip vortex, the axial induction factor remains at about -0.1 in three wake models; therefore, the assumption of the constant induced velocity is reasonable when the control point is moving.

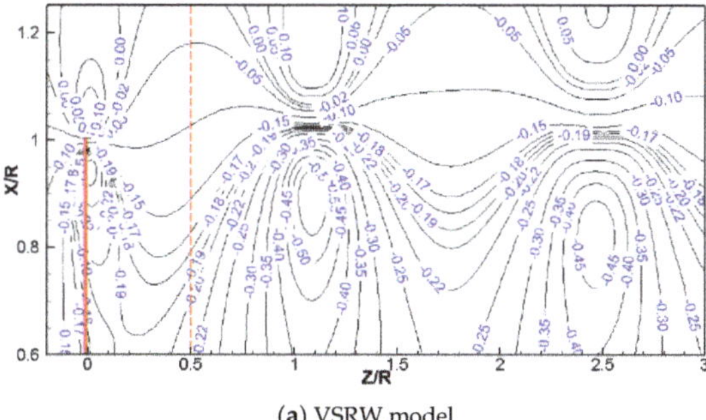

(a) VSRW model

Figure 15. *Cont.*

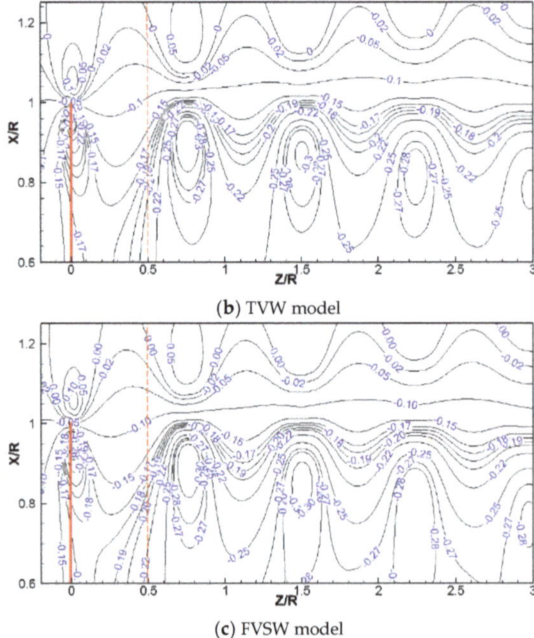

Figure 15. Contours of the axial induction factor.

6. Conclusions

In this study, a simplified FVW model was developed to rapidly calculate the aerodynamic performance of wind turbines in axial steady conditions. The helical tip vortex filament of the far wake in the traditional FVW model is replaced by several vortex rings. The proposed computing method of the vortex ring position and the proposed analytical formulas of the vortex ring-induced velocity greatly reduce the computational time. The length of the near wake was cut off at a 120° wake age angle to take into account both the accuracy and the calculation time of the VSRW model. The computational speed of the simplified FVW model is at least an order of magnitude faster than other two conventional models and the convergence and stability are good.

The simplified FVW model accurately predicts the blade aerodynamic load, including the LSST and the distributions of the span's normal and tangential force coefficients. The error of the low speed shaft torque obtained from the simplified FVW model is no more than 10% relative to the experiment at most wind speeds. The normal and tangential force coefficients obtained from the three models agree well with each other and with the measurement results at the low wind speed. At the high wind speed, the estimated values begin to differ from each other and from the measurement results because most of the blade is undergoing a stall flow

Except for differences in the axial induction factor field in the far wake, the axial induction factor field in the near wake computed by using the simplified FVW model agrees well with the other two FVW models and the radial expansion deformation of the wake is captured. The comparison of the far wake structure illustrates that the helical tip vortex filament, which is used in the TVW model, is more appropriate based on the physics of the flow.

Overall, in axial steady conditions, the computational time of the simplified FVW model is at least an order of magnitude less than that of traditional FVW models and the prediction accuracy is not affected significantly. On this basis, a study of the unsteady simplified FVW model will be conducted in the future.

Author Contributions: B.X. run the FVW codes and prepared this manuscript under the guidance of T.W. and Y.Y. Z.Z. and H.L. supervised the work and contributed in the interpretation of the results. All authors carried out data analysis, discussed the results and contributed to writing the paper.

Acknowledgments: This work was supported by the National Natural Science Foundation of China (grant number 51607058, 11502070); the National Basic Research Program of China (973 Program) (grant number 2014CB046200); and the Fundamental Research Funds for the Central Universities (grant number 2016B01514).

Conflicts of Interest: The authors declare no conflict of interest.

Nomenclature

Variables

a, b	Coefficients in the induced velocity equation (-)
c_i	Chord of the ith blade element (m)
C_l	Lift coefficient (-)
C_n	Normal force coefficient to the rotor disc (-)
C_t	Tangential force coefficients to the rotor disc (-)
h	Distance from the collocation point to the vortex-line segment (m)
i, k, n	Integer variables (-)
N_C	Number of vortex rings (-)
N_E	Number of blade elements (-)
N_T	Number of time steps of a circle (-), $N_T = 2\pi/\Delta\psi$
r	Radial location of the blade (m)
R	Rotor tip radius (m)
$\bar{r}_b$	Dimensionless radial location of the blade element boundary (-)
r_b	Radial location of the blade element boundary (m)
r_v	Position vector of the vortex collocation point (m)
r_A	Position vector from point A to point P (m)
r_B	Position vector from point B to point P (m)
$R_{cr(n)}$	Radial position of the nth vortex ring control point (m)
R_A	Radial position of the tip vortex release (m)
R_P	Radial position of point P (m)
V_{ind}	Induced velocity vector (m/s)
$v_{ind,r}$	Radial velocity at point P induced by the nth vortex ring (m/s)
$v_{ind,z}$	Axis velocity at point P induced by the nth vortex ring (m/s)
$V_{ind,R}^n$	Radial velocity at the nth vortex ring control point induced by all vortex field (m/s)
$V_{ind,R}^A$	Radial velocity at point A induced by all vortex field (m/s)
$V_{ind,Z}^n$	Axis velocity at the nth vortex ring control point induced by all vortex field (m/s)
$V_{ind,Z}^A$	Axis velocity at point A induced by all vortex fields (m/s)
V_0	Free stream velocity vector (m/s)
W_i	Resultant velocity at the control point of the ith blade element (m/s)
X	X axis in the coordinate system pointing right as viewed from the front (m)
Y	Y axis in the coordinate system pointing vertically downwards (m)
Z	Z axis in the coordinate system in the direction of wind flow (m)
$Z_{cr(n)}$	Axis position of the nth vortex ring control point (m)
Z_A	Axis position of the tip vortex release (m)
Z_P	Axis position of point P (m)
Γ	Vortex circulation (m²/s)
Γ_b	Bound circulation of the blade element (m²/s)
Γ_n	Vortex circulation of the nth vortex ring (m²/s)
$\Delta\zeta$	Discretization of the azimuthal angle (rad)
$\Delta\psi$	Discretization of the wake age angle (rad)
ζ	Vortex wake age angle (rad)
θ_A	Angle between vector r_A and vector $\mathbf{AB}$ (rad)
θ_B	Angle between vector r_B and vector $\mathbf{AB}$ (rad)
ψ	Azimuthal angle (rad)

Abbreviations

BEM	blade element momentum
CFD	computational fluid dynamics
CPU	Central Processing Unit
FVSW	full vortex sheet wake
FVW	free vortex wake
GPU	Graphics Processing Unit
LSST	low speed shaft torque
NASA	National Aeronautics and Space Administration
NREL	National Renewable Energy Laboratory
RMS	root mean square
TVW	tip vortex wake
VSRW	vortex sheet and ring wake
WInDS	Wake Induced Dynamics Simulator

References

1. Hansen, M.O.L.; Sørensen, J.N.; Voutsinas, S.; Sørensen, N.; Madsen, H.A. State of the art in wind turbine aerodynamics and aeroelasticity. *Aerosp. Sci.* **2006**, *42*, 285–330. [CrossRef]
2. Fernandez-Gamiz, U.; Zulueta, E.; Boyano, A.; Ansoategui, I.; Uriarte, I. Five megawatt wind turbine power output improvements by passive flow control devices. *Energies* **2017**, *10*, 742. [CrossRef]
3. Gohard, J.C. *Free Wake Analysis of Wind Turbine Aerodynamics*, Wind Energy Conversion, ASRL-TR-184-14; Massachusetts Institute of Technology, Department of Aeronautics and Astronautics, Aeroelastic and Structures Research Laboratory: Cambridge, MA, USA, 1978.
4. Arsuffi, G.; Guj, G.; Morino, L. Boundary element analysis of unsteady aerodynamics aerodynamics of windmill rotors in the presence of yaw. *J. Wind Eng. Ind. Aerodyn.* **1993**, *45*, 153–173. [CrossRef]
5. Garrel, A.V. *Development of a Wind Turbine Aerodynamics Simulation Module*; ECN-C-03-079; Energy Research Centre of The Netherlands: Petten, The Netherlands, 2003.
6. Sant, T.; Kuik, G.V.; Bussel, G. Estimating the Angle of Attack from Blade Pressure Measurements on the NREL Phase VI Rotor Using a Free Wake Vortex Model: Axial Conditions. *Wind Energy* **2006**, *9*, 549–577. [CrossRef]
7. Sebastian, T.; Lackner, M.A. Development of a free vortex wake method code for offshore floating wind turbines. *Renew. Energy* **2012**, *46*, 269–275. [CrossRef]
8. Rosen, A.; Lavie, I.; Seginer, A. A general free-wake efficient analysis of horizontal-axis wind turbines. *Wind Eng.* **1990**, *14*, 362–373.
9. Gupta, S. Development of a Time-Accurate Viscous Lagrangian Vortex Wake Model for Wind Turbine Applications. Ph.D. Dissertation, University of Maryland, College Park, MD, USA, 2006.
10. Qiu, Y.X.; Wang, X.D.; Kang, S.; Zhao, M.; Liang, J.Y. Predictions of unsteady HAWT aerodynamics in yawing and pitching using the free vortex method. *Renew. Energy* **2014**, *70*, 93–105. [CrossRef]
11. Xu, B.F.; Wang, T.G.; Yuan, Y.; Cao, J.F. Unsteady aerodynamic analysis for offshore floating wind turbines under different wind conditions. *Philos. Trans. R. Soc. A* **2015**, *373*, 20140080. [CrossRef] [PubMed]
12. Marten, D.; Lennie, M.; Pechlivanoglou, G.; Nayeri, C.N.; Paschereit, C.O. Implementation, optimization and validation of a nonlinear lifting line free vortex wake module within the wind turbine simulation. In Proceedings of the ASME Turbo Expo 2015: Turbine Technical Conference and Exposition, Montreal, QC, Canada, 15–19 June 2015.
13. Miller, R.H. The aerodynamics and dynamic analysis of horizontal axis wind turbines. *J. Wind Eng. Ind. Aerodyn.* **1983**, *15*, 329–340. [CrossRef]
14. Afjeh, A.A.; Keith, T.G.J. A vortex lifting line method for the analysis of horizontal axis wind turbines. *ASME J. Sol. Energy Eng.* **1986**, *108*, 303–309. [CrossRef]
15. Yu, W.; Ferreira, C.S.; van Kuik, G.; Baldacchino, D. Verifying the blade element momentum method in unsteady, radially varied, axisymmetric loading using a vortex ring model. *Wind Energy* **2017**, *20*, 269–288. [CrossRef]
16. Farrugia, R.; Sant, T.; Micallef, D. Investigating the aerodynamic performance of a model offshore floating wind turbine. *Renew. Energy* **2014**, *70*, 24–30. [CrossRef]

17. Elgammi, M.; Sant, T. Combining unsteady blade pressure measurements and a free-wake vortex model to investigate the cycle-to-cycle variations in wind turbine aerodynamic blade loads in yaw. *Energies* **2016**, *9*, 460. [CrossRef]
18. Turkal, M.; Novikov, Y.; Usenmez, S.; Sezer-Uzol, N.; Uzol, O. GPU based fast free-wake calculations for multiple horizontal zxis wind turbine rotors. *J. Phys. Conf. Ser.* **2014**, *524*, 012100. [CrossRef]
19. Weissinger, J. *The Lift Distribution of Swept-Back Wings*; NACA-TM-1120; National Advisory Committee for Aeronautics: Cleveland, OH, USA, 1947.
20. Du, Z.; Selig, M.S. *A 3-D Stall-Delay Model for Horizontal Axis Wind Turbine Performance Prediction*; AIAA-98–0021; American Institute of Aeronautics and Astronautics: Reston, VA, USA, 1998.
21. Bagai, A.; Leishman, J.G. Rotor free-wake modeling using a relaxation technique—Including comparisons with experimental data. *J. Am. Helicopter Soc.* **1995**, *40*, 29–41. [CrossRef]
22. Bhagwat, M.; Leishman, J.G. Stability, consistency and convergence of time marching free-vortex rotor wake algorithms. *J. Am. Helicopter Soc.* **2001**, *46*, 59–71. [CrossRef]
23. Grouse, G.L.; Leishman, J.G. A new method for improved rotor free vortex convergence. In Proceedings of the 31st AIAA Aerospace Sciences Meeting and Exhibit, Reno, NV, USA, 11–14 January 1993.
24. Wang, T.G.; Wang, L.; Zhong, W.; Xu, B.F.; Chen, L. Large-scale wind turbine blade design and aerodynamic analysis. *Chin. Sci. Bull.* **2012**, *57*, 466–472. [CrossRef]
25. Gupta, S.; Leishman, J.G. Accuracy of the induced velocity from helicoidal vortices using straight-line segmentation. *AIAA J.* **2005**, *43*, 29–40. [CrossRef]
26. Vatistas, G.H.; Kozel, V.; Minh, W. A simpler model for concentrated vortices. *Exp. Fluids* **1991**, *11*, 73–76. [CrossRef]
27. Sant, T.; del Campo, V.; Micallef, D.; Ferreira, C.S. Evaluation of the lifting line vortex model approximation for estimating the local blade flow fields in horizontal-axis wind turbines. *J. Renew. Sustain. Energy* **2016**, *8*, 023302. [CrossRef]
28. Gupta, S.; Leishman, J.G. Validation of a free vortex wake model for wind turbine in yawed flow. In Proceedings of the 44th AIAA Aerospace Sciences Meeting 2006, Reno, NV, USA, 9–12 January 2006; Volume 7, pp. 4529–4543.
29. Xu, B.F.; Yuan, Y.; Wang, T.G.; Zhao, Z.Z. Comparison of two vortex models of wind turbines using a free vortex wake scheme. *J. Phys. Conf. Ser.* **2016**, *753*, 022059. [CrossRef]
30. Xu, B.F.; Feng, J.H.; Wang, T.G.; Yuan, Y.; Zhao, Z. Application of a turbulent vortex core model in the free vortex wake scheme to predict wind turbine aerodynamics. *J. Renew. Sustain. Energy* **2018**, *10*, 023303. [CrossRef]
31. Bhagwat, M.J.; Leishman, J.G. Correlation of helicopter tip vortex measurements. *AIAA J.* **2000**, *38*, 301–308. [CrossRef]
32. Ananthan, S.; Leishman, J.G. The role of filament stretching in the free-vortex modeling of rotor wakes. *J. Am. Helicopter Soc.* **2004**, *49*, 176–191. [CrossRef]
33. Yoon, S.S. Heister, S.D. Analytical formulas for the velocity field induced by an infinitely thin vortex ring. *Int. J. Numer. Methods Fluids* **2004**, *44*, 665–672. [CrossRef]
34. Fukushima, T. Fast computation of complete elliptic integrals and Jacobian elliptic functions. *Celest. Mech. Dyn. Astron.* **2009**, *105*, 305–328. [CrossRef]
35. Hand, M.M.; Simms, D.A.; Fingersh, L.J.; Jager, D.W.; Cotrell, J.R.; Schreck, S.; Larwood, S.M. *Unsteady Aerodynamics Experiment Phase VI: Wind Tunnel Test Configurations and Available Data Campaign*; NREL/TP-500-29955; National Renewable Energy Laboratory: Golden, CO, USA, 2001.
36. Ananthan, S. Analysis of Rotor Wake Aerodynamics during Maneuvering Flight Using a Free-Vortex Wake Methodology. Ph.D. Dissertation, University of Maryland, College Park, MD, USA, 2006.
37. Breton, S.P.; Coton, F.N.; Moe, G. A study on rotational effects and different stall delay models using a prescribed wake vortex scheme and NREL Phase VI experiment data. *Wind Energy* **2008**, *11*, 459–482. [CrossRef]

 *applied sciences*

Article

Comparative Study of Dynamic Stall under Pitch Oscillation and Oscillating Freestream on Wind Turbine Airfoil and Blade

Chengyong Zhu and Tongguang Wang *

Jiangsu Key Laboratory of Hi-Tech Research for Wind Turbine Design, Nanjing University of Aeronautics and Astronautics, Nanjing 210016, China; rejoicezcy@nuaa.edu.cn
* Correspondence: tgwang@nuaa.edu.cn; Tel.: +86-025-8489-6138

Received: 15 June 2018; Accepted: 25 July 2018; Published: 27 July 2018

Abstract: This study aims to assess the dynamic stall of the wind turbine blade undergoing pitch oscillation (PO) and oscillating freestream (OF), respectively. Firstly, a thin-airfoil theoretical analysis was performed to differentiate between these two dynamic effects. During upstroke, PO results in a positive effective airfoil camber, while OF has an additional negative effective airfoil camber, and yet in contrast during downstroke, PO decreases the effective camber, while OF increases the effective camber. Secondly, the equivalence relation between PO and OF is investigated by numerically solving the unsteady Reynolds-averaged Navier-Stokes equations. The difference between PO and OF mainly exists in the linear part of the aerodynamic loads. Because the difference is great at high reduced frequencies or angle of attack (AOA) amplitudes, PO and OF should be treated separately for dynamic stall from different aerodynamic sources. Thirdly, the Beddoes-Leishman dynamic model coupled with Bak's rotational stall delay model was used to predict the yawed responses of the blade section. The obtained results show different aerodynamic responses between PO and OF, although consideration of rotational augmentation can greatly improve the accuracy of the lift and drag coefficients. To improve the understanding and coupling modeling of rotational augmentation and dynamic stall, an extended analysis of the coupled effect was performed as well.

Keywords: dynamic stall; pitch oscillation; oscillating freestream; rotational augmentation; wind turbine

---

1. Introductions

Horizontal axis wind turbines (HAWTs) often experience unsteady air loads, to which wind turbine failures, reduced machine life, and increased operating maintenance are all directly linked. However, deep understanding and accurate prediction of the unsteady blade air loads and rotor performance face many challenges [1], among which 'dynamic stall' is of particular significance. Dynamic stall is characterized by the shedding and passage of a strong vortical disturbance over the suction surface, inducing a highly nonlinear fluctuating pressure field [2]. The unsteady effects on HAWTs can be classified into two categories according to different aerodynamic sources. One is blade motion, such as blade pitching, elastic bending, and flapping. The other is varying flow field structure induced by atmospheric turbulence, wind shear, wind gust, yawed inflow, tower shadow, interactions among rotors in a wind farm, and so on. These sources ultimately act on the local angle of attack (AOA) and velocity field at the blade-section level, hence producing unsteady loads on the blades.

In the 1980s, McCroskey [2] and Carr [3] conducted a comprehensive review on the main physical features of dynamic stall of helicopter rotors. Then, Butterfield [4] verified the great effect of dynamic stall on HAWT performance. In order to consider dynamic stall in engineering, several empirical and semi-empirical models have been developed for rotor air load prediction and rotor design analysis.

These models often contain simplified representation of the essential physics using sets of linear and non-linear equations for the lift, drag, and pitching moment [5–7]. Leishman and Beddoes [6] developed the well-known Beddoes-Leishman (B-L) model based on proper considerations of flow mechanisms. An apparent advantage of the B-L model is that it yields good results with relatively few empirical coefficients, most of which can be derived from the static airfoil data. Despite the original application to helicopters (compressible flow), the B-L model has been introduced into wind turbines with several modifications [8–11]. Another promising method is the computational fluid dynamics (CFD) simulation achieved by numerically solving the Navier-Stokes equations with a suitable turbulence model. CFD methods have a very good capability of predicting both two-dimensional (2D) and three-dimensional (3D) dynamic stall events [12,13]. Ekaterinaris and Platzer [14] provided a comprehensive review on the computational prediction of dynamic stall. For Reynolds-averaged Navier-Stokes simulations (RANS), they found that the SST k-ω eddy viscosity model [15] performs better than any other two-equation models. However, their results are not yet satisfactory in the stalled regime and during flow reattachment. In this regard, they further found that the accurate prediction of transition from a laminar-to-turbulent boundary layer is a key element of improved hysteresis predictions.

As is well known, most of the information concerning dynamic stall is obtained by sinusoidal pitch oscillation about the quarter-chord axis, because this motion is considered to be adequate to represent the characteristics of dynamic stall. However, other types of motion, such as plunging oscillation, fore-aft motion, and oscillating freestream, can make a big difference in the dynamic-stall process [3,16]. In terms of relative motion, the blade motion and flow field variation may be further subdivided into the following three pairs (Figure 1):

- Firstly, the blade sections experience a time-varying tangential velocity (Figure 1a) under yawed inflow or in-plane gusts; as a counterpart, the blade sections undergo a fore-aft motion with edgewise vibration (Figure 1b).
- Secondly, unsteady inflow will cause a time-varying velocity component perpendicular to the rotor plane (Figure 1c); as a counterpart, the blade sections undergo a plunge motion during flapwise vibration (Figure 1d).
- Thirdly, a periodic AOA change (i.e., oscillating freestream, OF) results from the superposition of rotational velocity and the in-plane freestream velocity component (Figure 1e) under yawed inflow; as a counterpart, pitch oscillation (PO) occurs and affects the effective AOA under blade pitching or elastic bending (Figure 1f).

Van der Wall and Leishman [16] clarified the difference in the nature of the first pair, time-varying tangential velocity and fore-aft motion; the latter leads to a uniform perturbation velocity across the airfoil chord, while the former produces a perturbation velocity gradient across the chord. When the reduced frequency is low, both of them may be considered almost identical, which, however, is invalid for high reduced frequencies. Similarly, it can be speculated that time-varying freestream velocity will bring about a different effect on perturbation velocity from that of the plunging motion. For the third pair, Karbasian et al. [17] stated that OF can be referred to as PO. However, the equivalence relation between PO and OF still remains open from the point of view of Van der Wall and Leishman [16]. Practically, PO is generally used to conduct experimental studies and establish the engineering models. In order to improve the wind turbine unsteady aerodynamic prediction, it is necessary to clarify the relation between PO and OF.

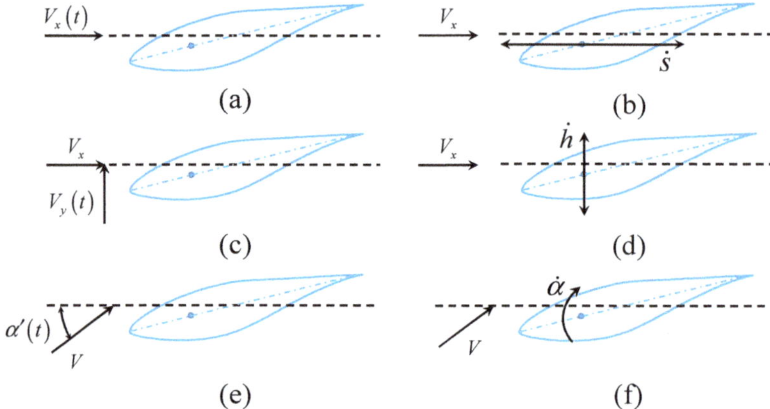

Figure 1. Schematic of different unsteady aerodynamic sources from a blade-section view. The left comes from varying flow field structures, and the right comes from the blade motion. The dashed line denotes the rotor plane. (**a**) Time-varying tangential velocity; (**b**) Fore-aft motion; (**c**) Time-varying normal velocity; (**d**) Plunge motion; (**e**) Oscillating freestream; (**f**) Pitch oscillation.

To investigate the dynamic stall of the blade sections under PO and OF, the problem of dynamic stall must be considered as fully 3D due to the rotational effect [1]. During the last few decades, both dynamic stall and rotational effects have captured significant attention but have been studied independently. The rotational effects play an important role in the boundary layer development due to the centrifugal and Coriolis forces. The centrifugal force results in radial flow, thinning the boundary layer of the blade. On the other hand, the Coriolis force adds a chordwise positive pressure gradient, flattening the separated zone and delaying the flow separation. Because the gross output power and sectional lift are clearly elevated with flow separation delayed, the rotational effect is also called a rotational augmentation. Several stall delay models have been developed to correct the 2D airfoil data by Snel et al. [18], Chaviaropoulos and Hansen [19], Du and Selig [20], Raj [21], Bak et al. [22], Corrigan and Schillings [23], and Lindenburg [24]. Breton et al. [25] evaluated these existing models on the Phase VI rotor with a prescribed wake vortex code and discussed their deficiencies and strengths. The 3D rotational effect is relatively strong on small-scaled stall-controlled HAWTs, and almost all of the correction models are established on the studies of stall-controlled HAWT aerodynamics. However, Zahle et al. [26] and Bangga et al. [27] noted that rotational augmentation is still prominent in the root region of a 10 MW modern pitch-controlled HAWT. Bangga et al. [28] also observed the unsteady effects on rotational augmented blade loads under a turbulent inflow.

A comparative analysis of the dynamic stall under PO and OF on the NREL S809 airfoil [29] and the Phase VI rotor [30] is presented in this paper. Firstly, the quasi-steady thin-airfoil theory is used to clarify the difference between PO and OF in the induced perturbation velocity across the chord. Secondly, the unsteady RANS method is used to assess the equivalence relation between PO and OF. The turbulence is modeled by the SST k-ω eddy viscosity model [15] incorporated with the γ-Re_θ transition model [31]. Thirdly, the classical B-L model is used to predict the yawed response of the Phase VI rotor under PO and OF. Bak's rotational stall delay model is incorporated into the dynamic stall model to consider rotational augmentation. The inverse BEM method is used to obtain the sectional AOAs and relative velocities from the experimental data. These extracted AOAs and velocities are then used to isolate the 3D rotational effect from the 3D unsteady aerodynamic data. All the predicted results agree well with the corresponding experimental data. It is found that PO and OF have an opposite effect on the effective airfoil camber. Therefore, it is necessary to separately consider the effects of PO and OF.

2. Methodology

The research objects are the Phase VI rotor and its basic airfoil, the NREL S809 airfoil. Here, PO denotes the sinusoidal pitch of the airfoil about the quarter-chord at a constant freestream velocity (Figure 1f), while OF denotes the sinusoidal change in the direction of inflow velocity at a constant magnitude to the stationary airfoil (Figure 1e). What is satisfied for both PO and OF is the same instantaneous angle between the freestream velocity and the chord as follows:

$$\alpha(t) = \alpha_m + A\sin(2\pi f t) \tag{1}$$

where α_m, A, and f are the mean AOA, the AOA amplitude, and the frequency of oscillation, respectively. The reduced frequency is often defined as $k = \pi f c / V$ with V being the freestream velocity and c the chord length.

2.1. Thin-Airfoil Theoretical Analysis

Allowing for a simple analysis of the flow mechanism, the airfoil is simplified as a flat plate with its camber and thickness neglected here, and then, a quantitative comparison can be made with the quasi-steady thin-airfoil theory. The governing equation for thin-airfoil theory is Laplace's equation. The idea is to place a vortex sheet singularity of unknown strength on the airfoil [32]. The strength of this vortex sheet is determined by satisfying the flow tangency on the camber line in conjunction with the Kutta condition at the trailing edge. Formulated in Fourier series, the general form of pressure distribution on the airfoil is as follows:

$$\Delta C_p(\alpha, \theta) = 4\left[A_0\left(\frac{1+\cos\theta}{\sin\theta}\right) + \sum_{n=1}^{\infty} A_n \sin n\theta \right] \tag{2}$$

where α is the AOA, and $\theta = \cos^{-1}(1 - 2x/c)$ (x is the chordwise location with the leading edge denoted by $x = 0$). The coefficients A_0 and A_n are given by following equations:

$$\begin{aligned} A_0 &= \alpha - \frac{1}{\pi}\int_0^\pi \frac{dy}{dx}d\theta \\ A_n &= \frac{2}{\pi}\int_0^\pi \frac{dy}{dx}\cos n\theta \, d\theta \end{aligned} \tag{3}$$

where $y(x)$ describes the camber line of the airfoil. Integrating Equation (2) along the chord, we can obtain the lift coefficient, $C_l = 2\pi(A_0 + 0.5A_1)$, and the pitching moment coefficient about the quarter-chord axis, $C_{m0.25} = -0.25\pi(A_1 - A_2)$. For PO, the pitch-rate term produces a linear variation in normal perturbation velocity [5], $w(x) = -\dot{\alpha}(x - 0.25c)$ (where $\dot{\alpha}$ is the pitch rate and positive during upstroke) so that the induced camber is a parabolic arc (Figure 2). For OF, the normal perturbation velocity can be formulated as $w(x) = V\alpha(t) - V\alpha(t-x/V)$. Because the quasi-steady condition means a low frequency, this formula can be linearized as $w(x) = \dot{\alpha}x$. Then, the term of dy/dx in Equation (3) can be replaced with $w(x)/V$. Hence, the lift coefficients are given by $C_l = 2\pi(\alpha + \dot{\alpha}c/2V)$ under PO and $C_l = 2\pi(\alpha - 3\dot{\alpha}c/4V)$ under OF, indicating an additional effective AOA, $\alpha_{eq} = \dot{\alpha}c/2V$ and $\alpha_{eq} = -3\dot{\alpha}c/4V$, respectively.

In short, PO and OF have an opposite effect on the effective airfoil camber. During upstroke, PO increases the effective camber, while OF decreases the effective camber. During downstroke, PO decreases the effective camber, while OF increases the effective camber. Therefore, the lift under PO is higher during upstroke and lower during downstroke compared with the lift under OF. Notice that this effect is directly related to the pitch rate $\dot{\alpha} = 2\pi f A\cos(2\pi f t)$. Consequently, the difference between PO and OF is negligible at low variation frequencies and substantial at high ones.

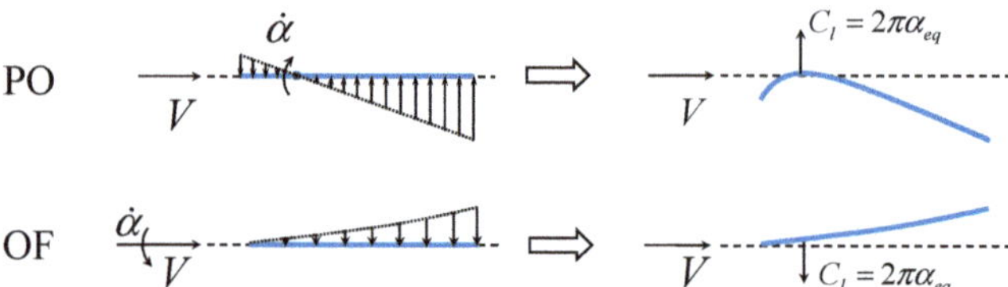

Figure 2. Induced effective camber for pitch oscillation (PO, upper) and oscillating freestream (OF, lower).

2.2. Numerical Modeling

For predictions of the dynamic stall, the commercial program ANSYS/FLUENT 16.0 (ANSYS, Inc., Canonsburg, PA, USA) [33] was used to solve the unsteady Reynolds-averaged Navier-Stokes equations. The mesh around airfoil geometry is generated in a structured O-type configuration to assure the wall orthogonality (Figure 3). The computational domain is exactly square. The first layer spacing off the wall is 10^{-5} c; this setting can ensure that the viscous sublayer is directly resolved with y^+ less than 1. There are 48 normal layers in the boundary layer mesh, with the growth rate being 1.08. After a proper grid independence study, the total size of the computational mesh is set to be 51,090 with 245 wrap-around points and 206 normal layers. A far-field distance of 20 c away from the airfoil is sufficient. The left, upper, and lower sides are all inlet boundary with velocity imposed. The right side is outlet boundary with pressure imposed (Figure 3). To simulate the dynamic stall under PO, the sliding-mesh method [33] was used. Therefore, the computational zone was divided into two subdomains, a rotating region (inner) and a stationary region (outer). These two regions interact by a circular sliding interface; this control circle of the sliding interface is 1.2 c in radius (Figure 3). To simulate the dynamic stall under OF, a user-defined function was implemented on the velocity inlet boundaries. The typical Reynolds number of 1×10^6 was selected in numerical simulations.

In order to obtain a good resolution, the third-order Monotone Upstream-Centered Schemes for Conservation Laws (MUSCL) convection scheme [33] is used for spatial discretization of the whole set of RANS and turbulence equations. This scheme was conceived of from the original MUSCL [34] by blending a central differencing scheme and a second-order upwind scheme. The second-order implicit scheme [33] is used for time differencing. Although it is theoretically valid that implicit methods are unconditionally stable with respect to the time step, nonlinearity effects would become prominent and oscillatory solutions may occur when the time step is increased. An operational cell convective courant number between 5 and 10 for viscous turbomachinery flows with an implicit scheme provides the best error damping properties [35]. On these bases, the number of inner iterations per time step is chosen to be 40 in this study to ensure the cell convective courant number is less than 10.

The pressure-based Coupled algorithm [33] is employed to handle the pressure-velocity coupling. Originally, the PISO algorithm [33] was preferable for transient problems, because it was derived from the SIMPLE algorithm [33] specifically for unsteady calculations. The PISO algorithm can yield accurate results with lower computational costs than the SIMPLE algorithm when time steps are sufficiently small. Both the SIMPLE and PISO algorithms, however, are known to converge slowly because the momentum equation and pressure-correction equation are solved separately. In the Coupled algorithm, the Navier-Stokes equations are directly solved through an implicit discretization of pressure in the momentum equations, with benefits in terms of robustness and convergence, especially with large time steps or with a poor-quality mesh. Balduzzi et al. [35] found that the obtained results with the Coupled algorithm are much closer to the experimental data than the SIMPLE and PISO algorithms. The lower sensitivity to the temporal discretization leads to the choice of the Coupled algorithm as the preferable and more robust formulation for the pressure-velocity coupling.

Because the incorporation of the transitional flow effects is a key element for an improved prediction of dynamic stall [14], the turbulence in the boundary layer is modeled by the four-equation SST k-ω eddy viscosity model [15] incorporated with the γ-Re$_\theta$ transition model [31].

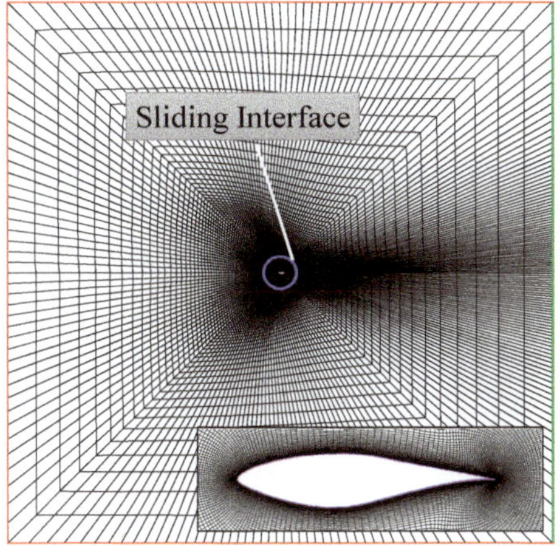

Figure 3. Computational mesh and boundary layer grids. The red line denotes the velocity inlet boundary, and the green line denotes the pressure outlet boundary.

2.3. Beddoes-Leishman Semi-Empirical Model

The classical B-L dynamic stall model [6] is capable of effectively assessing the unsteady loads with the flow physical features considered, and therefore, is fully implemented in the present work, including the leading-edge separation effect. Pereira et al. [10] have proven that it is unrealistic to disregard the leading-edge separation, because the wind turbine airfoils are thick. The main features and formulae of the B-L dynamic stall model may be summarized as follows.

2.3.1. Unsteady Attached Flow

Unsteady effects in attached flow conditions are simulated by the superposition of indicial aerodynamic responses for time step changes in the AOA. For a step change in the AOA, the normal force and pitching moment coefficients can be written as follows:

$$\Delta C_n = C_n^c + C_n^{nc}$$
$$\Delta C_m = C_m^c + C_m^{nc}$$

$$(4)$$

where the superscripts c and nc denote the circulatory part and non-circulatory part of the aerodynamic loads, respectively. The indicial response can be approximated empirically in terms of the exponential function. Hence, the circulatory loading is as follows:

$$C_n^c = C_{n\alpha}\alpha_e^c = C_{n\alpha}(\alpha(s) - X(s) - Y(s))$$

$$(5)$$

where $C_{n\alpha}$ is the normal force slope, and the non-dimensional time s represents the distance travelled at the relative velocity by the airfoil in semi-chords; that is, $s = 2Vt/c$. $X(s)$ and $Y(s)$ are deficiency functions written in a special finite difference approximation to Duhamel's integral as given by the following:

$$X(s) = X(s - \Delta s)\exp(-b_1 \Delta s) + A_1 \Delta \alpha \exp(-b_1 \Delta s/2)$$
$$Y(s) = Y(s - \Delta s)\exp(-b_2 \Delta s) + A_2 \Delta \alpha \exp(-b_2 \Delta s/2)$$

$$(6)$$

Here, A_1, A_2, b_1, and b_2 are the coefficients of the indicial functions and are a function of the airfoil and the Mach number. Then, the circulatory pitching moment coefficient is given by the following:

$$C_m^c = (0.25 - x_{ac}/c)C_n^c \tag{7}$$

where x_{ac} is the position of the airfoil aerodynamic center measured from the leading edge. A similar recurrence relation is used for the non-circulatory part of the air loads, C_n^{nc} and C_m^{nc} (see [6] for details).

2.3.2. Unsteady Separated Flow

The non-linear effects of the trailing-edge separation are implemented using a Kirchhoff flow model. The movement of the unsteady flow separation point position is related to the static separation position via a deficiency function. The onset of the leading-edge separation, and hence the dynamic stall, is identified using a criterion based on the attainment of a critical local leading-edge pressure, which is further related to the normal force.

Under unsteady conditions, there is a lag in the leading-edge pressure with the increasing AOA, which can be expressed as a first-order lag as given by the following:

$$C_n' = C_n^p - D_p \tag{8}$$

where

$$C_n^p = C_n^c + C_n^{nc}$$
$$D_p(s) = D_p(s - \Delta s)\exp(-\Delta s/T_p) + \left[C_n^p(s) - C_n^p(s - \Delta s)\right]\exp(-\Delta s/2T_p) \tag{9}$$

The constant T_p can be determined empirically from unsteady airfoil data. The effective AOA $\alpha_f = C_n'/C_{n\alpha} + \alpha_0$ after incorporating the unsteady pressure response is used to obtain the effective flow separation point f'. In this study, f' is determined by look-up-table interpolation from the static airfoil data. The additional effects of a boundary layer response are considered as a first-order lag as given by the following:

$$f'' = f' - D_f \tag{10}$$

where

$$D_f(s) = D_f(s - \Delta s)\exp\left(-\Delta s/T_f\right) + [f'(s) - f'(s - \Delta s)]\exp\left(-\Delta s/2T_f\right). \tag{11}$$

The constant T_f can be determined empirically from unsteady airfoil data. Then, the non-linear normal force coefficient is obtained by the Kirchhoff flow theory

$$C_n^f = C_{n\alpha}\left(\frac{1 + \sqrt{f''}}{2}\right)^2 (\alpha - \alpha_0) \tag{12}$$

where α_0 is the zero-lift AOA. Afterwards, the chordwise force coefficient is obtained by Gupa and Leishman's approximation [11] as given by the following:

$$C_n' \leq C_{n1} C_t^f = \begin{cases} C_{n\alpha}\sqrt{f''}\alpha\sin\alpha \\ K_1 + C_{n\alpha}\sqrt{f''}f''^\Phi\alpha\sin\alpha \quad C_n' > C_{n1} \end{cases} \tag{13}$$

where C_{n1} is the critical normal force value and the leading-edge separation is initiated when $C_n' > C_{n1}$. The parameter K_1 and Φ are obtained from the static test data [11]. The pitching moment coefficient is obtained by fitting the ratio C_m/C_n in a least-square method from the static airfoil data as given by the following:

$$C_m^f = C_{m0} + [k_0 + k_1(1 - f'') + k_2\sin(\pi f''^\mu)]C_n^f \tag{14}$$

where C_{m0} is the zero-lift pitching moment coefficient. The values of k_0, k_1, k_2, and μ can be adjusted for different airfoils.

2.3.3. Vortex Force

The induced vortex force and the associated pitching moment are represented empirically in a time-dependent manner during dynamic stall. The total accumulated vortex normal force coefficient under unsteady conditions is given by the following:

$$C_n^v = C_n^v(s - \Delta s)\exp(-\Delta s/T_v) + [C_v(s) - C_v(s - \Delta s)]\exp(-\Delta s/2T_v) \tag{15}$$

where

$$C_v(s) = C_n^c(s)\left[1 + \left(\frac{1 + \sqrt{f''}}{2}\right)^2\right]. \tag{16}$$

T_v is the vortex time decay constant. The center of pressure on the airfoil also varies with the chordwise position of the shedding vortex and will achieve a maximum value when the vortex reaches the trailing edge after a non-dimensional time period T_{vl}. A general representation of the center of pressure behavior (aft of the quarter-chord) is empirically formulated as follows:

$$(CP)_v = 0.25\left(1 - \cos\frac{\pi T_v}{T_{vl}}\right) \tag{17}$$

where T_v denotes the non-dimensional vortex time, $0 \le T_v \le 2T_{vl}$ (i.e., $T_v = 0$ at the onset of dynamic stall and $T_v = T_{vl}$ when the vortex reaches the trailing edge). Thus, the increment in the pitching moment about the quarter-chord due to dynamic stall is given by the following:

$$C_m^v = -(CP)_v \cdot C_n^v. \tag{18}$$

The total unsteady air loads on the airfoil is then obtained by superposition as follows:

$$\begin{aligned}
C_n &= C_n^f + C_n^{nc} + C_n^v \\
C_t &= C_t^f + (C_n^{nc} + C_n^v)\sin\alpha_e^c \\
C_m &= C_m^c + C_m^{nc} + C_m^f + C_m^v
\end{aligned} \tag{19}$$

Apart from dynamic stall, rotational augmentation is of crucial importance for wind turbine aerodynamic performance prediction. The rotational stall delay model of Bak et al. [22] is implemented. This model was developed on the difference in chordwise pressure distribution between the rotating blade section and the static airfoil. Although the obtained lift and drag coefficients were generally good, the modeled pitching moment coefficient indicated a poor agreement with the measurements, because the center of pressure was often underpredicted at large AOAs. Thus, a simple modification that the integrated location is changed from 0.25 to 0.1 was made to the pitching moment coefficient when fully separated flow occurs. Figure 4 implies an acceptable performance of this modification. It is generally acknowledged that the rotational augmentation is relatively apparent in separated flow; therefore, rotational augmentation mainly affects the nonlinear air loads.

To incorporate the rotational stall delay model into the dynamic stall model, the static airfoil data are corrected by Bak's model firstly. Then, the static separation point is obtained from the Kirchhoff formulation (Equation (12)). The dynamic separation point is calculated by a first-order lag of the static separation point (Equation (10)), while the nonlinear normal force is reconstructed by Equation (12). Afterwards, the chordal force coefficient is obtained by Equation (13), with an increment added to consider the rotational augmentation. Finally, the nonlinear pitching moment coefficient is obtained by Equation (14), and a rotational increment is added as well. According to the recommendations of Pereira et al. [10] and Gupa et al. [11], the set of corresponding empirical time constants, T_p, T_f, T_v, T_{vl}, is as follows: 1.7, 8.0, 1.0, and 7.0, where T_p and T_f are related to the first-order lag of the leading-edge separation and trailing-edge separation, respectively, and both T_v and T_{vl} are the non-dimensional time constants related to the movement of the dynamic vortex.

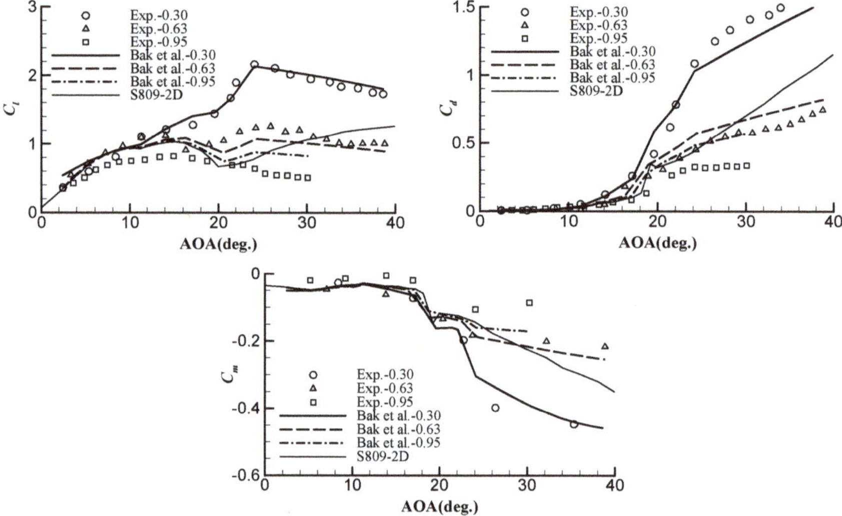

Figure 4. Three-dimensional (3D) corrected airfoil characteristics of different radial locations compared to two-dimensional (2D) measurements and measured 3D data.

The difference between PO and OF is addressed in the circulatory part of unsteady linear air loads, based on the thin-airfoil theory. To obtain input conditions, the sectional velocity components are extracted from the yawed measurements with the inverse BEM method [22,24,36–38]. This method has widely been used to determine the local AOA with the pre-determined local forces used to calculate the local induction factors and hence the induced velocities. In the inverse BEM method, Prandtl's tip correction is used to consider the tip loss effect.

3. Results and Discussion

3.1. Time-Varying Sectional Incident Velocity under Yawed Inflow

Schepers [39] provided a detailed discussion on two aerodynamic effects under yawed inflow. One is the skewed wake effect, inducing an unbalanced axial velocity, producing asymmetric air loads, and then resulting in yawing stability. The other is the advancing and retreating blade effect, making the in-plane tangential velocity vary with the azimuth. Figure 5 provides the definitions of yaw angle and azimuth angle. Here, the blade azimuth is defined as zero at the 6-o'clock position, where the rotor is assumed to rotate clockwise. Figure 6 indicates the variations of the local AOA and relative velocity at different blade spanwise locations. These variations can be approximately formulated in trigonometric functions due to the rotational periodicity. It can be found that the spanwise location strongly influences the mean AOA and the AOA amplitude; the highest unsteadiness can be observed on the inboard blade despite a low velocity magnitude. Therefore, dynamic stall is often severe at inboard sections, where the rotational augmentation has manifested.

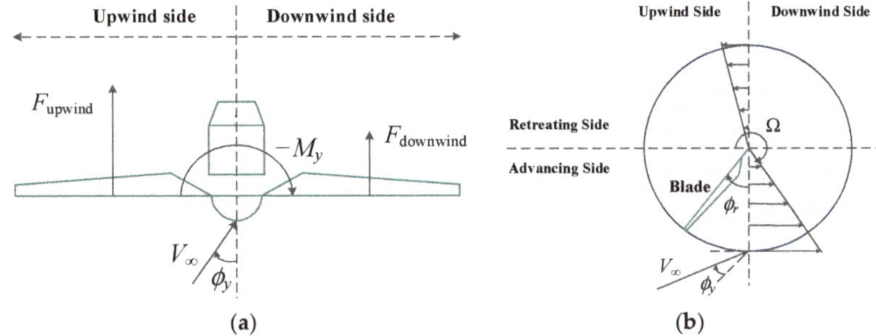

Figure 5. Schematic of a wind turbine in the yawed condition. (**a**) Top view; (**b**) Front view.

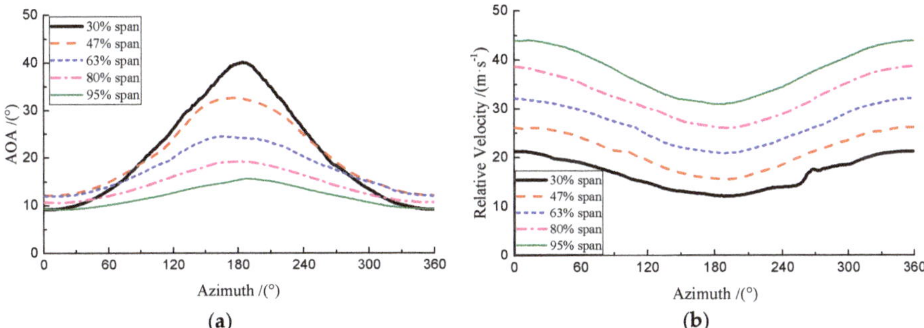

Figure 6. The variation of the angle of attack (AOA) and relative velocity with the azimuth at different spans. Wind speed is 13 m/s, and yaw angle is 30°. (**a**) AOA; (**b**) Relative velocity.

3.2. Validations of Numerical Modeling Methods

The pitch-oscillation case in the deep stall regime is selected for analyses of the grid independence and time step independence, where $\alpha_m = 14°$, $A = 10°$, and $k = 0.078$. Different grid resolutions are obtained by changing the numbers of wrap-around points on the airfoil and normal layers; then, three types of grids are generated with total numbers of 37,850, 51,090, and 70,525, respectively. Three different time steps of 0.002 s, 0.001 s, and 0.0005 s are used with the steps per cycle being 273, 546, and 1092, respectively. However, there is no clear difference in the lift coefficient among the test cases. The streamwise velocity profile is used to estimate the capability of capturing flow structures; this velocity profile is at mid-chord at the phase angle of 90° (i.e., $\phi = 90°$, where $\phi = 2\pi f t$). The coarsen grid and large time step can yield clear deviation (Figure 7). Finally, the grid in the size of 51,090 and the time step of 0.001 s are chosen as the computational settings.

In light stall, the numerical results agree well with the experimental data [40] as shown in Figure 8. An improvement over Karbasian, Esfahani, and Barati's CFD results [17] can be observed due to the consideration of the transition effect. The B-L model shows a reasonable representation of unsteady aerodynamics. Because of a lack of measurements under OF, the CFD results are compared with those of Gharali et al. in the same conditions [41]. First and foremost, it is necessary to clarify the determination of the AOA under OF. In steady flow, the AOA is defined as the angle between the chord and freestream velocity, and this instantaneous AOA (denoted as 'inflow angle') is assured at same value for both motions in the computations. Actually, the true AOA (denoted as 'local angle') is different, because the relative velocity direction to the airfoil can be influenced by the freestream time lag and flow disturbance of the airfoil. To obtain the local angle and compare it with the inflow angle, we extract the unsteady relative velocity vector at the location of 3 c before the airfoil stagnation point (Figure 9). An apparent lag between the two types of AOAs can be observed during downstroke. The variation amplitude of the local angle becomes slightly larger. Both the trend and the shape of the hysteresis loop agree well with the results of Gharali et al. [41], as shown in Figure 10. A notable feature is that the hysteresis loop is more apparent with respect to the inflow angle than to the local angle. To avoid ambiguity, the phase angle will replace the AOA as an independent variable afterwards.

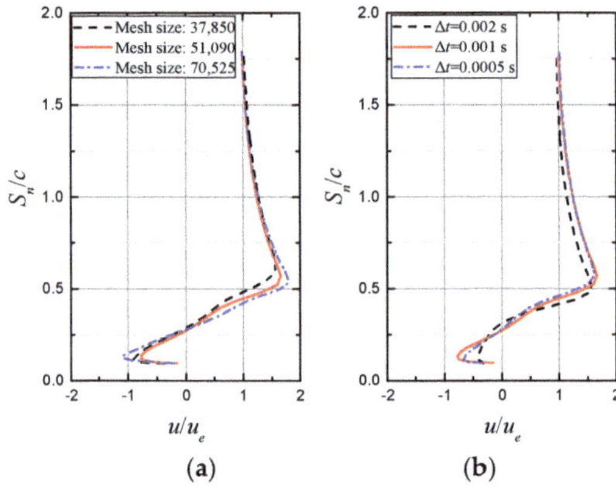

Figure 7. Effects of grid resolution and time steps on streamwise velocity profile at mid-chord at $\phi = 90°$. S_n is a normal distance away from the mid-chord, and u_e is the streamwise velocity of external flow. (**a**) Effect of grid resolution; (**b**) Effect of time steps.

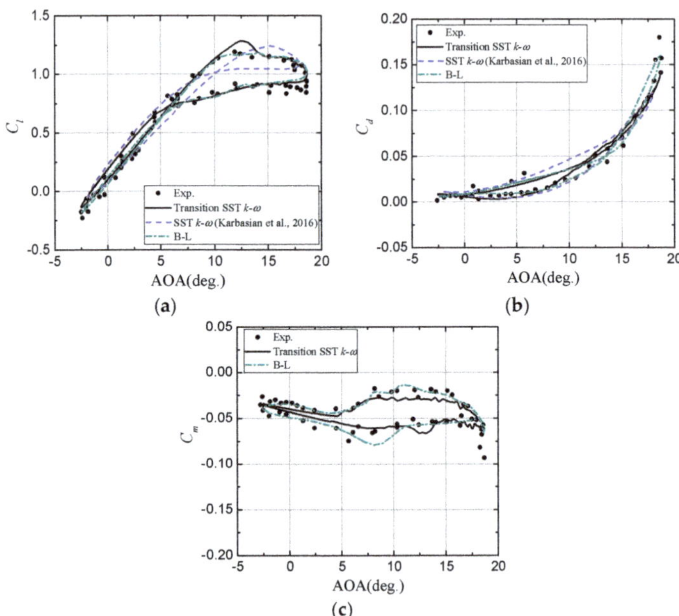

Figure 8. Comparison of aerodynamic coefficients under pitch oscillation. $\alpha_m = 8°$, $A = 10°$, and $k = 0.026$. (**a**) The lift coefficient; (**b**) The drag coefficient; (**c**) The pitching moment coefficient.

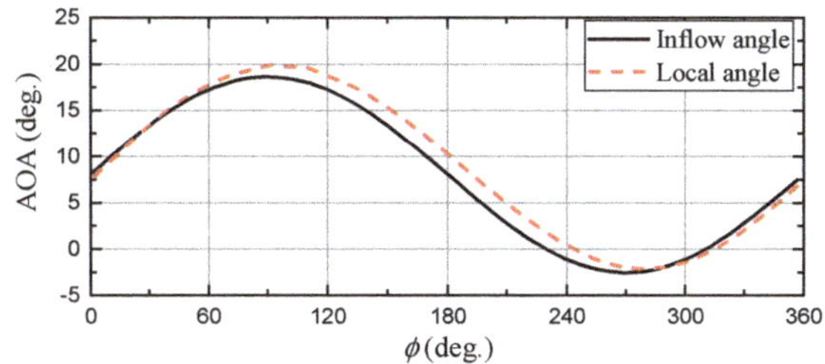

Figure 9. Comparison between the inflow angle and the local angle.

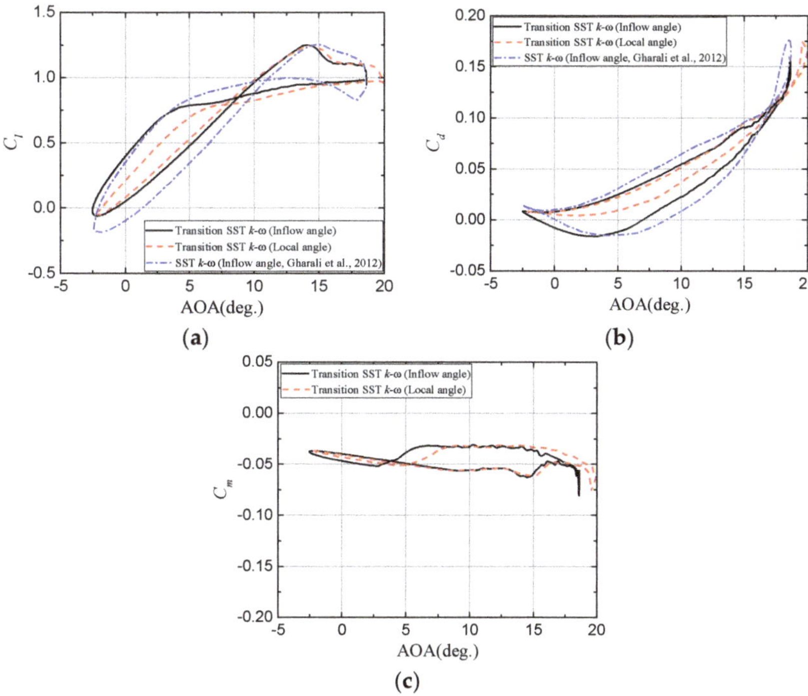

Figure 10. Comparison of aerodynamic coefficients under oscillating freestream. $\alpha_m = 8°$, $A = 10°$ and $k = 0.026$. (**a**) The lift coefficient; (**b**) The drag coefficient; (**c**) The pitching moment coefficient.

3.3. Equivalence Analysis between Pitch Oscillation and Oscillating Freestream

The aforementioned theoretical analysis indicates an opposite effect of PO and OF on the effective airfoil camber. Figure 11 shows the variations of the lift and the drag coefficients along one cycle. Notice that the difference in the nonlinear air loads (where ϕ is from 30° to 180°) is relatively negligible. A clear difference in the linear lift coefficient can be observed with a higher value under PO during upstroke and a higher value under OF during downstroke; the CFD results agree well with the thin-airfoil theory. From the vorticity contours in Figure 12, flow is attached to the airfoil and there is no obvious difference between the two motions from $\phi = 270°$ to $\phi = 360°$. When the AOA reaches the maximum at $\phi = 90°$, the flow separation is apparent under both PO and OF. The separated flow under PO is more persistent. In contrast, the vortex under OF rolls up and then sheds into wake. Despite a great difference in separated flow structure between the two motions, the associated aerodynamic loads do not show great differences. The possible reason is that the pressure distribution under PO possesses a sharp leading-edge suction peak, while the pressure distribution under OF experiences an elevated pressure on the suction side due to the separated vortex. These two factors may lead to an almost equal value in calculation. Figure 13 shows the unsteady pressure variations at different chordwise locations at $\alpha_m = 2°$, $A = 8°$, and $k = 0.078$. It is obvious that there is a stronger suction peak under PO than under OF. However, the pressure coefficient under OF decreases with a time lag.

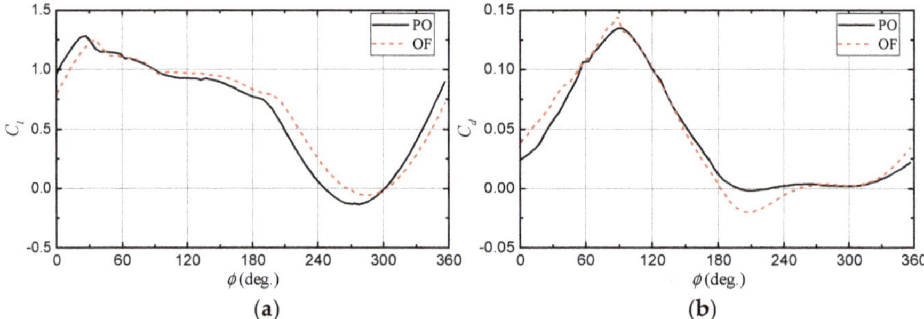

Figure 11. Comparison of lift coefficient and drag coefficient between pitch oscillation and oscillating freestream. $\alpha_m = 8°$, $A = 10°$, and $k = 0.026$. (**a**) The lift coefficient; (**b**) The drag coefficient.

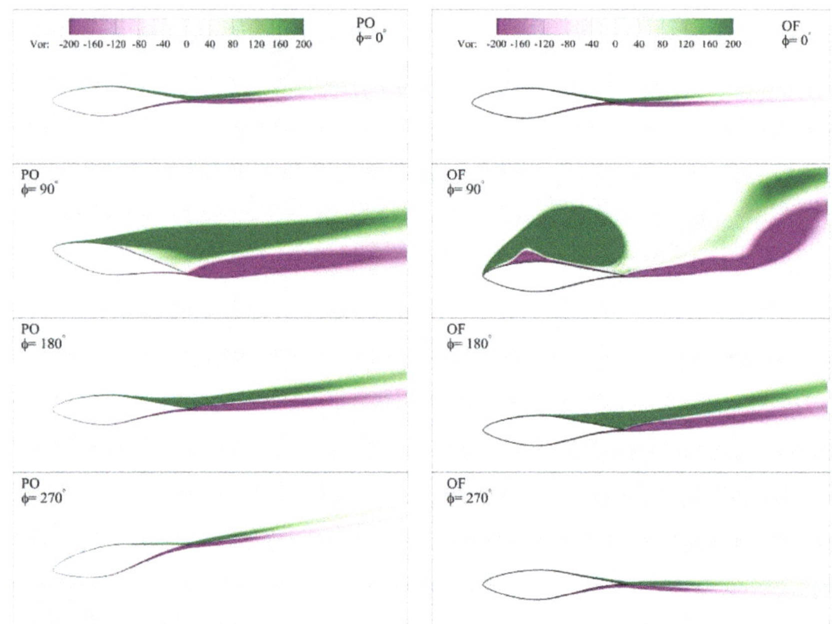

Figure 12. Vorticity contours of pitch oscillation and oscillating freestream at different phase angles. $\alpha_m = 8°$, $A = 10°$ and $k = 0.026$.

Because the difference between the two motions focuses on the linear part of the aerodynamic loads, the unsteady aerodynamic effect only at the low AOA is assessed. The mean AOA, AOA amplitude, and oscillation frequency are chosen as parameters to evaluate the difference. The deviation of the normal coefficient along one cycle between the two motions is defined as follows:

$$Dev. = \frac{max(|C_{n,PO}(\varnothing) - C_{n,OF}(\varnothing)|)}{C_{n,qs}} \tag{20}$$

where $C_{n,qs}$ is the quasi-steady normal coefficient at the mean AOA. Figure 14 indicates that the oscillation frequency f and the AOA amplitude A affect the deviation more strongly than the mean AOA. From the thin-airfoil theory, the key is the pitch rate $\dot{\alpha} = 2\pi f A\cos(2\pi f t)$. Because f and A are factors of the pitch rate, the deviation becomes great when f and A are high. Under yawed inflow, the oscillation frequency is directly determined by the rotor speed, and the AOA amplitude is seriously affected by the yaw angle. Consequently, it is necessary to consider the effects of PO and OF separately.

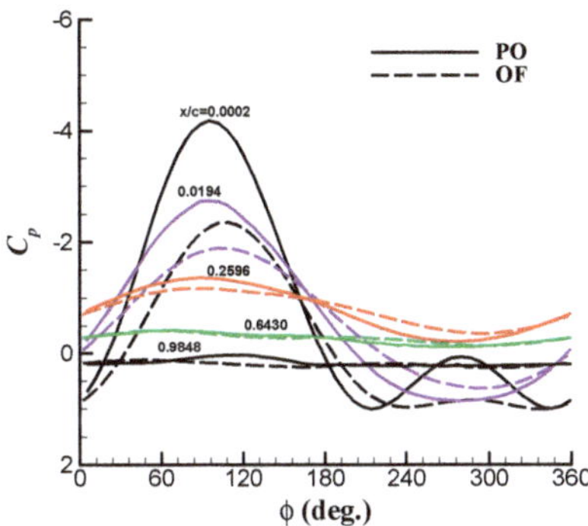

Figure 13. Comparison of unsteady pressures at different chordwise locations on the upper surface between pitch oscillation and oscillating freestream. $\alpha_m = 2°$, $A = 8°$, and $k = 0.078$. The same locations are in a same color.

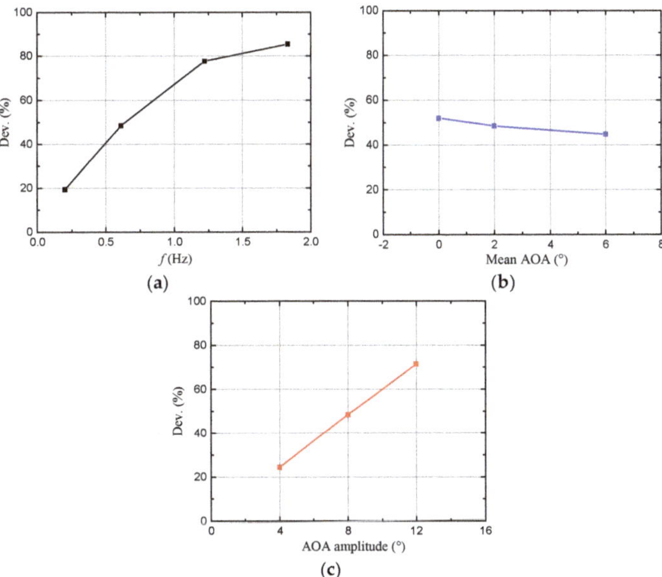

Figure 14. Effects of the oscillation frequency, the mean AOA and the AOA amplitude on the deviation between pitch oscillation and oscillating freestream. (**a**) Effect of the oscillation frequency; (**b**) Effect of the mean AOA; (**c**) Effect of the AOA amplitude.

3.4. Comparison of Dynamic Stall in Yawed Condition between the Two Motions

The 30% span of the Phase VI blade is investigated due to the manifest dynamic stall on the inboard blade. The wind speed is 10 m/s, and the yaw angle is 10° and 30°, respectively (the cases of S1000100 and S1000300 in NREL UAE Phase VI [30]). The B-L dynamic stall model incorporated with Bak's rotational stall delay model shows a good agreement with the experimental data for the yaw angle of 10°, as shown in Figure 15. A noticeable difference in both the lift coefficient and drag coefficient between the 2D and 3D results is observed, manifesting the 3D rotational effect.

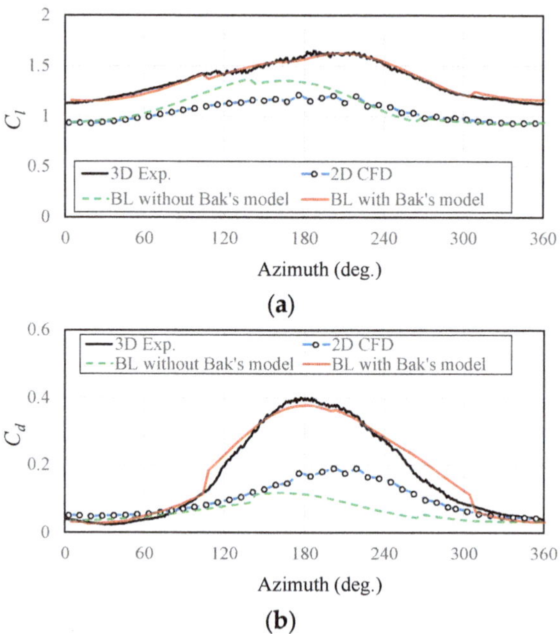

Figure 15. Validation of the corrected Beddoes-Leishman (B-L) dynamic stall model in lift coefficient and drag coefficient. The wind speed is 10 m/s, the yaw angle is 10°, and radial location is 30% span. (**a**) The lift coefficient; (**b**) The drag coefficient.

Figure 16 shows the aerodynamic force coefficients and pitching moment coefficient varying with azimuthal angle at the yaw angles of 10° and 30°, respectively. The calculated results are generally in a good agreement with the experimental data. When the yaw angle increases, the 3D dynamic stall becomes more intense with the high lift coefficient, but trends of air load variations are consistent. The difference between PO and OF is moderate. As aforementioned, in 2D cases, the aerodynamic coefficients under PO are higher during upstroke and lower during downstroke. A slight improvement can be observed in the prediction under OF; the possible reason is that dynamic stall under yawed inflow is essentially a velocity-varying problem without pitch motion and fore–aft motion. Notice that prediction of the pitching moment coefficients is quite poor in both amplitudes and trends. Given that the pitching moment depends not merely on the integrated normal force but also on the location of the center of pressure, it is somewhat not surprising that the predicted pitching moment is unsatisfactory under the dynamic condition, because 3D effects have an influence on the normal force and on the movement of the center of pressure, which depends on appropriate coupling of the dynamic stall and the 3D rotational stall delay models. This problem may be solved only if deep understanding and accurate modeling of both the dynamic stall and the 3D stall delay are achieved. Nevertheless, the results from the B-L dynamic stall model show a different effect between PO and OF when their different additional camber effects are included.

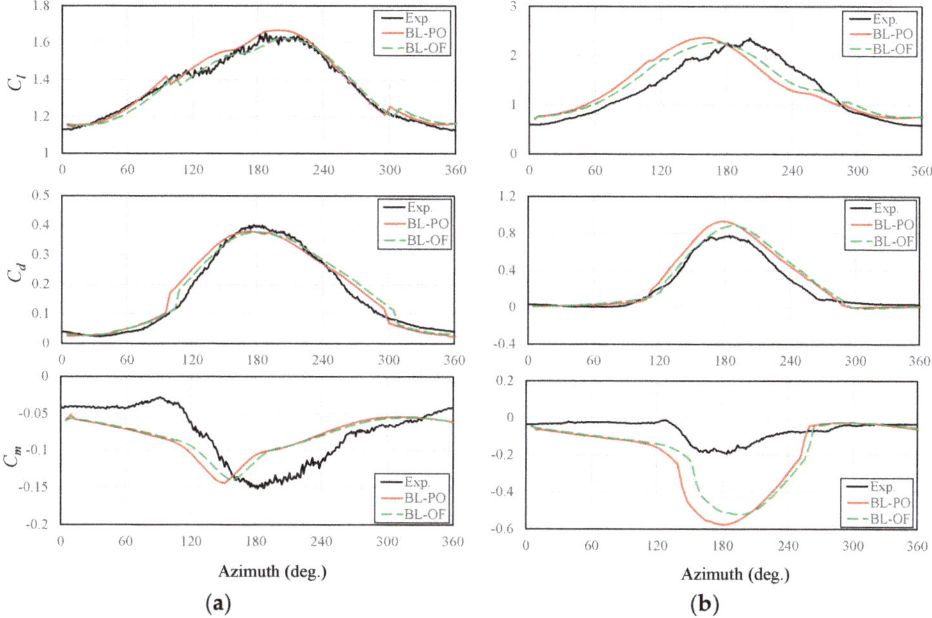

Figure 16. Effect of the yaw angle on the aerodynamic coefficients. The wind speed is 10 m/s, and the radial location is 30% span. (**a**) 10° yaw angle, and (**b**) 30° yaw angle.

3.5. Effect of Rotational Augmentation on the Dynamic Stall of the Inboard Blade

Because the rotational effect greatly suppresses the flow separation, the lift and nose-down pitching moment coefficients are increased. However, it is unclear whether the rotational stall delay effect under steady conditions is consistent with its performance under unsteady conditions. To provide an insight into the coupled effect of the rotational augmentation and dynamic stall, an extended analysis is conducted. As rotational augmentation is often isolated by making comparisons between 2D flow and 3D rotational flow [42], our strategy is to isolate the rotational augmentation by comparing the CFD results under an equivalent planer motion with yawed experimental data; the former contains only dynamic stall and the latter is coupled with rotational augmentation. The equivalent planer motion of a rotating blade section is obtained by the 2D time-varying incident velocity (2D TVIV, Figure 1a,c), including variations of the sectional AOA and the relative velocity magnitude.

Figure 17 shows the hysteresis loops of the aerodynamic coefficients at the 30% span. Because both the mean AOA and the AOA amplitude are low, the aerodynamic forces are generally attached to the unsteady linear part with almost no stall. Rotational augmentation apparently elevates the lift and drag coefficients, as under steady conditions. The value is even higher than the steady experimental data under axial inflow, meaning that the separation is more delayed in 3D rotational unsteady flow than in 3D rotational steady flow. Hence, both rotational augmentation and unsteady effects lead to a delayed flow separation and the elevated lift coefficient at a low yaw angle.

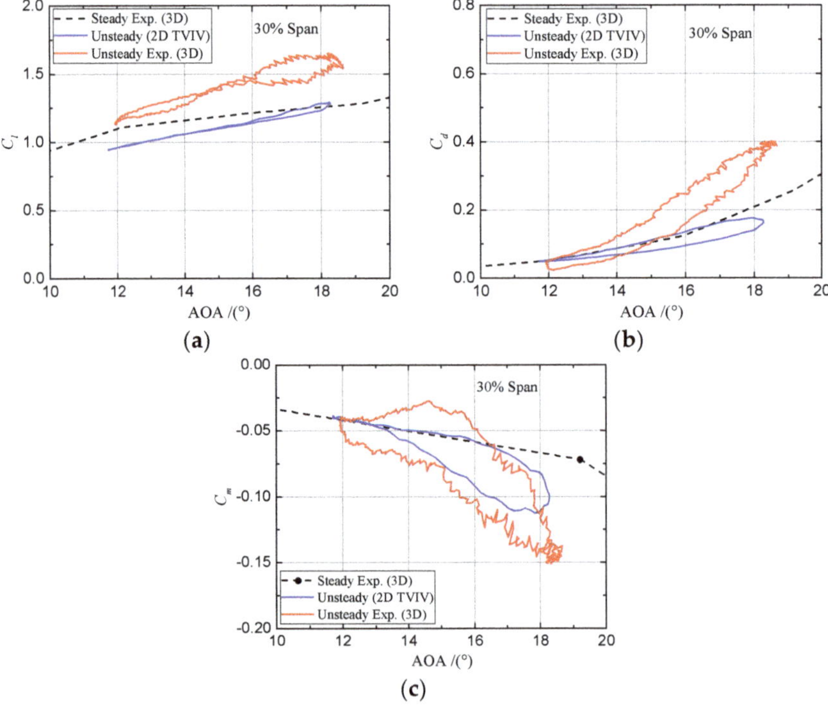

Figure 17. Hysteresis loops of the lift, drag and pitching moment coefficients at 30% spanwise location. The wind speed is 10 m/s, and the yaw angle is 10°. Steady Exp. (i.e., experimental) (3D) denotes the experimental data under axial inflow, while Unsteady Exp. (3D) denotes the experimental data under yawed inflow. (**a**) The lift coefficient; (**b**) The drag coefficient; (**c**) The pitching moment coefficient.

When the yaw angle is high, the air load variation is more severe with the large AOA amplitude. In comparison with Figure 17, Figure 18 shows hysteresis loops of the aerodynamic coefficients at 30° yaw angle at 30% span. It is rather remarkable that the 2D and 3D hysteresis loops of the lift coefficient are in the opposite direction. The 3D lift coefficient is even lower than the 2D value with the AOA increasing, meaning no rotational augmentation effect in the lift force. However, when the AOA decreases, the rotational augmentation dramatically advances the reattachment and results in a flattened hysteresis loop in an anticlockwise direction. In other words, the rotational augmentation effectively pushes the reattachment point rearward and reduces the hysteresis of the air loads on the downwind side (where the AOA is decreasing) (Figure 5b). Because of the unsteady effects, the unsteady 3D experimental data shows an apparent increase in the lift coefficient over the steady 3D experimental data. However, the 3D drag and pitching moment coefficients slightly vary around the steady data. With the AOA decreasing, the hysteresis loop under 2D TVIV indicates a notable additional fluctuation from secondary vortices in the pitching moment coefficient. In short, at a high yaw angle, the rotational augmentation is minor on the upwind side and effectively reduces the hysteresis on the downwind side, while unsteady effects greatly delay the flow separation and elevate the lift coefficient along the whole cycle.

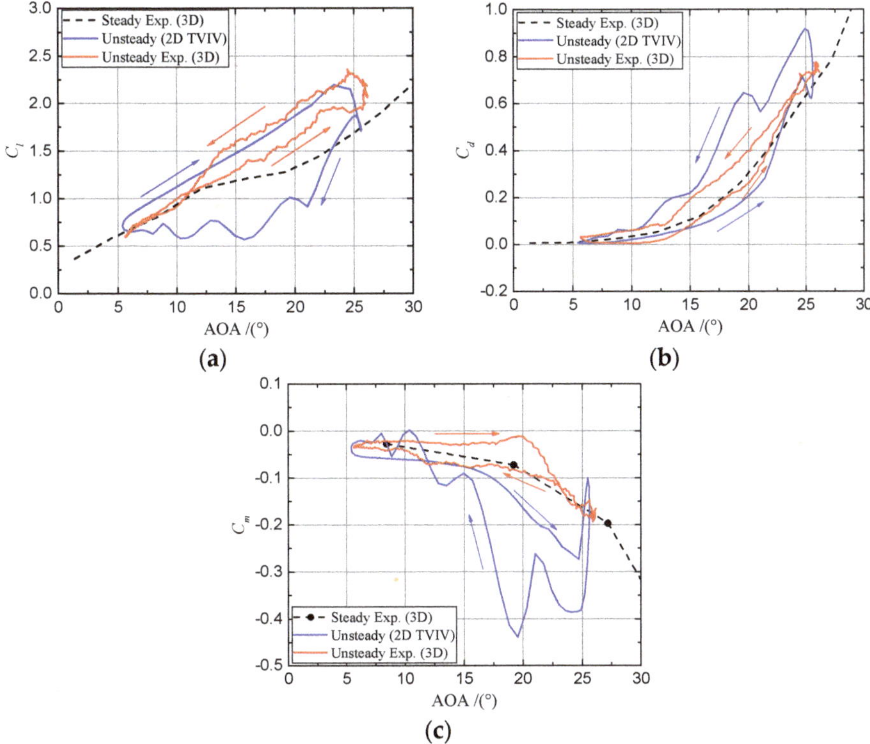

Figure 18. Hysteresis loops of the lift, drag, and pitching moment coefficients at 30° yaw angle. The wind speed is 10 m/s, and the spanwise location is 30%. (**a**) The lift coefficient; (**b**) The drag coefficient; (**c**) The pitching moment coefficient.

4. Conclusions

Dynamic stall is very common on HAWTs, resulting from many aerodynamic sources. Dynamic stall plays an important role in unsteady air loads. The accurate prediction of dynamic stall is extremely important for structural analysis. This study provides a comparative analysis of dynamic stall under pitch oscillation (PO) and oscillating freestream (OF) on wind turbine airfoil and blade, using the quasi-steady thin-airfoil theory, CFD method, and Beddoes-Leishman dynamic stall engineering model.

From the quasi-steady thin-airfoil theory, PO and OF have an opposite effect on the effective airfoil camber and the associated lift coefficient. The CFD results show that the difference between PO and OF mainly exists in the linear part of the aerodynamic loads. The deviation between these two motions is greatly influenced by the reduced frequency and the AOA amplitude. For the yawed response predictions, the Beddoes-Leishman dynamic stall model coupled with a rotational stall delay model also show a different effect between PO and OF. In addition, the effect of the rotational augmentation on the unsteady air loads is assessed. At a low yaw angle, the rotational augmentation apparently elevated the lift and drag coefficients along the whole cycle, as under steady conditions. At a high yaw angle, however, the rotational augmentation apparently flattens the hysteresis loop and its effect during which the AOA dynamically decreasing is more obvious than when the AOA is dynamically increasing.

It may be concluded from this work that dynamic stall behaviors under PO and OF are different at certain conditions and should separately be modeled in engineering.

Author Contributions: C.Z. and T.W. conceived of and designed the research; C.Z. conducted the data collection and wrote the paper; T.W. revised the paper.

Funding: This work is supported by the National Basic Research Program of China (973 Program) under grant No. 2014CB046200, the National Natural Science Foundation of China under grant No. 51506088, CAS Key Laboratory of Wind Energy Utilization under grant No. KLWEU-2016-0102, and the Priority Academic Program Development of Jiangsu Higher Education Institutions.

Conflicts of Interest: The authors declare no conflict of interest.

References

1. Leishman, J.G. Challenges in modelling the unsteady aerodynamics of wind turbines. *Wind Energy* **2002**, *5*, 85–132. [CrossRef]
2. McCroskey, W.J. *The Phenomenon of Dynamic Stall*; National Aeronautics and Space Administration: Washington, DC, USA, 1981.
3. Carr, L.W. Progress in analysis and prediction of dynamic stall. *J. Aircr.* **1988**, *25*, 6–17. [CrossRef]
4. Butterfield, C.P. Aerodynamic pressure and flow-visualization measurement from a rotating wind turbine blade. In Proceedings of the Eighth ASME Wind Energy Symposium, Houston, TX, USA, 22–25 January 1989; pp. 245–256.
5. Leishman, J.G. *Principles of Helicopter Aerodynamics*, 2nd ed.; Cambridge University Press: Cambridge, UK, 2006.
6. Leishman, J.G.; Beddoes, T.S. A Semi-Empirical Model for Dynamic Stall. *J. Am. Helicopter Soc.* **1989**, *34*, 3–17. [CrossRef]
7. Tran, C.T.; Petot, D. Semi-empirical model for the dynamic stall of airfoils in view of the application to the calculation of responses of a helicopter blade in forward flight. *Vertica* **1981**, *5*, 35–53.
8. Øye, S. *Dynamic Stall Simulated as Time Lag of Separation*; Technical University of Denmark: Lyngby, Denmark, 1991.
9. Hansen, M.H.; Gaunaa, M.; Madsen, H.A. *A Beddoes-Leishman Type Dynamic Stall Model in State-Space and Indicial Formulation*; Risø National laboratory: Roskilde, Denmark, 2004.
10. Pereira, R.; Schepers, G.; Pavel, M.D. Validation of the Beddoes-Leishman dynamic stall model for horizontal axis wind turbines using MEXICO data. *Wind Energy* **2013**, *16*, 207–219. [CrossRef]
11. Gupta, S.; Leishman, J.G. Dynamic stall modelling of the S809 aerofoil and comparison with experiments. *Wind Energy* **2006**, *9*, 521–547. [CrossRef]
12. Johansen, J. *Unsteady Airfoil Flows with Application to Aeroelastic Stability*; Risø National laboratory: Roskilde, Denmark, 1999.
13. Visbal, M.R. Numerical Investigation of Deep Dynamic Stall of a Plunging Airfoil. *AIAA J.* **2011**, *49*, 2152–2170. [CrossRef]
14. Ekaterinaris, J.A.; Platzer, M.F. Computational prediction of airfoil dynamic stall. *Prog. Aeosp. Sci.* **1998**, *33*, 759–846. [CrossRef]
15. Menter, F.R. Two-Equation Eddy-Viscosity Transport Turbulence Model for Engineering Applications. *AIAA J.* **1994**, *32*, 1598–1605. [CrossRef]
16. Van der Wall, B.G.; Leishman, J.G. On the Influence of Time-Varying Flow Velocity on Unsteady Aerodynamics. *J. Am. Helicopter Soc.* **1994**, *39*, 25–36. [CrossRef]
17. Karbasian, H.R.; Esfahani, J.A.; Barati, E. Effect of acceleration on dynamic stall of airfoil in unsteady operating conditions. *Wind Energy* **2016**, *19*, 17–33. [CrossRef]
18. Snel, H.; Houwink, R.; Bosschers, J. *Sectional Prediction of Lift Coefficients on Rotating Wind Turbine Blades in Stall*; Energy Research Center of the Netherlands: Petten, The Netherlands, 1993.
19. Chaviaropoulos, P.K.; Hansen, M.O.L. Investigating three-dimensional and rotational effects on wind turbine blades by means of a quasi-3D Navier-Stokes solver. *J. Fluids Eng. Trans. ASME* **2000**, *122*, 330–336. [CrossRef]
20. Du, Z.; Selig, M. A 3-D stall-delay model for horizontal axis wind turbine performance prediction. In Proceedings of the 36th AIAA Aerospace Sciences Meeting and Exhibit, 1998 ASME Wind Energy Symposium, Reno, NV, USA, 12–15 January 1998.
21. Raj, N.V. *An Improved Semi-Empirical Model for 3-D Post-Stall Effects in Horizontal Axis Wind Turbines*; University of Illinois: Urbana-Champaign, IL, USA, 2000.

22. Bak, C.; Johansen, J.; Andersen, P.B. Three-Dimensional Corrections of Airfoil Characteristics Based on Pressure Distributions. In Proceedings of the European Wind Energy Conference and Exhibition (EWEC), Athens, Greece, 27 February–2 March 2006.

23. Corrigan, J.J.; Schillings, J.J. Emprical Model for Stall Delay Due to Rotation. In Proceedings of the American Helicopter Society Aeromechanics Specialists Conference, San Francisco, CA, USA, 19–21 January 1994.

24. Lindenburg, C. *Investigation into Rotor Blade Aerodynamics*; Energy research Centre of the Netherlands Petten: Petten, The Netherlands, 2003.

25. Breton, S.P.; Coton, F.N.; Moe, G. A Study on Rotational Effects and Different Stall Delay Models Using i a Prescribed Wake Vortex Scheme and NREL Phase VI Experiment Data. *Wind Energy* **2008**, *11*, 459–482. [CrossRef]

26. Zahle, F.; Bak, C.; Guntur, S.; Soslashrensen, N.N.; Troldborg, N. Comprehensive aerodynamic analysis of a 10 MW wind turbine rotor using 3D CFD. In Proceedings of the 32nd ASME Wind Energy Symposium, National Harbor, MD, USA, 13–17 January 2014; p. 15.

27. Bangga, G.; Lutz, T.; Jost, E.; Kramer, E. CFD studies on rotational augmentation at the inboard sections of a 10 MW wind turbine rotor. *J. Renew. Sustain. Energy* **2017**, *9*, 023304. [CrossRef]

28. Bangga, G.; Yusik, K.; Lutz, T.; Weihing, P.; Kramer, E. Investigations of the inflow turbulence effect on rotational augmentation by means of CFD. *J. Phys. Conf. Ser.* **2016**, *753*, 022026. [CrossRef]

29. Somers, D.M. *Design and Experimental Results for the S809 Airfoil*; National Renewable Energy Laboratory: Golden, CO, USA, 1997.

30. Hand, M.M.; Simms, D.A.; Fingersh, L.J.; Jager, D.W.; Cotrell, J.R.; Schreck, S.J.; Larwood, S.M. *Unsteady Aerodynamics Experiment Phase VI: Wind Tunnel Test Configurations and Available Data Campaigns*; National Renewable Energy Laboratory: Golden, CO, USA, 2001.

31. Menter, F.R.; Langtry, R.B.; Likki, S.R.; Suzen, Y.B.; Huang, P.G.; Volker, S. A correlation-based transition model using local variables—Part I: Model formulation. *J. Turbomach. Trans. ASME* **2006**, *128*, 413–422. [CrossRef]

32. Glauert, H. *The Elements of Airfoil and Airscrew Theory*; Cambridge University Press: Cambridge, UK, 1947.

33. ANSYS Inc. *FLUENT Theory Guide, Release 16.0*; ANSYS Inc.: Canonsburg, PA, USA, 2015.

34. Van Leer, B. Towards the ultimate conservative difference scheme. V. A second-order sequel to Godunov's method. *J. Comput. Phys.* **1979**, *32*, 101–136. [CrossRef]

35. Balduzzi, F.; Bianchini, A.; Maleci, R.; Ferrara, G.; Ferrari, L. Critical issues in the CFD simulation of Darrieus wind turbines. *Renew. Energy* **2016**, *85*, 419–435. [CrossRef]

36. Guntur, S.; Sørensen, N.N. An evaluation of several methods of determining the local angle of attack on wind turbine blades. In *Science of Making Torque from Wind 2012*; Iop Publishing Ltd.: Bristol, UK, 2014; Volume 555.

37. Guntur, S.; Sørensen, N.N.; Schreck, S.; Bergami, L. Modeling dynamic stall on wind turbine blades under rotationally augmented flow fields. *Wind Energy* **2016**, *19*, 383–397. [CrossRef]

38. Guntur, S.; Bak, C.; Sørensen, N.N. Analysis of 3D Stall Models for Wind Turbine Blades Using Data from the MEXICO Experiment. In Proceedings of the 13th International Conference on Wind Engineering, Amsterdam, The Netherlands, 10–15 July 2012.

39. Schepers, J.G. *IEA Annex XX: Comparison between Calculations and Measurements on a Wind Turbine in Yaw in the NASA-Ames Wind Tunnel*; Energy Research Center of the Netherlands: Petten, The Netherlands, 2007.

40. Ramsay, R.F.; Hoffman, M.J.; Gregorek, G.M. *Effects of Grit Roughness and Pitch Oscillation on the S809 Airfoil*; National Renewable Energy Laboratory: Golden, CO, USA, 1995.

41. Gharali, K.; Johnson, D.A. Numerical modeling of an S809 airfoil under dynamic stall, erosion and high reduced frequencies. *Appl. Energy* **2012**, *93*, 45–52. [CrossRef]

42. Guntur, S.; Sorensen, N.N. A study on rotational augmentation using CFD analysis of flow in the inboard region of the MEXICO rotor blades. *Wind Energy* **2015**, *18*, 745–756.

 *applied*
sciences

Article

Accurate RANS Simulation of Wind Turbine Stall by Turbulence Coefficient Calibration

Wei Zhong *, Hongwei Tang, Tongguang Wang and Chengyong Zhu

Jiangsu Key Laboratory of Hi-Tech Research for Wind Turbine Design, Nanjing University of Aeronautics and Astronautics, No. 29 Yudao St., Nanjing 210016, China; thw1021@nuaa.edu.cn (H.T.); tgwang@nuaa.edu.cn (T.W.); rejoicezcy@nuaa.edu.cn (C.Z.)
* Correspondence: zhongwei@nuaa.edu.cn; Tel.: +86-025-8489-6132

Received: 17 July 2018; Accepted: 16 August 2018; Published: 23 August 2018

Abstract: Stall, a complex phenomenon related to flow separation, is difficult to be predicted accurately. The motivation of the present study is to propose an approach to improve the simulation accuracy of Reynolds Averaged Navier–Stokes equations (RANS) for wind turbines in stall. The approach is implemented in three steps in simulations of the S809 airfoil and the NREL (National Renewable Energy Laboratory) Phase VI rotor. The similarity between airfoil and rotor simulations is firstly investigated. It is found that the primary reason for the inaccuracy of rotor simulation is not the rotational effect or the 3-D effect, but the turbulence-related problem that already exists in airfoil simulation. Secondly, a coefficient of the SST turbulence model is calibrated in airfoil simulation, ensuring the onset and development of the light stall are predicted accurately. The lift of the airfoil in the light stall, which was overestimated about 30%, is reduced to a level consistent with experimental data. Thirdly, the calibrated coefficient is applied to rotor simulation. That makes the flow patterns on the blade properly simulated and the pressure distribution of the blade, as well as the torque of the rotor, are predicted more accurately. The relative error of the predicted maximum torque is reduced from 34.4% to 3.2%. Furthermore, the procedure of calibration is applied to the MEXICO (Model Experiments in Controlled Conditions) rotor, and the predicted pressure distributions over blade sections are better than the CFD (Computational Fluid Dynamics) results from the Mexnext project. In essence, the present study provides an approach for calibrating rotor simulation using airfoil experimental data, which enhances the potential of RANS in accurate simulation of the wind turbine aerodynamic performance.

Keywords: wind turbine; stall; NREL Phase VI; S809 airfoil; MEXICO; RANS

---

1. Introduction

Aerodynamics is one of the most important topics in wind turbine technology. Accurate calculation of the aerodynamic loads is essential for the power prediction and the structural design. A large number of computational tools has been developed based on the blade element momentum (BEM) theory [1–4], the prescribed/free vortex wake models (VWM) [5,6], or the computational fluid dynamics (CFD) method [7,8]. BEM and VWM are specially developed for rotary machinery, such as propellers, helicopter rotors, and wind turbine rotors. CFD is a general method that can be used to solve almost all common flow problems based on detailed numerical solutions of the Navier–Stokes equations and gives physical quantities of the whole computational domain. This brings many benefits to the computational research of wind turbines. For example, the rotational effect and the three-dimensional (3-D) effect are naturally included in CFD, and therefore the additional models are not required as in BEM. On the other hand, CFD still faces some problems, among which turbulence is the most intractable one, since it is too computationally expensive to be directly simulated.

Turbulence plays an important role in the aerodynamic performance of wind turbines. It exists in the wind, in the blade boundary-layer, and in the wake behind the rotor. There are mainly two ways to deal with turbulence in modern CFD applications. One is the Reynolds Averaged Navier–Stokes equations (RANS) simulation, the other is the Large Eddy Simulation (LES). Although LES provides a much better description of turbulence compared to RANS, it is much more computationally expensive. In order to make the computational cost acceptable, LES is usually used together with the actuator disk (AD) or the actuator line (AL) for wind turbine CFD. The methods of AD-LES and AL-LES have made great progress in the study of rotor wake [9,10]. However, they do not describe the geometry of rotors, and therefore the detailed flow around blades is not simulated. For simulations with blade geometry, RANS simulation is the most practical option.

The simplification of turbulence in RANS increases the inaccuracy of CFD, especially for phenomena directly related to turbulence. The prediction of stall, which is determined by the simulation of turbulence in the boundary-layer, becomes one of the most challenging problems for RANS. Stall is related to flow separation that occurs on the suction side of an airfoil (or blade) when the angle of attack (AoA) exceeds a certain value. It can be divided into two stages of light stall and deep stall, corresponding to trailing edge flow separation and complete flow separation, respectively. Light stall decreases the slope of lift curve, while deep stall causes the lift to drop sharply from the peak value.

A typical example of the inaccuracy of stall prediction is the RANS simulations of the S809 airfoil [11]. A large number of simulations using various turbulence models have been carried out since the end of the twentieth century. As early as 1994, Yang [12] made RANS simulation of the S809 airfoil using a 2-D incompressible solver and the Baldwin–Lomax algebraic turbulence model. The predicted lift coefficient significantly exceeds the experimental data when light stall occurs. Researchers later attempted to improve the simulation by using various turbulence models such as k-ω used by Yang [13], S-A, SST and SA-DES used by Benjanirat [14], and SST and RNG k-ε used by Guerri [15]. Their results repeatedly showed the same inaccuracy: predicting a late onset of stall and therefore overestimating the maximum lift coefficient. Instead of fully turbulent simulations, laminar/turbulent transitional simulations were carried out by other researchers. Wolfe [16] specified a transition point, while Bertagnolio [17] and Langtry [18] employed the Michel model and the γ-$Re\theta$ model [19], respectively. Their results showed that stall prediction depends on the simulation of transition, but it is still not accurate enough.

The problem also appears in RANS simulations of the well-known NREL Phase VI rotor [20] and MEXICO rotor [21]. A large number of simulations, including those in the blind comparison [22], were carried out for the NREL Phase VI rotor using various solvers such as OVERFLOW-D [23], elsA [24], EllipSys3D [25], CFX [26], FLUENT [27], THETA [28], etc. The MEXICO rotor was simulated by participants of the Mexnext project [29] and researchers using a variety of RANS codes [30–32]. Many of the above simulations failed to accurately predict the stall. The inaccuracy of stall prediction remains an unsolved problem [33,34].

Stall frequently occurs on stall-regulated wind turbines. Also, it may occur on variable-pitch wind turbines during operation in yaw or in extreme wind conditions, and particularly on the blade inboard sections where blade incidence angle is usually high. Therefore, the accurate stall prediction is of great significance to wind turbine load computation, as well as the optimal design of a wind turbine. The present study is devoted to achieving accurate stall prediction in wind turbine RANS simulations. Firstly, the similarity between airfoil simulation and rotor simulation is investigated, which makes it feasible to obtain accurate rotor simulation by making a calibration in airfoil simulation. Then, a coefficient β_∞^* of the SST turbulence model is calibrated according to airfoil experimental data. Finally, the calibrated β_∞^* is applied to rotor simulation, and more accurate prediction of stall is realized. The above approach enhances the potential of RANS in accurate simulation of the wind turbine aerodynamic performance.

2. Simulation Methods

In the present study, 2-D airfoil simulation and 3-D rotor simulation are carried out using ANSYS FLUENT on the S809 airfoil and the NREL (National Renewable Energy Laboratory) Phase VI rotor, respectively. The involved governing equations are the incompressible RANS equations. A comparison between an incompressible solver and a compressible solver was made by Länger-Möller [28] for the NREL Phase VI rotor. The results of the two solvers were almost identical. Therefore, it is reasonable to ignore the compressibility of air in the present study.

Two kinds of 2-D computational domains are generated for the airfoil simulation. One contains the walls of the wind tunnel's experimental section, the other does not. The outer boundaries of the domain without tunnel walls are set to more than 20 times the chord length away from the airfoil. In each domain, 430 grid nodes are properly distributed on the airfoil surface, and about 100,000 grid cells are employed. The initial height of the first grid layer measured from the airfoil surface is about 5×10^{-6} of the chord length. The values of y^+ (see Figure 1) are found less than 1.0 on most areas of the airfoil surface, indicating that the initial height of the first grid layer is small enough to capture the boundary layer. The computational settings of the airfoil simulation are summarized in Table 1.

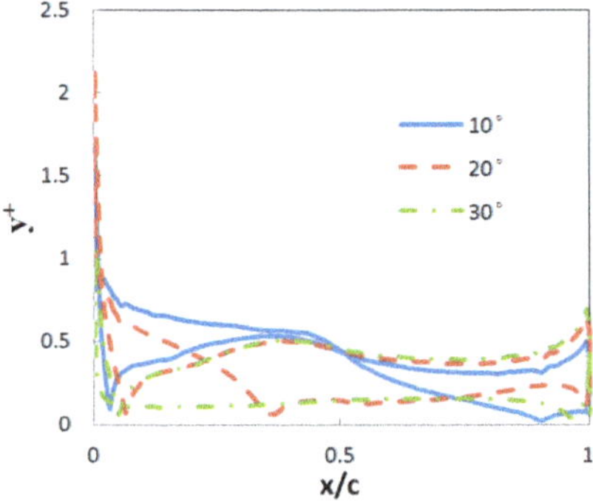

Figure 1. The wall y^+ distribution around the airfoil at angles of attack of $10°$, $20°$ and $30°$ ($Re = 1 \times 10^6$).

Table 1. Computational settings of the airfoil simulation.

Items	Options	Settings
Models	Space	2D
	Time	Steady
	Viscous	SST k-omega
Pressure-Velocity Coupling	Type	Coupled
	Courant Number	50
Discretization Scheme	Pressure	Second Order
	Momentum	Second Order Upwind
	Turbulent Kinetic Energy	Second Order Upwind
	Specific Dissipation Rate	Second Order Upwind

The computational domain for the rotor simulation is shown in Figure 2. A pair of rotational periodic surfaces is defined to reduce the computational domain to a half, taking advantage of the axial symmetry of the rotor geometry. The grid nodes on the two periodic surfaces would overlap one to one if one surface were rotated $180°$ around the axis. The far boundaries are set to more than 15 times the rotor diameter away from the rotor, with no consideration of wind tunnel walls. The computational settings of the rotor simulation are summarized in Table 2.

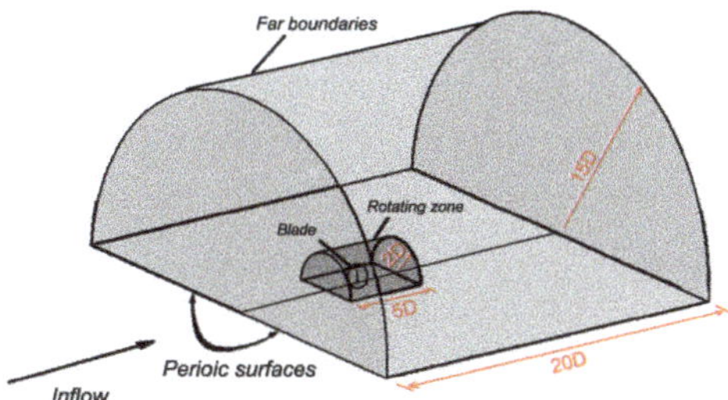

Figure 2. Computational domain for the rotor simulation ("D" is the rotor diameter).

Table 2. Computational settings of the rotor simulation.

Items	Options	Settings
Models	Space	3D
	Time	Steady
	Viscous	SST k-omega
	Motion type	Moving reference frame
	Periodic type	Rotationally Periodic
Pressure-Velocity Coupling	Type	Coupled
	Courant Number	50
Discretization Scheme	Pressure	Second Order
	Momentum	Second Order Upwind
	Turbulent Kinetic Energy	Second Order Upwind
	Specific Dissipation Rate	Second Order Upwind

A high-quality mesh with enough cells and properly distributed nodes is essential for a reliable numerical simulation. In the present study, the mesh independency has been tested for both the airfoil and the rotor simulations. The five meshes involved in the rotor test are shown in Table 3, where they are denoted by M01, M02, M03, M04, and M05, respectively. The initial height of the grid cells adjacent to the blade is set to 5×10^{-6} meter equally for all of the five meshes, according to the criterion of $y^+ < 1.0$. The simulation results of M03, M04, and M05 are found very close to each other, and the M03 mesh is preferred, since it has the smallest number of cells among the three. For a further verification of sufficient mesh resolution, the Richardson extrapolation [35,36] is applied to the predicted rotor torque from the simulations using M01, M02, and M03. Table 4 shows that the order of accuracy of the Richardson extrapolation is greater than 2 in conditions of various wind speeds. That means the M03 mesh (demonstrated in Figure 3) is fine enough for the present simulation based on the second order upwind scheme, and therefore it is finally adopted.

Table 3. The meshes involved in the test of mesh independency for the NREL Phase VI rotor.

Mesh ID	M01	M02	M03	M04	M05
Initial height of cells adjacent to blades (m)	5×10^{-6}	5×10^{-6}	5×10^{-6}	5×10^{-6}	5×10^{-6}
number of Nodes along the blade span	40	45	50	55	60
number of Nodes along the blade chord	140	190	240	290	320
Growth ratio of grid size away from blade	1.2	1.15	1.1	1.08	1.065
Total number of cells (million)	1.36	2.29	4.24	6.97	9.83

Table 4. Results of the mesh independency study based on Richardson extrapolation [1].

Wind Speed	Mesh ID			Richardson Extrapolation		
	M01	**M02**	**M03**	**p**	**R***	**TE**
7 m/s	744.2	756.5	757.4	4.18	0.073	757.5
10 m/s	1597.4	1778.6	1802.6	3.19	0.132	1806.5
15 m/s	850	1309.5	1408.8	2.36	0.216	1439.1
20 m/s	1047.6	1006.4	1001.4	3.33	0.121	1000.7

[1] TE represents the extrapolated torque, R* the ratio of errors, and p the order of accuracy.

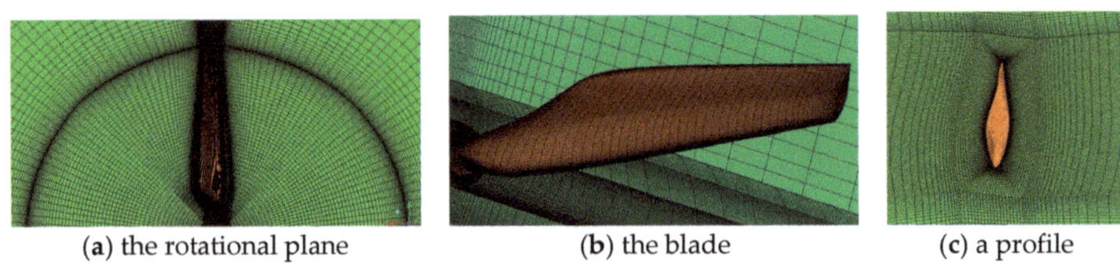

(a) the rotational plane   (b) the blade   (c) a profile

Figure 3. Demonstration of the mesh adopted for the rotor simulation. (a) The rotational plane; (b) the blade; (c) a profile.

3. Investigation of the RANS Simulation Error

3.1. Simulation of the S809 Airfoil

The wind tunnel experiments of the S809 airfoil, which were conducted in Delft University of Technology (DUT) [11], Ohio State University (OSU) [37], Colorado State University (CSU) [38], and University of Glasgow [39], respectively, cover a range of Reynolds number (Re) between 0.3×10^6 and 2×10^6. In the present study, full turbulence RANS simulation of the S809 airfoil is firstly made at Re = 1×10^6. It is close to the Reynolds number experienced by the blades of the NREL Phase VI rotor. (The Reynolds number varies between 0.7×10^6 and 1.4×10^6 at the blade root and between 1.0×10^6 and 1.1×10^6 at the blade tip, at the wind speeds between 7 m/s and 25 m/s [40].)

The simulation results of the airfoil lift coefficient (C_L) for a wide range of AoA between 0° and 40° are shown in Figure 4. The experimental data from two different sources are referenced in the figure. One is from the OSU [37] experiment, the other is the measured data of the r/R = 0.63 section of a parked NREL Phase VI blade [41]. The r/R = 0.63 section is close to the mid-span of the blade, and thus its performance can be approximately recognized equal to the 2-D S809 airfoil. The two sets of experimental data are consistent when AoA < 20° but deviate from each other when AoA > 20° where deep stall occurs. It is worth asking why such deviation occurs and which set of data is the appropriate reference for the simulation results in deep stall. Rooij [41] once made an analysis of the parked NREL Phase VI blade. The data collected in his article showed that the deviation does not become more notable at the r/R = 0.8 and r/R = 0.95 sections, although the 3-D effect is more significant there. Therefore, the primary reason for the deviation might not be attributed to the 3-D effect of the blade.

The present study pays attention to the tunnel blockage effect. The size of the test section of the NASA Ames wind tunnel, in which the NREL Phase VI experiment was conducted, is 24.4 m × 36.6 m. The chord length of the NREL Phase VI blade (between 0.355 m and 0.737 m) is very small compared to the wind tunnel, and therefore the tunnel blockage effect can be ignored in standstill cases. The wind tunnel experiment of the S809 airfoil was faced with a completely different situation. Figure 5 shows the present simulation result of the flow field around the airfoil at AoA = 40° in the wind tunnel of OSU. It is found that the actual blockage in deep stall is determined by the size of the vortices rather than the airfoil geometry. Therefore, the blockage effect correction applied to the experimental data was inaccurate, since only the size of the airfoil was considered [37].

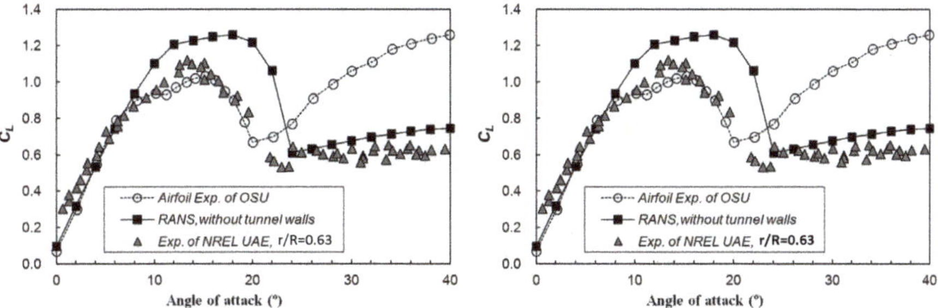

Figure 4. The computational and experimental lift coefficients of the S809 airfoil.

Figure 6 shows the lift coefficient from the airfoil simulation with tunnel walls compared with the measured data of the OSU experiment. The blockage effect correction has been applied to the simulation data, according to the method used in the experiment. That leads to an appropriate comparison between the simulation result and the experimental data. The shown curves can be divided into three stages according to the simulation error. At stage "A", the computational lift coefficient agrees with the measured data and increases linearly with the angle of attack. The lift coefficient is significantly overestimated by the simulation at stage "B" where light stall occurs, while it is underestimated at stage "C" where deep stall occurs. It is worth noting that the underestimation in deep stall is not as serious as that of the conventional comparison shown in Figure 4.

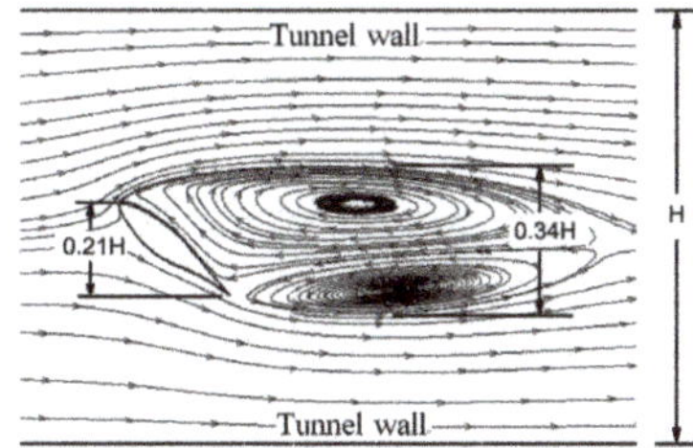

Figure 5. The simulated flow field of the S809 airfoil in the test section of the wind tunnel of OSU. (AoA = 40°, H = 1.4 m).

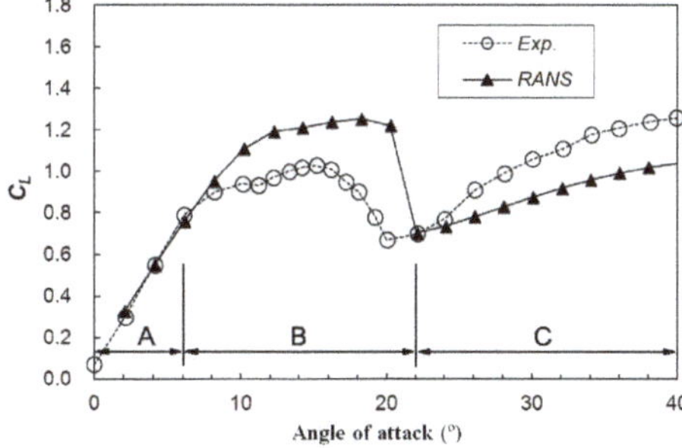

Figure 6. The lift coefficient from the airfoil simulation with tunnel walls, with a comparison to the measured data of the OSU experiment.

3.2. Simulation of the NREL Phase VI Rotor

Full turbulence simulation of the NREL Phase VI rotor in non-yawed condition is carried out at the wind speeds between 7 m/s and 25 m/s. Figure 7 shows the predicted rotor torque versus the wind speed, with a comparison to the measured data. The curves can also be divided into three stages. At stage "A", the predicted torque agrees with the measured data and increases linearly with the wind speed. The torque is overestimated at stage "B", while it is underestimated at stage "C". That is similar to the results of the airfoil simulation shown in Figure 6, which raises a question: Why is there such a similarity of simulation error between the airfoil lift coefficient (see Figure 6) and the rotor torque (see Figure 7)?

Figure 8 shows the simulated limiting streamlines on the suction side of the NREL Phase VI blade at different wind speeds. At 7 m/s, all streamlines are directed from the leading edge to the trailing edge, indicating the flow on the entire blade is attached and no stall occurs. At 10 m/s, the flow pattern on the blade is a combination of trailing edge flow separation (light stall) and attached flow (no stall). At 15 m/s, it becomes a combination of complete flow separation (deep stall) and trailing edge flow separation (light stall). When the wind speed increases to 20 m/s, complete flow separation (deep stall) covers the entire blade. The flow patterns at five sections of the blade are listed in Table 5. It is found that the simulation error of the rotor torque in Figure 7 is closely related to the flow patterns. At 7 m/s, no stall occurs, and the torque is well predicted. At 10 m/s, most sections are in light stall, and the torque is remarkably overestimated. At 15 m/s, three sections are under deep stall, while two sections are under light stall, and the overestimation of the torque is reduced. At 20 m/s, deep stall occurs on the entire blade, and the torque is underestimated. The above results indicate that the simulation accuracy of rotor torque is much affected by the flow patterns on the blade. The similarity of the simulation error between the airfoil lift coefficient (see Figure 6) and the rotor torque (see Figure 7) is raised by the fact that light stall leads to overestimation of the aerodynamic force, while deep stall leads to underestimation for both the airfoil and the blade sections.

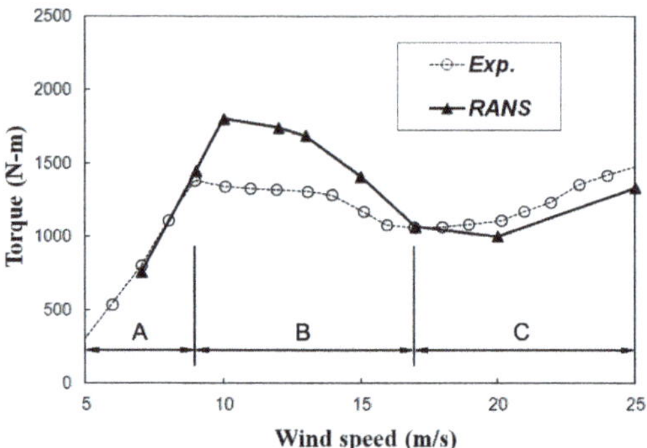

Figure 7. The predicted torque of the NREL Phase VI rotor, with a comparison to the measured data.

Table 5. The flow patterns of five sections of the NREL Phase VI blade at various wind speeds.

Wind Speed	r/R = 0.3	r/R = 0.47	r/R = 0.63	r/R = 0.8	r/R = 0.95
V = 7 m/s	No stall	No stall	No stall	No stall	No stall
V = 10 m/s	Light stall	Light stall	Light stall	Light stall	No stall
V = 15 m/s	Deep stall	Deep stall	Deep stall	Light stall	Light stall
V = 20 m/s	Deep stall	Deep stall	Deep stall	Deep stall	Deep stall

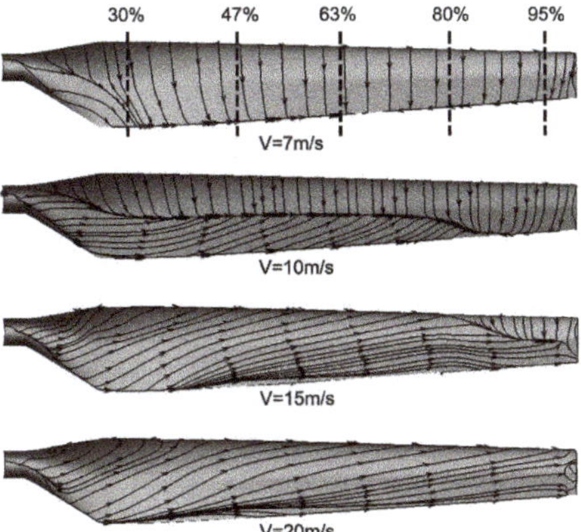

Figure 8. The simulated limiting streamlines on the suction side of the NREL Phase VI blade at various wind speeds.

Figure 9 shows the $C_t{'}$ versus the wind speed, in which $C_t{'}$ is the dimensionless coefficient of the aerodynamic force contributing positive torque of the blade. The shown three sections of $r/R = 0.3$, $r/R = 0.47$, and $r/R = 0.95$ represent the root, the mid-span, and the tip of the blade, respectively. The predicted curves exceed the experimental ones at the wind speeds between about 9 m/s and 17 m/s. The simulation error at the root section ($r/R = 0.3$) and the tip section ($r/R = 0.95$) is found not greater than that at the mid-span section ($r/R = 0.47$), although the flow at the root and the tip is more complex due to the rotational effect and the 3-D effect. It suggests that the rotational effect and the 3-D effect are not the primary cause of the inaccuracy of the rotor simulation.

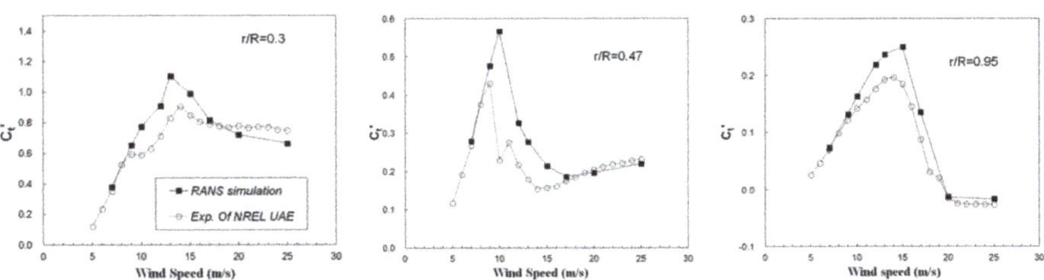

Figure 9. The predicted $C_t{'}$ versus wind speed at three typical blade sections, with a comparison to the measured data of the NREL Phase VI experiment.

3.3. The Similarity between Rotor Simulation and Airfoil Simulation

Figure 10 shows the $C_t{'}$ versus AoA at the three sections of $r/R = 0.3$, $r/R = 0.47$, and $r/R = 0.95$. For comparison, it also shows the corresponding curves of the S809 airfoil, which are obtained by replacing the aerodynamic force of each blade section with that of the airfoil. The shadow regions with sloping lines are determined by the positive simulation error (computational data subtracts experimental data). The points S (S′), E (E′), and P (P′) represent the starting, the ending, and the peak of the positive error, respectively. The shadow regions for both the airfoil and the blade sections are Λ-shaped, which is evidence of the similarity of simulation error between them. The shadow region for the root section ($r/R = 0.3$) is located at the upper right of the shadow for the airfoil, due to the rotational effect, which makes the stall delayed and thus makes the simulation error delayed as well. The two shadow regions at the mid-span section ($r/R = 0.47$) have large overlapping area, which is further evidence of the similarity of error between the airfoil and rotor simulations. The shadow region for the tip section (r/R = 0.95) becomes smaller than that for the airfoil, suggesting that the 2-D airfoil simulation may not be more accurate than the 3-D rotor simulation. Figure 11 shows the magnitude of the simulation error. The peak error of the blade sections is found lower than that of the airfoil. One reasonable explanation may be that the rotational effect and the 3-D effect produce a gentler slope in the process from light stall to deep stall.

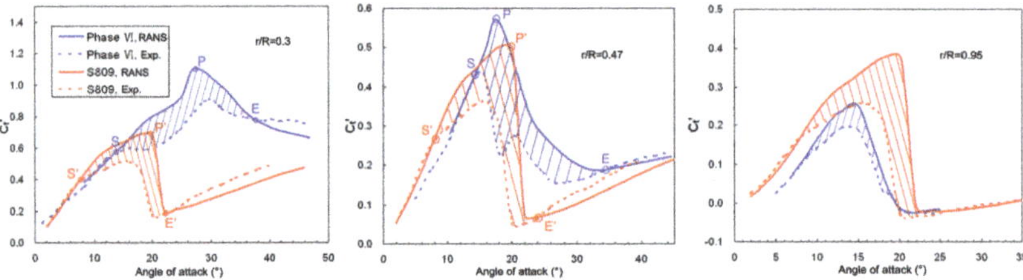

Figure 10. The C_t' versus AoA at different blade sections, with a comparison to the corresponding curves of the S809 airfoil.

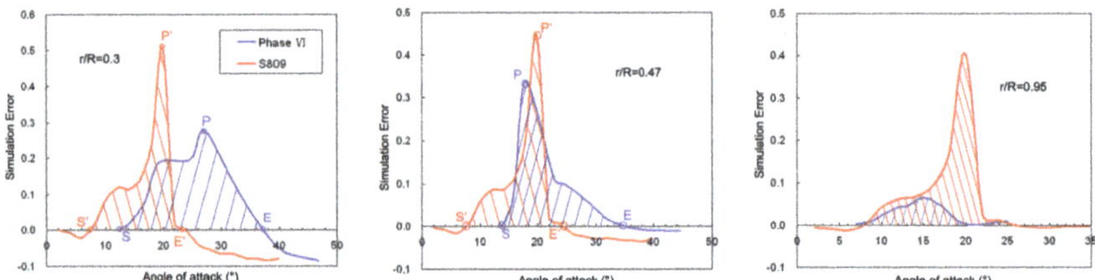

Figure 11. The magnitude of the simulation error at different blade sections, with a comparison to the corresponding curves of the S809 airfoil.

The above results suggest that there is a high degree of similarity of error between the airfoil and the rotor simulations, and the rotor simulation error is approximately at the same level of the airfoil simulation error. That leads to a conclusion that the main source of the rotor simulation error already exists in airfoil simulation. Therefore, it is feasible to achieve better rotor simulation by calibrating the RANS method in airfoil simulation.

4. Simulations with Turbulence Coefficient Calibration

4.1. Closure Coefficients of the SST Turbulence Model

Turbulence is a key determinant of flow separation and stall. The SST turbulence model used in the present study is recognized as one of the best linear eddy viscosity turbulence models providing a relatively good prediction of turbulence for both attached and separated flows. There are two transport equations for the turbulence kinetic energy k and the specific dissipation rate ω in the SST model. More than a dozen closure coefficients are involved in the equations. Their default values are

$$\sigma_{k,1} = 1.176,\ \sigma_{\omega,1} = 2.0,\ \sigma_{k,2} = 1.0,\ \sigma_{\omega,2} = 1.168,\ \alpha_1 = 0.31,\ \beta_{i,1} = 0.075,\ \beta_{i,2} = 0.0828,\ \alpha_\infty^* = 1.0,$$
$$\alpha_0 = 1/9,\ \beta_\infty^* = 0.09,\ R_\beta = 8.0,\ R_k = 6,\ R_\omega = 2.95,\ \xi^* = 1.5$$

Most of these values were determined by the observed turbulence properties in certain conditions. That limits the universality of the turbulence model. The turbulence simulation may be more suitable in some cases but less applicable in other cases. Calibrating the values of these coefficients is a way to achieve a better result for a given case. After a lot of computational tests, the coefficient β_∞^* is adopted to be calibrated in the present study. It is a key coefficient responsible for the dissipation of k and the production of ω in the transport equations. The variation of the value of β_∞^* eventually leads to a change of the turbulent viscosity in RANS simulations. Therefore, the calibration of β_∞^* can be considered as an appropriate adjustment of turbulent viscosity.

4.2. Airfoil Simulation with Different Values of β_∞^*

The airfoil simulation of the S809 airfoil has been made using various values of β_∞^*. The optimal value can easily be obtained by comparison with the airfoil experimental data. The simulation results show that $\beta_\infty^* = 0.11$ is the optimal value for the S809 airfoil at $\mathrm{Re} = 1 \times 10^6$. The predicted lift coefficient is shown in Figure 12 for $\beta_\infty^* = 0.09$ (the default value) and $\beta_\infty^* = 0.11$. The results for other values of β_∞^* are not displayed here for simplicity. In general, a higher β_∞^* leads to a lower lift coefficient in the range of light stall.

Table 6 lists the relative simulation error defined as

$$\delta = \frac{C_{L,sim} - C_{L,exp}}{C_{L,exp}} \times 100\% \tag{1}$$

in which $C_{L,sim}$ and $C_{L,exp}$ are the lift coefficients obtained from the simulation and the experiment, respectively. An averaged relative error is defined as following to measure the overall error level in a range of angles of attack,

$$\bar{\delta} = \frac{1}{n} \sum_{n=1}^{i} |\delta_i| \tag{2}$$

in which i is the index and n is the count.

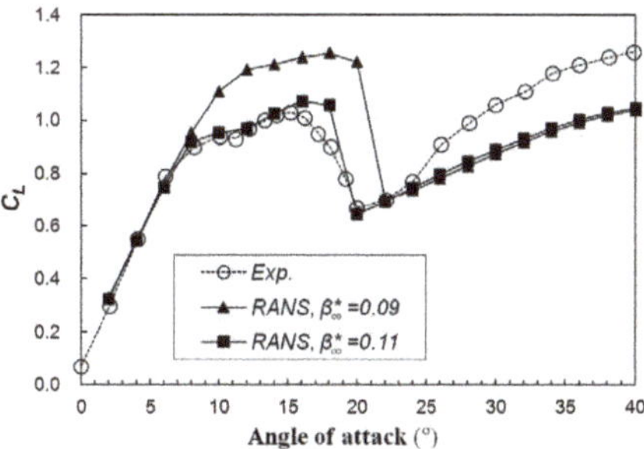

Figure 12. The simulation results of lift coefficient for different values of β_∞^* compared to experimental data. ($\mathrm{Re} = 1 \times 10^6$).

It can be seen from the figure and the table that $\beta_\infty^* = 0.11$ leads to significantly better results in the range of $8° < \mathrm{AoA} < 20°$ where light stall occurs. The averaged relative error is reduced to 4.5% from 30.0%. On the other hand, the calibration results in little change of lift coefficient in the range of $\mathrm{AoA} > 20°$ where deep stall occurs.

The pressure distributions over the airfoil at four angles of attack of 8°, 12°, 16°, and 20° are shown in Figure 13. According to the NREL Phase VI experiment, light stall is about to occur at 8°, is in progress at 12° and 16°, and converts to deep stall at 20°. The pressure distributions for $\beta_\infty^* = 0.11$ are consistent with the experimental data at all the four angles of attack. In contrast, $\beta_\infty^* = 0.09$ only performs well at 8° where no stall occurs. Figure 14 shows the simulated streamlines around the airfoil at angles of attack of 16° and 20°. At 16°, the flow patterns for $\beta_\infty^* = 0.09$ and $\beta_\infty^* = 0.11$ are both trailing edge flow separation (light stall). The difference is that $\beta_\infty^* = 0.11$ leads to a larger separation vortex. At 20°, $\beta_\infty^* = 0.11$ and $\beta_\infty^* = 0.09$ gives complete flow separation (deep stall) and trailing edge flow separation (light stall), respectively.

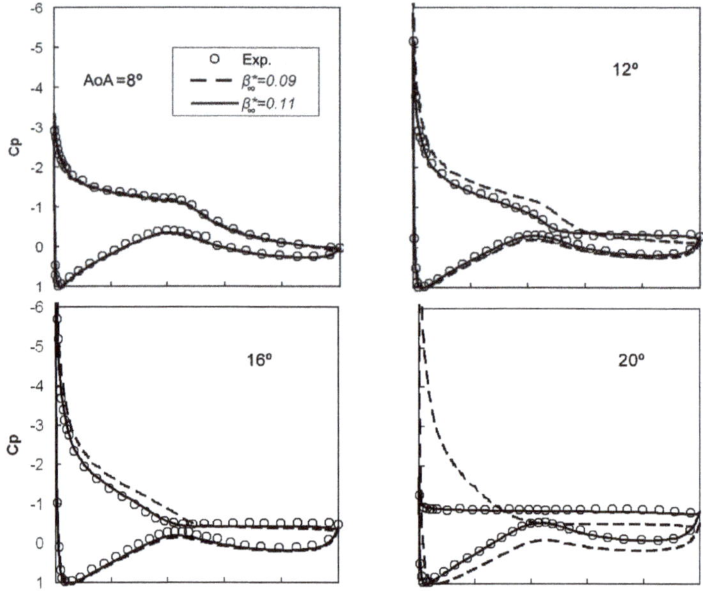

Figure 13. The predicted pressure distributions for different values of β_∞^*, with a comparison to experimental data (Re = 1 × 10^6).

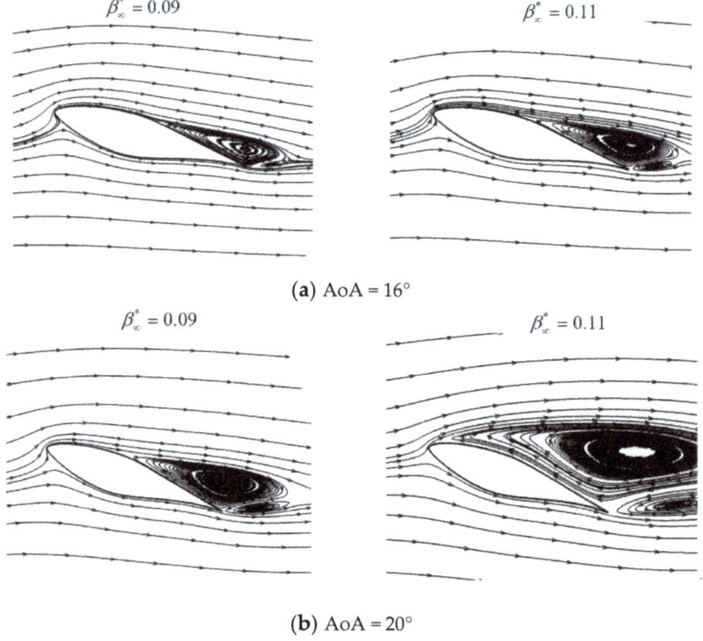

Figure 14. The simulated streamlines around the airfoil for different values of β_∞^* (Re = 1 × 10^6).

The airfoil simulation has also been carried out at the Reynolds numbers of 0.75 × 10^6 and 1.5 × 10^6, which are the lowest and the highest Reynolds number of the OSU experiment [37], respectively. The results are shown in Figure 15. $\beta^*_\infty = 0.11$ makes the simulation accuracy significantly improved at the both Reynolds numbers, suggesting that the optimal β^*_∞ calibrated at a certain Reynolds number stays in a relatively wide range of Reynolds numbers.

Table 6. The relative simulation error of lift coefficient for different values of β^*_∞.

AoA (°)	8	10	12	14	16	18	20	$\overline{\delta}$
$\beta^*_\infty = 0.09$	5.8%	18.1%	23.0%	18.9%	22.7%	39.4%	82.2%	30.0%
$\beta^*_\infty = 0.11$	2.2%	1.3%	−0.3%	0.5%	6.1%	17.5%	−3.9%	4.5%

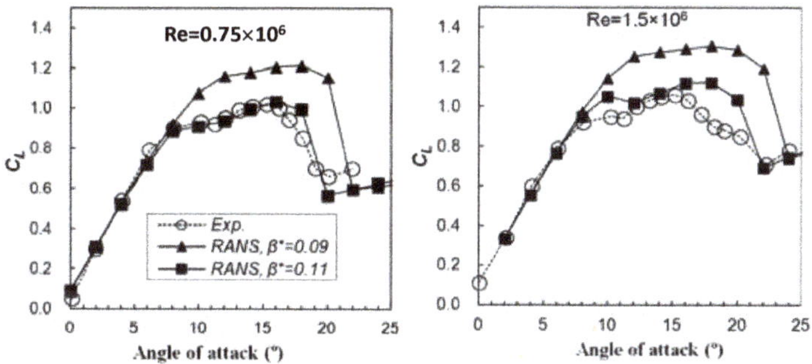

Figure 15. The simulation results of lift coefficient at different Reynolds numbers compared to experimental data.

4.3. Rotor Simulation Using the Calibrated Coefficient

$\beta^*_\infty = 0.11$, the optimal value for the S809 airfoil, is applied to the simulation of the NREL Phase VI rotor. The predicted rotor torque versus the wind speed is shown in Figure 16. Both the results for $\beta^*_\infty = 0.09$ and $\beta^*_\infty = 0.11$ are consistent with the experimental data when the wind speed is lower than 9 m/s. They diverge from each other at the wind speeds between 9 m/s and 20 m/s. $\beta^*_\infty = 0.09$ leads to a notable overestimation of the maximum torque, while $\beta^*_\infty = 0.11$ makes it accurately predicted. Table 7 shows the relative simulation error of the rotor torque. At 10 m/s, where the maximum torque is achieved for both the predicted curves, the relative error is reduced to 3.2% from 34.4% by the application of $\beta^*_\infty = 0.11$. At 17 m/s, the result for $\beta^*_\infty = 0.09$ looks better than that of $\beta^*_\infty = 0.11$. However, it is a mere coincidence due to the change from positive error to negative error of the curve for $\beta^*_\infty = 0.09$ at this wind speed. When the wind speed increases to 20 m/s or higher, the two curves overlap again and are lower than the experimental data. The maximum difference of the two predicted curves is located at the wind speed of 10 m/s. The pressure distributions of five typical blade sections at this wind speed are given in Figure 17. It is shown that the application of $\beta^*_\infty = 0.11$ leads to accurate prediction of the pressure distributions. Taking the section of r/R = 0.47 as an example, $\beta^*_\infty = 0.11$ makes the onset of deep stall accurately captured and properly gives the flat pressure distribution on the suction side of the blade.

Table 7. The relative simulation error of the rotor torque for different values of β^*_∞.

Wind Speed	9	10	13	15	17	20
$\beta^*_\infty = 0.09$	4.7%	34.4%	29.0%	20.2%	0.6%	−9.8%
$\beta^*_\infty = 0.11$	−2.3%	3.2%	−4.5%	−15.7%	−18.8%	−9.6%

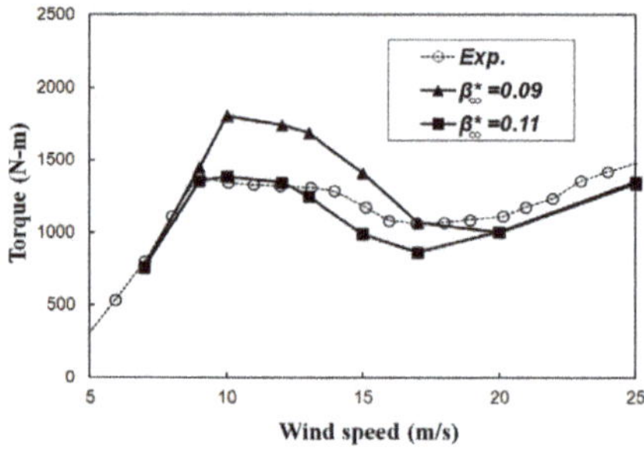

Figure 16. The predicted rotor torque for different values of β^*_∞ compared to the measured data of the NREL Phase VI experiment.

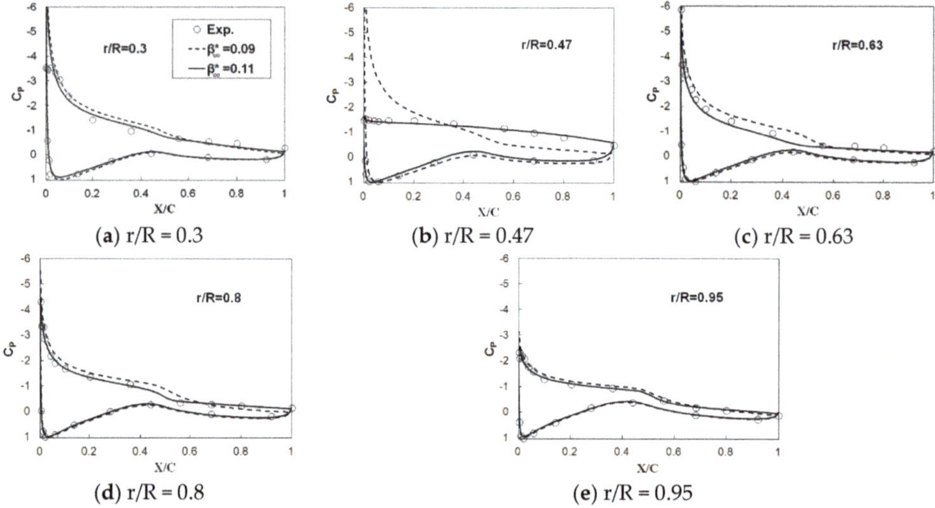

(a) r/R = 0.3

(b) r/R = 0.47

(c) r/R = 0.63

(d) r/R = 0.8

(e) r/R = 0.95

Figure 17. The simulation results of pressure distributions for different values of β^*_∞, with a comparison to the measured data of the NREL Phase VI experiment. (Wind speed: 10 m/s).

The limiting streamlines on the suction side of the blade at the wind speeds of 10 m/s, 15 m/s, and 20 m/s are shown in Figure 18. As compared to $\beta^*_\infty = 0.09$, the results for $\beta^*_\infty = 0.11$ have two differences at 10 m/s. Firstly, the area of trailing edge flow separation (light stall) is larger and the area of attached flow (no stall) is correspondingly smaller. Secondly, a small piece of complete flow separation (deep stall) occurs at about r/R = 0.47, which has been corroborated by the experimental pressure distribution at this section shown in Figure 17. At 15 m/s, a large part of the blade is under complete flow separation (deep stall) for both $\beta^*_\infty = 0.09$ and $\beta^*_\infty = 0.11$. The difference is that the area of deep stall is larger for $\beta^*_\infty = 0.11$. At 20 m/s, complete flow separation (deep stall) covers the whole blade, and there is little difference between the two results for $\beta^*_\infty = 0.09$ and $\beta^*_\infty = 0.11$. The above results suggest that the simulation using $\beta^*_\infty = 0.11$ predicts an earlier onset of trailing edge flow separation (light stall) and also an earlier conversion to complete flow separation (deep stall) compared to the simulation using $\beta^*_\infty = 0.09$.

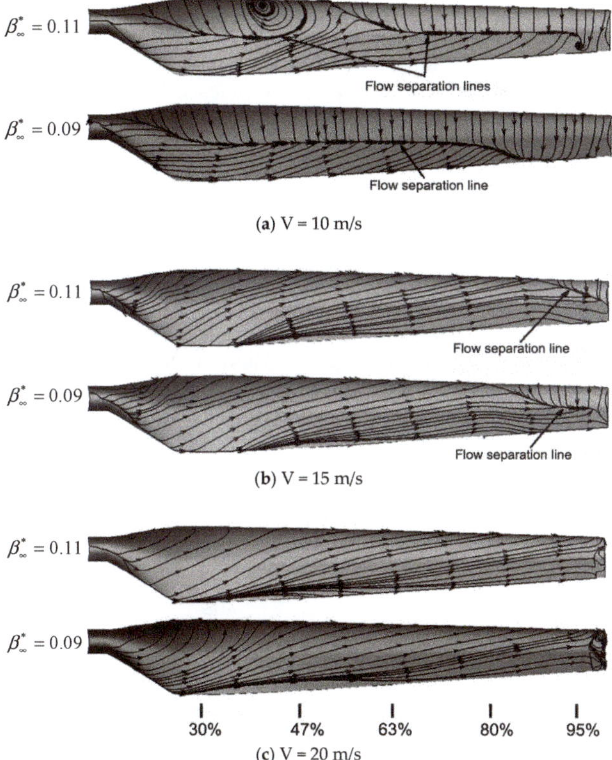

Figure 18. The simulation results of limiting streamlines on the suction side of the blade, for different values of β_∞^*.

5. Verification of the Calibration to the MEXICO Rotor

5.1. Introduction to the MEXICO Rotor

In order to study how robust the present approach of calibration is for different airfoils and rotors, the calibration is applied to RANS simulation of the turbine used in the MEXICO (Model Experiments in Controlled Conditions) project [21]. The MEXICO turbine consists of a three-bladed rotor with a diameter of 4.5 m. Three different airfoils of DU91-W2-250, RISØ-A1-21, and NACA 64-418 were used in the blade design. The DU91-W2-250 airfoil was applied from 20% to 45.6% span, the RISØ-A1-21 airfoil from 54.4% to 65.6% span, and the NACA 64-418 airfoil outboard to 74.4% span. The turbine was tested in the 9.5 × 9.5 m^2 open section of the largest European wind tunnel, the Large-Scale Low Speed Facility (LLF) of the German Dutch Wind Tunnels (DNW), in December 2006. The tests were performed at both non-yawed and yawed flows at different wind speeds and resulted in a database of combined blade pressure distributions, loads, and flow field measurements. As a follow up of the first MEXICO campaign, the New MEXICO measurement on the same turbine was carried out in the same wind tunnel between 20th June and 4th July 2014. Several open questions from the first campaign have been resolved, and good agreement has been found between the first and the new measurements [42].

An international research collaborative project (Mexnext) within the framework of IEA Task 29 was created in June 2008. It was carried out in three phases. The first phase from September 2008 until December 2011 included an assessment of the measurement uncertainties and a validation of different categories of aerodynamic models. In the second phase from January 2012 until the end of

2014, unexplored aerodynamic measurements on wind turbines (both in the wind tunnel as well as in the field) were analyzed from a wide variety of sources. In the third phase from January 2015 until December 2017, several rounds of comparisons were made between the New MEXICO measurements and the results from various codes. Some comparisons between CFD simulations and measurements are reported in [43–45].

5.2. Introduction to the Present Simulation

The present study investigates the performance of the SST turbulence model using different values of β_∞^* in a non-yawed case with a rotational speed of 424.5 rpm and a wind speed of 24 m/s. Flow separation occurs on the rotor blades in this case. The separation point location was generally predicted closer to the trailing edge than the experimental value, and the suction side pressure was over-predicted by most of the CFD codes used in the Mexnext project [29].

Pressure distributions were measured in the first and the New MEXICO experiments over five sections located at 25%, 35%, 60%, 82%, and 92% span of the blade. Information about the five sections and their Reynolds numbers in the present case (424.5 rpm, 24 m/s) are shown in Table 8.

Table 8. Information about the five sections and their Reynolds numbers in the present case.

r/R	r (m)	Chord (m)	Profile	Re
0.25	0.563	0.224	DU91-W2-250	0.38×10^6
0.35	0.788	0.193	DU91-W2-250	0.46×10^6
0.6	1.350	0.142	RISØ-A1-21	0.58×10^6
0.82	1.845	0.113	NACA 64-418	0.64×10^6
0.92	2.070	0.099	NACA 64-418	0.62×10^6

The detailed experimental aerodynamic coefficients of the airfoils DU91-W2-250, RISØ-A1-21, and NACA 64-418 were given in the project of Mexnext. The available experimental data of DU91-W2-250 and NACA 64-418 were measured at Reynolds numbers of 0.5×10^6 and 0.7×10^6, respectively. They are close to the operating Reynolds numbers of the corresponding sections shown in Table 8. However, the data of RISØ-A1-21 are available only at Reynolds number of 1.6×10^6, which is much higher than the operating Reynolds number of the corresponding section of r/R = 0.6. That may reduce the reliability of the present calibration for the blade segment based on the airfoil RISØ-A1-21.

The grid layout and density for the airfoils are identical to those for the S809 airfoil. Instead of another mesh independency study, the mesh for the MEXICO rotor is further refined compared to the mesh for the NREL Phase VI rotor. The number of grid cells is increased to about 13 million, which is more than three times that of the NREL Phase VI rotor. All other computational settings for the airfoils and the MEXICO rotor are identical to those for the S809 airfoil and the NREL Phase VI rotor, respectively.

5.3. Results and Disscussion

RANS simulations of the three airfoils of the MEXICO blade are carried out, and the resulted lift coefficients are shown in Figure 19. For DU91-W2-250 at Re = 0.5×10^6, the default β_∞^* of 0.09 gives results consistent with the measured data at a wide range of angles of attack. For RISØ-A1-21 at Re = 1.6×10^6, β_∞^* = 0.10 looks better. For NACA 64-418 at Re = 0.7×10^6, β_∞^* = 0.105 gives perfect prediction before deep stall. As a result, in the simulations of the MEXICO rotor, different values of β_∞^* should be applied at different span locations depending on the airfoil used there. That would be an inconvenience for application. A function might have to be established to give a continuous change of β_∞^* in the computational domain, especially along the blade span. The function could be applied to the SST model if an open source solver or a house code were used. However, it cannot easily have been realized in ANSYS FLUENT. As a compromise, three independent rotor simulations are carried out, in which the β_∞^* values of 0.09, 0.10, and 0.105 are applied, respectively.

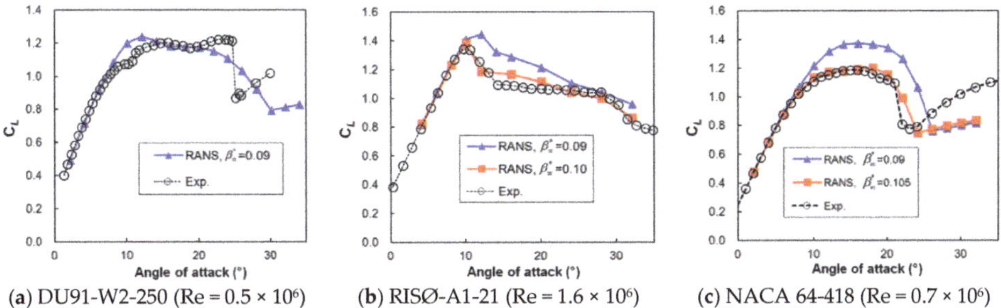

Figure 19. The predicted lift coefficients of the airfoils DU91-W2-250, RISØ-A1-21, and NACA 64-418 for different values of β_∞^*, with a comparison to experimental data.

The rotor simulation results of pressure distributions over the five blade sections of 25%, 35%, 60%, 82%, and 92% span are shown in Figure 20. Measured data from both the MEXICO and the New MEXICO experiments are shown for comparison. For the default β_∞^* value of 0.09, notable deviations between the computational and experimental data appear on the suction side at sections of r/R = 0.35, 0.82, and 0.92. The present simulation error of the three sections is at an average level of the CFD results from the Mexnext project [29]. The Figure 20c–e give the results of r/R = 0.6, 0.82, and 0.92 for the calibrated values of β_∞^*, in which the results for β_∞^* = 0.09 are also displayed. At r/R = 0.6, the prediction of = 0.09 is consistent with the New MEXICO data, while the result of β_∞^* = 0.10 is closer to the MEXICO data. At r/R = 0.82 and 0.92, the predicted pressure distributions on the suction side become much closer to the experimental data when β_∞^* = 0.105 is applied. Also, the present results are better than a collection of the CFD results from the Mexnext project [29]. The comparisons in the Figure 20d,e give strong evidence for the contribution of the present calibration to the accurate simulation of wind turbine stall.

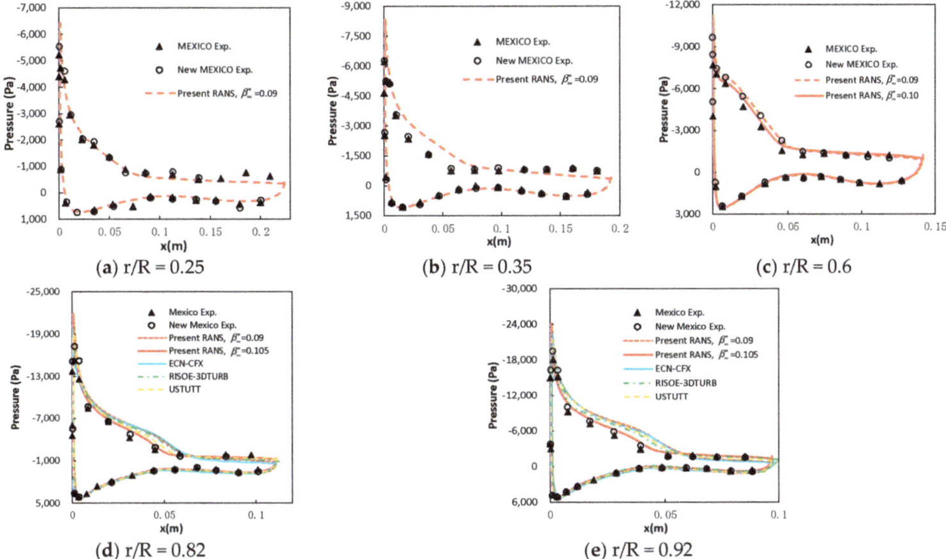

Figure 20. The predicted pressure distributions over blade sections for different values of β_∞^* compared to experimental data and a collection results of the Mexnext project [29].

6. Conclusions

An approach of turbulence coefficient calibration has been proposed for accurate simulation of wind turbine stall. The similarity of error between the rotor and airfoil simulations is investigated on the S809 airfoil and the NREL Phase VI rotor. It is an important foundation of the present work, which provides a way to improve rotor simulation using airfoil experimental data. The simulation of the NREL Phase VI rotor is significantly improved by making the calibration on the S809 airfoil. Furthermore, the procedure of calibration is applied to the MEXICO rotor, and it also significantly improves the simulation. The following conclusions are drawn from the present study:

- RANS simulations using the SST turbulence model tend to over-predict the aerodynamic force of airfoils and rotors in light stall. The present study involves four airfoils of S809, DU91-W2-250, RISØ-A1-21, and NACA 64-418, three of which face the problem of over-prediction at given Reynolds numbers. The over-prediction of the NREL Phase VI rotor and the MEXICO rotor is dependent on the airfoils of their blades.

- There is a high degree of similarity of error between the airfoil and the rotor simulations. In conditions of stall, the rotor simulation error is approximately at the same level of the airfoil simulation error. That means the rotor simulation error is not mainly caused by the rotational effect or the 3-D effect. For example, the range and amplitude of the simulation error at the mid-span section ($r/R = 0.47$) of the NREL Phase VI blade overlaps much with those of the S809 airfoil, and the peak error at the root section ($r/R = 0.3$) and the tip section ($r/R = 95$) is not higher than that of the airfoil.

- The closure coefficient β^*_∞ of the SST turbulence model has a great impact on the prediction of light stall for both airfoils and rotors. The predicted airfoil lift in light stall reduces with the increase of β^*_∞. For example, the lift coefficient of the S809 airfoil, which was over-predicted about 30% in light stall conditions, is reduced to a level consistent with experimental data as β^*_∞ is adjusted to 0.11. Also, the error of predicted torque of the NREL Phase VI rotor is reduced when the same adjustment is made.

- The optimal β^*_∞ obtained from the calibration in airfoil simulation can significantly improve the stall prediction in rotor simulation, and thus the rotor aerodynamic performance is predicted more accurately. For the NREL Phase VI rotor, the relative error of the predicted maximum torque is reduced from 34.4% to 3.2%. For the MEXICO rotor, the predicted pressure distributions over typical sections are better than the CFD results from the Mexnext project.

In essence, the present study provides an approach for calibrating the rotor simulation by using airfoil experimental data and enhances the potential of RANS in accurate simulations of the wind turbine aerodynamic performance. In future work, more simulations of various airfoils and rotors will be carried out to obtain a more comprehensive evaluation of the approach.

Author Contributions: Conceptualization, W.Z.; Data curation, H.T.; Funding acquisition, W.Z. and T.W.; Investigation, W.Z., H.T., and C.Z.; Methodology, W.Z.; Project administration, W.Z. and T.W.; Supervision, W.Z. and T.W.; Visualization, H.T. and C.Z.; Writing—original draft, W.Z.

Funding: This research was funded jointly by the National Natural Science Foundation of China (No. 51506088), the National Basic Research Program of China ("973" Program) under Grant No. 2014CB046200, the Priority Academic Program Development of Jiangsu Higher Education Institutions, and the Key Laboratory of Wind Energy Utilization, Chinese Academy of Sciences (No. KLWEU-2016-0102).

Acknowledgments: Part of the present study was once discussed in W.Z. Shen's research team, department of Wind Energy, DTU. The corresponding author wishes to express special acknowledgement to him and his team members.

Conflicts of Interest: The authors declare no conflict of interest.

References

1. Hansen, M.O.L. The classical blade element momentum method. In *Aerodynamics of Wind Turbines*, 3rd ed.; Routledge: Abingdon, UK, 2015; pp. 38–53.
2. Sørensen, J.N. Blade-element/momentum theory. In *General Momentum Theory for Horizontal Axis Wind Turbines*; Peinke, J., Ed.; Springer: Cham, Switzerland, 2016; pp. 99–122.
3. Hansen, M.O.L.; Sorensen, J.N.; Voutsinas, S.; Sorensen, N.; Madsen, H.A. State of the art in wind turbine aerodynamics and aeroelasticity. *Prog. Aerosp. Sci.* **2006**, *42*, 285–330. [CrossRef]
4. Fernandez-Gamiz, U.; Zulueta, E.; Boyano, A.; Ramos-Hernanz, A.J.; Lopez-Guede, M.J. Microtab Design and Implementation on a 5 MW Wind Turbine. *Appl. Sci.* **2017**, *7*, 536. [CrossRef]
5. Wang, T.G. Unsteady Aerodynamic Modeling of Horizontal Axis Wind Turbine Performance. Ph.D. Thesis, University of Glasgow, Glasgow, UK, 1999.
6. Cottet, G.H.; Koumoutsakos, P.D. Three-dimensional vortex methods for inviscid flows. In *Vortex Methods: Theory and Practice*; Cambridge University Press: Cambridge, UK, 2000; pp. 55–89.
7. Shen, W.Z.; Loc, T.P. Numerical method for unsteady 3d navier-stokes equations in velocity-vorticity form. *Comput. Fluids* **1997**, *26*, 193–216. [CrossRef]
8. Mikkelsen, R. Actuator Disc Methods Applied to Wind Turbines. Ph.D. Thesis, Technical University of Denmark, Copenhagen, Denmark, 2003.
9. Sanderse, B.; van er Pijl, S.P.; Koren, B. Review of computational fluid dynamics for wind turbine wake aerodynamics. *Wind Energy* **2011**, *14*, 799–819. [CrossRef]
10. Mehta, D.; van Zuijlen, A.H.; Koren, B.; Holierhoek, J.G.; Bijl, H. Large Eddy Simulation of wind farm aerodynamics: A review. *J. Wind Eng. Ind. Aerodyn.* **2014**, *133*, 1–17. [CrossRef]
11. Somers, D.M. Design and Experimental Results for the S809 Airfoil. Available online: https://www.nrel.gov/docs/legosti/old/6918.pdf (accessed on 17 August 2018).
12. Yang, S.L.; Chang, Y.L.; Arici, O. Incompressible Navier-Stokes computation of the NREL airfoils using a symmetric total variational diminishing scheme. *J. Sol. Energy Eng.* **1994**, *116*, 174–182. [CrossRef]
13. Yang, S.L.; Chang, Y.L.; Arici, O. Navier-Stokes computations of the NREL airfoil using a k-ω turbulent model at high angles of attack. *J. Sol. Energy Eng.* **1995**, *117*, 304–310. [CrossRef]
14. Benjanirat, S. Computaional Studies of Horizontal Axis Wind Turbines in High Wind Speed Condition Using Advanced Turbulence Models. Ph.D. Thesis, Georgia Institute of Technology, Atlanta, GA, USA, December 2006.
15. Guerri, O.; Bouhadef, K.; Harhad, A. Turbulent flow simulation of the NREL S809 airfoil. *Wind Eng.* **2006**, *30*, 287–302. [CrossRef]
16. Wolfe, W.P.; Ochs, S.S. *CFD Calculations of S809 Aerodynamic Characteristics*; AIAA 97-0973; American Institute of Aeronautics and Astronautics: Reston, VA, USA, 1997. [CrossRef]
17. Bertagnolio, F.; Sørensen, N.; Johansen, J.; Fuglsang, P. Wind Turbine Airfoil Catalogue. Available online: https://www.vindenergi.dtu.dk/english/Research/Research-Projects/Completed-projects/Wind-Turbine-Airfoil-Catalogue (accessed on 17 August 2018).
18. Langtry, R.B.; Gola, J.; Menter, F.R. *Predicting 2D Airfoil and 3D Wind Turbine Rotor Performance Using a Transition Model for General CFD Codes*; AIAA 2006-0395; American Institute of Aeronautics and Astronautics: Reston, VA, USA, 2006. [CrossRef]
19. Menter, F.R.; Langtry, R.B.; Volker, S. Transition Modelling for General Purpose CFD Codes. *Flow Turbul. Combust.* **2006**, *77*, 277–303. [CrossRef]
20. Hand, M.M.; Simms, D.A.; Fingersh, L.J.; Jager, D.W.; Cotrell, J.R.; Schreck, S.J.; Larwood, S.M. Unsteady Aerodynamics Experiment Phase VI: Wind Tunnel Test Configurations and Available Data Campaigns. Available online: https://www.nrel.gov/docs/fy02osti/29955.pdf (accessed on 17 August 2018).
21. Schepers, J.G.; Snel, H. Model Experiments in Controlled Conditions Final Report. Available online: https://www.ecn.nl/publicaties/PdfFetch.aspx?nr=ECN-E--07-042 (accessed on 17 August 2018).
22. Simms, D.; Schreck, S.; Hand, M. NREL Unsteady Aerodynamics Experiment in the NASA-AMES Wind Tunnel: A Comparison of Predictions to Measurements. Available online: http://citeseerx.ist.psu.edu/viewdoc/download?doi=10.1.1.452.974&rep=rep1&type=pdf (accessed on 17 August 2018).
23. Duque, E.P.N.; Burklund, M.D.; Johnson, W. Navier-Stokes and comprehensive analysis performance predictions of the NREL Phase VI experiment. *J. Sol. Energy Eng.* **2003**, *125*, 457–467. [CrossRef]

24. Pape, A.L.; Lecanu, J. 3D Navier-Stokes computations of a stall-regulated wind turbine. *Wind Energy* **2004**, *7*, 309–324. [CrossRef]
25. Sørensen, N.N. CFD modeling of laminar-turbulent transition for airfoil and rotor using the γ-Reθ model. *Wind Energy* **2009**, *12*, 715–733. [CrossRef]
26. Schmitz, S.; Chattot, J.J. A parallelized coupled Navier-Stokes vortex-panel solver. *J. Sol. Energy Eng.* **2005**, *127*, 475–487. [CrossRef]
27. Rooij, R.P.; Arens, E.A. Analysis of the experimental and computational flow characteristics with respect to the augmented lift phenomenon caused by blade rotation. *J. Phys. Conf. Ser.* **2007**, *75*, 012021. [CrossRef]
28. Länger-Möller, A.; Löwe, J.; Kessler, R. Investigation of the NREL phase VI experiment with the incompressible CFD solver THETA. *Wind Energy* **2017**, *20*, 1529–1549. [CrossRef]
29. Schepers, J.G.; Boorsma, K.; Cho, T.; Gomez-Iradi, S.; Schaffarczyk, A.; Jeromin, A.; Shen, W.Z.; Lutz, T.; Meister, K.; Stoevesandt, B.; et al. Final Report of IEA Task 29, Mexnext (Phase 1): Analysis of Mexico Wind Tunnel Measurements. Available online: http://www.mexnext.org/fileadmin/mexnext/user/documents/FinRep_Mexnext_v6_opt.pdf (accessed on 17 August 2018).
30. Herráez, I.; Stoevesandt, B.; Peinke, J. Insight into rotational effects on a wind turbine blade using Navier-Stokes computations. *Energies* **2014**, *7*, 6798–6822. [CrossRef]
31. Bechmann, A.; Sørensen, N.N.; Zahle, F. CFD simulations of the MEXICO rotor. *Wind Energy* **2011**, *14*, 677–689. [CrossRef]
32. Shen, W.Z.; Zhu, W.J.; Sørensen, J.N. Actuator line/Navier-Stokes computations for the MEXICO rotor: Comparison with detailed measurements. *Wind Energy* **2012**, *15*, 811–825. [CrossRef]
33. O'Brien, J.M.; Young, T.M.; O'Mahoney, D.C.; Griffin, P.C. Horizontal axis wind turbine research: A review of commercial CFD, FE codes and experimental practices. *Prog. Aerosp. Sci.* **2017**, *92*, 1–24. [CrossRef]
34. Xu, H.Y.; Qiao, C.L.; Yang, H.Q.; Ye, Z.Y. Delayed detached eddy simulation of the wind turbine airfoil S809 for angles of attack up to 90 degrees. *Energy* **2017**, *118*, 1090–1109. [CrossRef]
35. Almohammadi, K.M.; Ingham, D.B.; Ma, L.; Pourkashan, M. Computationa fluid dynamics (CFD) mesh independency techniques for a straight blade vertical axis wind turbine. *Energy* **2013**, *58*, 483–493. [CrossRef]
36. Fernandez-Gamiz, U.; Errasti, I.; Gutierrez-Amo, R.; Boyano, A.; Barambones, O. Computational modeling of rectangular sub-boundary layer vortex generators. *Appl. Sci.* **2018**, *8*, 138. [CrossRef]
37. Ramsay, R.R.; Hoffman, M.J.; Gregorek, G.M. Effects of Grit Roughness and Pitch Oscillations on the S809 Airfoil. Available online: https://wind.nrel.gov/airfoils/OSU_data/reports/3x5/s809.pdf (accessed on 17 August 2018).
38. Jonkman, J.M. Modeling of the UAE Wind Turbine for Refinement of FAST_AD. Available online: https://www.nrel.gov/docs/fy04osti/34755.pdf (accessed on 17 August 2018).
39. Sheng, W.; Galbraith, R.A.M.D.; Coton, F.N. On the S809 airfoil's unsteady aerodynamic characteristics. *Wind Energy* **2009**, *12*, 752–767. [CrossRef]
40. Sørensen, N.N.; Michelsen, J.A.; Schreck, S. Navier-Stokes predictions of the NREL Phase VI rotor in the NASA Ames 80 ft × 120 ft wind tunnel. *Wind Energy* **2002**, *5*, 151–169. [CrossRef]
41. Rooij, R.P. Analysis of the flow characteristics of two nonrotating rotor blades. *J. Sol. Energy Eng.* **2008**, *130*, 031015. [CrossRef]
42. Boorsma, K.; Schepers, J.G. New MEXICO Experiment: Preliminary Overview with Initial Validation. Available online: https://www.ecn.nl/docs/library/report/2014/e14048.pdf (accessed on 17 August 2018).
43. Sørensen, N.N.; Zahle, F.; Boorsma, K.; Schepers, G. CFD computations of the second round of MEXICO rotor measurements. *J. Phys. Conf. Ser.* **2016**, *753*, 022054. [CrossRef]
44. Imiela, M. CFD Simulations of the New MEXICO Rotor Experiment under Yawed Flow. *J. Phys. Conf. Ser.* **2018**, *1037*, 022031. [CrossRef]
45. Sarmast, S.; Shen, W.Z.; Zhu, W.J.; Mikkelsen, R.F.; Breton, S.P.; Ivanell, S. Validation of the actuator line and disc techniques using the New MEXICO measurements. *J. Phys. Conf. Ser.* **2016**, *753*, 032026. [CrossRef]

Article

Aerodynamic Performance of Wind Turbine Airfoil DU 91-W2-250 under Dynamic Stall

Shuang Li [1,2], Lei Zhang [1,3,4,*], Ke Yang [1,3,4], Jin Xu [1,2] and Xue Li [1,2]

[1] Institute of Engineering Thermophysics, Chinese Academy of Sciences, Beijing 100190, China; lishuang@iet.cn (S.L.); yangke@iet.cn (K.Y.); xujin@iet.cn (J.X.); lixue@iet.cn (X.L.)

[2] University of Chinese Academy of Sciences, Beijing 100049, China

[3] Key Laboratory of Wind Energy Utilization, Chinese Academy of Sciences, Beijing 100190, China

[4] National Research and Development Center of Wind Turbine Blade, Beijing 100190, China

* Correspondence: zhanglei@iet.cn

Received: 15 June 2018; Accepted: 29 June 2018; Published: 10 July 2018

Abstract: Airfoils are subjected to the 'dynamic stall' phenomenon in significant pitch oscillations during the actual operation process of wind turbines. Dynamic stall will result in aerodynamic fatigue loads and further cause a discrepancy in the aerodynamic performance between design and operation. In this paper, a typical wind turbine airfoil, DU 91-W2-250, is examined numerically using the transition shear stress transport (SST) model under a Reynolds number of 3×10^5. The influence of a reduced frequency on the unsteady dynamic performance of the airfoil model is examined by analyzing aerodynamic coefficients, pressure contours and separation point positions. It is concluded that an increasingly-reduced frequency leads to lower aerodynamic efficiency during the upstroke process of pitching motions. The results show the movement of the separation point and the variation of flow structures in a hysteresis loop. Additionally, the spectrum of pressure signals on the suction surface is analyzed, exploring the level of dependence of pressure fluctuation on the shedding vortex and oscillation process. It provides a theoretical basis for the understanding of the dynamic stall of the wind turbine airfoil.

Keywords: wind turbine airfoil; dynamic stall; boundary layer separation; aerodynamic characteristics

---

1. Introduction

Wind turbines operate in diverse environments under complex control strategies, suffering unsteady loads including aerodynamic forces. An important source of these forces comes from the blades when the relative velocity with respect to blade sections varies in magnitude and direction. Generally, the airfoils on the blades are considered as operating in dynamic stall [1]. The unsteady factors that cause dynamic stall are, e.g., wind shear and yaw, and the angles of attack of airfoils periodically vary with the rotation of blades [2–4]. Then, the stall characteristics are dynamically dependent on the operating route of airfoils with dramatically different lift and drag forces from those in static operations. The main disadvantage of dynamic stall for the safety of blades is that it causes fatigue loads and a deviation between design and actual running conditions. With the trend of the large-scale development of wind turbines, dynamic stall will have increasingly greater influence on the unsteady aerodynamic characteristics of wind turbine blades.

Dynamic stall has been one of the major interesting research subjects and has been studied over the latest few decades. The dynamic stall phenomenon was first observed in the 1930s [5]. It was first systematically studied in the aerospace field and was studied in a wider field later, such as for maneuverable jets wings, gas turbines and wind turbines [6]. Previous studies primarily concentrating on dynamic stall consist of experimental and numerical simulation. Experiments are important because experimental data provide a verification basis for the numerical simulation. However, due to the limitation of experimental conditions and other reasons, it is necessary to employ numerical simulation to grasp a broader and more general conclusion. With the gradual maturity of the theory, there are many analytical models to predict the effects of dynamic stall and analyze the load of power machines nowadays; such as the Boeing–Vertol model [7], the ONERA model [8], the Beddoes–Leishman (B–L) semi-empirical model [9], etc. The B-L model has been adopted most widely to analyze dynamic stall. It has been modified and improved with the increasing attention on unsteady aerodynamics [10–13]. The modified model shows better agreement than the original one. Although the semi-empirical models have the advantage of using less computational resources, they depend on the accuracy of the empirical factors greatly, and it is not possible to obtain detailed and accurate flow field information. With the development of computer technology, the computational fluid dynamics (CFD) method is used to study dynamic stall for the purpose of obtaining detailed information of dynamic stall. Some researchers have been dedicated to finding more accurate turbulence models. For example, Martinat et al. [14] used three turbulence models to investigate the dynamic stall of a NACA 0012 airfoil, and the unsteady Reynolds-averaged Navier–Stokes (URANS) $k - \varepsilon$ Chien model provided the best results. Wang et al. [15] compared the renormalization-group (RNG) $k - \varepsilon$ model and transition shear stress transport (SST) $k - \omega$ model used to simulate dynamic stall of an oscillating NACA 0012 airfoil at a relatively low Reynolds number. It was found that the SST $k - \omega$-based detached eddy simulation (DES) approach was superior. Some other researchers have focused on how typical factors affecting dynamic stall. These factors include the Reynolds number, mean angle of pitch oscillation, amplitude of pitch oscillation, reduced frequency, the position of pitch axis, turbulence intensity, etc. For instance, Akbari et al. [16] studied the effects of reduced frequency, mean angle of attack, pitch axis location and Reynolds number on the dynamic stall of a NACA 0012. It was shown that the reduced frequency was the most influential parameter on dynamic stall. Gharali et al. [17] investigated the effects of the oscillations of the free-stream velocity on pitch oscillation by the SST $k - \omega$ model; Kim et al. [18] studied the effects of free-stream turbulence on the aerodynamic characteristics of pitching airfoils by large eddy simulations. Gandhi et al. [19] investigated the influence of reduced frequency on the dynamic stall characteristics of a NACA 0012 airfoil pitching in a turbulent wake with direct numerical simulations. In addition, McCroskey and his partners made two efforts for decades to investigate the details of the dynamic stall phenomena [20–22]. The above results are extensive and continuous. However, the problem is that most of the research objects are wings or symmetrical airfoils. The literature lacks observations of specific asymmetric and blunt tail edge wind turbine airfoils, such as Delft University of Technology (DU) series airfoils. On the other hand, the flow field contour is a general method to explain the phenomenon of dynamic stall used by many previous researchers. It is summarized that a typical dynamic stall process consists of four major stages: attached flow at low angles of attack, the development of the leading edge vortex (LEV), the shedding of the LEV from the suction surface and reattachment of the flow to the suction surface [23]. However, there is a little attention given to the relationship between the frequency of shedding vortices and the frequency of airfoil pitching motion. The interpretation of the "delay" is always analyzed by the lift and drag coefficients or the qualitative expression of the flow field contours. The quantitative expression of the separation of the airfoil boundary layer is not obviously given.

In order to avoid flow separation and reduce the dynamic stall effects, various devices have been proposed for flow control. Depending on the operating principle, they can be classified as active or passive [24]. Passive control includes vortex generators, microtabs, serrated trailing edges, and so on. Active control includes trailing-edge flaps, synthetic jets, and so on. Johnson et al. [25] overviewed 15 different devices for wind turbine control. Lei Zhang et al. [26,27] made long-term efforts to investigate the effects of vortex generators. Some other researchers focused on microtabs [28,29] and obtained some good results. T.K. Barlas et al. [30] and Unai Fernandez-Gamiz et al. [31] reviewed some devices and discussed their effects. The flow control devices have good effects on avoiding flow separation, but the blades have a higher cost for their production. Based on the above, cases with different reduced frequencies (the most important influential factor) are compared regarding the aerodynamic forces and lift-to-drag ratio. In addition, the delay effect of dynamic stall is deeply analyzed from the perspective of two aspects: the pressure field and spectrum analysis.

2. Method

2.1. Numerical Method

The airfoil DU 91-W2-250 was developed by TU Delft, which has a relative thickness of 25% with a trailing edge gap of 0.65% chord [32]. The present study is based on a 2D simulation with ceiling and bottom boundary similarity in a wind tunnel. With this kind of set, the numerical method is easily validated by the authors' previous measurements [33]. In order to simulate the dynamic stall phenomenon, a rotational zone whose radius is three-times that of the model chord is placed at the pitching center, as shown in Figure 1. The rotational zone and airfoil model oscillate like a rigid body, and the movement is defined with a user-defined function (UDF) in ANSYS Fluent [22]. The rest of grid is connected to the rotational grid by interfaces. The block interfaces between the dynamic and static zones are non-conformal because, as the airfoil moves, the O-grid attached to the airfoil rotates rigidly (as a solid body) together with the airfoil.

The Reynolds number is 3×10^5, which confirmed the turbulence that appeared in the previous study. Flows under such a relatively low Reynolds number are highly non-linear [34]. Therefore, turbulence should be considered in the simulation. There are three main turbulence simulation methods, i.e., direct numerical simulation (DNS), Reynolds-averaged Navier–Stokes (RANS) and large eddy simulation (LES). A turbulence model should present a satisfactory representation of an important phenomenon while introducing the least amount of complexity [35]. In the present study, considering numerical simulation cost and acceptable accuracy, the unsteady Reynolds-averaged Navier–Stokes (URANS) method is seem to be the most suitable one to simulate the dynamic stall phenomenon [15]. Consequently, in the CFD software ANSYS Fluent, the transition SST $k - \omega$ model, which is capable of capturing the transition process, was chosen. Some researchers have also discussed that the accuracy of this model is satisfactory [15,17,36–40].

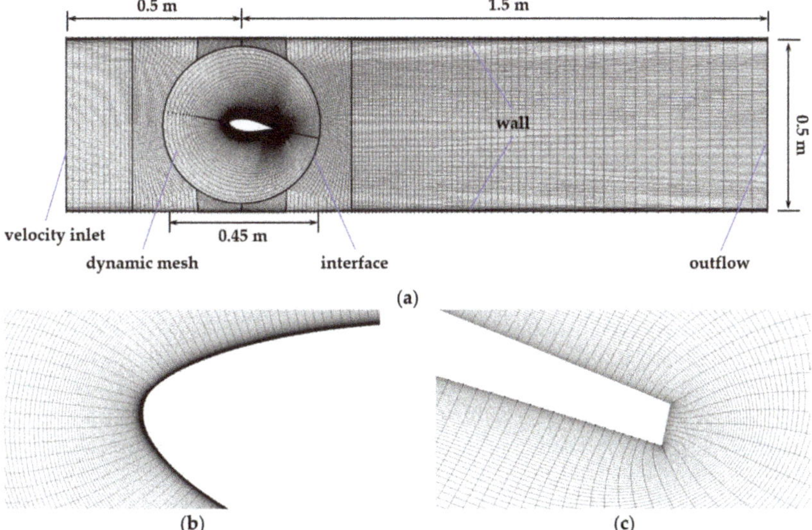

Figure 1. (**a**) Mesh topology of the whole domain; (**b**) mesh details near the leading edge; (**c**) mesh details near the trailing edge.

2.2. Oscillation of the Airfoil

The movement of the airfoil model is described as Equation (1) in a sinusoidal mode, as shown in Figure 2.

$$\alpha_t = \alpha_0 + \alpha_{amp}\sin(\omega t),\tag{1}$$

where,

α_t is the angle of attack at flow time t,

α_0 is the mean angle of attack,

α_{amp} is the pitch oscillation amplitude,

ω is the oscillation angular frequency.

The angular frequency ω is related to the reduced frequency k, which is an important parameter used to describe the 'degree of unsteadiness', as defined by Equation (2).

$$k = \frac{\omega c}{2U_\infty}\tag{2}$$

where U_∞ is the free-stream velocity and c is the chord length of the airfoil. It can be obtained from dimensionless Navier-Stokes (N-S) equations. J. G. Leishman has the empirical view that when $0 \leq k \leq 0.5$, the flow can be considered as quasi-steady flow, which means for some cases that the unsteady effects may be neglected completely [1]. Therefore, $k = 0.026$, 0.502 and 0.106 are set to be computed to find out whether the previous empirical view applies to the pitching motion of the DU 91-W2-250 airfoil. They correspond to the oscillating period T = 0.6 s, 0.3 s and 0.15 s, respectively.

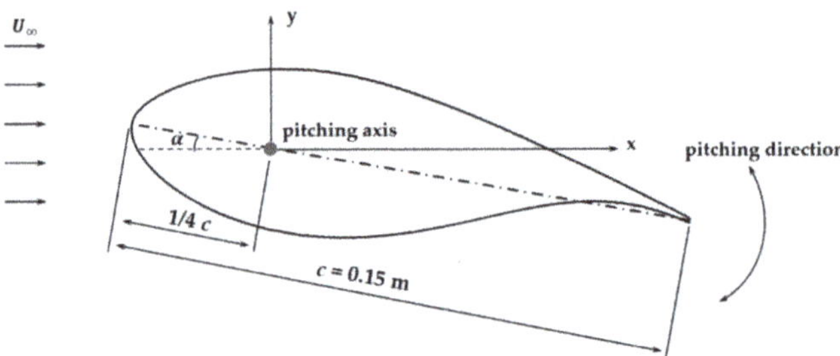

Figure 2. Schematic of the pitching movement of the airfoil model.

2.3. Grid Sensitivity

The independence of the solution from the grid is examined by different grid densities in Table 1. Only the circumferential grid densities are changed, but the radial grid densities are the same. The growth rate of grid is set as 1.05, and the maximum grid is 0.3c. It is important that the mesh near the wall be properly sized to ensure accurate simulation and no wall functions are used [41]. The height of the first mesh cell off the wall is 10^{-5} m, so that the non-dimensional wall-distance y^+ [42] is less than 1.0 at a Reynolds number of 3×10^5. It is computed by using flat-plate boundary layer theory [43]. The lift and drag coefficients during the 4th cycle of pitching motion are shown in Figure 3. The comparison shows that the numerical results for grid cases G3 and G4 do not differ considerably from other grid densities. In order to observe more detailed characteristics on the suction surface of the airfoil, the grid nodes on the suction side are increased to 200 and keep the same set of grids as G3 on the pressure side of the airfoil.

Table 1. Cases for grid sensitivity examination.

Grid	Suction Surface	Pressure Surface	Trailing Edge	Total Grids
G1	50 [1]	50	11	25,753
G2	100	100	11	45,387
G3	150	150	11	67,238
G4	200	200	11	89,070

[1] The number of cells on the indicated surfaces.

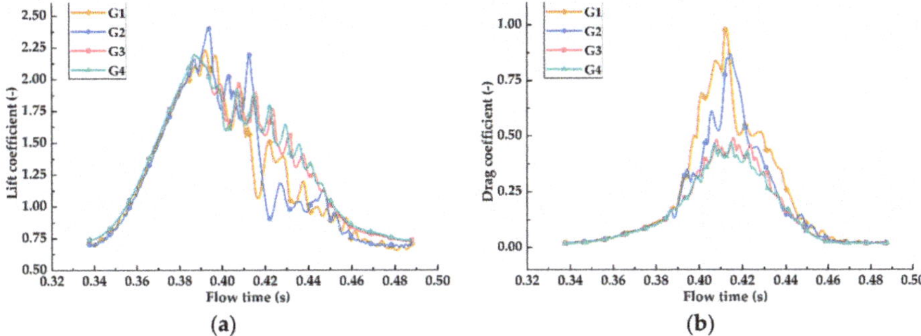

Figure 3. Comparison of computed (**a**) C_l and (**b**) C_d with different grid densities during the 4th cycle of pitching motion.

2.4. Time Step and Periodic Repeatability

To specify the optimum time step in the unsteady simulations, a series case with different time steps was simulated. A characteristic time T_c defined by Equation (3) is employed as base of the time step:

$$T_c = \frac{c}{U_\infty} \tag{3}$$

It has the same significance as the Courant, Friedrichs and Levy criterion (CFL) number in numerical simulations. The time steps of $0.1T_c$, $0.05T_c$, $0.025T_c$, $0.0125T_c$ and $0.00625T_c$ (denoted as dt1 to dt5, respectively) are used to examine the time step sensitivity.

As shown in Figure 4, when the time step dt4 = 6.25×10^{-5} s, the time cost and accuracy are acceptable for the present study. This is consistent with dynamic stall cases applied by other researchers [17,38].

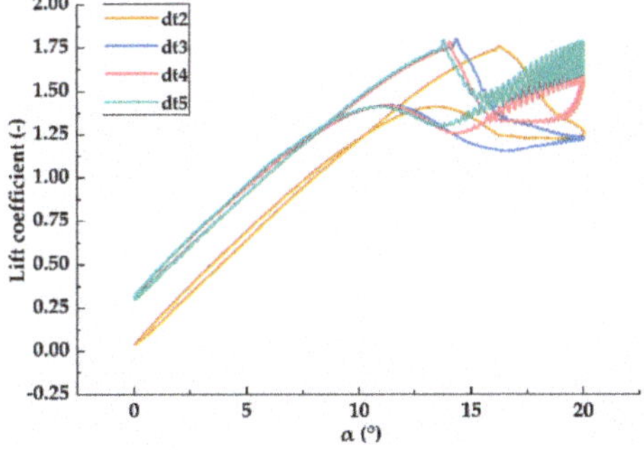

Figure 4. Comparison of computed C_l with different time steps.

Periodic repeatability is also examined, as shown in Figure 5. It shows that values tend to be well reproducible from the third cycle. That is why the other cases are computed only for 4 periods.

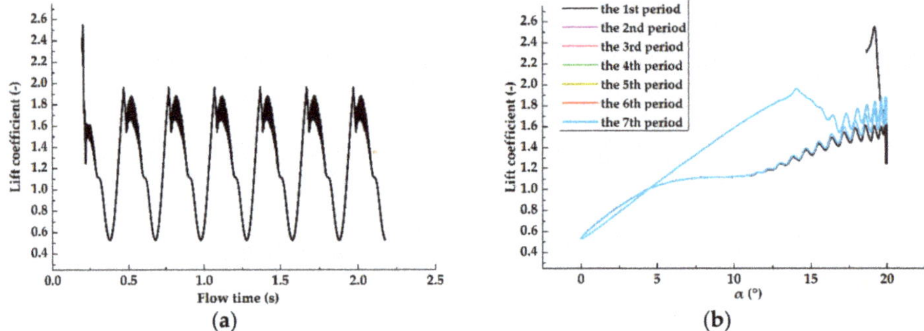

Figure 5. (**a**) history of the lift coefficient C_l; (**b**) the repeatability of the lift coefficient C_l with $k = 0.052$, $\alpha_0 = 10°$, $\alpha_{amp} = 10°$.

2.5. Comparison of the Static Result with Experiment and Vortex Panel Method

Due to the lack of experimental data for the dynamic stall of the DU 91-W2-250 airfoil at present, the static experimental data, which have been obtained from Jingyan Bai et al. [33], are used to examine the accuracy of the numerical method. The data computed by the software Xfoil and Rfoil, which are currently widely accepted, are also compared to each other, as shown in Figure 6.

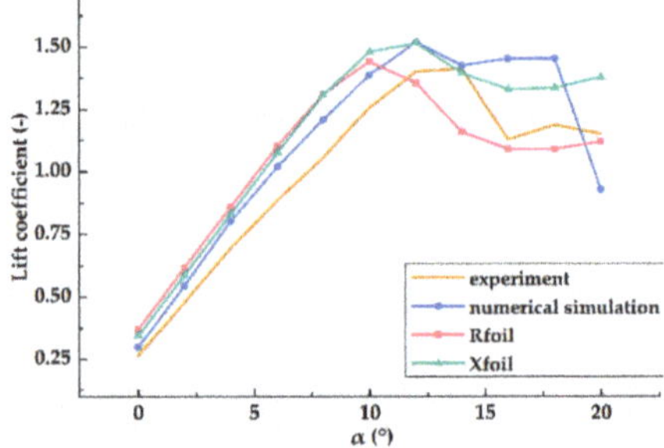

Figure 6. Comparison between the numerical results, Rfoil, Xfoil and experimental data [25].

Generally, Xfoil and Rfoil predict acceptable lift and drag coefficients for thin airfoils in attached flow. It is shown that the numerical simulation is more consistent with the data obtained by Xfoil. Because of the constraints of the experimental equipment and conditions in 2014, a large discrepancy appears between the experimental and CFD results. The accuracy of the zero angle of attack and the stiffness of the support structure may be the main factors, which usually cause the shift of the lift curve along the angle-axis and the variation of the slope of the linear section on the lift curve, respectively. Moreover, CFD greatly overestimates the lift coefficient at 18 and 20 with respect to that of Xfoil, Rfoil and the experiment. This is because the airfoil goes into a deep stall and the flow becomes fully separated at high angles of attack. Raffel et al. [44] showed that the 3D effects of separation flows should be more significant than those without separations at smaller angles of attack. Therefore, in the following analysis, static data calculated by Xfoil are considered as baseline data in the static condition.

3. Results and Discussions

In this part, the results of the cases defined in Section 2.2 are analyzed. The performance of airfoil DU 91-W2-250 under dynamic stall is illustrated with respect to three aspects, i.e.: (a) the influence of reduced frequency; (b) flow structures and separation points; and (c) spectrum characteristics.

3.1. Effects of Reduced Frequency on Aerodynamic Coefficients

Three cases when k = 0.026, 0.502 and 0.106 were simulated, and the results are shown in Figure 7. The dynamic characteristics of the airfoil undergoing pitch oscillation were strongly dependent on the reduced frequency. The hysteresis loop increased with the increasingly-reduced frequency. In a loop, as can be seen, the maximum lift coefficient in upstroke and the corresponding angle of attack were much higher than those in the static condition. Previous researchers have observed this phenomenon already as mentioned in Section 1. Contrarily, the lift coefficient in downstroke was smaller than that in upstroke and sometimes even smaller than that in the static condition. On the lift curves, the intersection point in a loop was named the "reunion point" by the authors. An increase in the reduced frequency made the angle of reunion point smaller. This also means that the effect of delay was more significant when the frequency reduction was greater. On the other hand, the increasingly-reduced frequency would cause the increase of the drag coefficient. This effect is clearly revealed in the lift-drag-ratio curve in Figure 7c.

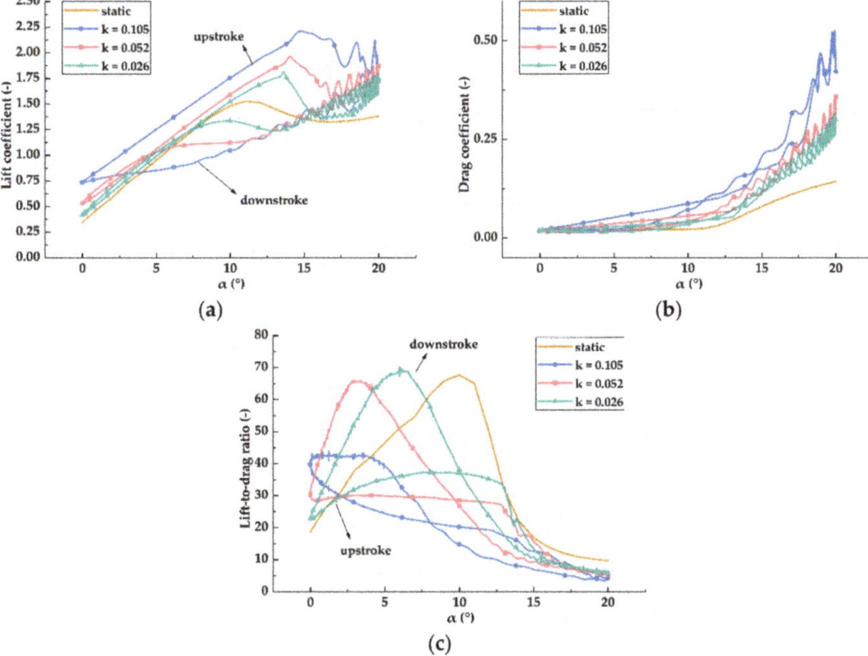

Figure 7. Comparison of the different reduced frequencies k = 0.105, 0.052, 0.026 of the computed (**a**) lift coefficient C_l, (**b**) drag coefficient C_d and (**c**) lift-to-drag ratio C_l / C_d with $\alpha_0 = 10°$, $\alpha_{amp} = 10°$.

The lift-to-drag ratio is an important parameter that indicates the aerodynamic efficiency of the airfoils. The aerodynamic performance of the airfoils is better with a larger lift-to-drag ratio. The angle of attack corresponding to the maximum lift-to-drag ratio is called a favorable angle of attack. In general, from the minimum angle to the favorable angle of attack, lift increases fast, but drag increases slowly. Therefore, the lift-to-drag ratio increases. Then, from the favorable angle to a critical stall angle of attack, the increase of lift is slow, and the drag increases rapidly. This makes the lift-to-drag ratio relatively small. After the critical stall angle of attack, the lift-to-drag ratio sharply decreases with the sharply increasing pressure drag. The lift-to-drag ratio of the airfoil during a complete hysteresis loop was quite different with a regular pattern. The peak of the lift-to-drag ratio curve in upstroke decreased and even disappeared with the increasingly-reduced frequency because of the faster increase of the drag coefficient. This effect was amplified in the lift-to-drag ratio and showed the tremendous difference between static and dynamic stall. Instead, the lift-to-drag ratio in the downstroke process showed a different form. Its shape was closer to the static form, which also had a peak. The angle of attack corresponding to the peak was smaller with increasingly-reduced frequency. This also shows the degree of the delay effect. Therefore, the aerodynamic efficiency was lower in dynamic stall process though the lift in upstroke was higher than that in the static condition.

3.2. Flow Development and Separation Points on the Suction Surface of the Airfoil

The reason for the variation of forces and lift-to-drag ratio can be found in flow field contours. One case was selected to investigate the characteristics as an example. The pressure fields and corresponding streamlines at different angles of attack through a full pitching cycle in the case with $k = 0.052$, $\alpha_0 = 10°$, $\alpha_{amp} = 10°$ are demonstrated in Figure 8. It was observed that the development trend of flow field structures of the DU 91-W2-250 was qualitatively in good agreement with the conclusions of other papers with different shapes of airfoils [15,45,46]. The flow was fully attached to the airfoil from the angles of attack $\alpha_t = 0°\sim10.31°$. The hysteresis of stall angles of attack in lift curves, which is the most obvious feature of dynamic stall, is observed clearly here. The lift dropped suddenly at about 14° in the hysteresis loop, while the static stall angle was 12° in the experiment. It was manifested as reversed flow generated near the trailing edge, as shown in Figure 8c. The reverse flow was due to the adverse pressure gradient caused by the rapid movement of the airfoil surface. As the angle of attack increased, the clockwise-rotating vortex on the suction surface moved forward to the leading edge, and the strength of the recirculation increased. There was a small counterclockwise-rotating vortex behind the trailing edge vortex in Figure 8c, implying that the vortex near the trailing edge started to move downstream slowly. This would lead to a decrease in the lift coefficient compared with the lift coefficient curves in Figure 7a. The effects of vortices rotating in two opposite directions on the lift coefficient are shown in Figure 9 in more detail. At Point A in Figure 9a, the clockwise-rotating vortex appeared. With the growth of the vortex, the lift coefficient was still increasing. The growing vortex increased the slope of the lift curves until the maximum value of C_l when the angle of attack was 14.07° at Point B; see Figure 9b. The reason that C_l decreased was the generation of the counterclockwise-rotating vortex, as shown in Figure 9c. C_l decreased with the growth of the counterclockwise-rotating vortex until the counterclockwise-rotating vortex traveled into the wake and the clockwise-rotating vortex appeared again; see Figure 9e. This process would repeat until the trend of the lift coefficient curve became flat when $\alpha_t = 10.31°$ in the downstroke process. The counterclockwise-rotating vortex did not appear from this moment. In other words, the growth and shedding of vortices in opposite directions led to the fluctuations of the lift coefficient curve. The lift coefficient C_l increased with the growth of the vortex, and it decreased with the shedding of the vortex. What needs to be noted here is that the range of the angles of attack in this case was from 0°–20°, so the airfoil did not experience a deep stall. Vortices on the suction surface of airfoil did not develop to reach the leading edge, but only reached 30% of the chord from the leading edge. Therefore, in this case with $k = 0.052$, $\alpha_0 = 10°$, $\alpha_{amp} = 10°$, the LEV observed in other papers [15] did not appear, but it appeared when $k = 0.052$, $\alpha_0 = 15°$, $\alpha_{amp} = 10°$. After the maximum angle of 20°, the airfoil was performing the downstroke motion. It is known that the lift coefficient of the downstroke process was much smaller than the upstroke process from Section 3.1. This is because the vortex still existed near the trailing edge and the pressure of pressure surface was lower at the same angle of attack by comparing Figure 8b,h and Figure 8c,g. As the airfoil attack angle became smaller, vortices eventually moved into the wake. The last vortex left the trailing edge at $\alpha_t = 6.61°$, and the flow completely reattached to the airfoil at $\alpha_t = 4.73°$. They were smaller than the angles respectively in upstroke with a similar structure. Therefore, the hysteresis of flow attachment during the downstroke process was considerable.

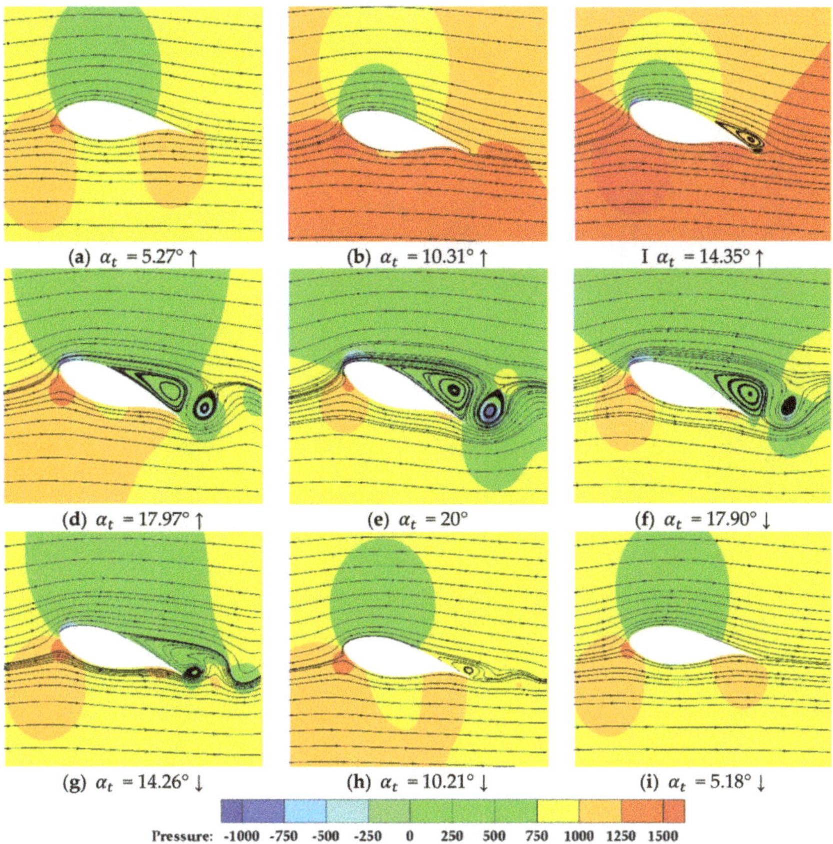

Figure 8. Pressure fields and streamlines for different angles of attack in the pitching cycle with $k = 0.052$, $\alpha_0 = 10°$, $\alpha_{amp} = 10°$.

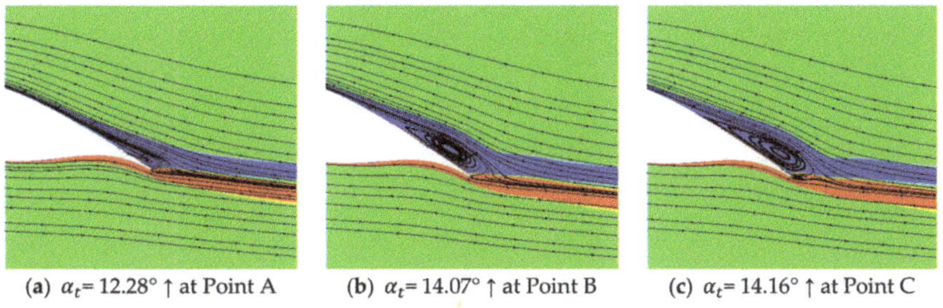

(a) $\alpha_t = 12.28° \uparrow$ at Point A   (b) $\alpha_t = 14.07° \uparrow$ at Point B   (c) $\alpha_t = 14.16° \uparrow$ at Point C

Figure 9. *Cont.*

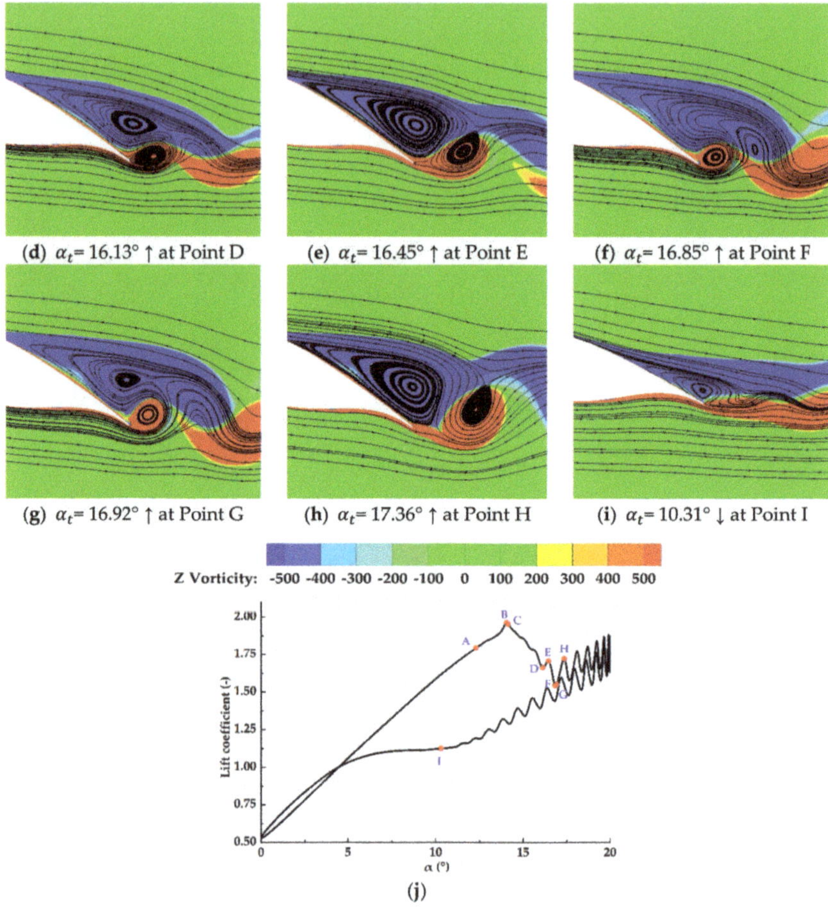

Figure 9. (**a–i**) Vorticity fields and streamlines for different points of angles of attack corresponding to (**j**) the lift coefficient curve.

It is not convenient to show separation situations at every angle of attack with contours. In order to examine dynamic stall delay quantitatively, the pressure coefficient and the wall shear stress around the airfoil were recorded at each time step. The separation points appeared when the wall shear stress abruptly became zero or the pressure signal showed a plateau. Figure 10 shows the position, i.e., x/c, from the leading edge of the separation point at each angle of attack. The delay to stall angle is observed clearly here, as well.

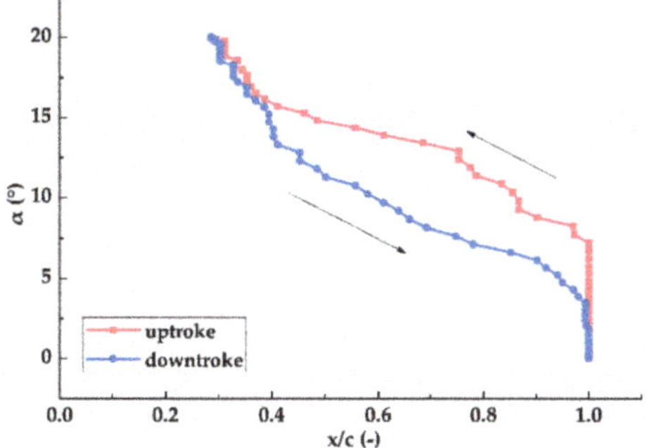

Figure 10. Positions of the separation point during dynamic stall with $k = 0.052$, $\alpha_0 = 10°$, $\alpha_{amp} = 10°$.

3.3. Frequency Spectrum Analysis of the Pressure Coefficient

The above pressure contours and streamlines can only be observed from a qualitative point of view. This requires multiple consecutive figures to obtain the characteristics of multiple cycles. Instead, spectrum analysis can capture the features of shedding vortices around the airfoil quickly and quantitatively. Spectrum analysis is a technique that converts a complex signal into a set of simple signals with different frequencies. It is a way to find the information (e.g., amplitude, power, intensity or phase, etc.) of a signal at different frequencies. In this paper, nine points are set on the suction surface of the airfoil to collect time series of pressure signals. The positions of the points are shown in Figure 11.

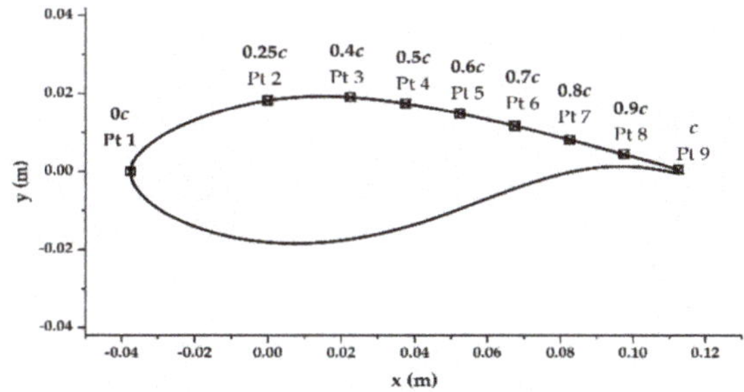

Figure 11. Positions of points to gather the history of pressure signals.

The fast fourier transform (FFT) of the pressure signals was performed with the software MATLAB under different calculations. The spectra of pressure signals were generated by two continuous period signals. Figure 12 shows an example of the time series signals and energy spectrum at Point 6. Figure 12b is the result of Figure 12a after FFT. It shows that the grid in the boundary layer was able to capture small-scale vortices properly, with the turbulence model in the present study.

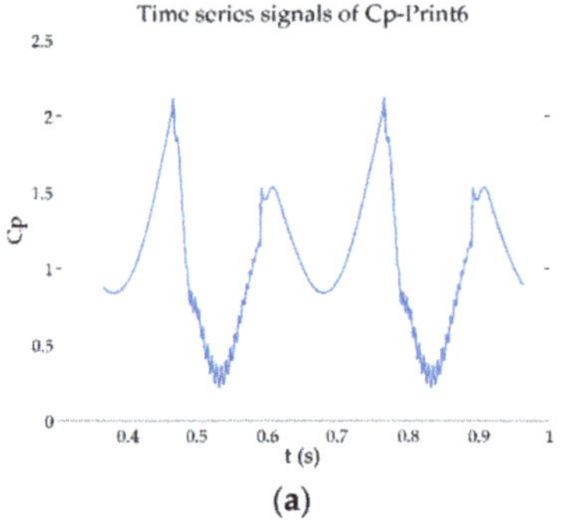
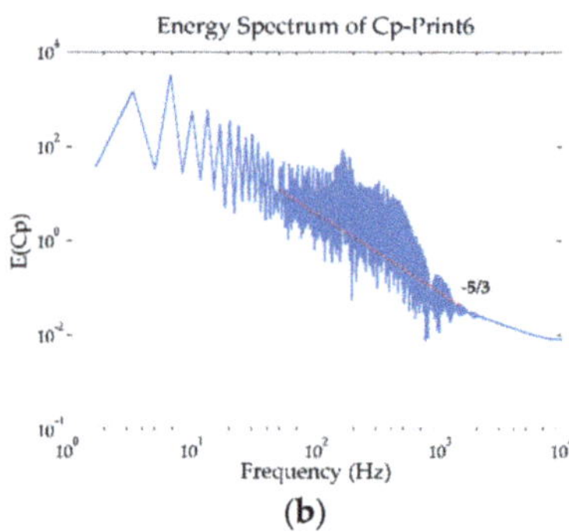

Figure 12. (a) time series signals and (b) energy spectrum of the pressure coefficient of Point 9 for two periods with $k = 0.052$, $= 10°$, $\alpha_{amp} = 10°$.

In order to show the differences between each frequency more clearly, the spectra of surface pressure at all nine points on one airfoil are shown in Figure 13 with the semi-log-x function. The abscissa value at the first peak corresponds to the pitch oscillation frequency $f = 3.3333$, which can be computed by Equation (4):

$$f = \frac{1}{T} = \frac{1}{2\pi/\omega} - \frac{\omega}{2\pi} - \frac{2U_\infty k}{2\pi c} = \frac{U_\infty k}{\pi c} \tag{4}$$

The abscissa value at the second peak corresponds to the vortices' shedding mode. The flow field around the pitching airfoil consists of large-scale motions and small-scale turbulence motions. The large-scale one has the same time scale as the unsteady pitch oscillations.

By observing the spectra of pressure at the different points on the airfoil in the same case, i.e., $k = 0.052$, $\alpha_0 = 10°$, $\alpha_{amp} = 10°$, it can be seen that most of the energy of the leading edge point, Pt, is concentrated in the first order frequency pitch oscillation. This is because that the pressure signals of the leading edge point basically were affected by the airfoil pitch oscillation and unaffected by shedding vortices. With the position of the points moving backward, the energy of pressure at the second order frequency became higher. This means that the pressure signals begin to be affected not only by the airfoil pitch oscillation, but also by the shedding of vortices. Due to the proximity of the location, the features of Pt 4 and Pt 5 were similar. This was the same for Pt 6 and Pt 7. It is worth noting that the energy of pressure at high frequency $f_v = 163.333$ at the trailing edge point Pt 9 was much higher than that of other points. This means that there were severe high-frequency fluctuations in the pressure signal at the trailing edge during pitching motion cycles.

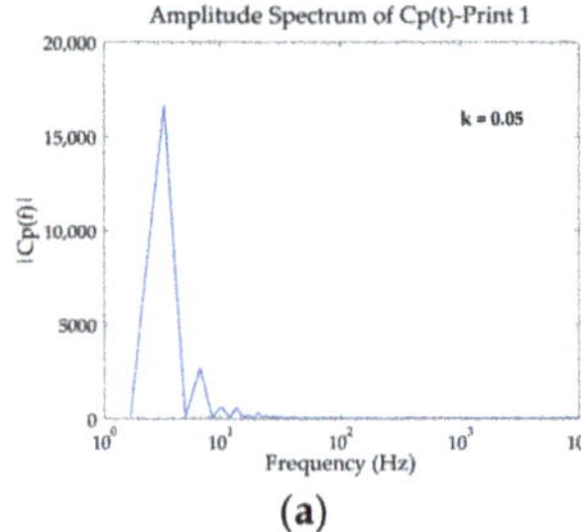
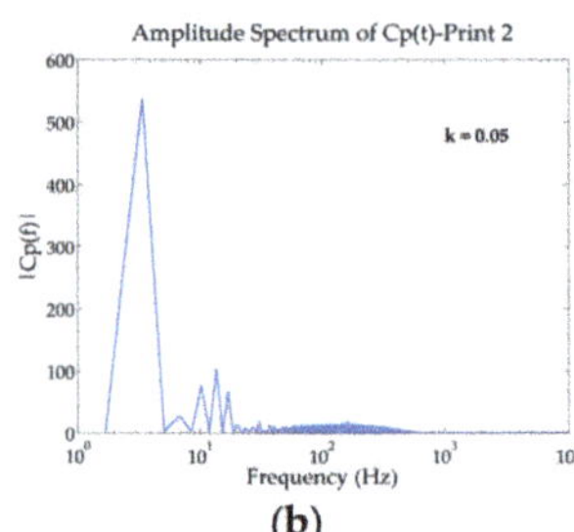
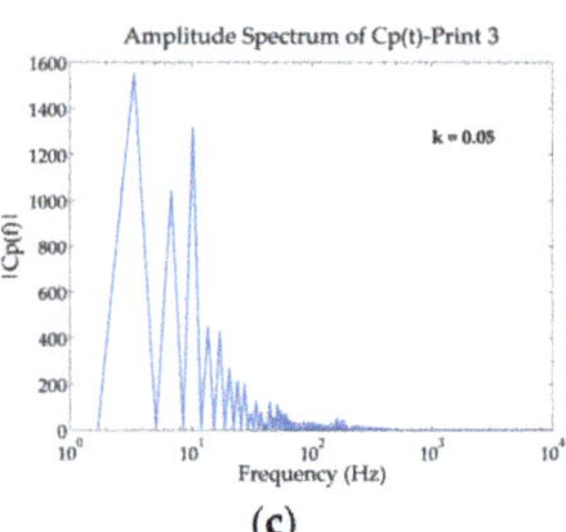

Figure 13. *Cont.*

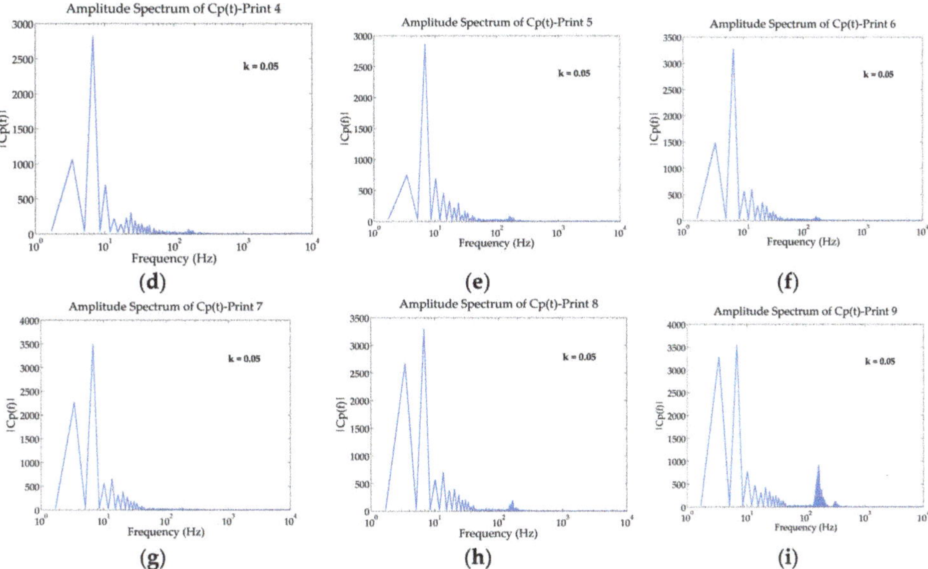

Figure 13. (a–i) Energy spectra of the pressure coefficient at different points on the suction surface of the airfoil in the semi-log-x coordinate system with $k = 0.052$, $\alpha_0 = 10°$, $\alpha_{amp} = 10°$.

By comparing the pressure spectra of the cases with different reduced frequencies, as shown in Figure 14, it can clearly be seen that the vortex shedding frequency increased with the increase of the reduced frequency. At the same time, pressure fluctuations at each point on the airfoil showed different characteristics dependent on reduced frequencies. At the leading edge of the airfoil, the proportion of energy for pressure at the first order of frequency became lower with the increasingly-reduced frequency. This shows that the proportion of pressure variation affected by pitch oscillation and shedding vortex changed. This implies that the increasingly-reduced frequency intensified the unsteady effects. Instead, the change of this proportion was reversed at the trailing edge point. Pitch oscillation played a dominant role in pressure fluctuations with increasingly-reduced frequency at the trailing edge point. The pitch oscillation dominated the fluctuations of pressure signals with $k = 0.1$, but did not with $k = 0.05$. This phenomenon provides new evidence of J.G. Leishman's conclusion that when k is smaller than 0.5, the flow can be considered as quasi-steady [1]. Besides the high-frequency fluctuations in the pressure signal, at the trailing edge during pitching motion, cycles had higher energy, but lower frequency with increasingly-reduced frequency; see Figure 14i.

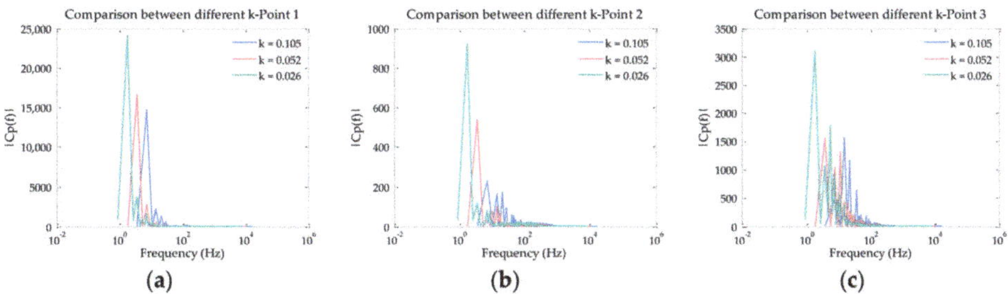

Figure 14. *Cont.*

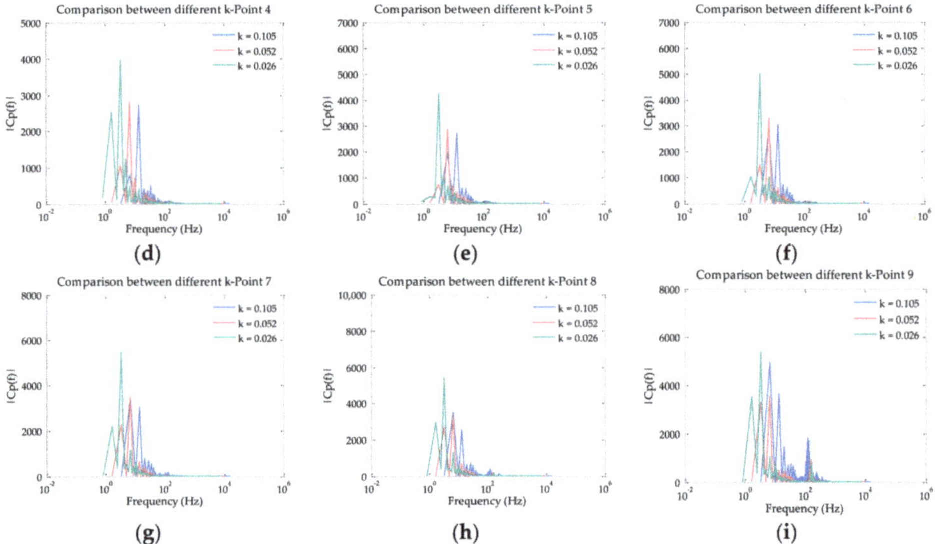

Figure 14. (a–i) Comparison of the amplitude spectrum of the pressure coefficient at typical points with different reduced frequencies *k*.

4. Conclusions

In this paper, in order to better understand the effects of dynamic stall on the aerodynamic characteristics, a CFD method was employed to investigate the influence of the pitch oscillations of the DU 91-W2-250 airfoil. Cases with different reduced frequency were studied by comparing aerodynamic forces and the lift-to-drag ratio. In addition, the hysteresis effect of dynamic stall was analyzed from the perspective of two aspects: the pressure field and spectrum analysis.

The effects of reduced frequency on dynamic stall were analyzed qualitatively and quantitatively. Aerodynamic forces and the lift-to-drag ratio during pitching motion were compared. During the hysteresis loop, the aerodynamic efficiency of the upstroke was lower than that of the downstroke. The effect of delay was more significant with the increased reduction of the frequency. Increasingly-reduced frequency led to lower aerodynamic efficiency during the upstroke process of pitching motions. The result that increasingly-reduced frequency aggravates dynamic stall effects was the same as the conclusions of other observations. Flow field structures and positions of separation points were used to explain the delay of lift stall from other aspects. There were a clockwise-rotating vortex and a counterclockwise-rotating vortex near the trailing edge of airfoil during pitching motions. The growth of the clockwise-rotating vortex increased the value of the lift coefficient, while the growth of the counterclockwise-rotating vortex decreased the value of the lift coefficient.

Frequency spectrum analysis of the pressure coefficient was used as a new method to find the relationship of shedding vortices and pitching motion. The vortex shedding frequency increased with the increase of the reduced frequency. The fluctuations of the pressure signals were the results of pitching motion and vortex shedding. The dominant factor of the pressure signal fluctuations was different at the leading edge and trailing edge. It was also different with different reduced frequencies. Pitch oscillation played a dominant role in pressure fluctuations with increasingly-reduced frequency at the trailing edge point.

In this paper, the dynamic stall characteristics of the DU 91-W2-250 airfoil at $Re = 3 \times 10^5$ were analyzed systematically. However, due to the large amount of computing resources and time cost, many problems need to be further studied in the project. For example, a conclusion of this paper was that increasingly-reduced frequency caused angles of attack of stall to increase. However, it is not known yet when will this trend would stop, i.e., what is the biggest reduced frequency to maintain this trend. In addition, dynamic experiments of the DU series airfoil should be carried out as soon as possible to obtain dynamic data for the purpose of verification.

Author Contributions: S.L. carried out numerical calculation, analyzed the data, and wrote the paper. L.Z. supervised the work, provided theoretical guidance and experimental data. K.Y. provided subject inspiration and writing comments. J.X. and X.L. provided writing help.

Funding: This research was funded by the National Natural Science Foundation of China (No.51776204).

Acknowledgments: The authors are grateful to Xingxing Li for several helpful discussions and suggestions.

Conflicts of Interest: The authors declare no conflict of interest.

Nomenclature/Abbreviations

c	chord of airfoil (m)
C_f	skin friction coefficient, $\tau_w/0.5\rho U_\infty^2$
C_l	lift coefficient, $L/0.5\rho U_\infty^2 c$
C_p	pressure coefficient, $(p - p_\infty)/0.5\rho U_\infty^2$
f	pitch oscillation frequency (Hz)
f_v	frequency of shedding vortices
k	reduced frequency, $\omega c/2U_\infty$
L	lift force (N)
p	pressure at one point (Pa)
t	time (s)
T	time period (s)
T_c	characteristic time (s)
U	velocity (m/s)
U_∞	free-stream velocity (m/s)
y	height of the first grid in the boundary layer of the airfoil (m)
y^+	dimensionless wall distance
α	angle of attack (deg)
α_t	angle of attack at flow time t (deg)
α_0	mean angle of attack (deg)
α_{amp}	amplitude of angle of attack (deg)
Δt	time step (s)
ρ	density (kg/m^3)
τ_{wx}	wall shear stress in the x direction (Pa)
ω	angular frequency (rad/s)
2D	two-dimensional
CFD	computational fluid dynamics
CFL	Courant, Friedrichs and Levy criterion
DES	detached eddy simulation
DNS	direct numerical simulation
dt1, dt2 ...	time step size 1, time step size 2 ...
DU	Delft University of Technology
FFT	fast fourier transform
G1, G2 ...	gird case 1, grid case 2 ...
LEV	leading edge vortex
N-S	Navier-Stokes
Pt1, Pt2 ...	point1, point2 ...
RANS	Reynolds-averaged Navier–Stokes
RNG	renormalization-group
SST	shear stress transport
UDF	user-defined function
URANS	unsteady Reynolds-averaged Navier–Stokes

References

1. Leishman, J.G. *Principles of Helicopter Aerodynamics*; Cambridge University Press: Cambridge, UK, 2006; p. 78, ISBN 9781107013353.
2. Holierhoek, J.; De Vaal, J.; Van Zuijlen, A.; Bijl, H. Comparing different dynamic stall models. *Wind Energy* **2013**, *16*, 139–158. [CrossRef]
3. Choudhry, A.; Leknys, R.; Arjomandi, M.; Kelso, R. An insight into the dynamic stall lift characteristics. *Exp. Therm. Fluid Sci.* **2014**, *58*, 188–208. [CrossRef]
4. Xiong, L.; Xianmin, Z.; Gangqiang, L.; Yan, C.; Zhiquan, Y. Dynamic response analysis of the rotating blade of horizontal axis wind turbine. *Wind Eng.* **2010**, *34*, 543–559. [CrossRef]
5. Ham, N.D. Aerodynamic loading on a two-dimensional airfoil during dynamic stall. *AIAA J.* **1968**, *6*, 1927–1934. [CrossRef]
6. Galvanetto, U.; Peiró, J.; Chantharasenawong, C. An assessment of some effects of the nonsmoothness of the leishman–beddoes dynamic stall model on the nonlinear dynamics of a typical aerofoil section. *J. Fluids Struct.* **2008**, *24*, 151–163. [CrossRef]
7. Tarzanin, F. Prediction of control loads due to blade stall. *J. Am. Helicopter Soc.* **1972**, *17*, 33–46. [CrossRef]
8. Tran, C.; Petot, D. Semi-Empirical Model for the Dynamic Stall of Airfoils in View of the Application to the Calculation of Responses of a Helicopter Blade in Forward Flight. *Vertica* **1981**, *5*, 35–53.
9. Leishman, J.G.; Beddoes, T. A semi-empirical model for dynamic stall. *J. Am. Helicopter Soc.* **1989**, *34*, 3–17. [CrossRef]
10. Larsen, J.W.; Nielsen, S.R.; Krenk, S. Dynamic stall model for wind turbine airfoils. *J. Fluids Struct.* **2007**, *23*, 959–982. [CrossRef]
11. Hansen, M.H.; Gaunaa, M.; Madsen, H.A. *A Beddoes–Leishman Type Dynamic Stall Model in State-Space and Indicial Formulation*; Technical University of Denmark: Lyngby, Denmark, 2004; ISBN 8755030890.
12. Gupta, S.; Leishman, J.G. Dynamic stall modelling of the s809 aerofoil and comparison with experiments. *Wind Energy* **2010**, *9*, 521–547. [CrossRef]
13. Sheng, W.; Galbraith, R.A.M.; Coton, F.N. A modified dynamic stall model for low mach numbers. *J. Sol. Energy Eng.* **2008**, *130*, 653. [CrossRef]
14. Martinat, G.; Braza, M.; Harran, G.; Sevrain, A.; Tzabiras, G.; Hoarau, Y.; Favier, D. *Dynamic Stall of a Pitching and Horizontally Oscillating Airfoil*; Springer: Berlin, Germany, 2009; pp. 395–403.
15. Wang, S.; Ingham, D.B.; Ma, L.; Pourkashanian, M.; Tao, Z. Numerical investigations on dynamic stall of low reynolds number flow around oscillating airfoils. *Comput. Fluids* **2010**, *39*, 1529–1541. [CrossRef]
16. Akbari, M.H.; Price, S.J. Simulation of dynamic stall for a naca 0012 airfoil using a vortex method. *J. Fluids Struct.* **2003**, *17*, 855–874. [CrossRef]
17. Gharali, K.; Johnson, D.A. Dynamic stall simulation of a pitching airfoil under unsteady freestream velocity. *J. Fluids Struct.* **2013**, *42*, 228–244. [CrossRef]
18. Kim, Y.; Xie, Z.T. Modelling the effect of freestream turbulence on dynamic stall of wind turbine blades. *Comput. Fluids* **2016**, *129*, 53–66. [CrossRef]
19. Gandhi, A.; Merrill, B.; Peet, Y.T. Effect of Reduced Frequency on Dynamic Stall of a Pitching Airfoil in a Turbulent Wake. In Proceedings of the 55th AIAA Aerospace Sciences Meeting, Grapevine, TX, USA, 9–13 January 2017; p. 0720. [CrossRef]
20. Martin, J.; Empey, R.; McCroskey, W.; Caradonna, F. An experimental analysis of dynamic stall on an oscillating airfoil. *J. Am. Helicopter Soc.* **1974**, *19*, 26–32. [CrossRef]
21. Mccroskey, W.J. Unsteady airfoils. *Annu. Rev. Fluid Mech.* **2003**, *14*, 285–311. [CrossRef]
22. McCroskey, W.J. *The Phenomenon of Dynamic Stall*; NASA Ames Research Center: Mountain View, CA, USA, 1981. [CrossRef]
23. Wernert, P.; Geissler, W.; Raffel, M.; Kompenhans, J. Experimental and numerical investigations of dynamic stall on a pitching airfoil. *AIAA J.* **1996**, *34*, 982–989. [CrossRef]
24. Aramendia, I.; Fernandez-Gamiz, U.; Ramos-Hernanz, J.A.; Sancho, J.; Lopez-Guede, J.M.; Zulueta, E. Flow control devices for wind turbines. In *Energy Harvesting and Energy Efficiency*; Springer: Berlin, Germany, 2017; pp. 629–655.
25. Johnson, S.J.; Baker, J.P.; Van Dam, C.P.; Berg, D. An overview of active load control techniques for wind turbines with an emphasis on microtabs. *Wind Energy* **2010**, *13*, 239–253. [CrossRef]

26. Zhang, L.; Li, X.; Yang, K.; Xue, D. Effects of vortex generators on aerodynamic performance of thick wind turbine airfoils. *J. Wind Eng. Ind. Aerodyn.* **2016**, *156*, 84–92. [CrossRef]

27. Yang, K.; Zhang, L.; Jianzhong, X.U. Simulation of aerodynamic performance affected by vortex generators on blunt trailing-edge airfoils. *Sci. China Technol. Sci.* **2010**, *53*, 1–7. [CrossRef]

28. Tsai, K.-C.; Pan, C.-T.; Cooperman, A.M.; Johnson, S.J.; Van Dam, C. An innovative design of a microtab deployment mechanism for active aerodynamic load control. *Energies* **2015**, *8*, 5885–5897. [CrossRef]

29. Fernandez-Gamiz, U.; Zulueta, E.; Boyano, A.; Ramos-Hernanz, J.A.; Lopez-Guede, J.M. Microtab design and implementation on a 5 mw wind turbine. *Appl. Sci.* **2017**, *7*, 536. [CrossRef]

30. Barlas, T.K.; Van Kuik, G. Review of state of the art in smart rotor control research for wind turbines. *Prog. Aerosp. Sci.* **2010**, *46*, 1–27. [CrossRef]

31. Fernandez-Gamiz, U.; Zulueta, E.; Boyano, A.; Ansoategui, I.; Uriarte, I. Five megawatt wind turbine power output improvements by passive flow control devices. *Energies* **2017**, *10*, 742. [CrossRef]

32. Timmer, W.; Van Rooij, R. Summary of the delft university wind turbine dedicated airfoils. *J. Sol. Energy Eng.* **2003**, *125*, 488–496. [CrossRef]

33. Bai, J.Y.; Zhang, L.; Xingxing, L.I.; Yang, K. Analyzing the effect of wind tunnel wall on the aerodynamic performance of airfoils. *Sci. Sin.* **2016**, *46*, 124707. (In Chinese)

34. Poirel, D.; Métivier, V.; Dumas, G. Computational aeroelastic simulations of self-sustained pitch oscillations of a naca0012 at transitional reynolds numbers. *J. Fluids Struct.* **2011**, *27*, 1262–1277. [CrossRef]

35. Blazek, J. *Computational Fluid Dynamics: Principles and Applications*, 2nd ed.; Elsevier: New York, NY, USA, 2005; pp. 1–4.

36. Hand, B.; Kelly, G.; Cashman, A. Numerical simulation of a vertical axis wind turbine airfoil experiencing dynamic stall at high reynolds numbers. *Comput. Fluids* **2017**, *149*, 12–30. [CrossRef]

37. Liu, X.; Lu, C.; Liang, S.; Godbole, A.; Chen, Y. Vibration-induced aerodynamic loads on large horizontal axis wind turbine blades. *Appl. Energy* **2017**, *185*, 1109–1119. [CrossRef]

38. Wang, S.; Ingham, D.B.; Ma, L.; Pourkashanian, M.; Tao, Z. Turbulence modeling of deep dynamic stall at relatively low reynolds number. *J. Fluids Struct.* **2012**, *33*, 191–209. [CrossRef]

39. Gharali, K.; Johnson, D.A. Numerical modeling of an s809 airfoil under dynamic stall, erosion and high reduced frequencies. *Appl. Energy* **2012**, *93*, 45–52. [CrossRef]

40. Karbasian, H.R.; Esfahani, J.A.; Barati, E. Effect of acceleration on dynamic stall of airfoil in unsteady operating conditions. *Wind Energy* **2016**, *19*, 17–33. [CrossRef]

41. *Ansys Fluent 14 User's Guide*; Ansys Inc.: Canonsburg, PA, USA, 2011.

42. Durbin, P.A.; Reif, B.P. *Statistical Theory and Modeling for Turbulent Flows*; John Wiley & Sons: Hoboken, NJ, USA, 2011; ISBN 1119957524.

43. White, F.M. *Fluid mechanics*, 7th ed.; McGraw-Hill: New York, NY, USA, 2011; ISBN 978-0-07-352934-9.

44. Raffel, M.; Favier, D.; Berton, E.; Rondot, C.; Nsimba, M.; Geissler, W. Micro-piv and eldv wind tunnel investigations of the laminar separation bubble above a helicopter blade tip. *Meas. Sci. Technol.* **2005**, *17*, 1–13. [CrossRef]

45. Barakos, G.N.; Drikakis, D. Computational study of unsteady turbulent flows around oscillating and ramping aerofoils. *Int. J. Numer. Methods Fluids* **2010**, *42*, 163–186. [CrossRef]

46. Martinat, G.; Braza, M.; Hoarau, Y.; Harran, G. Turbulence modelling of the flow past a pitching naca0012 airfoil at 10^5 and 10^6 reynolds numbers. *J. Fluids Struct.* **2008**, *24*, 1294–1303. [CrossRef]

Article

Impact of Economic Indicators on the Integrated Design of Wind Turbine Systems

Jianghai Wu, Tongguang Wang *, Long Wang and Ning Zhao

Jiangsu Key Laboratory of Hi-Tech Research for Wind Turbine Design, Nanjing University of Aeronautics and Astronautics, Nanjing 210016, China; jianghaiwu@nuaa.edu.cn (J.W.); longwang@nuaa.edu.cn (L.W.); zhaoam@nuaa.edu.cn (N.Z.)

* Correspondence: tgwang@nuaa.edu.cn; Tel.: +86-139-1391-7623

Received: 8 July 2018; Accepted: 8 September 2018; Published: 15 September 2018

Abstract: This article presents a framework to integrate and optimize the design of large-scale wind turbines. Annual energy production, load analysis, the structural design of components and the wind farm operation model are coupled to perform a system-level nonlinear optimization. As well as the commonly used design objective levelized cost of energy (LCoE), key metrics of engineering economics such as net present value (NPV), internal rate of return (IRR) and the discounted payback time (DPT) are calculated and used as design objectives, respectively. The results show that IRR and DPT have the same effect as LCoE since they all lead to minimization of the ratio of the capital expenditure to the energy production. Meanwhile, the optimization for NPV tends to maximize the margin between incomes and costs. These two types of economic metrics provide the minimal blade length and maximal blade length of an optimal blade for a target wind turbine at a given wind farm. The turbine properties with respect to the blade length and tower height are also examined. The blade obtained with economic optimization objectives has a much larger relative thickness and smaller chord distributions than that obtained for high aerodynamic performance design. Furthermore, the use of cost control objectives in optimization is crucial in improving the economic efficiency of wind turbines and sacrificing some aerodynamic performance can bring significant reductions in design loads and turbine costs.

Keywords: wind turbine; wind turbine design; optimization; blade length; economic analysis

---

1. Introduction

Wind turbine design is a complex task comprising multiple disciplines, requiring a trade-off between many conflicting objectives. Many research articles have been published to achieve an optimal turbine design. Some use a single objective such as maximum annual energy production (AEP) or maximum AEP per turbine weight to carry out a single-objective optimization [1]. Others use multi-objective methods [2,3] or a multi-level system design [4] to accomplish a balance between different conflicting objectives, often drawn from different scientific and economic disciplines. The objective functions of wind turbine design can be divided in four main categories: Maximization of the energy production, minimization of the blade mass, minimization of the cost of energy, and multi-objective optimization [5].

Multi-objective optimization offers a set of Pareto Optimal design solutions and places the burden of choice on the shoulders of the decision maker. In contrast, a single objective lumps all different objectives into one and provides a unique design result to the decision maker, which seems to be more practical. The difficulty is that the objective must reflect the nature of the problem. Inappropriate design goal will lead to unfeasible results. For example, maximization of the power coefficient results in larger root chords and a very high blade twist [5], while minimization of mass/AEP may overemphasize the role of the tower [6]. Hence, the selection of objective is crucial in wind turbine design.

It is noticed that the development of a wind farm is essentially an investment activity and wind turbine design is an upstream stage of wind farm development. In this case, turbine optimization also needs to be performed from an engineering economics point of view. Many economic functions such as levelized cost of energy (LCoE), net present value (NPV), internal rate of return (IRR) and discounted payback time (DPT) have been used to evaluate the profitability of wind farms [7,8]. The LCoE represents the minimum energy price that meets the desired interest rate by the designers, NPV defines the total profit of the wind farm and takes into account the price of energy, IRR is the interest rate that sets the NPV function equal to zero and enables checking if a minimum rate of return set by the designers is met, and DPT determines the time required to cover the initial investment while taking into account the time value of the money [9,10]. In these economic indicators, LCoE is the most common one used as an objective in wind turbine design [11–13]. But sometimes LCoE alone is not a sufficient measure to determine a project's profitability or competitiveness. Investors need other parameters as inputs to make investment decisions [10]. However, there has been no research published using other economic indicators as objectives to perform a wind turbine design. Therefore, besides LCoE, economic functions like NPV, IRR and DPT will also be utilized as design objectives in this study. Insights into the differences between these metrics and the impact of these economic functions when used as optimization objectives will be assessed.

To carry out this research, a system-level optimization framework is used to perform an integrated design of a wind turbine. Blade length, the geometry of the blade and the tower height are selected as design variables. The purpose of this study is not to develop a design methodology, or present exact parameters for an optimized wind turbine, but rather to assess the effect of different objective functions on the properties of optimal wind turbines.

2. Methodology

2.1. Calculation Tools

Recent developments in the methodology of wind turbine design have mainly focused on simultaneous evaluation of aerodynamic and structural design [14–17]. In this paper a series of automated calculation models are coupled to perform an integrated optimization. The integrated methodology mainly builds upon that previously described in References [6,16], which includes the rotor aerodynamic analysis, blade structure design and cost model. These models are selected as a compromise between computational effort and calculation accuracy. The integrated design strategy has been successfully used in the design of a 1.5 MW stall regulated rotor [14], BONUS 1 MW and WM 600 wind turbine [18], and the investigation of 5 MW upwind and downwind turbines [16], proved to be effective in the wind turbine optimization.

In the next subsection we describe how the integrated modelling tool is implemented, giving details of how the models are coupled and how the optimization process is sequenced.

2.1.1. Power Production

Aerodynamic analysis is based on ordinary blade element/momentum (BEM) theory. A detailed description of the process with an extensive explanation of its fundamental equations can be found in Reference [19]. AEP is computed using a Rayleigh distribution in which mean speed varies with the hub height. An exponential law is used to calculate the average wind speed at different hub heights as follows:

$$V(h) = V(h_0)\left(\frac{h}{h_0}\right)^{\alpha} \tag{1}$$

where h_0 is the reference height, $V(h_0)$ is the average wind speed at h_0 and α is the wind shear exponent.

2.1.2. Loading

Load calculation is an important discipline within wind turbine technology. Design loads serve as inputs for the structural design of rotor blades and other turbine components, and essentially determine the cost of producing a wind turbine. Typically, load analysis is performed using dynamic simulation software such as "DNV Bladed" according to the IEC61400-1 standard [20] during the real-world wind turbine design process. However, this load estimation procedure is difficult to carry out in the optimization period for the following reasons:

1. For an accurate estimation of design loads, there are thousands of design load cases (DLCs) which must be simulated, causing a significant computational cost. Design optimization of a wind turbine requires estimations of hundreds of thousands of different turbine configurations. For this reason, a full IEC loads analysis is not computationally feasible.
2. The output loads are influenced by the controller algorithm. For example, coefficients in the PID controller and pitch rates under different operational conditions can have a significant effect. Typically, we must modify many parameters of the controller to find the best controller algorithm and configuration for a given wind turbine. This is another computationally expensive aspect which limits the usefulness of automated design load analysis.

Some approaches have been described to deal with this problem. For example, some essential load cases can be selected to reduce the computational cost [18]. It is also possible to use the static forces of rotor and tower at some critical conditions, with an amplification factor to correct for dynamic effects [16]. In accordance with the latter approach, we explored a series of load calculation results from commercial wind turbines, correlated the conditions a wind turbine experiences when ultimate loads occur (rotor speed, wind speed, pitch angle, azimuth angle, etc.), and then identified the appropriate static load cases (SLCs) to perform the loads estimation, which are summarized in Table 1.

Table 1. Static conditions for load estimation.

Static Load Cases	Wind Speed (m/s)	Rotor Speed (rpm)	Pitch Angle (deg)	Yaw Angle (deg)	Azimuth Angle (deg)
SLC 1.1	$V_r + 3\sigma$	$1.1\,\omega_r$	0~10	−8~+8	0~90
SLC 1.2	$V_{out} + 3\sigma$	ω_r	10~20	−8~+8	0~90
SLC 2.1	V_{e1}	0	90	90,270	0
SLC 2.2	V_{e50}	0	90	30,330	0

Note: V_r: rated wind speed; V_{out}: cut-out wind speed; σ: standard deviation of the wind speed, relates to short term turbulence; ω_r: rated rotor speed; V_{e1}: 10-min average extreme wind speed with a recurrence period of 1 year; Ve50: 10-min average extreme wind speed with a recurrence period of 50 years.

SLC 1.1 represents an extremely gusty wind condition. The ultimate loads on a wind turbine are more likely to happen in gusty conditions near the rated wind speed, typically due to a rapid increase of wind speed which the pitch angle can't catch up with. SLC 1.2 examines the operational condition IEC DLC 1.3, with an extreme turbulence model. In this case the maximum wind speed will increase to a very high value in a short period due to extreme turbulence at the cut-out wind speed. SLC 2.1 and 2.2 represent the parked survival cases of wind turbines under extreme conditions which occur every 1 year and every 50 years, respectively.

The predicted static thrust from the blade requires a dynamic amplification factor of 1.45 to fall within the range of dynamic loadings from "DNV Bladed" simulations. To check the feasibility of this assumption, we carried out several load estimations for different wind turbines with various blade lengths. Figure 1 shows the comparison of blade root flap-wise moments (Mf) between the SLCs and a more rigorous DLCs calculation procedure.

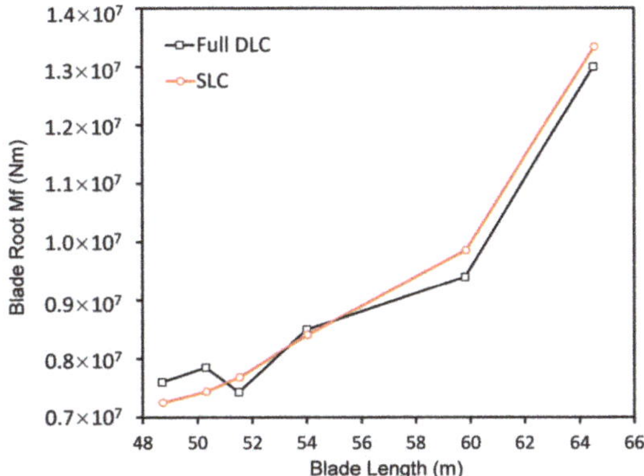

Figure 1. Blade root flap-wise moments of SLC and full DLC calculations.

Although there is some deviation between the calculated loads, the errors are within an acceptable range. Usually the output loads on a turbine blade will increase with the blade length, however, the load evaluation of real-world turbines shows that sometimes longer blades can have lower output loads. This is mostly due to variations in the turbine configuration and control algorithms. If this influence is eliminated, we can expect that the error between the load trends in the DLC and SLC calculations would be reduced. Moreover, since the load calculations as well as the cost model described in the next section inevitably need to be simplified, the primary goal of this study is not to present an exact geometry for an optimized wind turbine, but to assess the effect of different objective functions on the properties of optimal wind turbines. For this purpose, the SLC method is more suitable.

2.1.3. Cost Model

Besides AEP, capital expenditure (CAPEX), operational expenditures (OPEX) and financing are needed as inputs to perform overall economic analysis. These are calculated using the models described in this section.

The CAPEX herein represents the overall cost of a single wind turbine. It includes turbine manufacturing cost and the sharing portion of wind farm construction and maintenance costs. The manufacturing cost uses the individual component masses for the turbine to evaluate the costs of individual components. The blade mass is derived by a structural model using beam finite element theory and classical laminate theory to determine the effect of the design loads [2]. The blade structure is formed from a double web I-beam. The material used for the bulk of the blade is glass fiber with reinforced polyester and foam.

A set of physics-based models are used to estimate the sizes and costs of a subset of the major load-bearing components (the main shaft, gear box, bedplate, gearbox, tower) and parametric formulations representative of current wind turbine technology are used to evaluate sizes and costs for the remaining components (the hub and yaw system). For example, the hub is treated as a thin-walled, ductile, cast iron cylinder with holes for blade root openings and main shaft flanges. The bedplate module is separated into two distinct front and rear components. The rear frame is modelled as two parallel steel I-beams and the front frame is modelled as two parallel ductile cast iron I-beams. The physics-based models have internal iteration schemes based on system constraints and design criteria, which are described in Reference [21].

The balance-of-station costs vary with the rotor diameter and tower height, and the OPEX are accounted for by taking a small, fixed percentage of the capital cost as the annual maintenance cost [18]. Decommissioning costs are also considered in this model.

2.2. Objective Function

The appropriate choice of an objective function is critical to an optimization process. In order to perform the wind turbine optimization from an economic point of view and ensure the designs correctly assess fundamental trade-offs (primarily between energy capture and overall cost) we chose LCoE, NPV, IRR and DPT as design metrics. The models employed in this research are proposed in Reference [10] and shown in Table 2. Dynamic evaluation of these economic functions can also be found in References [22,23].

Table 2. Economic functions for the wind turbine optimization problem.

Function	Equation	Parameters
LCoE ($/kWh)	$\frac{1}{AEP}\left(\frac{CAPEX}{a} + OPEX\right)$	a—annuity factor $\left(a = \frac{1-(1+r)^{-n}}{r}\right)$, r—interest rate, n—wind farm lifetime
NPV ($)	$(AEP \cdot p_{kWh} - OPEX)a - CAPEX$	p_{kWh}—market energy price
IRR (%)	$(AEP \cdot p_{kWh} - OPEX)\frac{1-(1+r_{IRR})^{-n}}{r_{IRR}} = CAPEX$	r_{IRR}—interest rate that zeroes the NPV equation
DPT (years)	$\frac{n \cdot CAPEX}{(AEP \cdot p_{kWh} - OPEX) \cdot a}$	

In the economic models, the total utilized energy output and the total cost over the lifetime of the wind turbine are both discounted to the start of operation.

2.3. Design Variables and Constraints

There are a total of 15 design variables used in the optimization process. B-Spline curves are used to create smooth distributions of chord, twist and relative thickness of the blade. Five control points are used to shape the chord distribution and a further four points are used to describe both twist and relative thickness distribution, as shown in Figure 2.

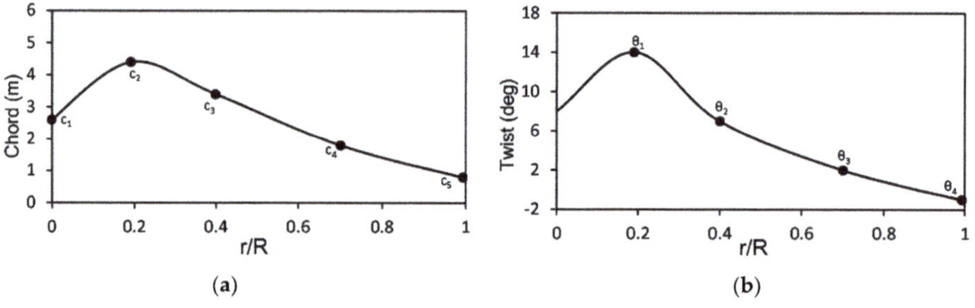

Figure 2. Design variables of blade geometry. (**a**) Chord distribution; (**b**) Twist distribution.

Blade length and tower height are also set as variables for the purpose of determining the most suitable rotor diameter and tower height for a specific wind speed site. Upper and lower limits for these variables are implemented to define the bounds of the design space.

The DU and NACA 64 series airfoils are employed in this study, as shown in Table 3. These airfoils are optimized for high speed wind condition with an advantage of high maximum lift coefficient and very low drag over a small range of operation condition. They have been used by various wind turbine manufacturers worldwide in over 10 different rotor blades for turbines with rotor diameters ranging from 29 m to over 100 m [24].

Blade structure design is set to meet the needs of strength and deflection constraints. The maximum velocity of the blade tip is 80 m/s. The rated rotational speed and the gearbox ratio change with the blade length in proportion to keep the generator operational speed the same. To have a safe blade-tower-clearance, a constraint for the blade out-of-plane deflection is used and the clearance between the ground and the blade tip is constrained by a minimum of 25 m [25].

Table 3. Airfoil family.

Airfoil Name	*t/c* Ratio
DU99-W-405	40%
DU99-W-350	35%
DU97-W-300	30%
DU91-W2-250	25%
DU93-W-210	21%
NACA 64-618	18%

To examine wind turbine designs at low speed sites, the average wind speed is set as 6 m/s at a reference height of 70 m. The wind shear exponent α is set to be 0.2 for normal wind profile model according to the IEC61400-1 standard [20]. The power transfer efficiency is 92% and losses such as those caused by wake interference, electrical grid unavailability and air density correction are estimated using an array loss factor and an availability factor. An overall correction factor of 0.64 is used in this research.

2.4. Genetic Algorithm

The genetic algorithm (GA) is a search procedure based on genetics and natural selection mechanisms, which has proved to be efficient and robust for wind turbine system [26], wind turbine layout in wind farms [27], and offshore wind turbine support structures [28].

In this study, multi island genetic algorithm (MIGA) is employed for the optimization. In MIGA, the population is divided into several subpopulations staying on isolated "islands," whereas traditional genetic algorithm operations are performed on each subpopulation separately. A certain number of individuals between the islands migrate after a certain number of generations. Thus, MIGA can prevent the problem of "premature" by maintaining the diversity of the population [29]. Table 4 presents the main parameters of MIGA.

Table 4. Main parameters of MIGA.

Item	Value
Sub-Population Size	10
Number of Islands	5
Number of Generations	150
Rate of Crossover	0.8
Rate of Mutation	0.01
Rate of Migration	0.3
Interval of Migration	5

3. Description of the Optimization Process

All sequential programming methods and the optimization processes are formulated in a framework. Figure 3 shows the full optimization methodology adopted in this study. Parameterization, geometry generation, AEP and load evaluation, structural design, mass and cost estimation, and economic analysis are presented in the flow chart.

Objectives such as LCoE and NPV are specified at the beginning of the optimization process. The AEP and design loads are evaluated by BEM theory, followed by calculation of the component masses and costs with the consideration of the design loads. The turbine capital cost and AEP together with the wind farm operational model are used to obtain the economic characteristic during the whole wind turbine life cycle as the final design target. The variables are changed in the optimization depending on the design objective. The maximum number of iterations is the stopping criterion. When stopping criterion is reached, the design result is picked out by the algorithm.

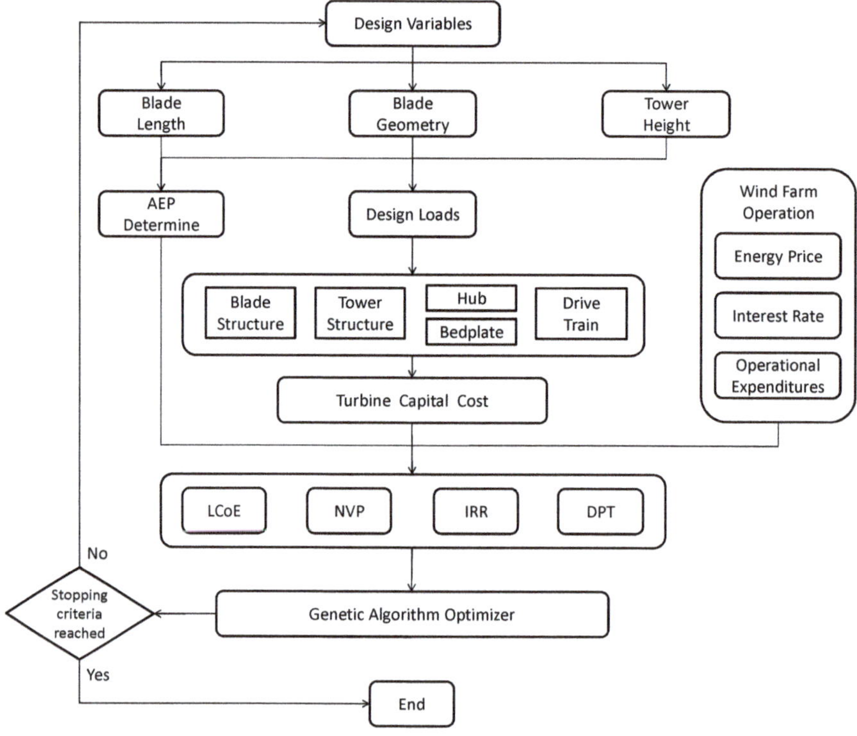

Figure 3. Flow chart of the optimization methodology.

4. Results and Discussion

4.1. 2 MW Case Study

Recently, increasing numbers of low wind speed sites have been developed, and the 2 MW turbine is dominating the market (In 2017, the average rated power of the new installed wind turbine in China is 2.1 MW [30] and China alone accounted for 37% of the world's new installed capacity [31]). Thus, this paper will focus on the optimization of 2 MW low wind speed turbines to capture the industrial trends of the state-of-the-art wind farm.

As the market energy price, labor costs, material costs, etc. vary between countries, the parameters considered for this case study are mainly derived from the industry's status in China. For example, the NPV is calculated with a market energy price of 8.8 cents/kWh, the weight unit price of the blade is set to be $5.88/kg and $1.54/kg for the tower. A discount rate of 8% and an economic lifetime of 20 years are employed in this study.

The results of the optimization are presented in the following sections. Optimization outputs using different design targets are examined. Fundamental differences and observations from each case are discussed below.

4.2. Effects of Blade Length

Increasing blade length will increase energy production, but it will also increase the costs of material, labor and delivery. Generally, the initial capital cost and maintenance cost will also increase.

To analyze how blade length impacts the economic effectiveness of a design, we recorded the trends in energy production, loading, mass, and costs from the properties of all the intermediate cases during the optimization. These data points are results for all metrics put together, show clearly the relationships between various turbine properties and the blade length.

The annual capacity factor is commonly used as an indicator of energy performance. It is defined as the energy generated during the year divided by wind turbine rated power multiplied by the number of hours in the year. Figure 4 shows the capacity factor against blade length. Each point in the scatter plot represents an individual turbine configuration evaluated during the genetic algorithm. Although wind energy is proportional to the square of blade length, due to the limitation of generator capacity, the output electrical power is not permitted to exceed the rated power beyond the rated wind speed. The total energy production increases with the blade length while the effect is reduced as the blade length increases. The capital expenditure also increases with blade length, however in this case the effect increases as blade length increases (Figure 5).

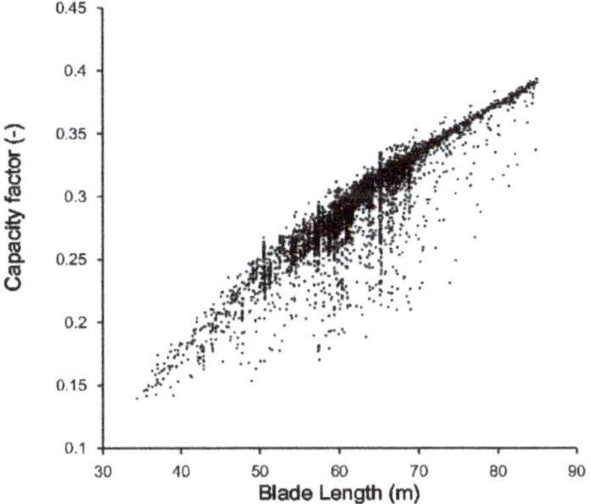

Figure 4. Equivalent full-loaded hours against blade length.

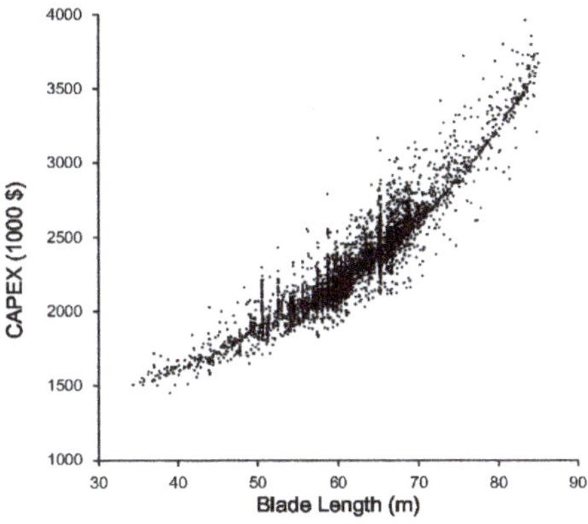

Figure 5. Initial capital cost against blade length.

The same comparison was made for blade length against blade root loading and against blade mass, as shown in Figures 6 and 7 respectively. Power law relationships are fitted to this data to quantify the correlations. The flap-wise bending moment (Mf) is the critical load in blade structure design. This research shows that the flap-wise bending moment of blade root increases with length, L, as $L^{2.84}$, while the mass of the blades scales with $L^{2.52}$. The enveloping functions on the outer border of the scattered points are also presented to indicate the boundary of the design results.

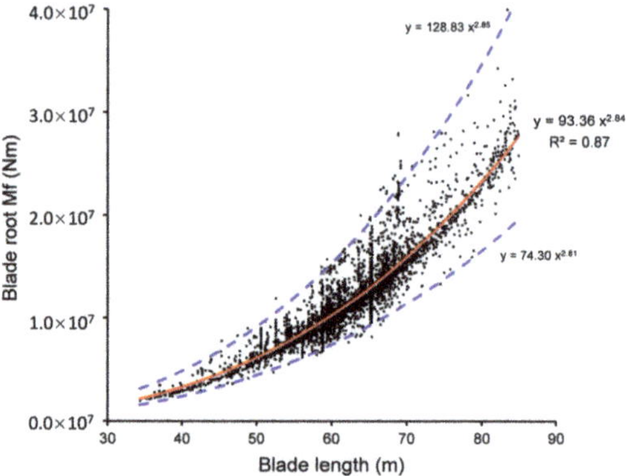

Figure 6. Blade root Mf against blade length.

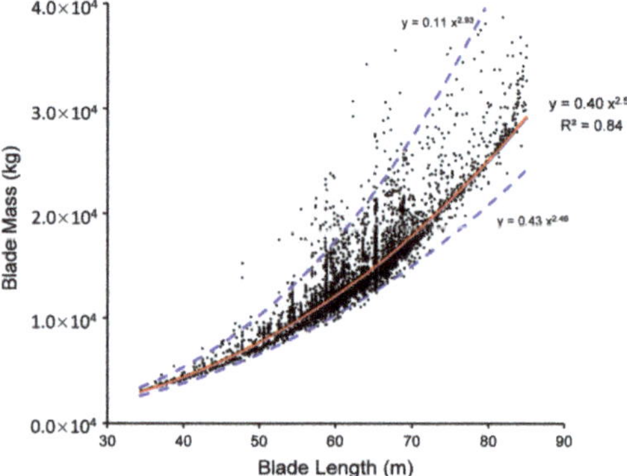

Figure 7. Blade mass against blade length.

Torsional moment (Mx) under fixed hub coordinates greatly influences the gearbox design so this is examined closely in this work. The optimization data trend predicts that Mx increases linearly with L as depicted in Figure 8. Meanwhile, Figure 9 shows the gearbox mass scales as $L^{1.78}$. The loading and mass of the remaining components such as the low speed shaft (Figure 10) and the bed-plate (Figure 11) each have different scale exponents as described in the figures.

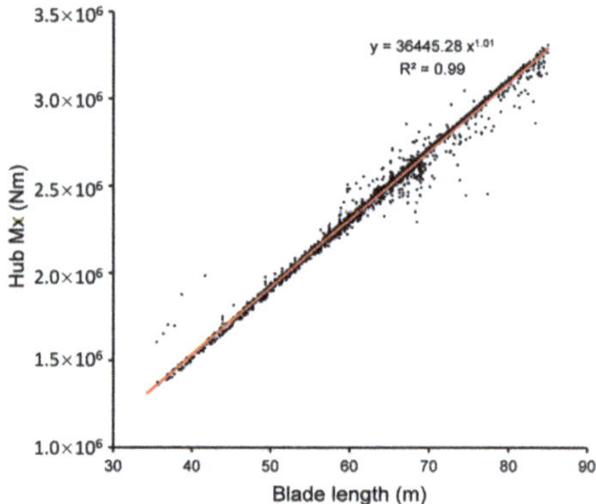

Figure 8. Hub Mx against blade length.

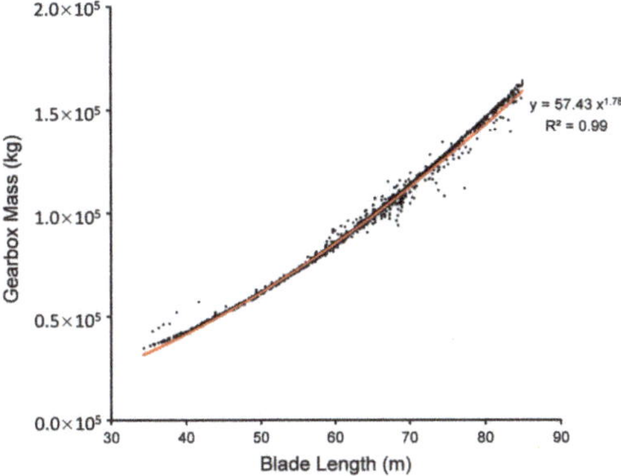

Figure 9. Gearbox mass against blade length.

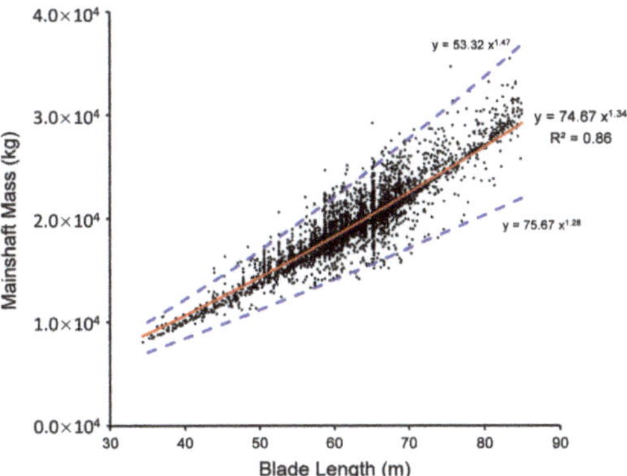

Figure 10. Mainshaft mass against blade length.

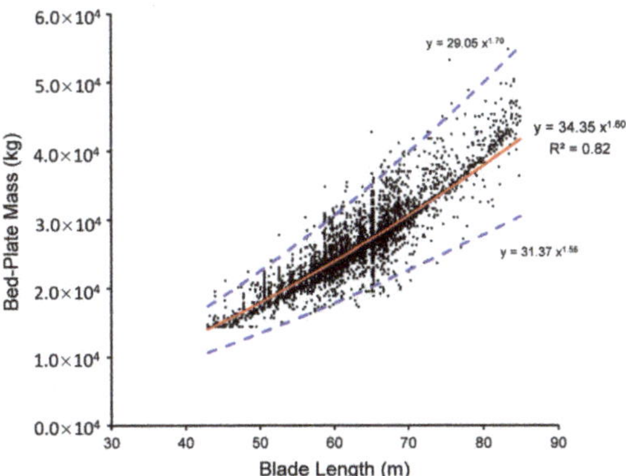

Figure 11. Bed-plate mass against blade length.

In the scatter plots of economic indicators as a function of blade length, the boundaries of these scatter cloud appear in the form of a parabola. The data points to an optimal blade length for each of these economic metrics are shown in Figures 12–15.

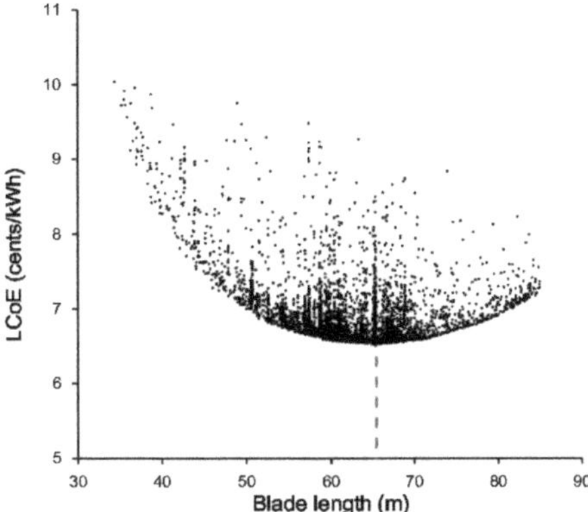

Figure 12. LCoE against blade length.

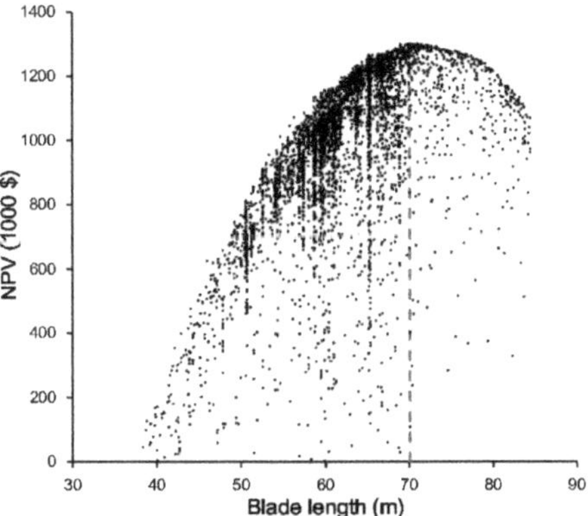

Figure 13. NPV against blade length.

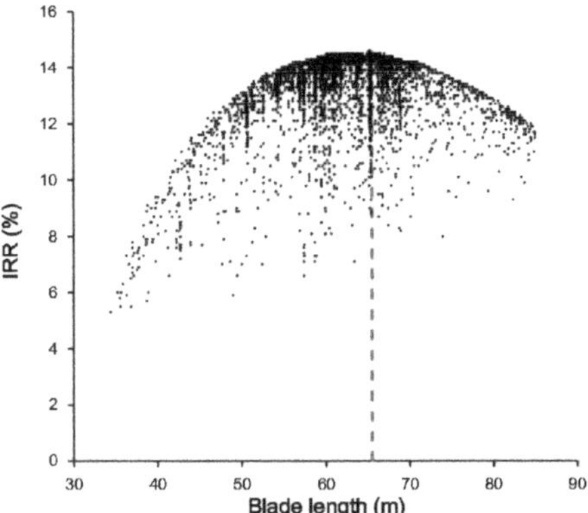

Figure 14. IRR against blade length.

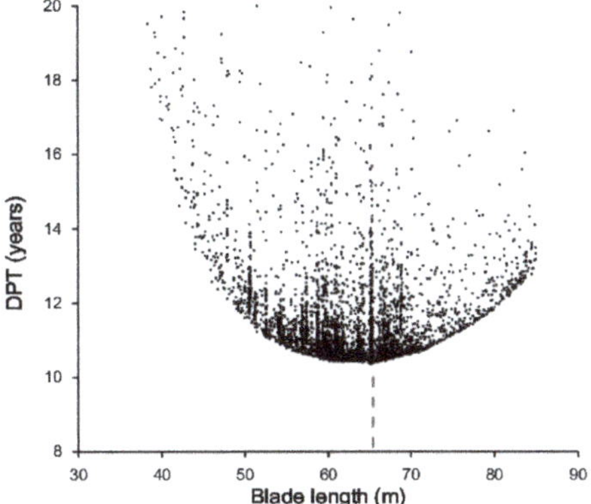

Figure 15. DPT against blade length.

A large proportion of the initial capital costs are unrelated to the rotor diameter or design loads, for example the generator, converter and control system costs. This leads to a higher LCoE and lower NPV when blade length is comparatively short. For example, a 30 m–38 m blade is uneconomical for a 2 MW turbine since the energy production is inadequate compared with the costs (Figure 13). As the blade length increases beyond 38 m, the benefits outweigh the economic costs. The economic efficiency improves until it reaches a maximum at a certain blade length. Beyond this length, the turbine may produce more energy, but due to higher manufacturing effort and increased loading on many components, the increase in cost exceeds the improvement in AEP, resulting in a lower economic efficiency.

As a result, LCoE first decreases then increases with blade length. The characteristic of NPV is just the opposite. Optimal blade length and corresponding values for LCoE and NPV can be observed. LCoE achieves a minimal value of 6.54 cents/kWh at a blade length of 65.4 m, while the NPV reaches a maximum value of $1,295,000 when the blade is 70.1 m long. The same configuration that had the lowest LCoE also presented the best values for the IRR and DPT. The highest IRR is 14.6% and the shortest DPT is 10.4 years.

4.3. Economic Functions

Based on the equations presented in Table 3, as previously assumed, OPEX are accounted for by taking a fixed percentage of the capital cost ($OPEX = f \cdot CAPEX$), after some mathematical transformation, it can be found that the task of minimization of LCoE, maximization of IRR and minimization of DPT can be considered as a task to minimize the ratio between CAPEX and the AEP:

$$LCoE|_{\min} \Rightarrow \frac{CAPEX}{AEP}(\frac{1}{a} + f)|_{\min} \tag{2}$$

$$IRR|_{\max} \Rightarrow \frac{AEP \cdot p_{kwh}}{CAPEX} - f|_{\max} \tag{3}$$

$$DPT|_{\min} \Rightarrow \frac{1}{\frac{AEP \cdot p_{kwh}}{CAPEX} - f}|_{\min} \tag{4}$$

Consequently, optimum LCoE, IRR and DPT occur at the same turbine configuration while optimum NPV is achieved at an entirely different design point. Maximum NPV is achieved when the margins between total discounted income and total discounted costs are maximized. As seen in Figures 12 and 13, the optimum NPV occurs at a larger blade length than the optimum LCoE.

Figure 16 illustrates the variation of total discounted incomes of different wind turbines as a function of their total discounted costs. The dashed line in the middle is the break-even line, while turbines above the line are economic and those below the line are not. Two points have been highlighted on the cost-income front. Point A, with total discounted costs of $3.9 m, has the highest net profit margin between incomes and costs which implies the maximum NPV is achieved. The investment returns of designs with costs higher than point A decline. Point B, with total discounted costs of $3.5 m has the maximum slope which indicates that the ratio between incomes and costs is maximized. This point corresponds to minimum LCoE. In this optimization problem, a wind turbine designed with the objective of minimum LCoE tends to choose a smaller-scale investment compared with the objective of maximum NPV.

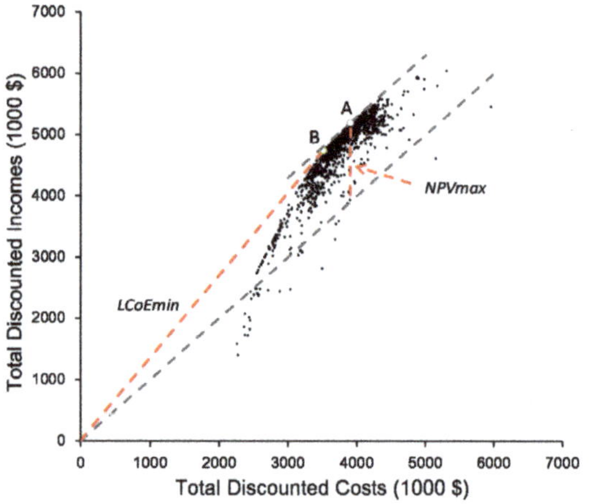

Figure 16. Total discounted incomes against total discounted costs.

The choice between LCoE and NPV is a complex economic problem. Investment size and incremental internal rate of return may also be taken into consideration to make the decision that most meets the desire of the designers. Further exploration is beyond the scope of this research. Nonetheless, the result obtained herein, that the optimal blade length of a certain turbine should be situated between the value of optimal LCoE and optimal NPV, may be valuable to wind turbine manufacturers.

4.4. Discussion of the Optimization Blades

In addition to the different blade lengths obtained from these metrics, the detailed turbine configurations show the effect of several other variables. In the following subsection, optimization results are presented to provide insight about the impact of blade geometry to the integrated turbine performance. Blade shapes of optimum NPV and optimum LCoE are plotted in Figure 17, and the turbine properties with respect to these blades are summarized in Table 5.

The design providing an optimal NPV has a larger chord distribution along the full blade span, but the difference remains almost constant up to the blade tips, revealing a certain degree of similarity between these two blades. Both blades have a relatively high thickness-to-chord ratio along the blade span. The minimum relative thickness is 21.6% for optimal NPV and 23% for optimal LCoE. Though large relative thickness will undoubtedly reduce blade aerodynamic performance, the beneficial aspect is that larger section thickness can increase the section moment of inertia, which is helpful in increasing both strength and stiffness, consequently reduces the blade mass. This configuration is mainly the result of a trade-off between aerodynamic performance and blade structure. Additionally, the reduced lift characteristic of high relative thickness airfoils may also be helpful in controlling the aerodynamic loads.

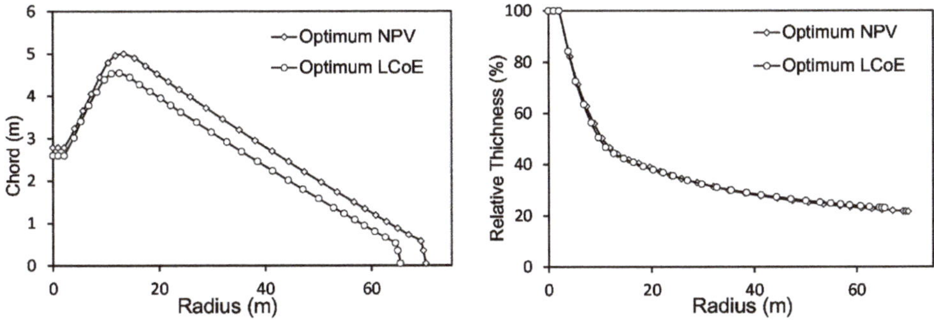

Figure 17. Design results of optimum NPV and optimum LCoE.

Generally, neither optimum LCoE design nor optimum NPV designs tend to pursue a maximum electricity production. The maximum power coefficients (Cp) of both blades are about 0.46, which is a relatively low value by current industrial standards. Note that the optimization procedure can achieve a Cp of 0.487, as shown by the High Cp design in Table 5. One of the main reasons for this is that the increase of Cp is always associated with an increase in design loads and consequently the initial capital cost.

Table 5. Turbine properties of designs produced for optimum NPV, LCoE and Cp.

Design Results	Blade Length (m)	Cp (-)	Capacity Factor (-)	Blade Root Mf (Nm)	Blade Mass (kg)	Hub Fx (Nm)
Optimum NPV	70.1	0.463	0.342	1.58×10^7	17,930	7.44×10^5
Optimum LCoE	65.4	0.457	0.312	1.27×10^7	14,552	6.80×10^5
High Cp	65.4	0.487	0.321	1.40×10^7	17,303	8.63×10^5

Design Results	Tower Height (m)	Tower Mass (tonne)	CAPEX (1000 $)	LCoE (cents/kWh)	NVP (1000 $)
Optimum NPV	105	369	2725	6.62	1295
Optimum LCoE	93	270	2377	6.54	1223
High Cp	93	346	2625	6.84	1096

In order to elucidate why optimization designs based on economic analysis lead to relatively lower AEP, we chose the High Cp design from Table 5, which shares the same blade length and tower height as optimum LCoE. A commercial 58 m blade with a maximum Cp of 0.49 is also presented for comparison, as is shown in Figure 18. This commercial blade holds a geometry layout more similar to High Cp design, shows that some industrial practices are using the design strategy of maximum aerodynamic performance.

The turbine performance of High Cp design differs greatly from the optimum LCoE design as presented in Table 5. Generally, the design loads and the mass of structural components rise considerably.

The minimum LCoE design has a much larger relative thickness distribution along most of the blade but a smaller chord distribution at the blade tip compared with the High Cp design. This is because the minimum LCoE sacrifices rotor performance to reduce thrust and decrease turbine costs. As a result, the design loads are significantly reduced. For example, the flap-wise moment of the blade root is reduced by 9%. Together with a larger blade thickness at the mid span, which is beneficial to the structural integrity of the blade, the blade mass is reduced from 17.3 tonnes to 14.5 tonnes, a decrease of 16%. The expense is that the maximum power coefficient is reduced from 0.487 to 0.457.

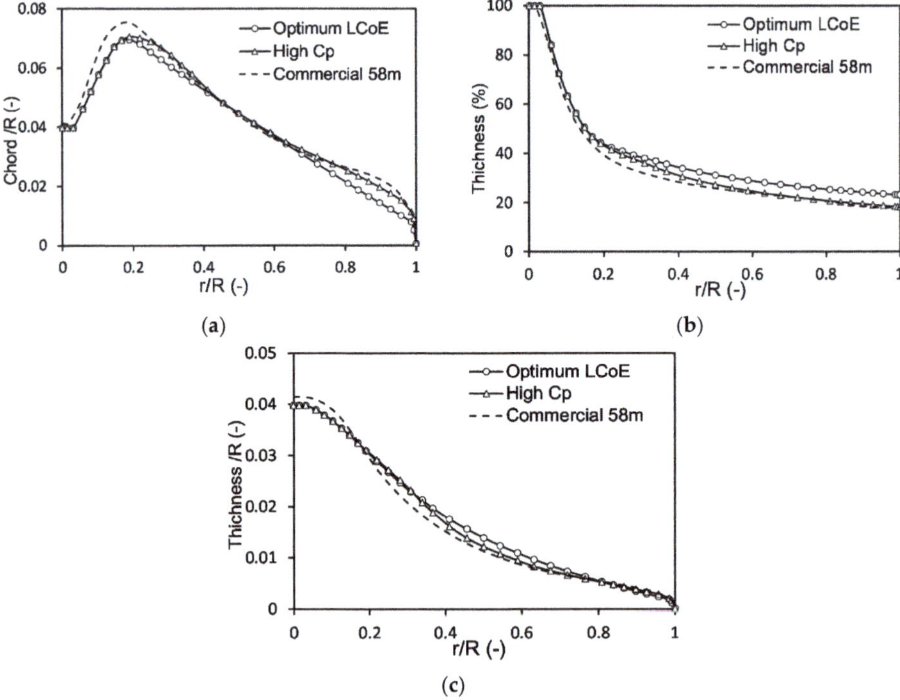

Figure 18. Comparison between Optimum LCoE, High Cp and Commercial 58 m blade. (**a**) Chord/R distribution; (**b**) Relative thickness distribution; (**c**) Thickness/R distribution.

0.457 may seem as unacceptable compared with 0.487, a decrease of 6.2%. However, the AEP is only degraded by 3%. Meanwhile, the design loads and structure masses were reduced significantly, accounting for a reduction in capital expenditure by 9.3%. The result produces 4.1% less LCoE and the NPV is increased. The LCoE design could be a more reasonable choice from the economic point of view.

4.5. Optimization of the Tower Height

At a certain wind shear exponent, the quality of the wind improves with increasing tower height. However, the higher tower results in a heavier structure which increases the cost. Therefore, the tower height of the turbine should match the site and rotor to achieve maximum economic efficiency [32,33]. An analysis was conducted to examine the impact of the tower height on the LCoE and NPV. Both are depicted graphically in Figures 19 and 20.

We took the rotors obtained from the optimization for examination. One is equipped with a 65.4 m blade and the other with 70.1 m blade. There are local optimum values of LCoE and NPV as a function of the tower height. The reason for this is similar to that which describes the influence of the blade length. The economic efficiency increases first and then decreases with the tower height, due to an increase in the costs of rotor, tower and the balance-of-station. Although the increased tower height can produce more energy, the increase in cost eventually outweighs the increase in profits. The optimum NPV design results in a higher tower than the optimum LCoE design, showing again that a minimum LCoE design is more geared towards controlling the cost.

Interestingly, we observed that a lower tower is a better match for the larger rotor in both cases. For example, NPV is optimized at a 110 m tower height for a 65.4 m blade rotor, but at a 105 m tower for a 70.1 m blade rotor, as shown in Figure 21. This trend is repeated in the optimal LCoE design.

Therefore, it seems from this result that, unlike the case for smaller rotor, the cost of the increase in tower height for larger wind turbine may overweight the increase in AEP due to its higher rotor thrust. The result may be influenced by the value selected for the wind shear exponent, but it is certain that more attention should be paid in upgrading the tower height of large rotors.

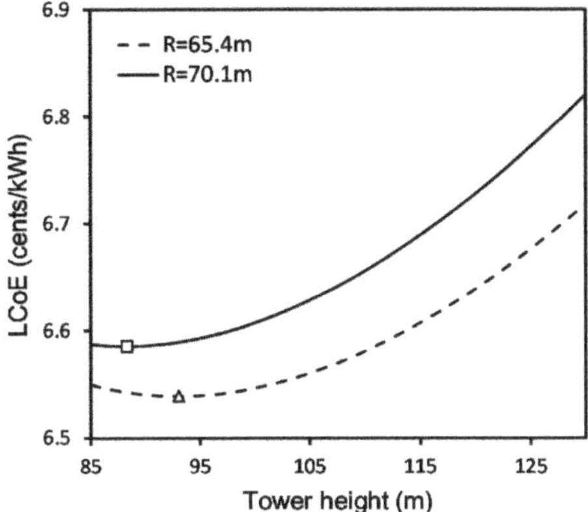

Figure 19. LCoE against tower height.

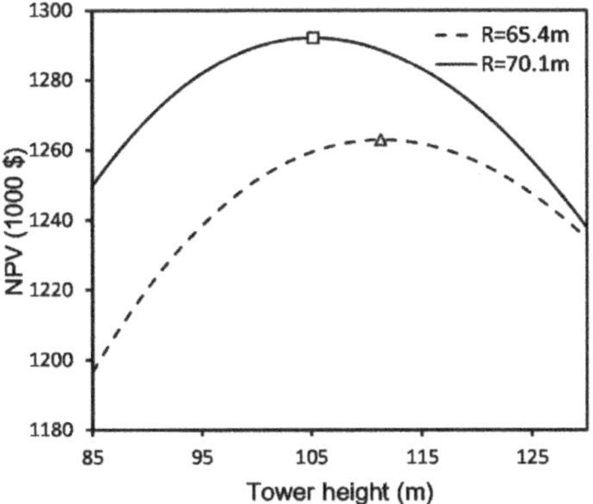

Figure 20. NPV against tower height.

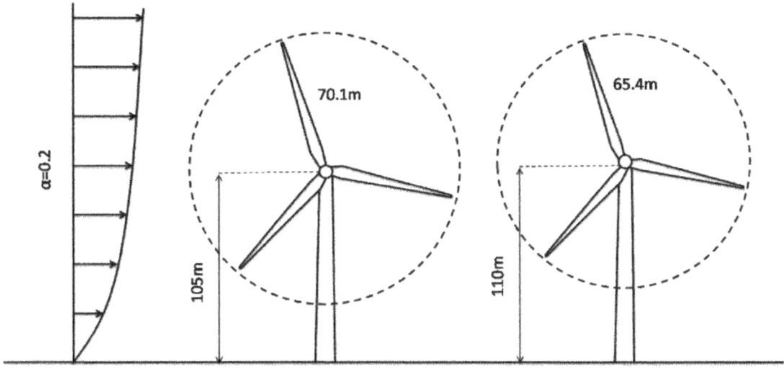

Figure 21. Optimal tower heights of different rotors.

5. Conclusions

In this paper, a system-level optimization analysis is used to perform an integrated design of a 2 MW low wind speed turbine. Besides LCoE, the economic functions NPV, IRR and DPT are also applied as objectives in wind turbine optimization. Theoretical analysis and design results indicate that in this study IRR and DPT effect the turbine design in the same way as LCoE, all leading to a minimization of the ratio of capital expenditures to AEP. However, optimum NPV implies the largest margin between incomes and costs, resulting in a longer blade than optimum LCoE. Though more economic factors should be included to make the final decision, the optimal blade length for a certain turbine should be situated between the value of optimal LCoE and optimal NPV according to the economic analysis in this study.

The blades obtained from these economic objectives seem to be aerodynamically uncompetitive due to their larger relative thickness and smaller chord distribution compare with high Cp design. However, the sacrifice of aerodynamic performance brings significant reduction in design loads and turbine costs. This optimization shows that sometimes control of construction and maintenance costs is more crucial than optimal aerodynamics in improving the economic efficiency of wind turbines.

Further developments in this work could consider other economic metrics, such as investment size and incremental internal rate of return, to achieve more practical results. Adding modal properties, local and global buckling in the structural design constraints for the blade and tower could also be considered for future work. Finally, through a more accurate estimation of the design loads and turbine capital costs, the economic performance of a wind turbine could be more accurately evaluated.

Author Contributions: Writing-Original Draft Preparation, J.W.; Writing-Review & Editing, T.W.; Supervision, L.W. and N.Z.

Funding: This research was funded by the National Basic Research Program of China (973 Program) [grant number 2014CB046200], the National Nature science Foundation [grant number 51506089], CAS Key Laboratory of Wind Energy Utilization [grant number KLWEU-2016-0102], Jiangsu Key Laboratory of Offshore wind Turbine Blade Design and Manufacture Technology [grant number BM2016024].

Conflicts of Interest: The authors declare no conflict of interest.

References

1.	Johansen, J.; Madsen, H.A.; Gaunaa, M.; Bak, C.; Sørensen, N.N. Design of a wind turbine rotor for maximum aerodynamic efficiency. *Wind Energy* **2009**, *12*, 261–273. [CrossRef]

2.	Wang, L.; Wang, T.G.; Wu, J.H.; Chen, G.P. Multi-objective differential evolution optimization based on uniform decomposition for wind turbine blade design. *Energy* **2017**, *120*, 346–361. [CrossRef]

3.	Benini, E.; Toffolo, A. Optimal design of horizontal-axis wind turbines using blade-element theory and evolutionary computation. *J. Sol. Energy Eng.* **2002**, *124*, 357–363. [CrossRef]

4.	Maki, K.; Sbragio, R.; Vlahopoulos, N. System design of a wind turbine using a multi-level optimization approach. *Renew. Energy* **2012**, *43*, 101–110. [CrossRef]

5.	Chehouri, A.; Younes, R.; Ilinca, A. Review of performance optimization techniques applied to wind turbines. *Appl. Energy* **2015**, *142*, 361–388. [CrossRef]

6.	Ning, A.; Damiani, R.; Moriarty, P.J. Objectives and constraints for wind turbine optimization. *J. Sol. Energy Eng.* **2014**, *136*, 041010. [CrossRef]

7.	Savino, M.M.; Manzini, R.; Della Selva, V.; Accorsi, R. A new model for environmental and economic evaluation of renewable energy systems: The case of wind turbines. *Appl. Energy* **2017**, *189*, 739–752. [CrossRef]

8.	Herbert-Acero, J.F.; Probst, O.; Réthoré, P.E.; Larsen, G.C.; Castillo-Villar, K.K. A review of methodological approaches for the design and optimization of wind farms. *Energies* **2014**, *7*, 6930–7016. [CrossRef]

9.	Short, W.; Packey, D.J.; Holt, T. *A Manual for the Economic Evaluation of Energy Efficiency and Renewable Energy Technologies*; Technical Report NREL/TP-462-5173; National Renewable Energy Laboratory: Golden, CO, USA, 1995.

10. Rodrigues, S.; Restrepo, C.; Katsouris, G.; Pinto, R.T.; Soleimanzadeh, M.; Bosman, P.; Bauer, P. A multi-objective optimization framework for offshore wind farm layouts and electric infrastructures. *Energies* **2016**, *9*, 216. [CrossRef]

11. Mirghaed, M.R.; Roshandel, R. Site specific optimization of wind turbines energy cost: Iterative approach. *Energy Convers. Manag.* **2013**, *73*, 167–175. [CrossRef]

12. Ashuri, T.; Zaaijer, M.B.; Martins, J.R.; van Bussel, G.J.; van Kuik, G.A. Multidisciplinary design optimization of offshore wind turbines for minimum levelized cost of energy. *Renew. Energy* **2014**, *68*, 893–905. [CrossRef]

13. Sun, Z.Y.; Sessarego, M.; Chen, J.; Shen, W.Z. Design of the OffWindChina 5MW Wind Turbine Rotor. *Energies* **2017**, *10*, 777. [CrossRef]

14. Fuglsang, P.; Madsen, H.A. Optimization method for wind turbine rotors. *J. Wind Eng. Ind. Aerodyn.* **1999**, *80*, 191–206. [CrossRef]

15. Dykes, K.; Platt, A.; Guo, Y.; Ning, A.; King, R.; Parsons, T.; Petch, D.; Veers, P. *Effect of Tip-Speed Constraints on the Optimized Design of a Wind Turbine*; Technical Report NREL/TP-5000-61726; National Renewable Energy Laboratory: Golden, CO, USA, 2014.

16. Ning, A.; Petch, D. Integrated design of downwind land-based wind turbines using analytic gradients. *Wind Energy* **2016**, *19*, 2137–2152. [CrossRef]

17. Ashuri, T.; Zaaijer, M.B.; Martins, J.R.; Zhang, J. Multidisciplinary design optimization of large wind turbines—Technical, economic, and design challenges. *Energy Convers. Manag.* **2016**, *123*, 56–70. [CrossRef]

18. Fuglsang, P.; Bak, C.; Schepers, J.G.; Bulder, B.; Cockerill, T.T.; Claiden, P.; Olesen, A.; van Rossen, R. Site-specific Design Optimization of Wind Turbines. *Wind Energy* **2002**, *5*, 261–279. [CrossRef]

19. Hansen, M.O.L. *Aerodynamics of Wind Turbines*, 2nd ed.; Earthscan: London, UK, 2008.

20. *International Electrotechnical Committee IEC 61400-1: Wind turbines Part 1: Design Requirements*, 3rd ed.; IEC: Geneva, Switzerland, 2005.

21. Guo, Y.; Parsons, T.; King, R.; Dykes, K.; Veers, P. *An Analytical Formulation for Sizing and Estimating the Dimensions and Weight of Wind Turbine Hub and Drivetrain Components*; Technical Report NREL/TP-5000-63008; National Renewable Energy Laboratory: Golden, CO, USA, 2015.

22. Díaz, G.; Gómez-Aleixandre, J.; Coto, J. Dynamic evaluation of the levelized cost of wind power generation. *Energy Convers. Manag.* **2015**, *101*, 721–729. [CrossRef]

23. Tang, S.L.; Tang, H.G. The variable financial indicator IRR and the constant economic indicator NPV. *Eng. Econ.* **2003**, *48*, 69–78. [CrossRef]

24. Timmer, W.A.; van Rooij, R.P.J.O.M. Summary of the Delft University wind turbine dedicated airfoils. *J. Sol. Energy Eng. Trans. ASME* **2003**, *125*, 488–496. [CrossRef]

25. Ning, A.; Dykes, K. Understanding the benefits and limitations of increasing maximum rotor tip speed for utility-scale wind turbines. *J. Phys. Conf. Ser.* **2014**, *524*, 012087. [CrossRef]

26. Diveux, T.; Sebastian, P.; Bernard, D.; Puiggali, J.R.; Grandidier, J.Y. Horizontal axis wind turbine systems: Optimization using genetic algorithms. *Wind Energy* **2001**, *4*, 151–171. [CrossRef]

27. Gao, X.; Yang, H.; Lu, L. Optimization of wind turbine layout position in a wind farm using a newly-developed two-dimensional wake model. *Appl. Energy* **2016**, *174*, 192–200. [CrossRef]

28. Gentils, T.; Wang, L.; Kolios, A. Integrated structural optimisation of offshore wind turbine support structures based on finite element analysis and genetic algorithm. *Appl. Energy* **2017**, *199*, 187–204. [CrossRef]

29. Zhang, J.J.; Xu, L.W.; Gao, R.Z. Multi-island Genetic Algorithm Optimization of Suspension System. *Telkomnika* **2012**, *10*, 1685–1691. [CrossRef]

30. China Wind Energy Association. *China Wind Power Industry Map 2017*; CWEA: Beijing, China, 2018.

31. Global Wind Energy Council. *Global Wind Report*; GWEC: Brussels, Belgium, 2018.

32. Abdulrahman, M.; Wood, D. Investigating the Power-COE trade-off for wind farm layout optimization considering commercial turbine selection and hub height variation. *Renew. Energy* **2017**, *102*, 267–278. [CrossRef]

33. Alam, M.M.; Rehman, S.; Meyer, J.P.; Al-Hadhrami, L.M. Review of 600–2500 kW sized wind turbines and optimization of hub height for maximum wind energy yield realization. *Renew. Sustain. Energy Rev.* **2011**, *15*, 3839–3849. [CrossRef]

 *applied sciences*

Article

Wind Turbine Optimization for Minimum Cost of Energy in Low Wind Speed Areas Considering Blade Length and Hub Height

Han Yang, Jin Chen * and Xiaoping Pang

State Key Laboratory of Mechanical Transmissions, Chongqing University, Chongqing 400044, China;
yh098@126.com (H.Y.); pangxp@cqu.edu.cn (X.P.)
* Correspondence: chenjin413@cqu.edu.cn

Received: 8 July 2018; Accepted: 20 July 2018; Published: 23 July 2018

Abstract: In recent years, sites with low annual average wind speeds have begun to be considered for the development of new wind farms. The majority of design methods for a wind turbine operating at low wind speed is to increase the blade length or hub height compared to a wind turbine operating in high wind speed sites. The cost of the rotor and the tower is a considerable portion of the overall wind turbine cost. This study investigates a method to trade-off the blade length and hub height during the wind turbine optimization at low wind speeds. A cost and scaling model is implemented to evaluate the cost of energy. The procedure optimizes the blades' aero-structural performance considering blade length and the hub height simultaneously. The blade element momentum (BEM) code is used to evaluate blade aerodynamic performance and classical laminate theory (CLT) is applied to estimate the stiffness and mass per unit length of each blade section. The particle swarm optimization (PSO) algorithm is applied to determine the optimal wind turbine with the minimum cost of energy (COE). The results show that increasing rotor diameter is less efficient than increasing the hub height for a low wind speed turbine and the COE reduces 16.14% and 17.54% under two design schemes through the optimization.

Keywords: wind turbine optimization; low wind speed areas; cost of energy; particle swarm optimization

---

1. Introduction

Wind energy is renewable and clean, which can help mitigate global climate change. Wind farms with high quality wind resources are limited. The wind farms with low quality wind resources are far more plentiful than high-quality ones and have some advantages such as being closer to the existing electrical grid. The design and development of wind turbines in low wind speed areas faces several technical and financial challenges related to maximizing energy conversion efficiency and minimizing cost of energy (COE). The classical wind turbine literature mainly deals with acquiring ideal blade geometry or structural design and improving structural properties of the tower. The research objective mainly aims to maximize annual energy production (AEP), minimize COE, minimize blades mass, or a combination of these.

Each wind turbine is designed for specific wind conditions. The IEC 61400-1 standard [1] defines wind classes according to wind speed. Currently, the Classes III, II and I of wind turbines correspond to low, medium and high wind speed locations, respectively. In the global wind energy market, wind turbines installed in high wind speed sites (Class I) have progressively lost market share for the past few years in favor of wind turbines in low wind speed locations (Class III). The Asian wind energy market has been dominated by low wind speed wind turbines during the last decade mainly due to the lower quality wind resources in most places in China and India [2].

There has been a moderate amount of literature regarding low wind speed wind turbines, mainly focusing on: low wind resources [3], blade aerodynamic performance [4,5] or structural performance [6,7], tower structural design or analysis [8,9]. In general, the rotor of a wind turbine designed for low wind speed areas has a higher aspect ratio with longer blades, which aims to acquire more wind power and reduce the cost. The costs of larger rotor diameter and higher tower height have a strong impact on the total cost of the wind turbine system. With the additional loads acting on a tower's top owing to the increasing thrust when increasing the blade length and hub height to attract more wind power, the initial capital cost of a wind turbine will also increase. Thus, the wind turbine optimization design should trade-off the hub height and rotor diameter, especially in low wind speed sites. As previously described, because of a prominent increase in the low wind speed wind turbine market, wind turbine manufacturers have invested considerable effort to design and develop wind turbine blades operating in low wind speed sites. Hence, with the wind turbine scale increased, reducing the cost of energy is important. The objective of the optimization model by Xudong et al. [4] is the minimum cost of energy. The design variables in their study are the blade geometric parameters including chord, twist and thickness distribution with fixed rotor diameter. In the present work, the entire system of a wind turbine is considered.

The literature about wind turbines for low wind speed areas is very limited. Some studies work on the shape optimization of low wind speed wind turbine blades, but the object models are always the micro wind turbine [10–13]. Some low wind speed blade designs in the literature [5,6] focus on large scale wind turbines. The work in [5] focuses on aerodynamic performance analysis and AEP enhancement of 3 MW wind turbine blade for low wind speed areas. The optimization objective is to improve AEP and to minimize the thrust based on blade element momentum (BEM) theory. The design variables such as blade length and rated wind speed are selected by existing wind turbine blade groups. The work in [6] proposes a method to quantitatively compare wind turbine operating in wind Class I and wind Class III. Their results show that the traditional design method for low wind speed blade structures is not efficient, and the research is used to improve the design process for low wind speed blades.

The minimized cost of energy is acquired by altering the wind turbine system parameters in the study [14]. They use a multi-level optimization approach to minimize the cost of energy by maintaining AEP and reducing blade root loads. However, their optimization only considers the geometric shape and structure of the blade, and the costs of other components are estimated by the cost and scaling model in reference [15]. Because the design model selects a 1 MW wind turbine for Class I wind resources, the variations of rotor diameter and hub height have less influence on the total system. The optimization process only considers the hub height to evaluate the tower mass by scaling with the product of the swept area and hub height. This cannot ensure accuracy when the tower has a height much greater than 80 m and the wind turbine operates in low wind speed areas [15]. Bortolotti [16] is concerned with the holistic optimization of wind turbines. A multi-disciplinary optimization procedure considering rotor diameter and tower height is presented. The methodology is applied to a commercial 2.2 MW onshore and a conceptual 10 MW offshore wind turbine used in high wind speed locations.

To our knowledge, no studies have focused on a wind turbine model for low wind speed areas with a full scale system taken into account. In this study, the main work is to reduce the COE with rotor and tower integrated optimization. An aero-structural design optimization for a commercial 2.1 MW wind turbine for low wind speed locations is performed. The BEM theory is adopted to evaluate the aerodynamic performance and the classical laminate heory (CLT) is used to estimate the structural performance for the wind turbine blade. The aerodynamic loads acting on blade and tower would change according to the rotor diameter and hub height. The COE model comprises the overall wind turbine system and is a function of rotor diameter and tower height. Most researchers employ the model proposed in reference [15], but this reference indicates that when the tower is much greater than 80 m, the scaling relationships should be used carefully because the cost of the tower will have a major impact on design. In addition, in order to estimate an accurate cost of the tower with much greater height, the matched tower of the commercial wind turbine is used as a reference.

2. Wind Turbine Model

A commercial 2.1 MW HAWT for low wind speed areas is selected as a case study to verify the efficacy of the design method. The length of the blade is 59.8 m and its mass is 13,240 kg. The corresponding tower is 95 m in height and the hub is considered as a concentrated mass. The blade is made of glass fiber reinforced plastic (GFRP). The core materials consist of balsa wood and polyvinyl chloride (PVC). The blade baseline is described by the chord distributions, twist distributions and selected airfoils distributions. Details of each parameter are shown in Tables 1 and 2. The geometric and material properties of the wind turbine tower are presented in Table 3. The material of the tower consists of Q345 steel, and the steel density is 7820 kg/m^3.

Table 1. Baseline blade geometry.

Span Location (m)	Chord (m)	Twist (°)	Airfoil	Thickness (%)
0	2.40	18.0	circle	100
1.0	2.40	18.0	circle	100
13.4	3.42	9.6	DU00-W2-401	40
16.2	3.17	6.9	DU00-W2-350	35
24.9	2.36	2.7	DU97-W-300	30
49.8	1.02	−1.56	DU91-W2-250	25
59.8	0.02	−1.6	DU93-W-210	21

Table 2. Wind turbine characteristics.

Rated power	2.1 MW
Rated wind speed	9 m/s
Cut in wind speed	3 m/s
Cut out wind speed	20 m/s
Number of blades	3
Design tip speed ratio	10.6
Rotor diameter	122.3m
Control type	Variable speed—variable pitch
Maximum power coefficient	0.483

Table 3. Geometric and material properties of the tower.

Tower height (m)	95
Tower top outer diameter (m)	3.15
Tower base outer diameter (m)	5.72
Tower top wall thickness (mm)	24
Tower base wall thickness (mm)	40
Density (kg/m^3)	7820
Young's modulus (GPa)	200
Poisson ratio	0.3
Yield strength (MPa)	345

2.1. Materials

The main material of the blade is GFRP, consisting of leading/trailing edge reinforcement, panel sections, two shear webs and a spar cap. The spar cap is made of unidirectional glass fiber composite materials to withstand the aerodynamic loads flap-wise. The shear webs are composed of sandwich panels to withstand the shear force. The core materials comprise of glass fiber composite materials with balsa wood and PVC. The material properties (E_1 is the longitudinal modulus, E_2 is the transverse modulus, G_{12} is the shear modulus, v_{12} is the Poisson's ratio and the thickness is the laminate thickness of each ply) are listed in Table 4.

Table 4. The material properties.

Stack ID	Materials	E1 (GPa)	E2 (GPa)	G12 (GPa)	v12	ρ (kg/m³)	Thickness (mm)
1	Unidirectional	39.18	11.69	3.5	0.3	1940	0.892
2	Bi-axial	11.5	11.5	10.8	0.48	1970	0.555
3	Triax	25.8	13.59	7.36	0.36	1842	0.906
4	Gelcoat	3.44	3.44	1.38	0.3	1235	0.6
5	Balsa	1.321	0.03	0.006	0.39	151	25.4
6	Polyvinyl chloride (PVC)	0.035	0.035	0.022	0.3	60	5

2.2. Original Blade Layup Schedule

Figure 1 shows the distributions of the layup layers along the blade span-wise. The first two meters of span is circumferential reinforcement for the blade root. The main material for this section is triaxial composite materials with large layers. The first 13 m of span is inner and outer surface reinforcement for the blade root. Sections outboard of 5 m have thick caps and trailing edge layup by unidirectional materials. The airfoil with thickness of 40% is set at the blade span location of 13.4 m. The root region of the blade experiences the highest bending moments, and cylindrical reinforcement is used to withstand the moments. Thus, the thickness of cylindrical reinforcement is also considered as a design variable in the optimization process. The structural design variables are set as layup layers for spar caps and trailing edge reinforcement. The configuration of the blade section is also considered. The details of the design variables are described in Section 4.1.

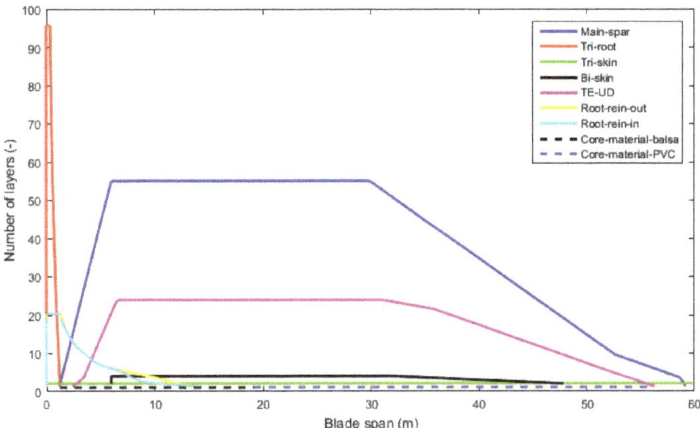

Figure 1. Distribution of section's layup layers along the blade radial coordinate.

2.3. Cost of Energy (COE) Model

The study [17] has shown the design objective of maximizing annual energy production or using sequential aerodynamic and structural optimization to be significantly suboptimal compared to the aero-structural integrated methods. For variable rotor diameter and hub height, the optimal rotor diameter may be misleading because the tower mass dominates the total mass of turbine. Thus, minimizing COE is a much better metric for the full-scale wind turbine optimization objective than minimizing the ratio of turbine mass to annual energy production. The main objective is to minimize the COE of a wind turbine used in low wind speed areas. Following the study [15], COE is calculated as:

$$COE = \frac{FCR \times ICC + AOE}{AEP} \tag{1}$$

where FCR is the fixed charge rate, ICC is the initial capital cost, AOE is the annual operating expenses, and AEP is annual energy production.

As wind turbines become more sophisticated in terms of aerodynamic shape and structural configuration with increasing size, the design of each wind turbine component is not always clear. The COE model is used to assess the true impact of the technical change and various designs. The FCR, ICC and AOE are obtained from the National Renewable Energy Laboratory (NREL) wind turbine design cost and scaling model [15]. The BEM code and the Weibull probability distribution with annual average wind speed of 6 m/s at hub height are used to calculate AEP. The blade stiffness and mass distributions are computed by the classical laminate theory code named PreComp [18]. The total cost of a wind turbine is defined by adding the value of the initial capital cost multiplied by the fixed charge rate to the annual operating expenses. The fixed charge rate is the annual amount per dollar of initial capital cost used to cover the capital cost and other fixed charges. The FCR value is set to 0.1158 per year as suggested in reference [15].

3. Theoretical Models and Verifications

3.1. Aerodynamic Model

In this work, the combination of blade element theory and momentum theory is used to estimate the aerodynamic performance and calculate the AEP. The blade element theory aims to calculate the aerodynamic force acting on each blade element. The blade momentum theory refers to analysis of both axial and tangential induced velocity by introducing axial and angular induction factors. The induced velocity in each blade section influences the angle of attack of the airfoil and therefore affects the loads acting on the blade sections. With an iterative procedure, BEM theory provides a method to calculate the aerodynamic loads on each blade element. For a rotor with a finite number of blades, the vortex system in the wake is different from that of a rotor with an infinite number of blades [19]. By introducing the Prandtl's tip loss factor, the BEM theory can simulate a wind turbine rotor with a finite blade number.

The aerodynamic performance of the total wind turbine blade can be obtained from each blade section. Thus, it is necessary to acquire the lift and drag coefficient of the adopted airfoil. However, the experimental polar of airfoils are not often available. In this work, RFoil [20] is used to estimate the aerodynamic characteristics of the airfoil. The RFoil code is an improved version of XFoil [21] and has many advantages such as better corrections after stall.

The annual average wind speed is the main source of wind power. Due to the friction of the plants and constructions, better wind resources are located at higher heights. The friction coefficient n could affect the average wind speed at hub height, and the different average wind speed at hub height could have an impact on the AEP of wind turbines. In this work, the friction coefficient is set as 0.23 which corresponds to the roughness of the countryside. The wind speed is varied depending on the hub height according to the following equation:

$$V_{hub} = V_{ref} \left(\frac{H_{hub}}{H_{ref}} \right)^{n} \tag{2}$$

where V_{hub} is the wind speed at the hub; H_{hub} is the hub height; V_{ref} is the reference wind speed at the reference height; H_{ref} is the reference height; and n is the friction coefficient and set as 0.23 for low wind speed condition.

3.2. Blade Element Momentum (BEM) Code Validation

In this work, a commercial wind turbine for Class III with blade length 59.8 m and 2.1 MW of rated power is used as a reference. The aerodynamic performance of the blade is calculated by considering a tip speed ratio of 10.6 with a range of wind velocities between 3 m/s and 20 m/s. Air density is set to 1.225 kg/m^3 and the cone and tilt angle of the wind turbine are set to zero degrees. The RFoil is used to calculate the aerodynamic characteristics of Delft University (DU) series airfoils for a specific Reynolds number according to their locations. The power coefficient and the power curves calculated by BEM code are implemented in MATLAB software (MATLAB 2014b; The MathWorks Inc.: Natick, MA, USA, 2014). In order to validate the BEM code, the aerodynamic performance data from the wind turbine manufacturer, based on simulations by WINDnovation Engineering Solutions GmbH, are compared with the BEM code. The numerical simulations and the manufacturer's data show an excellent agreement as shown in Figure 2. The power curve calculated by BEM code shows that the rated power is achieved with wind velocity of 9 m/s. Therefore, the aerodynamic performance prediction of the wind turbine blade calculated by BEM code for this study can be considered acceptable.

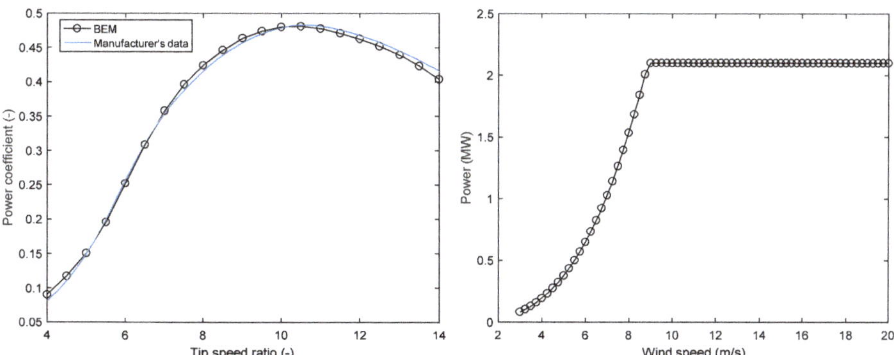

Figure 2. Comparison between the power coefficient and power curves obtained by blade element momentum (BEM) code and wind turbine manufacturer's data.

3.3. Structural Model

The structural performance can be represented by the cross-sectional composite properties of each blade section. In order to evaluate the cross-sectional composite properties of the blade section, classical lamination theory [22] is applied during the optimization procedure. For the laminated plates, the CLT model is an extension of classical plate theory [23] which has improvements in reasonable validation and computational efficiency. Thus, the CLT model can confirm the overall effective performance of composite laminated structure of wind turbine blade. Bir et al. [18] developed a FORTRAN code named PreComp (pre-processor for computing composite blade properties) to calculate the blade structural properties based on CLT. PreComp requires input files of the blade geometric shape, interior structural layout and materials properties. In addition, PreComp allows various internal geometry configurations and a general layup of composite laminates. The CLT model can transfer a complex three-dimensional problem into calculating two-dimensional blade section properties. In this paper, PreComp is applied to evaluate the structural properties of wind turbine blade sections. The structural parameters like blade geometry and layup schedule are set the same as the actual blade. After that, the stiffness and the mass per unit length distributions can be calculated. Figure 3 shows that the calculated stiffness values are in good agreement with the manufacturer's data based on simulations by WINDnovation Engineering Solutions GmbH.

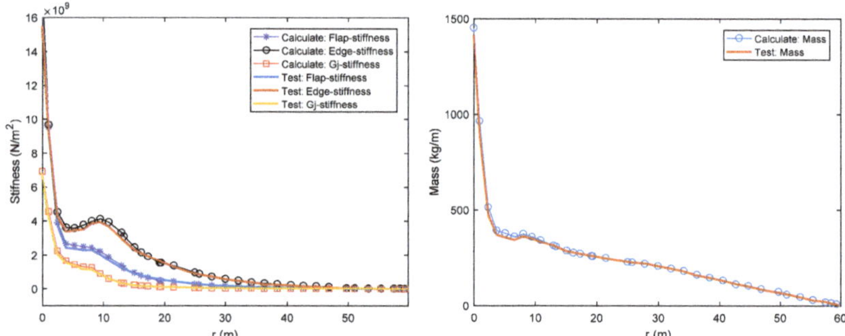

Figure 3. Comparison between the stiffness and mass per unit length distributions obtained by CLT code and actual blade.

4. The Optimization Model

In general, according to the wind turbine blades used in Class I, the same turbine family used in Class III has a length of 130~140% [6]. When the length of the blades increases, the blade mass and tip deflections will also increase. For structural and aerodynamic performance, the fitness function f in this paper is to minimize the cost of energy:

$$\text{maximize} \quad f = 1/\text{COE} \tag{3}$$

4.1. Design Variables

The geometric shape of the blade can be generated by chord distribution, twist distribution and the airfoil series. Eight control points are used for the thickness distribution, and the upper and lower bounds of the control points are set with adequate space, as shown in Figure 4. Nine control points are used for the chord and twist distributions, as shown in Figure 5. The first control point of chord distributions is at the root, the third control point is at the maximum chord station, and the ninth control point is at the blade tip. The thickness distribution is defined with a 9th order Bezier curve, and the chord and twist distributions are defined with a 10th order Bezier curve. For large scale wind turbine, a lower tip speed ratio would result in less efficient aerodynamic performance but lower cost of fabrication because of less aerodynamic loads by lower blade centrifugal stresses. The lower tip speed ratio can also reduce the tower cost because of the lower aerodynamic loads from the rotor. Thus, the tip speed ratio is set as a design variable in the design process.

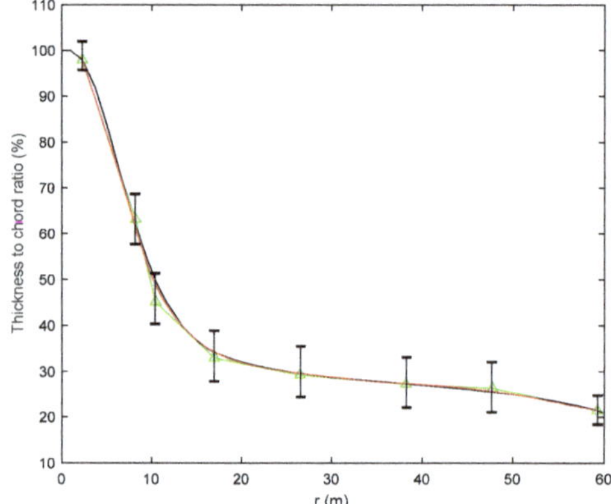

Figure 4. Thickness distribution of the blade.

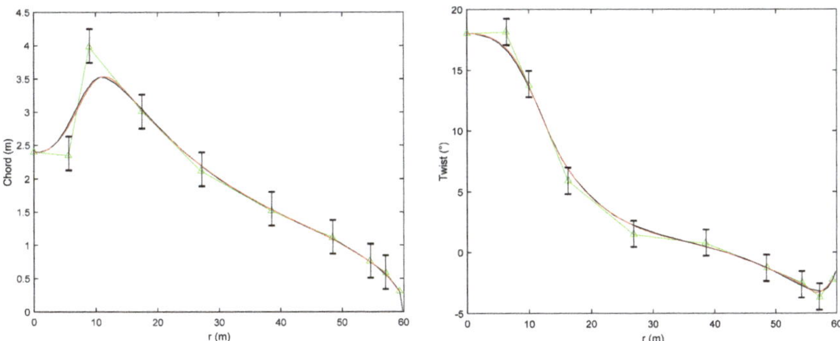

Figure 5. Chord and twist distributions of the blade.

The typical airfoil and its internal structural configuration of a typical modern wind turbine blade are shown in Figure 6 (c is the chord length). The internal structure mainly consists of spar cap and shear webs. The spar caps at the maximum thickness resist the flap-wise loads and the two shear webs resist the torsional loads. In this work, the maximum thicknesses of the spar cap and trailing edge reinforcement are the design variables for the blade layup schedule which are shown in Figure 7. The shear webs distributions are also considered in the structural optimization design. The locations of each web at blade root and tip are set as design variables.

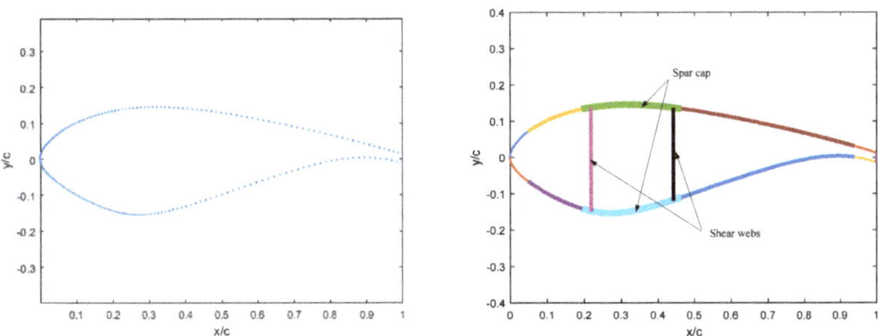

Figure 6. Typical structural parameters for section of the blade.

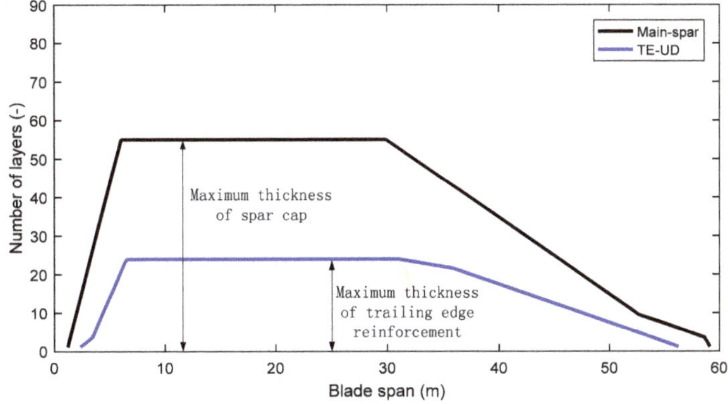

Figure 7. Design variables for spar cap and trailing edge reinforcement.

Figure 8 presents the schematic of the tower structure. With increasing tower height, the optimization design needs to trade off between the energy production and the cost of the structure. The wind turbines used in low wind speed areas always have higher hub height to obtain better wind sources than those used in high wind speed areas. The tower heights are always much greater than 80 m in low wind speed areas. Thus, the relationship of the tower mass and cost scaling are not efficient [15]. Therefore, an actual 2.1 MW commercial tower is used in an optimization process instead of using an empirical equation for the tower cost. Foundations were scaled as a function of hub height and rotor swept area, which is directly proportional to the tower overturning moment. The top and base diameters of the tower and their thickness are taken as design variables. The variations of diameters and thickness across the length of the tower are assumed as a linear distribution.

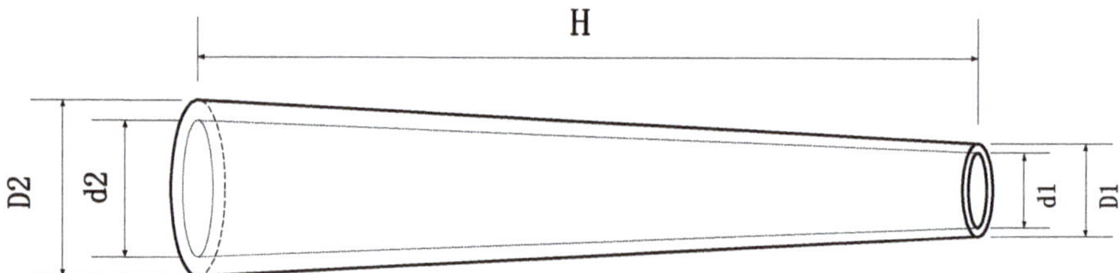

Figure 8. Schematic of tower structure.

4.2. Design Constraints

A real wind turbine must be designed to meet plenty of constraints. For the aerodynamic design, the blade root bending moment is always [14,17] set as a constraint for the blade design individually, but it is unnecessary for whole wind turbine design. The full scale design considers the whole structural stability instead of local aerodynamic performance. In this work, representative cases are used to meet the structural constraints. The wind turbine needs to meet the criteria of the structural strength in the process of operation. The ultimate strength analysis of the blade and the tower is performed in an extreme load condition. The 50-year extreme wind condition is defined as $V_{e50} = 1.4V_{ref}$ ($V_{ref} = 37.5$ m/s for Class III wind resources).

4.2.1. Blade Design Constraints

In order to verify the structural stability of the blade, the structural design needs to withstand the extreme wind condition. Some representative sections are selected to calculate the strain and verify the strength. These sections are at the blade root and the airfoils' locations along the blade span on both the upper and lower surfaces. A maximum strain condition is used as [17]:

$$\varepsilon_{50} \leq \frac{1}{\gamma_f \gamma_m} \varepsilon_{ult} \tag{4}$$

where the ε_{50} is the strain at 50-year extreme wind condition, the ε_{ult} is ultimate strain, the partial safety for loads (γ_f) is set as 1.35 and the partial safety factor for materials (γ_m) is set as 1.1 according to IEC 61400-1 requirement. The representative numbers of ultimate strain is set as 0.03. The maximum tip deflection of the blade under extreme loading is added to ensure adequate stiffness. In this work, considering the blade length is set as a variable, the maximum tip deflections with various blade lengths are set as:

$$\delta_{opt} \leq 1.05 \cdot \delta_{orig} \cdot \frac{R_{opt}}{R_{orig}} \tag{5}$$

where δ_{opt} is the maximum tip deflection of the optimal blade, the δ_{orig} is the maximum tip deflection of the original blade, R_{opt} is the length of the optimal blade, and R_{orig} is the length of the original blade.

4.2.2. Tower Design Constraints

The tower is the main support component of the wind turbine to withstand the aerodynamic loads and the total mass of the other components. The structural properties of the tower have a significant impact on the total cost of the wind turbine. An efficient, safe and economic design of the tower that meets all the design criteria is needed to minimize the total wind turbine cost. The loads acting on the tower mainly come from the aerodynamic loads on the rotor, and the wind loads on tower itself are also considered. The thrust force on the rotor is transferred to the tower top and causes tower deformation. In order to ensure structural stability and avoid large deformation, the maximum deformation at the top of the tower should not exceed the allowable deformation:

$$d_{\max} \leq d_{allow} \tag{6}$$

The d_{allow} is the allowable deformation and it can be determined according to the reference [18]. In order to ensure the tower has an adequate structural performance during optimization process, the allowable deformation is set as [8]:

$$d_{allow} \leq \frac{H_{opt}}{H_{orig}} d_{orig} \leq 1.25 \frac{H_{opt}}{100} \tag{7}$$

where H_{opt} is the height of the optimal tower, H_{orig} is the height of the original tower, and d_{orig} is the top deformation of the original tower.

In order to verify the structural stability of the tower, the structural design needs to withstand thrust from the wind turbine rotor. For a tapered tower, the location of the critical stress is at about the middle of the tower (the bending moment is largest at the base, but so is the moment of inertia) [17]. In this work, we assumed that the critical location was at the middle of the tower. Then, the stress at that location can be calculated as:

$$\sigma_{cr} = \frac{(m_{RNA} + m_{tower\_mid\_up})g}{A_{mid}} + \frac{THr_{mid}}{2I_{mid}} \tag{8}$$

where σ_{cr} is the critical stress, m_{RNA} is the mass of rotor and nacelle assembly, $m_{tower\_mid\_up}$ is the tower mass above the middle of the tower, A_{mid} is the area of the middle of the tower, T is the rotor thrust, H is the tower height, and I_{mid} is the area moment of inertia at the middle of the tower.

The stress σ_{cr} generated by the loads cannot exceed the allowable stress:

$$\sigma_{cr} < \sigma_{allow} \tag{9}$$

The σ_{allow} is the allowable stress and the value of it is given by:

$$\sigma_{allow} = \sigma_{ys} / \gamma_{tm} \tag{10}$$

where σ_{ys} is the yield strength and γ_{tm} is the material safety factor of the steel. The yield strength of the steel is 345 MPa for the wall thickness between 16 mm and 40 mm. The material safety factor is set as 1.1 according to IEC 61400-1.

4.3. The Optimization Algorithm

The purpose of this paper is to investigate the cost of energy for a low wind speed turbine considering rotor diameter and hub height as variables. The optimization progress aims to find the ideal blade shape parameters such as chord, twist, thickness distributions and the blade length, and to find the structural performance like webs locations, layer-up distributions and the tower parameters (height, taper and wall thickness distributions). The design process interfaces the particle swarm optimization (PSO) algorithm, BEM code, PreComp code and the COE model. There are three modules integrated in the PSO algorithm by using MATLAB: (1) the aerodynamic performance analysis based on BEM theory; (2) the structural performance analysis by PreComp code; and (3) the design constraints by beam theory and strength theory. The basic parameters of the PSO algorithm are defined as: the population size (p = 80), maximum number of generations (g = 200), learning factors ($c_1 = c_2 = 0.5$), variable dimensions ($n = n_{aerodynamic} + n_{structural} = 36$), and the inertia factor (0.85).

Figure 9 shows the flowchart of the optimization process. The aerodynamic and structural variables for blades and the tower are randomly generated as the initial population. The aerodynamic variables are used to control the blade shape, the structural variables are used to control the webs locations and number of layers along the blade, and the variables for the tower are used to control the tower height and structural parameters. BEM theory is applied to calculate the annual energy production and the aerodynamic loads which will cause the blade deflection and tower deformation. Moreover, the PreComp code is adopted to evaluate the mass and stiffness of the blade sections. The requisite parameters are airfoil profiles, chord length and twist angle, materials characteristics, internal configurations of blade sections and the lay-up distributions. When a wind turbine is acquired during the optimization process, the blade and tower constraints are validated. By analyzing the results, a new wind turbine design with minimum COE will be outputted.

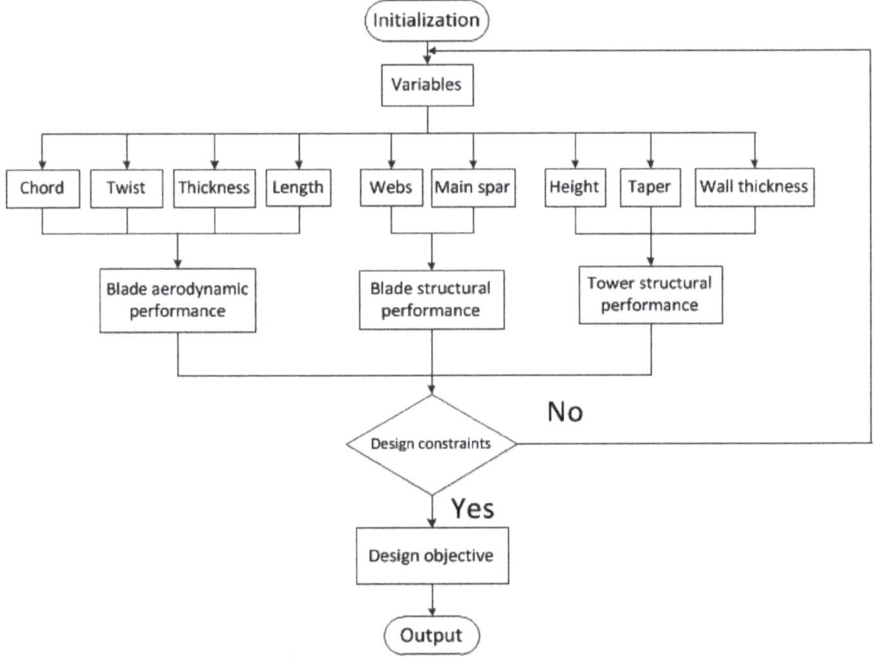

Figure 9. Flowchart of the optimization process.

5. Results and Discussion

In this section, the results are presented and analyzed. Most of the components are a function of the parameters such as rotor diameter and rated power, as shown in Figure 10. The relationship shows that the rated power also has a strong impact on COE. Thus, the optimization procedures considering rated power or not (set as 2.1 MW) are shown.

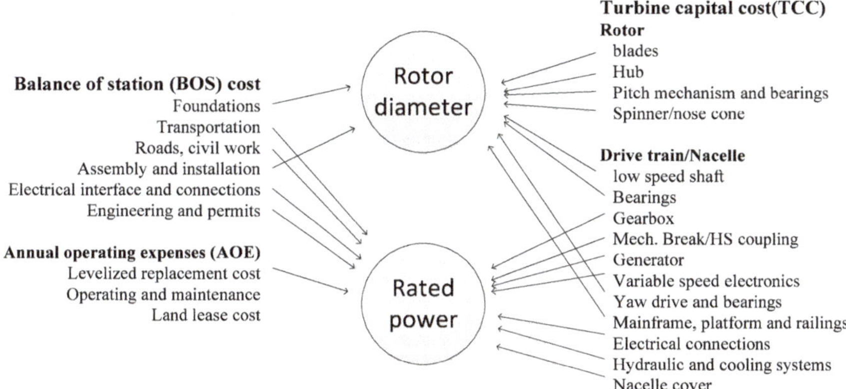

Figure 10. Relationship of rotor diameter and rated power to the components of wind turbine.

After the design procedure is finished, the iterative course of the optimization process by considering rated power is shown in Figure 11. It can be seen that the optimal results converge when the number of iteration steps reach 80.

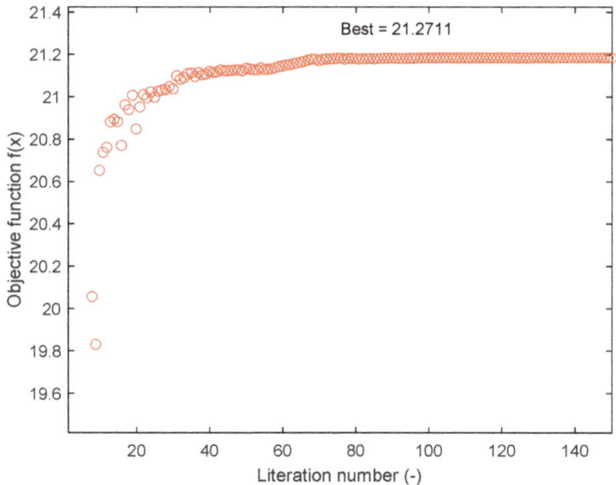

Figure 11. The iterative course for the optimization process considering rated power.

5.1. Blade Optimization Results

The chord and twist distributions of the new blade acquired by the optimization process are shown in Figure 12. The optimization process considering rated power is referred to as "CRP" for brevity. The chord distributions of the new blades decrease slightly, and the optimal chord distribution considering rated power is less than that without. The reason is that the optimal blade length acquired from the design considering rated power is less than that of the original blade and the optimal one without considering the rated power. Due to differences in aerodynamic loads, the structural requirements of the shorter blade are less than those of the longer one. The reduction of the chord distribution can result in a lighter blade mass and a lower thrust at the blade root. The twist distributions of the new blade are almost the same as the original one, the main reason is that the airfoil series adopted in this work are fixed, and the aerodynamic characteristics of the airfoils are the same. Another reason is that the twist distribution does not have a significant impact on COE compared to the chord distribution. The thickness distributions of the new blade acquired by the optimization process are shown in Figure 13. The main reason why the thickness distributions do not change much could be that the lengths of the optimal blades do not change significantly. In addition, when the thickness reduces, the stiffness would reduce too. In order to meet the design constraints, the thickness distributions do not change much.

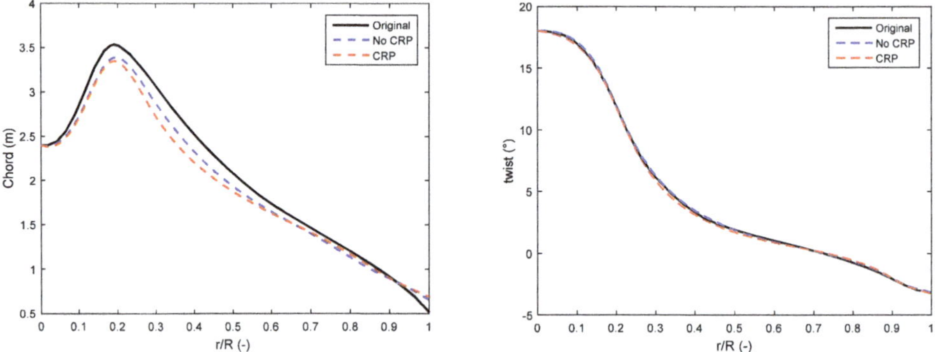

Figure 12. The comparison of chord and twist distributions between the optimal blades and the original one.

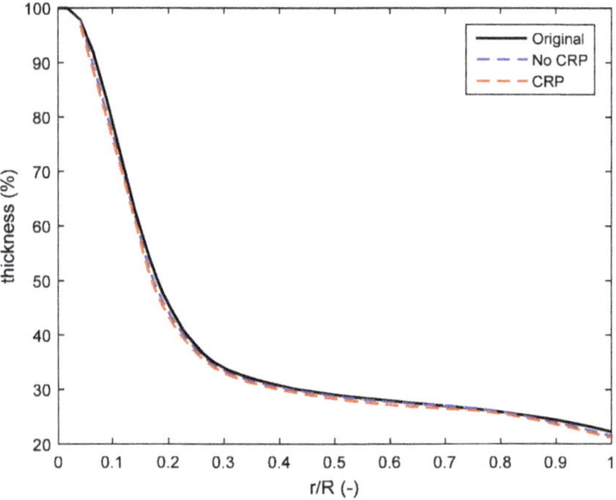

Figure 13. The comparison of thickness distributions between the optimal blades and the original one.

The Weibull distributions with annual average wind speed according to corresponding hub height (95, 117.2, 119.8 m) are shown in Figure 14. The annual average wind speed of 6 m/s of Weibull scaling factor (A = 6.8) and form factor (k = 1.91) determined from meteorological data of inland China is set as the annual average wind speed at hub height of 95 m. The annual average wind speed at optimal hub heights considering rated power (119.8 m) or not (117.2 m) are 6.3288 m/s and 6.2969 m/s, respectively. The Weibull form factor is assumed to be fixed as 1.91. The Weibull distributions show that the optimal hub heights have more probability with higher wind speed.

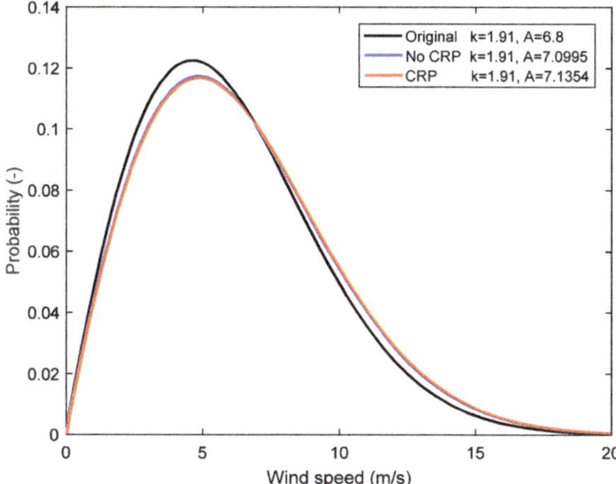

Figure 14. Weibull probability distribution of wind speed (6 m/s, 6.2969 m/s and 6.3288 m/s).

Figure 15 shows the comparison of the annual energy production between the optimal blades and the original one. Due to the differences of the Weibull distribution of the average wind speed at the hub height, the shape curve of the energy production between the original wind turbine and the optimal ones is not proportional. The energy production of the optimal wind turbine is slightly less than the original blade with a range of 3 m/s to about 8 m/s wind speed and is much higher from a speed of about 8 m/s to cut out speed (20 m/s). The results indicate that the AEP of the wind turbine (6.0921 GW) without considering rated power is higher than the original one (5.783 GW), but the one considering rated power (5.8364 GW) is similar to the original one.

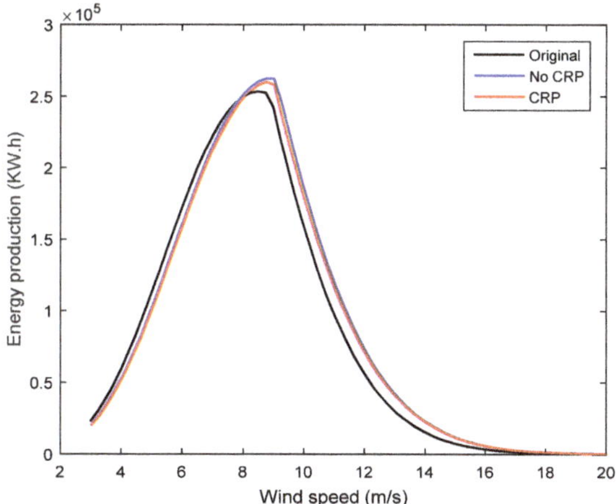

Figure 15. The comparison of the annual energy production (AEP) between the optimal blades and the original one.

Figure 16 shows that the comparison of the power curves between the optimal blades and the original one. The results indicate that the new blade without considering the design value of rated power as the variable can achieve rated power of 2.1 MW at the rated wind speed of about 9.2 m/s. The design value of rated power of the optimal wind turbine by considering rated power as the variable is 1.9 MW, and can achieve rated power at the rated wind speed of about 9.1 m/s. Both designs need higher wind speeds to achieve the rated power compared to the rated wind speed of 9 m/s of the original wind turbine. The main reason is that the lengths of the new blades are shorter than those of the original. This will result in less capacity in energy capture.

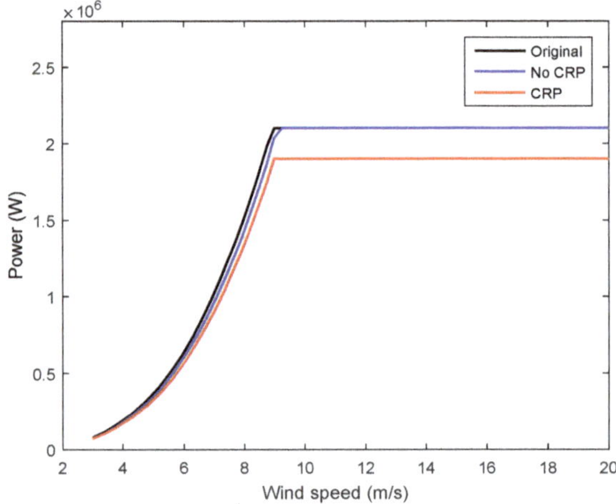

Figure 16. The comparison of the power curves between the optimal blades and the original one.

Figure 17 shows the distributions for spar caps and trailing edge reinforcement. The thickness of the spar caps and trailing edge reinforcement has increased compared to that of the baseline blade. The main reason for this is that the optimal blade needs to meet the structural constraints due to the large decrease of chord length distributions.

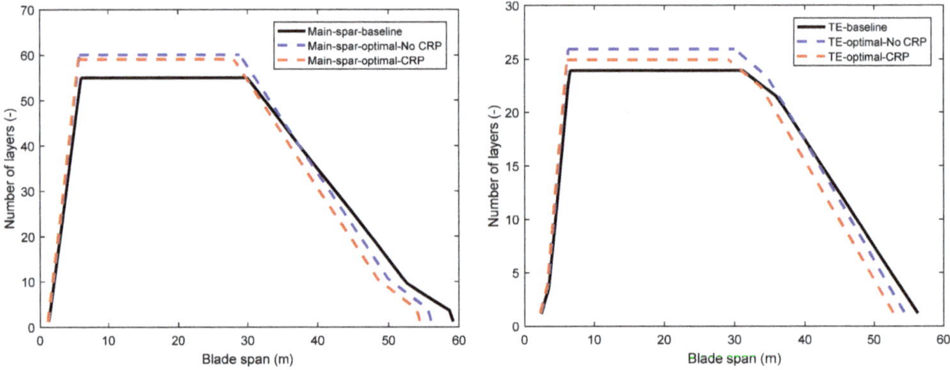

Figure 17. Structural thickness distributions of the baseline and the optimized blades.

The detailed results of the design objectives of minimizing the cost of energy are shown in Table 5. The longest blade is the original one and the shortest blade is the optimization considering rated power. The optimal wind turbines have COE of 0.0478 $/kWh and 0.047 $/kWh (CRP), compared to that of 0.057 $/kWh of the initial one.

Table 5. Optimization design results of the blade.

Parameter	Initial Value	Optimal Value	Optimal Value (Rated)
Length (m) /Mass (kg)	59.8/13,240	57.84/12,540	56.2/11,620
Tip deflection (m)	7.576	6.92	6.31
AEP (GW)	5.7830	6.0921	5.8324
Cost of energy ($/kWh)	0.057	0.0478	0.047
Tip speed ratio	10.6	9.8	9.6

5.2. Tower Optimization Results

The distributions of outer diameters and wall thicknesses along the tower height are shown in Figure 18. Both of the optimal towers have larger outer diameter and thickness distributions to meet the structural constraints. Table 6 shows the comparison of the optimization results between the optimal towers and the original one. Both of the optimal towers are taller than the original one. With the increase of tower height, the annual average wind speed also increases. Hence, the rotor can capture more wind power to improve the AEP. However, with the increase of wind speed, the aerodynamic loads acting on the rotor and hence the thrust will also increase. The tower's structural design needs to consider that an increase in thrust on the tower top will cause a larger deformation. The tower mass increases when the tower height increases. Compared to the original tower, the two optimal towers have an increase in mass of about 43.65% and 42.49% (CRP). The main reason is that the tower height with 118m and 120m has higher annual average wind speed, and the greater rotor thrust requires a better structural performance with thicker walls. The height of the optimal tower considering rated power is higher but its mass is less than the one without considering rated power. The main reason for this is that the wind turbine with longer blades has larger thrust acting on the tower top.

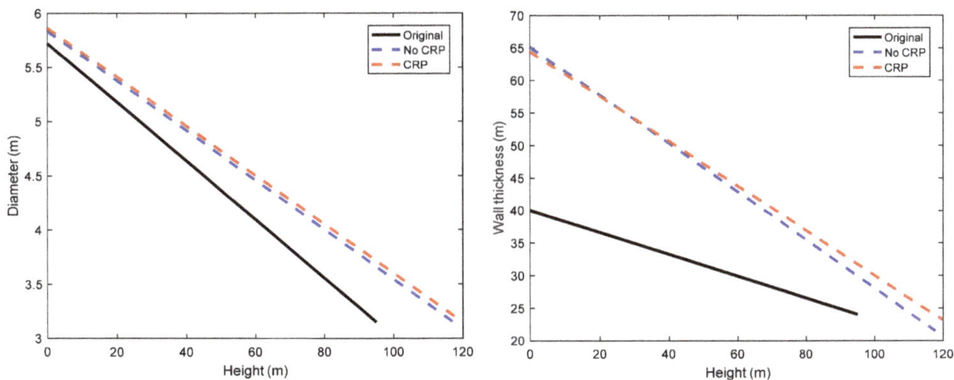

Figure 18. Tower outer diameter and wall thickness distributions of the baseline and the optimized towers.

Table 6. Optimization design results of the tower.

Parameter	Initial Value	Optimal Value	Optimal Value (Rated)
Height (m) /Mass (kg)	95/186,780	117.2/268,320	119.8/266,140
Top deformation (m)	0.65	0.828	0.816

Table 7 shows the costs of rotor, tower, and foundation in the cases of the original design, the optimal design (without CRP) and the optimal design (with CRP). The rotor costs of the optimal wind turbines are reduced compared with that of the original one. The main reason is that the optimal blades have a shorter length and lighter blade mass. The tower costs of the optimal wind turbines are increased compared with the original one due to the higher height and thicker wall distributions. Because of the shorter blade length and significant decrease in chord length distributions, the rotor thrust of the optimal wind turbine is less than that of the original one. Thus, the foundation costs do not change much, although the tower of the optimal wind turbine is higher than that of the original one.

Table 7. Costs of rotor, tower, and foundation.

Parameter	Rotor Cost ($1000)	Tower Cost ($1000)	Foundation Cost ($1000)
Original wind turbine	866.56	280.17	83.79
Optimal design (without considering rated power (CRP))	787.81	402.48	88.84
Optimal design (with CRP)	723.92	399.21	87.62

6. Conclusions

In this work, a wind turbine optimization for low wind speed areas is performed to investigate the relationship between rotor diameter and hub height. With the characteristic of high aspect ratio of the blade used in regions of wind Class III, the deflection of the longer blade is larger than the blade used in regions of wind Class I. According to this characteristic, the design needs to pay more attention to structural performance. In order to minimize the COE, the yield by increasing rotor diameter is less efficient than that by adding hub height, despite the fact that there is a large initial capital cost for the tower.

The AEP is evaluated by the determination of the electric energy generation as a function of average annual wind speed. The capacity of power generation is related to the aerodynamic performance of the rotor which is calculated by BEM theory and the annual average wind speed based on the hub height. The structural performance of the blade is calculated by the CLT method and the blade stability is estimated by the imposed constraints. The hub height has an impact on the annual average wind speed as well as AEP. A wind turbine with higher hub height has better wind resources but needs more efficient structural design. Although the larger rotor diameter and higher tower can generally produce more energy, the capital cost for that is not worthwhile. In low wind speed areas, it is not suitable to increase the rotor diameter alone and a solution needs to combine the whole system. For the purpose of minimizing the COE, increasing the rotor diameter for more annual energy production is not sufficient to outweigh increasing the cost of the wind turbine. The rated power is set as a variable to investigate its impact on COE. The results show that the design value of the rated power has a strong impact on COE.

Author Contributions: J.C. conceived and designed the project; H.Y. performed the computations and analyzed the results; H.Y., J.C. and X.P. wrote the paper.

Funding: This work is supported by National Natural Science Foundation of China (No.: 51175526).

Conflicts of Interest: The authors declare no conflict of interest.

References

1. International Electrotechnical Commission. *IEC 61400-1: Wind Turbines Part 1: Design Requirements*; International Electrotechnical Commission: Geneva, Switzerland, 2005.
2. Cristina, V.H.; Thomas, T.; Anahí, V.P. *JRC Wind Energy Status Report—2016 Edition*; Market, Technology and Regulatory Aspects of Wind Energy; Publications Office of the European Union: Luxembourg, 2017; pp. 10–11.
3. Wichser, C.; Klink, K. Low wind speed turbines and wind power potential in Minnesota, USA. *Renew. Energy* **2008**, *33*, 1749–1758. [CrossRef]
4. Wang, X.; Shen, W.-Z.; Zhu, W.-J.; Sørensen, J.-N.; Jin, C. Shape optimization of wind turbine blades. *Wind Energy* **2009**, *12*, 781–803.

5. Lee, J.; Lee, K.; Kim, B. Aerodynamic optimal blade design and performance analysis of 3 MW wind turbine blade with AEP enhancement for low-wind-speed-sites. *J. Renew. Sustain. Energy* **2016**, *8*, 1–12. [CrossRef]

6. Barnes, R.H.; Morozov, E.V.; Shankar, K. Improved methodology for design of low wind speed specific wind turbine blades. *Compos. Struct.* **2015**, *119*, 677–684. [CrossRef]

7. Barnes, R.H.; Morozov, E.V. Structural optimisation of composite wind turbine blade structures with variations of internal geometry configuration. *Compos. Struct.* **2016**, *152*, 158–167. [CrossRef]

8. Nicholson, J.C. Design of Wind Turbine Tower and Foundation Systems: Optimization Approach. Master's Thesis, University of Iowa, Iowa City, IA, USA, 2011.

9. Wang, L.; Kolios, A.; Luengo, M.M. Structural optimisation of wind turbine towers based on finite element analysis and genetic algorithm. *Wind Energy Sci. Discuss.* **2016**. [CrossRef]

10. Wright, A.K.; Wood, D.H. The starting and low wind speed behaviour of a small horizontal axis wind turbine. *J. Wind Eng. Ind. Aerod.* **2004**, *92*, 1265–1279. [CrossRef]

11. Pourrajabian, A.; Ebrahimi, R.; Mirzaei, M. Applying micro scales of horizontal axis wind turbines for operation in low wind speed regions. *Energy Convers. Manag.* **2014**, *87*, 119–127. [CrossRef]

12. Singh, R.K.; Ahmed, M.R. Blade design and performance testing of a small wind turbine rotor for low wind speed applications. *Renew. Energy* **2013**, *50*, 812–819. [CrossRef]

13. Salih, N.A.; Mohammed, A.H.; Talha, A.; Kamel, A.K. Experimental and theoretical investigation of micro wind turbine for low wind speed regions. *Renew. Energy* **2018**, *116*, 215–223.

14. Maki, K.; Sbragio, R.; Vlahopoulos, N. System design of a wind turbine using a multi-level optimization approach. *Renew. Energy* **2012**, *43*, 101–110. [CrossRef]

15. Fingerish, L.; Hand, M.; Laxson, A. *Wind Turbine Design Cost and Scaling Model*; Technical Report; National Renewable Energy Laboratory: Golden, CO, USA, 2006.

16. Bortolotti, P.; Bottasso, C.L.; Croce, A. Combined preliminary–detailed design of wind turbines. *Wind Energy Sci.* **2016**, *1*, 71–88. [CrossRef]

17. Ning, A.; Damiani, R.; Moriarty, P.J. Objectives and constraints for wind turbine optimization. *J. Sol. Energy Eng.* **2014**, *136*, 041010. [CrossRef]

18. Bir, G.S. *User's Guide to PreComp: Pre-Processor for Computing Composite Blade Properties*; National Renewable Energy Laboratory: Golden, CO, USA, 2005.

19. Hansen, M.O.L. *Aerodynamics of Wind Turbines*, 2nd ed.; Earthscan: Abingdon, UK, 2008.

20. Timmer, W.A.; van Rooij, P.R.J.O.M. Summary of the Delft University wind turbine dedicated airfoils. *J. Sol. Energy Eng.* **2003**, *125*, 48–96. [CrossRef]

21. Drela, M. *XFoil: An Analysis and Design System for Low Reynolds Number Airfoils*; Technical Report; MIT Dept. of Aeronautics and Astronautics: Cambridge, MA, USA, 1989.

22. Reddy, J.N. *Mechanics of Laminated Composite Plates and Shells: Theory and Analysis*; CRC: New York, NY, USA, 2003.

23. Timoshenko, S.; Woinowsky-Krieger, S.; Woinowsky, S. *Theory of Plates and Shells*; McGraw-Hill: New York, NY, USA, 1959.

Article

Optimize Rotating Wind Energy Rotor Blades Using the Adjoint Approach

Lena Vorspel [1,], Bernhard Stoevesandt [2] and Joachim Peinke [1,2]

[1]   ForWind, University of Oldenburg, Ammerländer Heerstr. 114-118, 26129 Oldenburg, Germany; peinke@uni-oldenburg.de
[2]   Fraunhofer IWES, Küpkersweg 70, 26129 Oldenburg, Germany; bernhard.stoevesandt@iwes.fraunhofer.de
*   Correspondence: lena.vorspel@forwind.de; Tel.: +49-441-798-5056

Received: 5 June 2018; Accepted: 6 July 2018; Published: 10 July 2018

Abstract: Wind energy rotor blades are highly complex structures, both combining a large aerodynamic efficiency and a robust structure for lifetimes up to 25 years and more. Current research deals with smart rotor blades, improved for turbulent wind fields, less maintenance and low wind sites. In this work, an optimization tool for rotor blades using bend-twist-coupling is developed and tested. The adjoint approach allows computation of gradients based on the flow field at comparably low cost. A suitable projection method from the large design space of one gradient per numerical grid cell to a suitable design space for rotor blades is derived. The adjoint solver in OpenFOAM is extended for external flow. As novelty, we included rotation via the multiple reference frame method, both for the flow and the adjoint field. This optimization tool is tested for the NREL Phase VI turbine, optimizing the thrust by twisting of various outer parts between 20–50% of the blade length.

Keywords: rotor blade optimization; blade parametrization; computational fluid dynamics; OpenFOAM; gradient-based; adjoint approach

---

1. Introduction

The generated electric power from wind turbines increases linearly with the swept rotor area. Therefore, wind turbines grow larger and larger. Otherwise, limits in maximal mountable rotor weight and constraints coming from material science restrict the growth of rotor blades. Further limiting factors are alternating wind loads, gusts, turbulences and rotational forces. As wind turbines are normally designed for a life time of 20–25 years, the estimated total amount of rotations is about 150×10^6, which leads to 5×10^8–10×10^8 load changes. Periodic loads due to the vertical wind velocity gradient and the tower shadow, each one of them proportional to the blade rotation frequency, add up to this [1]. Added to those load changes due to the rotational movement are the wind loads, which were shown to be highly intermittent [2]. At the same time, most favorable sites for wind turbines with little turbulence intensity and large average wind speeds are already exploited, such that either repowering or worse sites are chosen in order to increase the generated wind power [3]. For slow average wind speeds, slender blades are favorable, which react even on small wind velocities due to a large aerodynamic efficiency. This enforces the trend towards long, slender blades. Such blades are prone to elastic deformations and even more influenced by wind load changes.

In return, this might lead to critical fatigue of the material, both from blades, but also from bearings and support structures. However, flexibility and light structures offer the possibility of passive control. The pitch systems of wind turbine blades are too slow to react on highly intermittent load fluctuations within the wind field, the use of structural bend twist coupling, which reacts instantly on aerodynamic loads, is a promising concept to reduce the amount of load changes and therefore of fatigue by a pre-designed deformation of the blade [4]. Especially adaptive changes in the twist of the blade can reduce the angle of attack and thus lead to a significant load reduction, which results in a longer life time or less cumbersome maintenance of the wind turbine blades and therefore in the end in lower specific costs of energy [5,6].

Based on the actual development in rotor design, we present in this work an optimization tool for the aerodynamics of rotating rotor blades within the open-source toolbox for numerical simulations OpenFOAM. This tool is tested for the optimization of the twist distribution in order to reduce the overall thrust force. Gradient-based optimization offers on the one hand the large advantage of a high-quality search direction for optima. On the other hand, only local optima are found. Thus, this method is suitable if there is not more than one optimum or if the initial point is not too far away from the global optimum. In our case, the aerodynamics of wind turbine rotor blades can be considered as sophisticated, so that the mentioned disadvantage of the gradient-based method can be neglected, while the advantage of a high-quality search direction is utilized. The continuous adjoint approach is used for the computation of gradients. In contrast to other methods, the adjoint approach is marked by the independence of the computational cost on the amount of design parameters, thus no full simulation per each single design parameter is required [7]. This qualifies the adjoint approach for the combination with computationally expensive Computational Fluid Dynamics (CFD). The optimization is based on the solution of steady-state flow equations of the current blade geometry. Obtained gradients are then evaluated to find the optimum. The implementation of the continuous adjoint equations in OpenFOAM was firstly done by Ohtmer [8] for ducted flows. Several publications show the applicability of the adjoint approach in combination with CFD, both for external and ducted flows [9–12]. However, these works showed the application either for low Reynolds numbers, for non-rotating fields or using porosity of numerical cells to impact the shape of ducts. Thus, the usage of the adjoint approach in wind energy needed further development and investigation. The adjoint solver for ducted flows could be used in the field of diffuser-augmented wind turbines. The generalized actuator disc theory, proposed by Jamieson [13] and extended by Lui and Yoshida [14], could be used within OpenFOAM to compute optimal difuser geometries with maximal effective diffuser efficiency. In this work, we focus on the optimization of rotor blade geometries of wind turbines in open flow. Therefore, the adjoint equations in OpenFOAM were adopted for external flow and the resulting gradients were verified against finite differences by Schramm [15,16]. It was shown in Schramm [16] that the adjoint approach can be used in wind energy relevant cases, in that work shown for the optimization of an airfoil slat. In a previous work, the adjoint optimization in OpenFOAM has been tested against other optimization strategies, like finite differences and Nelder-Mead. The independence from the amount of design parameters was shown there, as well as the applicability for wind energy relevant profiles [17].

In this work, the adjoint approach is used for rotating geometries and large Reynolds numbers. To the authors' knowledge, this combination poses a novelty and offers new areas of application. It should be noted that the rotational flow around rotor blades is complex and it is generally already challenging to achieve low residuals for such flows using stationary flow solvers. In addition, the NREL Phase VI turbine is stall regulated and the chosen velocity already corresponds to the rated velocity. Thus, the flow is separated at the inner blade region, leading to poor convergence behavior of the flow field. The adjoint field is driven by the flow field so that the errors in the flow field transfer directly to the adjoint field and decrease the convergence quality of it. This challenging set-up is used purposely to thoroughly test the optimization tool, its convergence and applicability.

The paper is structured as follows. First, the theory of the underlying optimization is given, including the gradient estimation via the adjoint approach and their processing is explained in Section 2. In Section 3, the simulation set-up in OpenFOAM is shown, as well as the optimization set-up and cases. Finally, results of the optimization are given in Section 4 followed by a conclusion in Section 5.

2. Optimization Method

To design a blade using bend-twist-coupling as a passive control mechanism, a favorable deformed blade shape that ensures the optimal load reductions has to be determined. The deformed blade shape can then be integrated into the structural blade design where layer lay-up or geometrical sweep are options to achieve the favorable deformations. To find those favorable deformations, an optimization tool for the aerodynamic shape of the rotor blades based on the continuous adjoint approach is implemented in OpenFOAM. The gradients are then projected to central nodes, adopting geometry updates like in Fluid–Structure Interaction (FSI) based on a beam. In this work, the optimization is directly coupled with the CFD toolbox, allowing direct communication and data usage. This tool is used to find deformations that reduce loads under certain load conditions. Our approach is that the general geometrical shape of the 2D profiles is constant, and only elastic deformations are allowed, such as changing the twist angle along the blade. In order to test the optimization tool and its performance, the optimizations are conducted for the same blade under the same inflow conditions but with a variable movable part of the blade and different aimed values. Therefore, between 20% and 50% of the outer blade are allowed to be changed with respect to its twist distribution. This investigation allows an assessment of the effect of complex lay-up within large to short parts of the blades, respectively. The optimization in this work is based on the reduction of the thrust that effects the blade root bending, tower bending and other forces. Although no constraints were used, it is possible to include limiting ranges for other forces in further applications of this tool.

In addition, the developed optimization tool allows other definitions of design parameters and cost functions.

Details regarding the adjoint approach are discussed next.

2.1. Gradient Estimation by the Adjoint Approach

The adjoint approach allows for computing the gradient of a cost function with respect to the design parameters independently from the amount of design parameters. The main idea is to build a Lagrangian function, which includes both the cost function to be optimized and the constraint and thus can be used for constrained optimization problems. For optimization based on CFD, the flow equations have to be satisfied in each optimization point. Thus, the constraining functions are the Navier–Stokes equations (NSE)

$$\mathbf{R_u} = (\mathbf{u} \cdot \nabla)\,\mathbf{u} + \nabla p - \nabla \cdot (2\nu D\,(\mathbf{u})), \tag{1}$$

$$\mathbf{R}_p = -\nabla \cdot \mathbf{u}, \tag{2}$$

with $\mathbf{u}$ and p being the flow velocity and pressure, respectively. The rate of strain tensor for any given vector x is defined as $D\,(\mathbf{x}) = \frac{1}{2}\left(\nabla x + (\nabla x)^T\right)$. This notation is used for the velocity $\mathbf{u}$ and adjoint velocity $\mathbf{u_a}$ in the flow and adjoint field, respectively. The viscosity ν is a sum of the molecular and turbulent viscosity.

The Lagrangian function

$$L := I + \int_\Omega (\mathbf{u_a}, p_a) \cdot \mathbf{R}\, d\Omega \tag{3}$$

combines the cost function I with the NSE $\mathbf{R}$ in the flow domain Ω. The Lagrangian multipliers, which allow the combination of the cost function with the NSE, are the adjoint velocity $\mathbf{u_a}$ and adjoint pressure p_a.

The adjoint field equations

$$-2D\,(\mathbf{u_a})\,\mathbf{u} = -\nabla p_a + \nabla \cdot (2\nu D\,(\mathbf{u_a})) - \frac{\partial I_\Omega}{\partial \mathbf{u}}, \tag{4}$$

$$\nabla \cdot \mathbf{u} = \frac{\partial I_\Omega}{\partial \mathbf{p}}, \tag{5}$$

result from the variation of the Lagrangian function with respect to the flow and design variables. A step-by-step derivation of the adjoint field is shown in the work of Othmer [18].

In this work, the flow equations are solved following the SIMPLE algorithm [19]. The continuous adjoint equations are similar to the ones from the flow field, such that similar numerical solving strategies can be applied. The frozen turbulence approach is applied, meaning that the viscosity from the flow field is also applied for the adjoint field, without additional adjoint turbulence equations [7,20].

In the following, flow gradients refer to changes in the flow field over distance, whereas gradients are the gradients of the cost function w.r.t. the design variables.

The two discussed fields have to be solved once, respectively, to derive all gradients at each numerical grid node on the blade geometry surface. Thus, the additional cost for the gradient information at any amount of design parameters equals the cost of solving a second set of partial differential equations that are derived from the flow field equations. One consequence from this is that the adjoint field is driven by the flow field on which it is based. This can also be seen in Equations (4) and (5) and leads to a transfer of the error made in the flow field computations to the adjoint field. Thus, the adjoint field is less stable. The large advantage of the adjoint method compared to other gradient estimation methods is that the amount of necessary solutions of the flow field doesn't scale with the amount of design parameters. This is especially desirable when working with gradient based optimization based on fully resolved simulations.

2.2. Gradient Projection and Evaluation

For an optimization procedure, it is an essential point to work out suitable parameters. The definition of the rotor blade's shape can be done in various ways. It was shown in [17] that the free parametrization, i.e., all numerical grid points are design parameters, is not suitable for optimizing rotor blades. This results from large flow field gradients in the chordwise direction and in the spanwise direction, which transfer to the computed gradient. In areas of large gradients, the deformation is much larger, which leads to bumps and uneven surfaces. Instead of the use of all grid points, one can select parameters that define the blade geometry, like the twist distribution, thickness of airfoils along the blade and so forth, which define the blade geometry. For the developed optimization tool, an automatized routine is implemented that allows the projection of the gradients at each numerical grid cell to a reduced design space.

The flow around airfoils and rotor blades is highly sheared with large flow gradients both in velocity and pressure. Therefore, the numerical grid has to be appropriately fine at those areas in order to catch the flow phenomena. Typical values for the amount of cells in the chordwise direction vary between 250 and 350. The adjoint approach leads to gradient information on each of these cells. To use a parametrization with less design parameters, a projection from these 250–350 gradients on less design parameters is implemented. The main idea for the parametrization used in the optimization tool comes from FSI using a finite beam. The gradients are treated like forces acting on the blade and therefore projected on central nodes. The principle of the parametrization is shown in Figure 1.

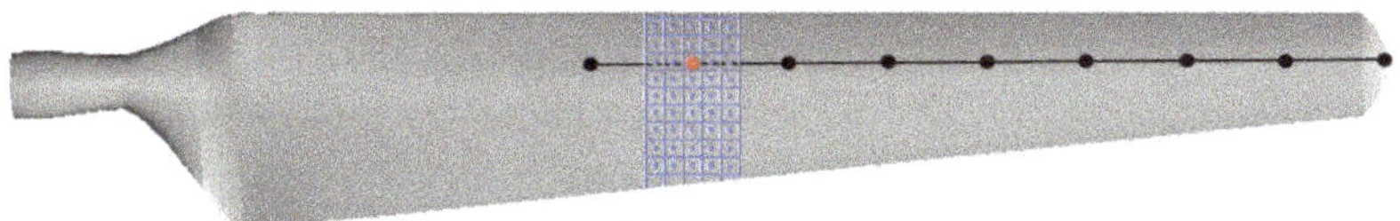

Figure 1. Principle of gradient projection for a wind turbine rotor blade.

As a pre-processing step, the blade is split into a selected amount of elements. Each of these elements has a central node, which is represented by a big red dot in Figure 1. All big dots represent one final design parameter each. The amount of design parameters is chosen beforehand by the user. The number can be controlled by adapting a dictionary that is defined for the usage of the developed tool.

In this exemplary case, there are nine design parameters. For one exemplary element, the structured grid is shown with thin blue lines. In the figure, the represented grid has been coarsened for the sake of clarity. Within each cell of this grid, the small blue dot represents the local gradient of each grid cell, computed via the adjoint approach. The adjoint gradients are then transferred onto the nodes depending on the distance of the local cell to the central node of the element. Evaluating the projected gradients then leads to a movement of the nodes, which act like beam nodes. The difference to FSI is that the structure is not solved, and it is assumed that any movement given by the gradients can be realized. This assumption is necessary for the presented optimization. The final, optimized geometries can be used within the blade design process as an input to the design of the structure, for example when passive control shall be applied. Therefore, the structure is not final when the optimization takes place. In addition, the movement of the blade due to a structural response might interfere with the movement given by the gradients.

The pre-condition for this projection is that an elastic deformation shall be guaranteed, where the blade cross sections are constant. However, even this parametrization on a reduced design space takes into account the gradient information of each numerical grid cell, as each local gradient is used within the projection. This makes the usage of the adjoint approach convenient.

The gradients of each numerical cell that belongs to one central node of an element are transferred to two vectors. One is acting like a force, that can be compared to the bending force in FSI and one is acting like a torsional moment that can be compared to a moment of force responsible for twisting in FSI. After projection, each element of the blade has these two gradient vectors. In a second step, for each of these two vector types, the maximal value among all elements is used for normalization, respectively. This leads to a similar range of values, with a maximal magnitude of one, for both. With this, weighting factors can be introduced that allow for augmenting one of the movements, when needed. Optional smoothing by neighboring elements is implemented and can be switched on and off via a dictionary. In between the elements, an interpolation is used to avoid kinks on the blade surface. This interpolation as well as the mesh update is taken from an in-house FSI solver by Dose [21].

The projected gradients can then be used within an optimization algorithm for gradient based optimization. The most direct way to use the gradients is via the steepest descent algorithm, which is also used in this work. According to the direction and dimension of the gradient, changes in the geometry are performed for all selected design parameters. Depending on the slope of the design space, the optimal search direction (pointing to the minimum) can differ largely from the gradient direction. In these cases, the steepest descent search results in a highly inefficient, but typical zigzag pattern. Consequently, the method of steepest descent has found little applicability in practice, since its performance is in most cases largely inferior to second-order methods, like the Quasi-Newton method. For high-dimensional design spaces, though, the method of steepest descent can be advantageous, since it does not require the estimation of the inverse Hessian, which scales with the number of design variables [22].

3. Simulation and Optimization Set-Ups

This is a conceptual work in which we want to show a new optimization approach for wind energy applications. Therefore, the optimizations are conducted for the NREL Phase VI turbine. This turbine is a stall-regulated, two-bladed turbine with a rated power of 19.8 kW. It was used for experiments in the NASA/Ames 24.4 m × 36.6 m wind tunnel [23] and is often used for validation of CFD. The S809 airfoil with 21% thickness was used for the construction of the blade [24]. In this work, the cone angle of the rotor is $0°$ and the pitch angle of the blade is $3°$.

To include rotational effects in steady-state simulations, the Multiple Reference Frame (MRF) is used. The rotation is modelled via a source force, that adds a rotational component to each cell. This way, the rotor does not have to move within the mesh reducing the computational effort, as no mesh update is necessary to include the rotation. The MRF method is also adopted to the adjoint field, so that the used schemes and solution methods for both fields stay similar.

The mesh used in this work was created by the bladeBlockMesher (BBM) developed by Rahimi et al. [25] for the automatic mesh generation for wind turbine rotor blades. The grid quality and validation based on the NREL Phase VI turbine is also shown in their publication. In this work, a grid with 5.2 million cells is used with a y^+-value below 1. A symmetry plane is used to exploit the symmetry of the two bladed rotor. In Figure 2 the domain and the mesh around a cut along the chord of the blade is shown.

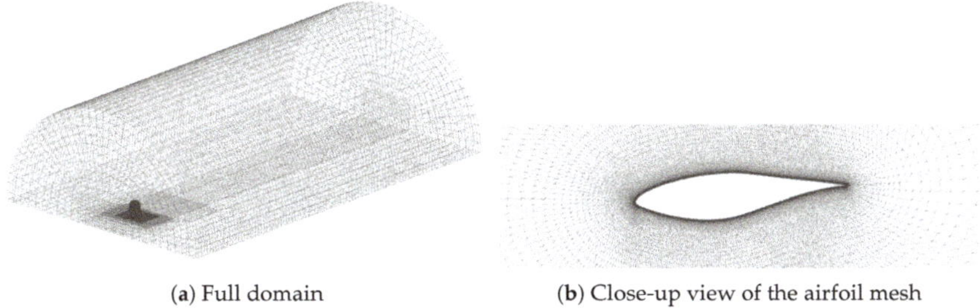

(**a**) Full domain   (**b**) Close-up view of the airfoil mesh

Figure 2. Numerical grid. The flow is coming from the left in (**a**). In (**b**), the o-type mesh around a cut through the blade is shown.

A rounded tip is created and meshed via the BBM. The k-ω-SST turbulence model is chosen for the flow field, whereas for the adjoint field the frozen turbulence approach is applied. This approach was used in various publications and it was shown that the gradient quality is adequate [7,18]. Standard boundary conditions are applied, similar to those in Rahimi [25]. The inflow velocity is 10 m/s and the rotational speed is 7.54 rad/s, i.e., 72 rpm and the density equals 1.225 kg/m^3. This velocity leads the stall regulated turbineto already partially separate flow at large parts of the blade. Steady state CFD computations of this turbine at separated flow are known to differ from the experimental data at those separated areas [25]. As it was shown above, the adjoint field is driven by the flow field and therefore it is expected that the adjoint field converges even less as good as the flow field. Still, this velocity was chosen in this work, as even small changes in AoA have a large impact on the forces, when separation is reduced.

As the adjoint field moves upstream, the boundary conditions are switched compared to the flow field. The adjoint velocity has a Neumann boundary condition at the inlet, and a Dirichlet boundary condition at the outlet and at the blade. The adjoint pressure has a Dirichlet boundary condition at the inlet and Neumann boundary condition at the outlet and blade. For a higher stability, the adjoint velocity at the hub is put to zero. This was necessary, as the transition between hub and blade root led to large instabilities in the adjoint field.

It was shown in Section 2.1 that the adjoint field is driven by the primal field. Therefore, preconverged solutions of the flow field are used in order to stabilize the adjoint equations. A convergence of the velocity up to a residual in the range of 10^{-7} is aimed before the adjoint field equations are switched on. An initial study using the same numerical grid based on the initial geometry was conducted to find adequate amounts of CFD iterations to reach good convergence of both fields. As a result, 25,000 iterations are used solving the flow field based on the SIMPLE algorithm [19]. An additional 15,000 iterations are used for solving the adjoint and flow field before the optimization tool is started. These pre-converged fields are used for all optimizations, as the initial blade geometry and flow boundary conditions are identical. Starting from the total amount of 40,000 iterations used for pre-convergence of the fields, the optimizer uses an additional 500 iterations after each geometry update for a pre-convergence of the flow field starting from the last flow field data before the geometry update. An additional 2500 iterations are then used solving both fields. After a total of an additional 3000 iterations, the gradients are evaluated. As the allowed movement of the design parameters, leading to an update of the geometry, is not too large, the flow and adjoint field before and after a geometry update are also rather similar. Thus, using the variables of the last step before a geometry update leads to a good pre-converged solution and a reduced amount of needed iterations between two optimization steps compared to the iterations needed for the first step. The convergence of the optimization, in this case the evaluation of the thrust force, is tested after the 500 iterations solving the flow field.

The general definition of the minimization problem

$$\min f\left(s\right) \quad s \in R^{N}, \tag{6}$$

with s being the N design parameters of the cost function f in R^{N}, is specified in this work for the thrust force F_T. The cost function $f(s)$ is defined as a least square function of the thrust

$$f(s) = \tfrac{1}{2}\left(F_T\left(s\right) - F_{T*}\right)^2. \tag{7}$$

This allows to set an aimed value of the thrust F_{T*} which the optimizer then tries to reach. As the thrust force has an impact on blade root bending moments and the loads transferred on the wind turbine, this force is a good candidate to be one key element when it comes to passive control.

The initial value $F_T(s_0)$ of the original blade with initial design parameters s_0 and the above given inflow and rotational speed equals 958.9 N. Two different aim values F_{T*} are chosen, namely 925 N and 900 N. This equals a reduction by 3.54% and 6.14%, respectively. Two different convergence limits are applied for the simulations, i.e., an interval of $\pm1.5\%$ and $\pm2.5\%$ around the aimed thrust force. Whenever the new geometry leads to a thrust within the given range around the aimed thrust, the optimization is ended. This can lead, for example, to a final value only a bit below of $1.025 \cdot F_{T*}$ in the case of the $\pm2.5\%$ range. This convergence criterion is chosen due to the instable flow phenomena transferring to the convergence of the gradients. Three different radial parts r of the blade are chosen to be optimized. The tip region is left out because the tip vortex leads to more complex flows and therefore leads to inferior gradient quality. From the blade length of 5 m, the following radial parts r are chosen, with 0 m being the blade root and 5 m the blade tip: $r_1 = 2$ m–4.5 m, $r_2 = 3$ m–4.5 m and $r_3 = 3.75$ m–4.75 m. The short range r_3 is split into five elements, while, for the middle and long range, nine elements are used. An overview of the optimization cases is given in Table 1.

Table 1. Optimization set-ups for different parts of the blade, aimed values and convergence interval around the aimed value.

Case	F_{T*} (N)	Convergence Interval %	Convergence Interval (N)	Radial Part r (m)	Blade Part %	Length (m)
1				2–4.5	50	2.5
2		1.5	886.5–913.5	3–4.5	30	1.5
3	900			3.75–4.75	20	1
4				2–4.5	50	2.5
5		2.5	877.5–922.55	3–4.5	30	1.5
6				3.75–4.75	20	1
7				2–4.5	50	2.5
8		1.5	911.125–938.875	3–4.5	30	1.5
9	925			3.75–4.75	20	1
10				2–4.5	50	2.5
11		2.5	901.875–948.125	3–4.5	30	1.5
12				3.75–4.75	20	1

4. Results

In this section, the performance and results of the developed optimization tool are presented. The implementation of the rotating adjoint field is one major step towards an optimization tool for wind energy application. First, the flow and adjoint fields around the blade are shown and analysed with focus on the adjoint field and the output of the MRF method applied to it. Then, the optimization tool is used for the test cases mentioned above. The additional twist deformations for various cases are compared, as well as the convergence behavior and stability of the optimization tool.

4.1. Flow and Adjoint Field

In some engineering applications, the adjoint field is used as a sensitivity map within the design process. For the design of complex geometries, like cars, the solution of the adjoint field offers the information where the surface should be changed to achieve a more aerodynamical efficient behavior. For this type of application, the adjoint field is only solved once and the surface is not adopted automatically within an optimization. However, it is also possible to compute gradients from the adjoint field, which then are used for optimization. This is also done in this work. To get an impression of the adjoint field, it is shown for the initial geometry at a cut through the blade, showing the airfoil geometry, at an exemplary radial position of 3 m of the span. In Figure 3, cuts through the blade are colored by the velocity (left) and adjoint velocity (right).

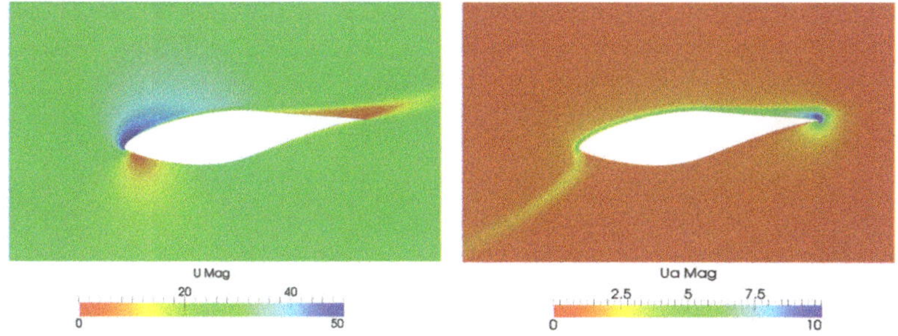

Figure 3. Magnitude (Mag) of the velocity U and adjoint velocity Ua around the blade at a cut at 3 m of the span.

In the left plot, typical flow around rotor blades can be seen. The stagnation point is below the leading edge, which shows a positive angle of attack. From the stagnation point onwards along the first part of the suction side, the flow is accelerating and finally the flow is slightly separated at the last third of the suction side. The adjoint velocity in the right figure shows an upstream movement, which is also expected from the theory. It is high along the suction side, where the flow gradients are also large. Then, the adjoint velocity moves upstream from the stagnation point on and follows a path around the angle of attack.

The trailing edge of an airfoil is a discontinuity that impairs the convergence of the adjoint field. The flow is separated in the rear part of the blade, thus turbulent effects are dominant in this region. At this part of the blade, the frozen turbulence approach decreases the convergence behavior of the adjoint field. This can also be seen at the adjoint velocity plot, showing a peak at the trailing edge. A used workaround for this problem in literature is to exclude the cells close to discontinuities from the evaluated gradients [16]. This would imply an additional processing of the gradients. For this work, such an exclusion of certain blade areas is not done, as this could overlap with the effects of the chosen projection of gradients. Testing this projection was one of the main goals in this work, which is why overlapping effects are avoided. However, it could easily be added for further applications of the optimization tool using the existing utilities available in OpenFOAM.

In order to show the flow around the blades in its rotational component, the Q-criterion in terms of the vorticity was used in Figure 4.

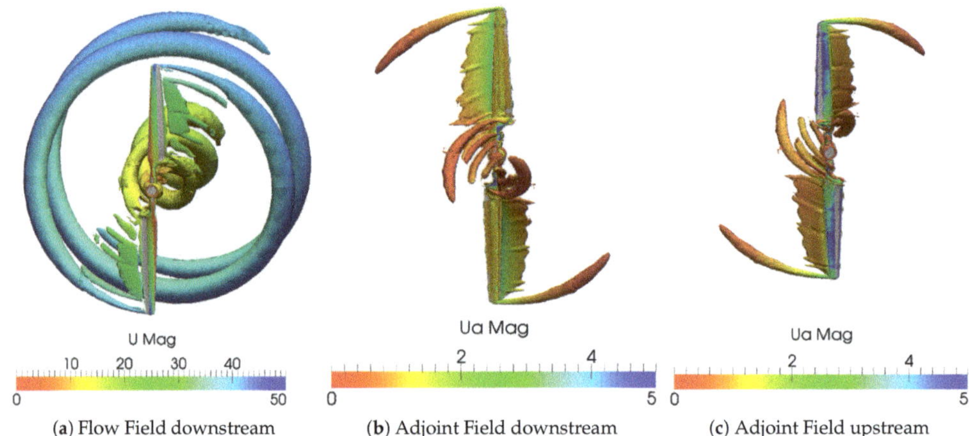

(a) Flow Field downstream (b) Adjoint Field downstream (c) Adjoint Field upstream

Figure 4. Field visualisation based on the Q-criterion in terms of vorticity. The full rotor for the flow field in (a) and for the adjoint field in (b,c) is shown. For good comparability, the adjoint field is shown in (b) in the same direction, as the flow field, i.e., downstream. As it moves upstream, it is secondly shown in the direction of the propagation of the adjoint field in (c).

The shown flow structures then are colored by the magnitude of the flow velocity and the adjoint velocity, respectively. The computations were performed for both the flow and adjoint velocity. As the adjoint field moves upstream, both views of the rotor are shown for the adjoint field. One view is in the downstream direction, according to the view for the flow field and one is in the upstream direction according to the propagation of the adjoint field. In Figure 4a, the tip and root vortex of the rotor can be seen, as well as low velocities close to the stagnation point and increasing velocities towards the tip of the blade. The tip and root vortex impair the flow field convergence at the root and tip area of the blade, respectively. The computed flow variables are less stable and reach inferior convergence in stationary simulations due to the instationary flow phenomena. As the gradients are evaluated along the movable part of the blade, which are according to Table 1 starting at 2 m spanwise and reach up to 4.75 m of the span, the influence of the vortices is supposed to be small for the gradient computation. In Figure 4b, it can be seen that similar vortex structures are built for the adjoint field, whereas they are moving upstream and rotating in the other direction compared to the flow field. Figure 4c shows the adjoint field looking at the suction side of the rotor. The large adjoint velocities at the suction side can be seen. In addition, the influence of the tip and root vortex, leading to a reduced adjoint velocity at the tip and root of the blade, respectively, is evident. Large and small adjoint velocities transfer to large and small gradients, respectively. Thus, large adjoint velocities indicate areas of the shape, where changes of the geometry are the most immediate. Overall, the implemented MRF approach for the adjoint field works in a stable manner and allows the computation of adjoint approach based gradients for rotating geometries, and the implemented version shows results expected from theory.

4.2. Convergence Behavior of the Optimizations

In this work, different types of convergence are used. Numerical convergence of the computed fields is investigated as well as convergence of the optimization tool. Convergence criteria for the flow field are mentioned in Section 3. The latter is defined depending on the ratio between current cost function value and aimed cost function value. The convergence behavior of the various optimizations for different aimed values and varying lengths of the twistable part of the blade is compared in Figure 5a–d.

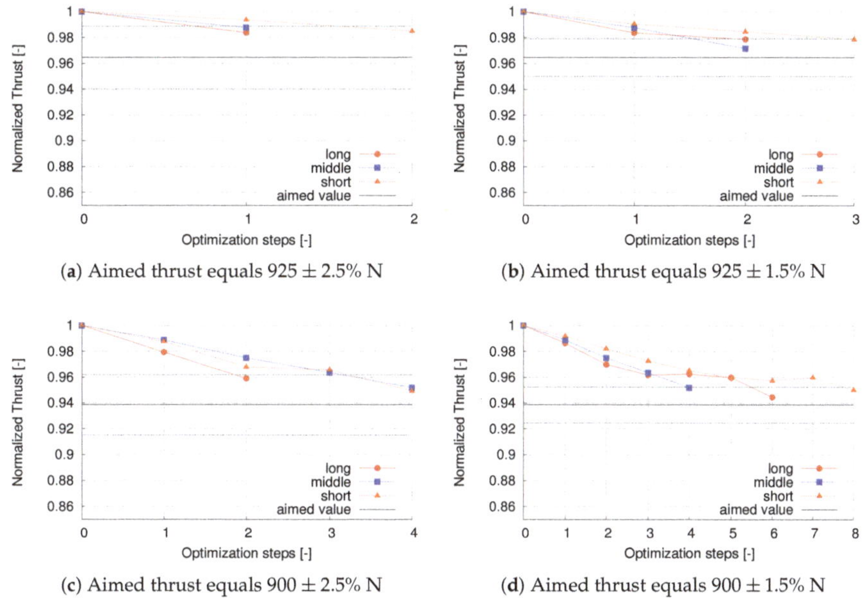

(**a**) Aimed thrust equals $925 \pm 2.5\%$ N

(**b**) Aimed thrust equals $925 \pm 1.5\%$ N

(**c**) Aimed thrust equals $900 \pm 2.5\%$ N

(**d**) Aimed thrust equals $900 \pm 1.5\%$ N

Figure 5. Development of the normalised thrust force over the optimization steps for design parameters within varying parts of the blade. Long: 2 m–4.5 m radial position, middle: 3 m–4.5 m radial position and short: 3.75 m–4.75 m radial position. The aimed values and the limits of the allowed range are shown with black and grey lines, respectively.

For all plots, the linear lines between the points are added to ease the following of the trend, whereas only the dots represent computed values of the actual geometries of the blade during the optimization. The thrust is normalized by the initial value of 958.9 N to ensure comparability of the different aimed values and convergence ranges. The aimed values equal 925 N for the upper plots (a) and (b) and 900 N for the lower plots (c) and (d). The left plots are for the wider allowed range of 2.5% around the aimed value, whereas the right plots are for a range of 1.5% around the aimed value.

All optimizations, besides the short and long blade parts for an aimed value of $900 \pm 1.5\%$ N, show a similar behavior. With each optimization step, the thrust is reduced until the upper limit of the allowed aim range is met. The two named exceptions show a zig-zagging behavior close to the bound, which is a known problem of steepest descent.

Comparing all optimizations, it can be seen that more optimization steps are needed when the aim is further away from the initial value and when the bounds are closer. Furthermore, longer movable blade parts lead to faster convergence, except for the case shown in Figure 5d. In this case, the middle length converges faster than the long length case. It was expected that longer blade parts lead to faster convergence of the optimization as larger parts of the blade are moved relatively to the inflow angle, which then has larger influence on the forces, respectively. For the case shown in Figure 5d, the sensitivity of the optimization process with respect to the user's input, like initial step-size and pre-convergence of the flow and adjoint field, was found to be the largest. This explains the outlier of the long radial section r_1 compared to r_2.

It has to be remembered that each optimization step represents a larger amount of CFD iterations, which are necessary to compute reliable gradients via the adjoint approach. For a convergence after four optimization steps, e.g., shown for the short and middle length in Figure 5c and the given pre-convergence iterations and iterations between two optimization steps specified in Section 3, this sums up to 54,500 CFD iterations per optimization. Roughly 35,000 of these iterations include solving two fields. These numbers are still very convincing for gradient based optimization based on CFD.

When the optimizations for these cases were conducted, a dependence of the convergence, appearance of zig-zagging and initial step-size were observed. As mentioned in Section 2.2, the projected gradients are normalized by the current maximal value. These normalized gradients are then multiplied by the step size, such that the maximal gradient leads to a movement according this step size and the further design points move less. Once converged, optimizations could be sped up by changing the step-size. This is shown in Figure 6 for the long blade part when the aimed thrust equals $900 \pm 1.5\%$ N.

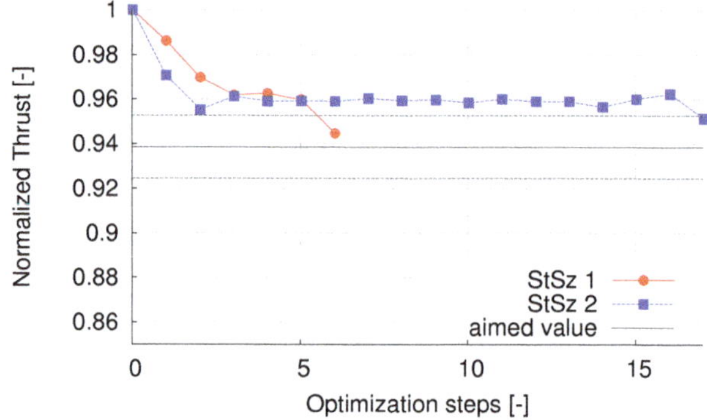

Figure 6. Development of the normalised thrust force over the optimization steps for the long blade part. The aimed force is $900 \pm 1.5\%$ N and the initial step sizes are 10^{-2} (red dots, StSz1) and 2×10^{-2} (blue squares, StSz2).

The two step sizes 10^{-2} (red) and 2×10^{-2} (blue) are compared. The larger step size leads to a faster decrease of the thrust within the first steps, but then zig-zagging happens for 14 optimization steps before convergence is reached. The case with the lower step size shows zig-zagging for two steps, before convergence is reached.

This underlines one major drawback of the continuous adjoint approach, also reported in other publications [26,27]. The user input, e.g., initial step size and set-up of the underlying flow simulation, is highly important for these optimizations, whereas other methods are more robust. Still, the shown amount of needed CFD iterations also underlines the advantage of this approach, which is its independence from the amount of design parameters. This still holds for the large Reynolds numbers and rotating geometry investigated in this work.

4.3. Additional Twist Computed by the Optimization

The final additional twist of the blade for the various optimization cases is shown in Figure 7a–c. The dots represent the moving element nodes where the gradients of each element are projected on (see Figure 1). The lines represent the smoothed movement between the nodes, as the geometry movement is also smoothed by the optimization tool. This plotted twist is then added to the initial twist of the according blade. The shown twist are the final ones, summed up over all optimization steps.

All twist values increase from the first element node towards a maximal value close to a spanwise position of 4 m. Afterwards, the additional twist decreases and even reaches negative values for the last element. The positive values in this set-up lead to a reduced angle of attack.

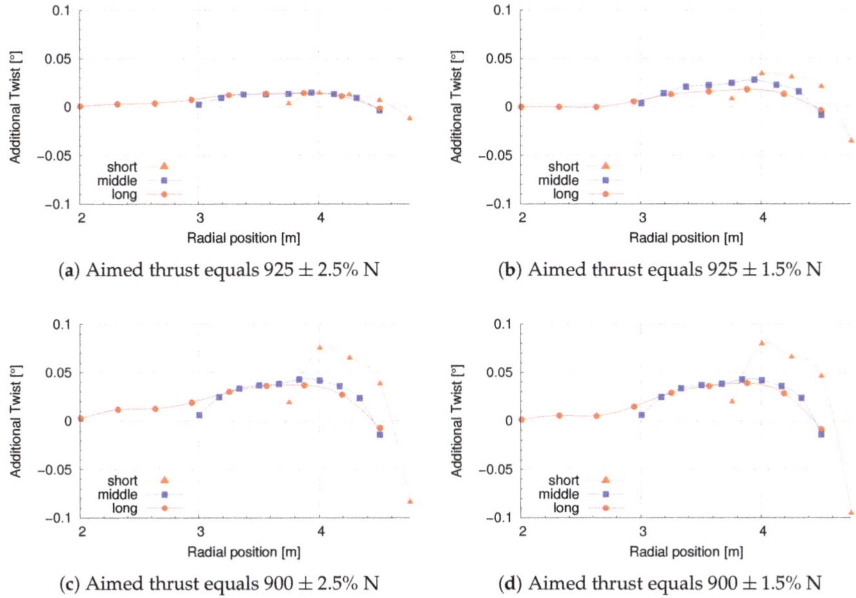

(**a**) Aimed thrust equals $925 \pm 2.5\%$ N

(**b**) Aimed thrust equals $925 \pm 1.5\%$ N

(**c**) Aimed thrust equals $900 \pm 2.5\%$ N

(**d**) Aimed thrust equals $900 \pm 1.5\%$ N

Figure 7. Development of the normalized thrust force over the optimization steps for design parameters within varying deformable parts of the blade. Long: $r_1 = 2$ m–4.5 m radial position, middle: $r_2 = 3$ m–4.5 m radial position and short: $r_3 = 3.75$ m–4.75 m radial position.

In general, those plots all show two trends: the shorter the movable section, the larger the additional twist. Furthermore, the more restrictive the aimed value, the more deformation is necessary to reach this value. This verifies the developed tool for optimizing rotating rotor blades.

To investigate this behavior further, the spanwise production of thrust of the initial NREL Phase VI blade is plotted in Figure 8.

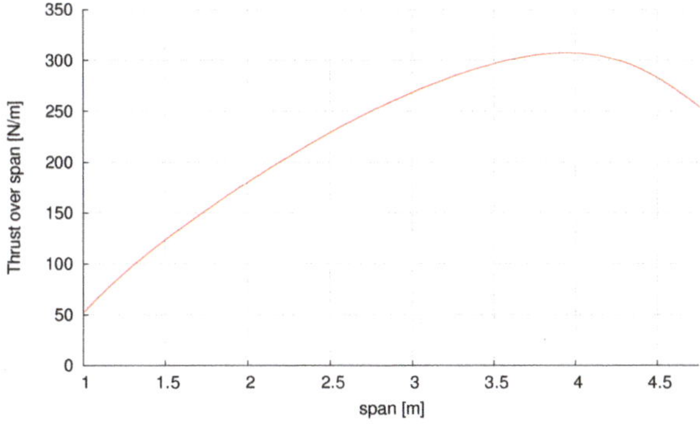

Figure 8. Spanwise production of thrust of the original NREL Phase VI blade from 1 to 4.75 m in radial direction.

It can be seen that the thrust production increases in the spanwise direction until a peak at a radial position of $r = 4$ m. Afterwards, the production of thrust decreases. Thus, the gradients follow the trend of production of thrust in size and direction, leading to a larger deformation around $r = 4$ m and decreased deformation towards the tip.

Larger movements in the outer blade part, which are aerodynamically more efficient, also have a larger impact on the thrust. This is apparently also represented by the gradients leading to a larger deformation in the outer regions. Close to the tip of the blade, the vortices seem to destabilize the adjoint field and thus also the accuracy of the computed gradients. This is also underlined by the larger negative twist values, which where not expected beforehand, of the short deformable blade part r_3 in comparison to the two longer ones. The deformable blade part r_3 ends at 4.75 m, whereas the other two deformable blade parts r_1 and r_2 end at 4.5 m. Thus, the influence from the tip vortex seems to be stronger for r_3, which reaches 0.25 m further towards the tip.

The conducted optimizations are chosen to test the developed optimization tool. It is therefore important to reduce the occurring effects as far as possible and avoid converse effects that could arise when using cost functions based on various force components and constraints. For future applications of the developed tool, it is a challenge to extend this method to more important design constraints and more elaborate definitions of the cost function.

5. Conclusions

An optimization tool for rotating wind energy rotor blades at realistic Reynolds numbers was developed and implemented based on the adjoint approach in OpenFOAM. In this work, it was tested for the NREL Phase VI blade. The rotation is represented using the MRF method, both for the flow and adjoint field. A projection for gradients to nodes similar to beam nodes in FSI tools was developed and implemented. This optimization tool is very flexible with respect to the cost function and other optimization settings, using the same dictionary-based input as OpenFOAM.

In this work, the cost function is a least square of the thrust and various aimed values and allowed ranges for the cost function were defined. The different test cases were used to test the stability and convergence behavior of the developed optimization tool. Defining a rather easy cost function allows for interpreting the optimization results based on rotor blade theory.

The 12 test cases all converged within a reasonable amount of CFD iterations. It was found that pre-convergence of both the flow and adjoint field is absolutely necessary in order to achieve usable gradient information. Still, a large influence of the chosen step size for the evaluation of the gradients was observed. This effect got more evident with aimed values further away from the initial value and tighter ranges around this aimed value. An initial study for a range of step sizes, which leads to good convergence behavior, might help to avoid zig-zagging. In the future, gradients can be used in more efficient optimization algorithms like SQP or Quasi-Newton. These algorithms also avoid zig-zagging that can appear when purely following the gradient direction. In addition, known implementations of these algorithms already have an automatized adjustment of the step sizes in run time.

In addition, an adjoint turbulence model is a promising approach to increase the quality of the gradients in future work and thereby the optimization convergence and stability. Using this tool for a blade of a turbine that is not stall regulated could perform better with respect to convergence of the flow and adjoint field.

The applicability of the developed tool, even for a challenging test case based on a rotating, stall regulated turbine at high Reynolds numbers, was shown in this work. When the suggested improvements of the gradient estimation are applied, the optimization tool is expected to perform in a more stable manner. This will allow the usage within the design process—for example, using more elaborate cost functions and being under consideration of constraints.

Author Contributions: Conceptualization, L.V.; Methodology, L.V. and B.S.; Software, L.V.; Writing—Original Draft Preparation, L.V.; Writing—Review and Editing, J.P.; Supervision, J.P.; Project Administration, B.S.

Funding: This research received no external funding.

Acknowledgments: For the simulations, we use the high performance computing cluster Eddy of the University of Oldenburg [28].

Conflicts of Interest: The authors declare no conflict of interest.

References

1. Gasch, R.; Twele, J. *Wind Power Plants*, 2nd ed.; Springer: Berlin/Heidelberg, Germany, 2012; ISBN 3-642-22938-1.
2. Mücke, T.; Kleinhans, D.; Peinke, J. Atmospheric turbulence and its influence on the alternating loads on wind turbines. *Wind Energy* **2011**, *14*, 301–316. [CrossRef]
3. Kumar, Y.; Ringenberg, J.; Depuru, S.; Devabhaktuni, V.K.; Lee, J.W.; Nikolaidis, E.; Andersen, B.; Afjeh, A. Wind energy: Trends and enabling technologies. *Renew. Sustain. Energy Rev.* **2016**, *53*, 209–224, doi:10.1016/j.rser.2015.07.200. [CrossRef]
4. Fedorov, V.; Berggreen, C. Bend-twist coupling potential of wind turbine blades. *J. Phys. Conf. Ser.* **2014**, *524*, 012035. [CrossRef]
5. Berry, D.T.A. *Design of 9-Meter Carbon-Fiberglass Prototype Blades: CX-100 and TX-100*; Sandia National Laboratories: Livermor, CA, USA, 2007.
6. Ashwill, T. Passive Load Control for Large Wind Turbines. In Proceedings of the 51st AIAA/ASME/ASCE/AHS/ASC Structures, Structural Dynamics, and Materials Conference, Orlando, FL, USA, 12–15 April 2010.
7. Soto, O.; Löhner, R. On the computation of flow sensitivities from boundary integrals. In Proceedings of the 42nd AIAA Aerospace Sciences Meeting and Exhibit, Reno, NV, USA, 5–8 January 2004.
8. Othmer, C.; de Villiers, E.; Weller, H. Implementation of a continuous adjoint for topology optimization of ducted flows. In Proceedings of the 18th AIAA Computational Fluid Dynamics Conference, Miami, FL, USA, 25–28 June 2007.
9. Othmer, C. Adjoint methods for car aerodynamics. *J. Math. Ind.* **2014**, *4*, 6, doi:10.1186/2190-5983-4-6. [CrossRef]
10. Soto, O.; Löhner, R.; Yang, C. A stabilized pseudo-shell approach for surface parametrization in CFD design problems. *Commun. Numer. Methods Eng.* **2002**, *18*, 251–258. [CrossRef]
11. Anderson, W.; Venkatakrishnan, V. Aerodynamic design optimization on unstructured grids with a continuous adjoint formulation. In Proceedings of the 35th Aerospace Sciences Meeting and Exhibit, Reno, NV, USA, 6–9 January 1997.
12. Grasso, F. Usage of Numerical Optimization in Wind Turbine Airfoil Design. *J. Aircr.* **2011**, *48*, 248–255. [CrossRef]
13. Liu, Y.; Yoshida, S. An extension of the Generalized Actuator Disc Theory for aerodynamic analysis of the diffuser-augmented wind turbines. *Energy* **2015**, *93*, 1852–1859. [CrossRef]
14. Jamieson, P. Generalized Limits for Energy Extraction in a Linear Constant Velocity Flow Field. *Wind Energy* **2008**, *11*, 445–457. [CrossRef]
15. Schramm, M.; Stoevesandt, B.; Peinke, J. Lift Optimization of Airfoils using the Adjoint Approach. In Proceedings of the EWEA 2015, Europe's Premier Wind Energy Event, Paris, France, 17–20 November 2015.
16. Schramm, M.; Stoevesandt, B.; Peinke, J. Simulation and Optimization of an Airfoil with Leading Edge Slat. *J. Phys. Conf. Ser.* **2016**, *753*. [CrossRef]
17. Vorspel, L.; Schramm, M.; Stoevesandt, B.; Brunold, L.; Bünner, M. A benchmark study on the efficiency of unconstrained optimization algorithms in 2D-aerodynamic shape design. *Cogent Eng.* **2017**, *4*, doi:10.1080/23311916.2017.1354509. [CrossRef]
18. Othmer, C. A continuous adjoint formulation for the computation of topological and surface sensitivities of ducted flows. *Int. J. Numer. Methods Fluids* **2008**, *58*, 861–877, doi:10.1002/fld.1770. [CrossRef]
19. Ferziger, J.; Peric, M. *Computational Methods for Fluid Dynamics*; Springer:Berlin/Heidelberg, Germany, 2001.
20. Papadimitriou, D.; Giannakoglou, K. A continuous adjoint method with objective function derivatives based on boundary integrals, for inviscid and viscous flows. *Comput. Fluids* **2007**, *36*, 325–341. [CrossRef]
21. Dose, B.; Rahimi, H.; Herráez, I.; Stoevesandt, B.; Peinke, J. Fluid-structure coupled computations of the NREL 5MW wind turbine blade during standstill. *J. Phys. Conf. Ser.* **2016**, *753*, 022034. [CrossRef]

22. Nocedal, J.; Wright, S. *Numerical Optimization*; Springer: Berlin/Heidelberg, Germany, 2006.
23. Hand, M.M.; Simms, D.; Fingersh, L.; Jager, D.; Cotrell, J.; Schreck, S.; Larwood, S. *Unsteady Aerodynamics Experiment Phase VI: Wind Tunnel Test Configurations and Available Data Campaigns*; National Renewable Energy Laboratory: Golden, CO, USA, 2001.
24. Tangler, J.L.; Somers, D.M. *NREL Airfoil Families for HAWTs*; Citeseer: Pennsylvania, PA, USA, 1995.
25. Rahimi, H.; Daniele, E.; Stoevesandt, B.; Peinke, J. Development and application of a grid generation tool for aerodynamic simulations of wind turbines. *Wind Eng.* **2016**, *40*, 148–172, doi:10.1177/0309524X16636318. [CrossRef]
26. Giles, M.B.; Duta, M.C.; Müller, J.D.; Pierce, N.A. Algorithm Developments for Discrete Adjoint Methods. *AIAA J.* **2003**, *41*, 198–205. [CrossRef]
27. Müller, J.D.; Cusdin, P. On the performance of discrete adjoint CFD codes using automatic differentiation. *Int. J. Numer. Methods Fluids* **2005**, *47*, 939–945, doi:10.1002/fld.885. [CrossRef]
28. HPC Cluster EDDY of the University of Oldenburg. Available online: https://www.uni-oldenburg.de/en/school5/sc/high-perfomance-computing/hpc-facilities/eddy/ (accessed on 9 July 2018).

Article

Development of a CFD-Based Wind Turbine Rotor Optimization Tool in Considering Wake Effects

Jiufa Cao [1], Weijun Zhu [1,2,*], Wenzhong Shen [2], Jens Nørkær Sørensen [2] and Tongguang Wang [3]

[1] School of Hydraulic Energy and Power Engineering, Yangzhou University, Yangzhou 225009, China; jfcao@yzu.edu.cn
[2] Department of Wind Energy, Technical University of Denmark, 2800 Kgs. Lyngby, Denmark; wzsh@dtu.dk (W.S.); jnso@dtu.dk (J.N.S.)
[3] Jiangsu Key Laboratory of Hi-Tech Research for Wind Turbine Design, Nanjing University of Aeronautics and Astronautics, Nanjing 211100, China; tgwang@nuaa.edu.cn
* Correspondence: wjzhu@yzu.edu.cn

Received: 31 May 2018; Accepted: 25 June 2018; Published: 28 June 2018

Abstract: In the present study, a computational fluid dynamic (CFD)-based blade optimization algorithm is introduced for designing single or multiple wind turbine rotors. It is shown that the CFD methods provide more detailed aerodynamics features during the design process. Because high computational cost limits the conventional CFD applications in particular for rotor optimization purposes, in the current paper, a CFD-based 2D Actuator Disc (AD) model is used to represent turbulent flows over wind turbine rotors. With the ideal case of axisymmetric flows, the simulation time is significantly reduced with the 2D method. The design variables are the shape parameters comprising the chord, twist, and relative thickness of the wind turbine rotor blades as well as the rotational speed. Due to the wake effects, the optimized blade shapes are different for the upstream and downstream turbines. The comparative aerodynamic performance is analyzed between the original and optimized reference wind turbine rotor. The results show that the present numerical optimization algorithm for multiple turbines is efficient and more advanced than conventional methods. The current method achieves the same accuracy as 3D CFD simulations, and the computational efficiency is not significantly higher than the Blade Element Momentum (BEM) theory. The paper shows that CFD for rotor design is possible using a high-performance single personal computer with multiple cores.

Keywords: wind turbine blade optimization; computational fluid dynamic; actuator disc; wake effect; Non-dominated Sorting Genetic Algorithm (NSGA-II)

---

1. Introduction

To reduce the cost of material and improve the aerodynamic efficiency of wind energy conversion systems, aerodynamic optimization of wind turbine rotors is important. Modern horizontal axis wind turbine (HAWT) rotors are aerodynamically optimized to efficiently extract wind energy. The physical limit to extract energy from wind is known as the Betz limit. The wind energy extraction process is through the reduction of wind speed passing the wind turbine rotor. The zero wake flow scenario will yield full energy absorption from wind, which certainly cannot be achieved. Since a wake always exists behind any wind turbine, and wind turbines are mostly clustered in a wind farm, the design and optimization of wind turbines in the wake situation are indeed necessary to represent some real-world wind farm cases.

From the theoretical to practical rotor aerodynamic design, the final products suffer from an accumulation of many aerodynamic and mechanical losses, such as tip loss, root loss, blade surface roughness change loss, blade shape manufacturing accuracy loss, drive train efficiency loss and wake effect loss. Despite of mechanical and manufacturing imperfections, most of aerodynamic optimizations have considered all the above-mentioned facts. The effect of wind turbine wake has been a very popular topic which is essentially important for wind farm layout optimization. However, in terms of rotor design itself, the wake effect is seldom included due to the inherent property of aerodynamic design tools. For example, the Blade Element Momentum (BEM) method is fast for optimization of single rotors, but it cannot deal with wake flow problems; the CFD-based methods are more accurate but practically not used for rotor optimization purposes.

In the field of wind turbine design and optimization, various simulation methods are involved for single rotors and for wind farms. For single rotor aerodynamic force calculations, the wind turbine aerodynamic modeling relies mainly on the following approaches: BEM [1–6]-based methods, Vortex Wake (VW)-based methods [7–12], and CFD-based methods [13–16]. The CFD methods either apply body-fitted grid or AD/AL/AS (Actuator Disc, Actuator Line, Actuator Surface) techniques. BEM has been the most efficient method which has been widely used to solve engineering problems. As the most efficient tool, BEM coupling with an optimization algorithm is mainly adopted as the method of wind turbine rotor design [13,17,18]. The VW methods are more flexible in the sense that it deals with the turbulent wake state that cannot be done with BEM. A study of BEM and VW methods are seen in [19] where general agreements are achieved for most of the considered cases but not for the high-thrust cases. However, VW methods are still not efficient for optimization purposes, and the accuracy in wake flow predictions affects the wind turbine performance to some extent. It is worthwhile to point out that the method to be used is relied on the matter of interest. From the multi-wind turbine optimization point of view, it is important to focus on both blade aerodynamics and turbulent wake. Therefore, BEM, VM and CFD based on curvilinear grid are commonly used for load calculations. However, rotor optimization requires a high computational demand, and so far CFDs are popularly used for wind turbine aerodynamic analysis. On the other hand, rotor aerodynamics has never been isolated with wake dynamics. However, existing models are individually used either for rotor design or wake study. Due to the inherent limitations of the BEM method, it cannot be directly applied to study wind turbine wakes. The more accurate methods, such as VM and CFD methods can be good choices to investigate both rotor aerodynamics and wake dynamics, but VM is not designed for long distance wake simulation and CFD is very expansive to study long range wake propagations, such that it is very hard to combine rotor aerodynamics with wake effects for rotor optimization purpose.

For a wind turbine cluster or wind farm, the velocity deficit causes a momentum loss in its wake before reaching the downstream wind turbines. Optimal design of the most upstream turbine will certainly increase the power product of the first turbine or perhaps the entire first row of a wind farm. However, for the second and third turbines downwind in the wake of the first turbine, the original design may not be the best configuration. Although this problem is realized through wind farm wake studies, but a practical design of wind turbines with wake effect is difficult either due to the model accuracy or high computational demand. The current work relates the rotor aerodynamics to the wake dynamics of multi-wind turbines. Therefore, modeling the velocity deficit is significantly important. Numerous studies were carried out in connection of large wind farm developments [20,21]. More recent work of Göçmen et al. [22] presented a good overview of various wake models that were developed at DTU. Some selected wake modelling approaches are listed in Table 1. The methods are divided into four groups according to their computational time demand. The model of Jensen [23] and Larsen [24,25] belong to class 1 that are the fastest engineering models. Some modifications of class 1 models become necessary to better represent the flow physics. For example, Tian et al. [26,27] proposed a new Jensen model that uses a cosine wake shape instead of the top-hat shape for the velocity deficit. The VW models and its modifications are adopted from aircraft aerodynamics and well-suited for wind turbine aerodynamics in terms of accuracy and efficiency [8,9,28]. Efforts were also made to reduce the computational time using CFD-based approaches [29,30], where the source equations are highly simplified. From the CFD point of view, many turbulence models are investigated and improved either for the purpose of Atmospheric Boundary Layer (ABL) simulations or of wake dynamics. Direct Numerical Method (DNS) is not yet applicable for rotor simulations, instead, the Reynolds Averaged Navier-Stokes/Large Eddy Simulation/Detached Eddy Simulation (RANS/LES/DES) and the hybrid turbulence models are widely used [31–34]. For wake investigations, in particular wake meandering and wake stability analysis, the traditional CFD with body-fitted mesh is very expensive even for one- or two-rotor analysis. So far, the best configuration to achieve a good quality and simulation speed is the so-called AL/AD techniques [35–37] which still apply the above-mentioned turbulence models but relatively faster. Using the similar mesh topology as the AL/AD techniques, the Lattice Boltzmann Method (LBM) [38] and Immersed Boundary Method (IBM) [39,40] have an advantage of reducing the computational time, but still demand a large grid number for high Reynolds number flows.

Table 1. Selected wake models.

Class	Methodology	References
Class 1	Jensen Dynamic Wake Meandering	Jensen (1983) Larsen (2007, 2009)
Class 2	Vortex Wake Method	Bareiss (1993), Voutsinas (1993), Wang (1998)
Class 3	Fuga Eddy Viscosity Model	Ott (2011) Ainslie (1988)
Class 4	RANS/LES + curvilinear RANS/LES + AD/AL/AS LBM/IBM	Sørensen (2014), Thé (2017), Allah (2017), Abdulqadir (2017) Sørensen, Shen et al. (1998, 2002, 2009, 2010, 2012), Troldborg (2014) Deiterding (2015), Angelidis (2015), Zhu (2013)

The main purpose of the current paper is to develop a wind turbine rotor optimization tool that has a numerical accuracy of Class 4 and has a computational efficiency better than Class 2. More specifically, the current study will not only focus on single wind turbine rotor optimization but also for two or more aligned wind turbine rotors. In this scenario, the 3D CFD with RANS/AL/AD approaches are suitable methods to perform rotor aerodynamic analysis, but too computationally heavy to perform aerodynamic optimization. Benefiting from the axisymmetric characteristic of a HAWT, the 3D RANS/AD equations can be reduced into a 2D formulation. The axisymmetric boundary condition applied for a 2D simulation is a good assumption in particular for rotor optimization of ideal flow cases, such as no wind shear, no yawing effects. The novelty of such method is to achieve the same accuracy as using 3D CFD simulation with a significantly higher efficiency. Then, based on the multi-objective and multi-variable optimization algorithm, the design variables are the shape parameters comprising the chord, twist, and relative thickness of wind turbine rotor blade. Finally, the blade optimization is achieved based on an in-house developed flow solver combining with an optimization algorithm. A multi-objective optimization model is proposed with maximizing the overall power efficiency for multi-wind turbines. The non-dominated sorting genetic algorithm-II (NSGA-II) is used to handle the complex process. The novel part of the present study is to combine the CFD solver with an efficient optimization algorithm, such that the single or multiple wind turbine rotors are optimized at the same time, which finally obtain a global high aerodynamic efficiency.

The paper is organized as follows: Section 2 describes the governing equations of the aerodynamic models; Section 3 presents numerical validations of the current wake model; in Section 4, the aerodynamic methods are applied for rotor optimization. Conclusions are given in the final section.

2. Numerical Methods

The recent research development on wind turbine flow modeling has largely focused on CFD approaches. With the assumption of incompressible flows over rotor blades, the NS equations are solved together with RANS, LES, DES etc. The rotor blades can either be modeled as a solid body surface or with AD/AL/AS techniques. In addition to that, more numerical issues are involved such as inflow turbulence, and atmospheric boundary layer which lead to possible modifications of turbulence models.

2.1. Governing Equations

The incompressible RANS equations in 2D form reads

$$\frac{\partial u_i}{\partial x_i} = 0 \ \ (i = 1, 2) \tag{1}$$

$$\frac{\partial u_i}{\partial t} + u_j \frac{\partial u_i}{\partial x_j} = -\frac{1}{\rho} \frac{\partial P}{\partial x_i} + \frac{\partial}{\partial x_j}(v \frac{\partial u_i}{\partial x_j} - \overline{u'_i u'_j}) + f_{ext} \tag{2}$$

$$-\rho \overline{u'_i u'_j} = \mu_t \left(\frac{\partial u_i}{\partial x_j} + \frac{\partial u_j}{\partial x_i} \right) - \frac{2}{3}\rho k \delta_{ij} \tag{3}$$

where x_i and u_i are the displacement and velocity components in the inertial coordinate system. t is the time, ρ is the density of fluid, P is the static pressure, v is the kinematic viscosity coefficient, μ_t is the turbulent eddy viscosity δ_{ij} is the Kronecker delta function, k is the turbulent kinetic energy, f_{ext} is the momentum source terms representing possible external forces.

2.2. Turbulence Model

Herein the turbulence model is discussed with the aim to model the nonlinear Reynolds stresses terms in Equations (2) and (3). Accurate predictions of turbulent kinetic energy and dissipation are the key factors in turbulent wake modelling. In the present work, the modified k-ω turbulence model of Menter [41] is used. Menter's baseline model combines the k-ω model by Wilcox [42] in the inner region of the boundary layer with the standard k-ε model in the outer region and the free steam outside the boundary layer. The modified transport equations are written as follows:

$$\frac{\partial k}{\partial t} + \frac{\partial (u_j k)}{\partial x_j} = \frac{1}{\rho}P_k + \frac{1}{\rho}\frac{\partial}{\partial x_j}[(\mu + \sigma_k \mu_t)\frac{\partial k}{\partial x_j}] - \beta^* k\omega \tag{4}$$

$$\frac{\partial \omega}{\partial t} + \frac{\partial (u_j \omega)}{\partial x_j} = \frac{\gamma}{\rho v_t}P_k + \frac{1}{\rho}\frac{\partial}{\partial x_j}[(\mu + \sigma_\omega \mu_t)\frac{\partial \omega}{\partial x_j}] + 2(1 - F_1)\frac{\rho \sigma_{\omega 2}}{\rho \omega}\frac{\partial k}{\partial x_j}\frac{\partial \omega}{\partial x_j} - \beta \omega^2 \tag{5}$$

where the original model constants for the inner region are:

$$\sigma_{k1} = 0.5, \ \sigma_{\omega 1} = 0.5, \ \beta_1 = 0.075, \ \beta^* = 0.09, \ \kappa = 0.41, \ \gamma_1 = 0.55$$

and for the outer region

$$\sigma_{k2} = 1.0, \ \sigma_{\omega 2} = 0.856, \ \beta_1 = 0.0828, \ \beta^* = 0.09, \ \kappa = 0.41, \ \gamma_2 = 0.44 \tag{6}$$

The re-formulation of the original turbulence model contains two parts: (1) change of model constants; (2) added the source terms to maintain the turbulence level.

(1) The modified model constants are based on the work of Prospathopoulos et al. [43] in which most of the constants remain the same except for the following ones

$$\beta_1 = 0.033, \ \beta^* = 0.025, \ \gamma_1 = 0.37$$

The new constants were suggested according to measurements. With the decrease of dissipation effect, the kinetic energy towards the rotor is effectively increased, which improves the numerical accuracy for wind turbine wake flow modeling [43].

(2) For RANS solvers, the turbulence level is usually specified from the inlet of a computational domain. For the k-ω models, it is the initial k and ω values that are responsible for the turbulence level

$$k = \frac{3}{2}(u_0 I_0)^2 \tag{7}$$

$$\omega = \frac{k}{\mu}\left(\frac{\mu_t}{\mu}\right)^{-1} \tag{8}$$

where I_0 represents the ambient turbulence level and μ_t/μ is the eddy viscosity ratio. A similar approach is implemented in some commercial CFD programs such as the ANSYS Fluent software. In another approach proposed by Richards and Hoxey [44], the boundary values of u_0, k and ω are basically a function of an aerodynamic roughness height corresponding to a specific site. This approach does not apply to the current work where an axisymmetric flow is assumed, and no wall boundary is specified. No matter what boundary values are given at inlet, the free decay of turbulence intensity cannot be avoided with the current set of RANS equations. In the free-stream flow, the mean velocity gradient is zero such that the turbulence quantities decay. Therefore, the production terms and the diffusion terms in Equations (4) and (5) can be neglected. The k and ω equations are reduced to

$$\frac{\partial(u_j k)}{\partial x_j} = -\beta^* k\omega \tag{9}$$

$$\frac{\partial(u_j \omega)}{\partial x_j} = -\beta\omega^2 \tag{10}$$

Solving the partial differential equations for k and ω, the turbulence decay in the free-stream condition is obtained as a function of downstream distance

$$k = k_{in}\left(1 + \frac{\omega_{in}\beta x}{u_0}\right)^{-\beta^*/\beta} \tag{11}$$

$$\omega = \omega_{in}\left(1 + \frac{\omega_{in}\beta x}{u_0}\right)^{-1} \tag{12}$$

where k_{in} and ω_{in} are the initial values and x is the downstream distance. It is desired that these turbulence quantities specified at inflow boundary can be maintained without unphysical decay. In the work of Spalart and Rumsey [45], measurements of turbulent airfoil flows in wind tunnel were numerically reproduced with RANS computations. Even though the turbulence intensity is very low in the wind tunnel, the numerical simulations were not in perfect agreement with measurements. Inspired from the experimental and numerical work done in [45], the idea of maintaining turbulence quantities is extended to the current external wind turbine flows. The modified transport equations are

$$\frac{\partial k}{\partial t} + \frac{\partial(u_j k)}{\partial x_j} = \frac{1}{\rho}P_k + \frac{1}{\rho}\frac{\partial}{\partial x_j}[(\mu + \sigma_k \mu_t)\frac{\partial k}{\partial x_j}] - \beta^* k\omega + \beta^* k_{amb}\omega_{amb} \tag{13}$$

$$\frac{\partial \omega}{\partial t} + \frac{\partial(u_j \omega)}{\partial x_j} = \frac{\gamma}{\rho v_t}P_k + \frac{1}{\rho}\frac{\partial}{\partial x_j}[(\mu + \sigma_\omega \mu_t)\frac{\partial \omega}{\partial x_j}] + 2(1 - F_1)\frac{\rho\sigma_{\omega 2}}{\rho\omega}\frac{\partial k}{\partial x_j}\frac{\partial \omega}{\partial x_j} - \beta\omega^2 + \beta\omega_{amb}^2 \tag{14}$$

Different from the original model, the added source terms are $\beta^* k_{amb}\omega_{amb}$ and $\beta\omega_{amb}^2$ with the subscript "*amb*" indicates the ambient values [46]. It is quite straight forward to see that the destruction terms can be exactly cancelled if the inflow values of k and ω are set equal to the ambient values, which means in the free-stream flow case, turbulence quantities will remain unchanged throughout the computational domain. An example is given in Figure 1 that clearly shows the effect of using the added source terms. In the figure, the downstream distance x is non-dimensionalized to the rotor size D. In the free stream condition, the decay of turbulent kinetic energy using different approaches is compared. The original k-ω turbulence model agrees well with the theory (Equation (10)), where the modified model maintains the turbulent kinetic energy everywhere in the computational domain.

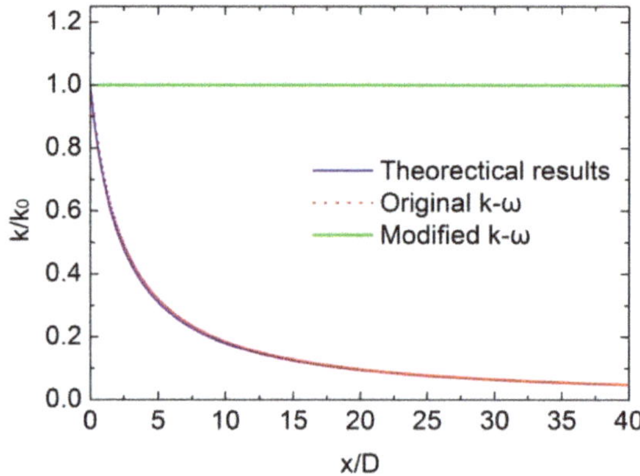

Figure 1. Comparisons of the normalized turbulence kinetic energy.

2.3. Actuator Disc

As the disc is assumed permeable, the mass conservation equation remains unchanged. In Equation (2), an external force f_{AD} is added that represents aerodynamic loads of wind turbine blades. To avoid the numerical singularity, each volume force needs to be smoothly re-distributed. A practical way of smearing the force is to use the 1D Gaussian function such that the actual forces in the disc is redistributed along one direction. In the current model, each force element is smeared locally in the blade normal direction with a distance d away from the disc using the convolution and the smeared force f' is computed as [14,47]

$$f' = f_{AD} \otimes \eta^{1D} \tag{15}$$

$$\eta^{1D}(d) = 1/\left(\varepsilon\sqrt{\pi}\right) \exp\left[-(d/\varepsilon)^2\right] \tag{16}$$

The parameter use in the smearing function is $\varepsilon = 3\Delta z$ where Δz is the reference grid size in the axis direction. Based on the hypothesis of blade elements, the body forces of rotor blades can be achieved through airfoil data and the flow field information.

The actuator disc solves the entire axisymmetric flow field and the induced velocity is naturally included into the formulation i.e., the relative velocity V_{rel} and flow angle ϕ are determined from

$$\phi = \tan^{-1}\left(\frac{V_0 - W_z}{\Omega r + W_\theta}\right), \quad V_{rel}^2 = (V_0 - W_z)^2 + (\Omega r + W_\theta)^2 \tag{17}$$

As shown in Figure 2, the z-axis and θ-axis represent the rotor normal and rotational directions, respectively. The relative velocity V_{rel} seen by the blade element is a combination of axial velocity and rotational velocity, as shown in Equation (17). The lift force L and the drag force D are perpendicular and parallel to V_{rel}, respectively, as sketched in Figure 2. The local angle of attack is given by $\alpha = \phi - \gamma$, where γ is the local pitch and twist angle, the axial velocity $(V_0 - W_z)$, where V_0 is the inflow wind speed and W_z is the axial induced velocity, the tangential velocity $(\Omega r + W_\theta)$, where Ω is the rotational speed and r is the radius of wind turbine and W_θ is the tangential induced velocity. Lift and drag forces per spanwise length are found from tabulated airfoil data and the velocity triangles shown in Figure 2.

$$L = \frac{1}{2}\rho V_{rel}^2 cBC_l, \quad D = \frac{1}{2}\rho V_{rel}^2 cBC_d \tag{18}$$

Therefore, projecting the force in the axial and tangential directions to the rotor gives the aerodynamic force components

$$F_z = L \cos \phi + D \sin \phi, \ F_\theta = L \sin \phi - D \cos \phi \tag{19}$$

The forces are calculated in every time-step such that the momentum equation produces a new velocity field which updates the flow angle of attack as shown in Figure 2.

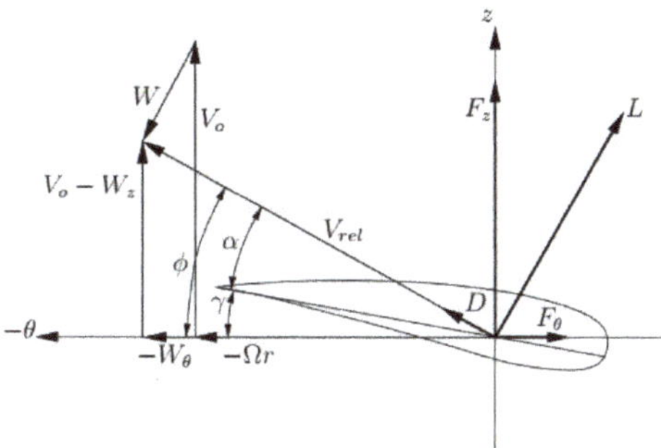

Figure 2. Sketch of velocity triangles of a blade element.

2.4. General Flow Solver

The RANS equations are solved by the *EllipSys* code [48,49], which is a general-purpose incompressible Navier-Stokes code based on a second-order multi-block finite volume method. The code solves the velocity-pressure coupled flow equations employing the SIMPLE/SIMPLEC/PISO (Semi-Implicit Method for Pressure-Linked Equations/SIMPLE Consistent/Pressure Implicit with Split Operator) methods and a multi-grid strategy. The momentum equations are first solved with a guessed pressure as a predictor. Next, the continuity equation is used as a constraint to obtain an equation for the pressure correction. In the predictor step, the momentum equations are solved by a second-order accurate backward differential scheme in time. For the spatial discretization a central difference scheme is employed for the diffusive terms and the QUICK (Quadratic Upstream Interpolation for Convective Kinematics) upwind scheme is utilized for the convective terms. In the corrector step, an improved Rhie-Chow interpolation technique is applied to suppress numerical oscillations from the velocity-pressure decoupling [50]. Furthermore, an improved SIMPLEC scheme developed for collocated grids is used to ensure that the solution does not depend on the value of relaxation parameters and the time-step [51]. The Navier-Stokes equations are discretized using a finite-volume scheme and all the information from the grid geometry is directly transferred into the discretized terms.

3. Results and Discussion

One of the major difficulties for engineering wind turbine wake models is to simulate the wake superposition effect. In this section, the 2D axisymmetric flows over two rotors are numerical validated against experimental data or LES results. The Danish Nibe wind turbines and Horns Rev wind turbines are selected for the numerical study.

3.1. Computational Setup

As sketched in Figure 3a, the two rotors are placed in tandem. The uniform inflow is applied at inlet, and the distance to the first rotor is about 10D (rotor diameter). The distance between downstream wind turbine and outlet is about 30D. The inlet and outlet width are chosen as 10D. The distance between the two rotors is not necessarily fixed which provides flexibility to be further optimized.

With a non-dimensional form, the uniform flow at the inlet and far field is Vo = 1. The change of wind turbine operational condition is through the Tip Speed Ratio (TSR). For the turbulence inflow conditions, the initial k and ω are computed through Equations (6) and (7) which are determined by the selected turbulence intensity. The mesh points are clustered around the rotor as seen in Figure 3b where only half the symmetrical plan is displayed. The mesh is block-structured with a total number of 14 blocks and 64 × 64 points in each block. The thick white lines indicate the two rotors whereas the mesh lines are clustered towards the rotors. With an OpenMPI (Message Passing Interface), the parallel simulation (on a Dell work station with Intel Xeon CPU E5 series) takes about 30 s for each simulation depending on the convergence rate. The computational time varies according to different TSRs as well as initial k and ω values. The simulation time using an AD method on a full 3D mesh will be more than hundred times slower than the 2D axisymmetric solver. On the other hand, the 2D CFD method is still slower than the classical steady BEM, but the 2D CFD method provides detailed flow field around the wind turbines.

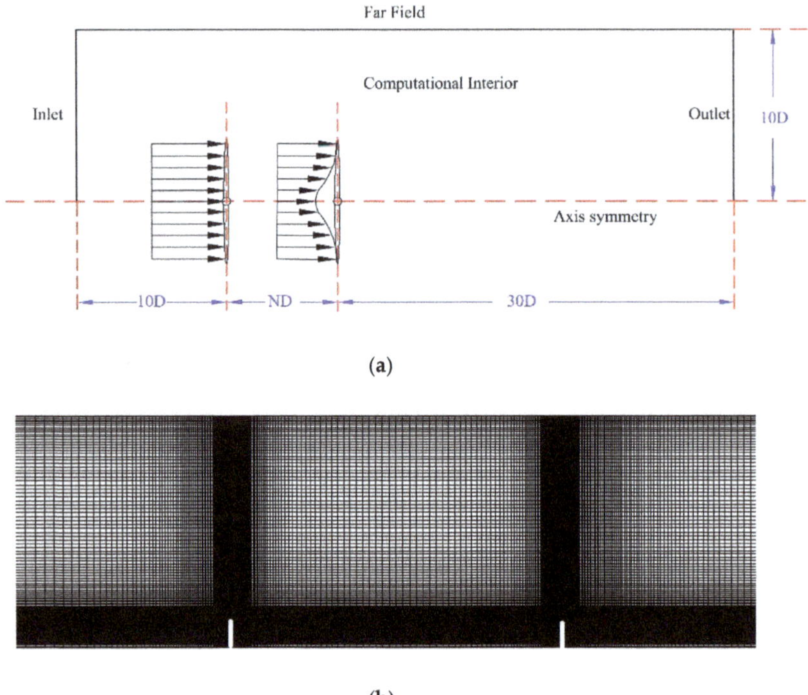

Figure 3. (**a**) Sketch of the boundary conditions and turbine positions; (**b**) The mesh configuration. The horizontal and vertical directions are defined as x and y for later reference.

3.2. Nibe Wind Turbine Case

To validate the aerodynamic model accuracy, the full-scale measurements [52] of the Nibe wind turbines are shown in Figure 4 where the two turbines are seperated with an axial distance of 5D. The Nibe wind turbines operate at a rated power of 630 KW, and the wind turbine has a rotor diameter of D = 40 m, and the hub height is H = 45 m. The simulation is performed at an inflow wind speed of 8.55 m/s and a turbulence intensity of 10%.

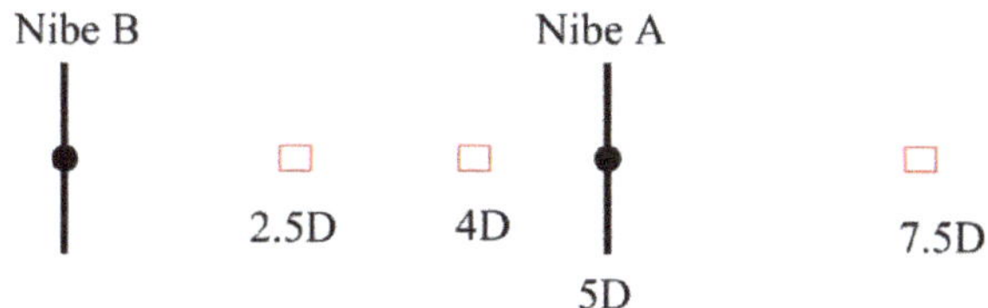

Figure 4. Sketch of the two Nibe wind turbines and the wake measurement locations.

The comparisons between the measured and simulated wake velocity profiles are shown in Figure 5. The measurement data are avialable for a single wind turbine (Nibe B) at down-stream positions of 2.5D, 4.0D and 7.5D as depicted in Figure 4. It is worth noting that for the present single wake study, Nibe A is not operating during the measurement period. The detailed wake deficit is clearly observed from Figure 5a–c which ranges from the near wake to the far wake region. In terms of maximum wake deficit and turbulence intensity, the new model shown relatively better agreement with the measured data. At the near wake region, x = 2.5D, the new model does not produce a better velocity deficit than the original model. At x = 4D and 7.5D, the new model shows better results, which is also the normal spacing between wind turbines.

For the double wake case, the streamwise velocity contour plot is presented in Figure 6. The axial velocity is largely reduced as the flow first passes through the Nibe B wind turbine and than the Nibe A wind turbine. The comparisons between the measured and simulated wake profiles for the two wind turbines at down-stream positions of 2.5D, 4.0D and 7.5D are shown in Figure 7. In this case, both Nibe A and Nibe B operate at a wind speed of 8.55 m/s. At the downstream locations 2.5D and 4D, the wake profiles are expected to be similar as that shown in Figure 6. As expected, only tiny differences are seen from the simulated results, which are due to the flow feedback from the Nibe A wind turbine, but the effect is very small. On the other hand, the experimental data showed a wider spread at 2.5D and 4D which is caused by different measurement conditions, Figure 7c shows the more complicated double wake case where the wake flow of two wind turbines are superimposed. At x = 7.5D, wake data are both measured behind the Nibe B and Nibe A wind turbines. The velocity deficit is seen to be relatively small as previously observed in Figure 5c. However, when the Nibe A wind turbine is operating, the x = 7.5D position is only 2.5D apart from the Nibe A wind turbine and thus it is considered to be the near wake position of turbine Nibe A. As seen from Figure 7a,c, wake profiles are similar at positions x = 2.5D and x = 7.5D. However, at x = 7.5D, a smaller velocity deficit is observed, which is due to the wake mixing effect introduced by the Nibe B wind turbine. At x = 7.5D, it is also seen that the fluctuation of turbulence intensity is quite large which cannot be captured by any RANS model.

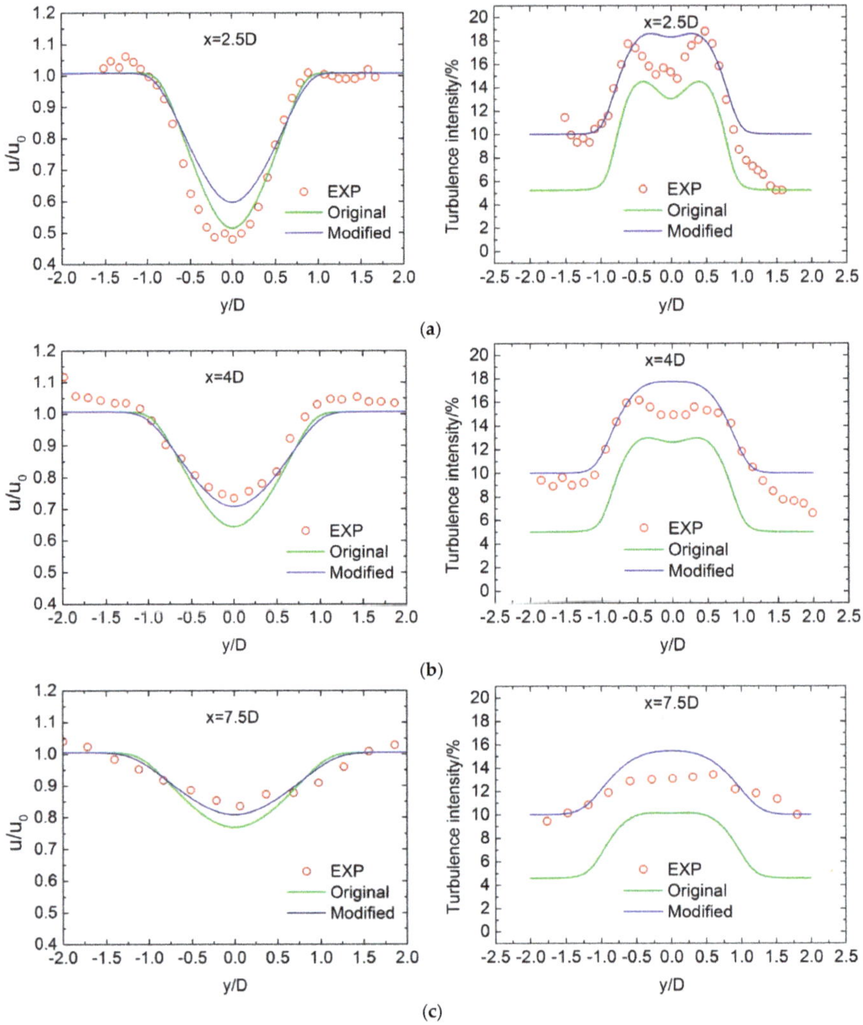

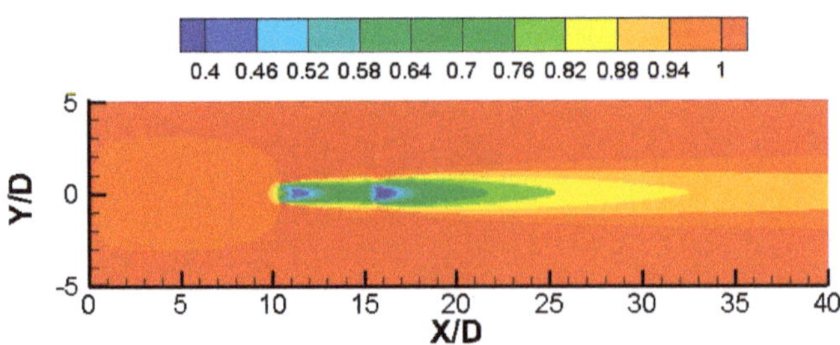

Figure 5. Comparisons between the measured and simulated wake data of a single wind turbine (Nibe B) at down-stream positions of 2.5D, 4.0D and 7.5D. (**a**) 2.5D; (**b**) 4D; (**c**) 7.5D.

Figure 6. Contours of the time-averaged stream-wise velocity for two Nibe wind turbines.

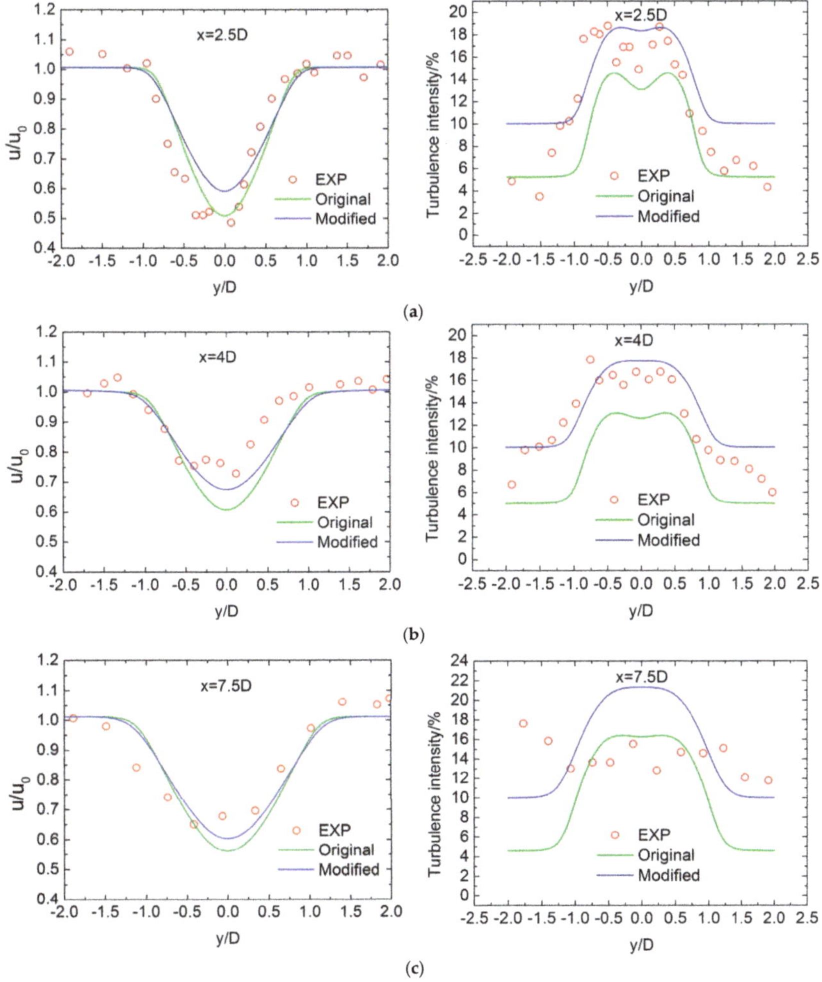

Figure 7. Comparisons between the measured and simulated wake data of the two wind turbines at down-stream positions of 2.5D, 4.0D and 7.5D. (**a**) 2.5D; (**b**) 4D; (**c**) 7.5D.

3.3. Horns Rev I Wind Farm Case

The Horns Rev I offshore wind farm is in the eastern North Sea near Denmark. Horns Rev wind farm consists of 80 Vestas V80 2 MW wind turbines within an area of about 20 km^2, as shown in Figure 8. The wind turbines are arranged in a regular array of 8 by 10 turbines which is very suitable to the study of multiple wake effects. The wind turbines are installed with an internal spacing along the main directions of 7D [53]. The wind turbines have a hub height of 70 m, a rotor diameter of D = 80 m, and a rated rotational speed of 18 rpm. The detailed pitch angle, rotational speed and blade geometry data are provided from the reference paper [54]. The 2 MW V80 wind turbine blade is composed of NACA63-series airfoils between the blade tip and the mid-span, and the FFA W3-series airfoils are used between the mid-span towards the blade inboard part. To validate the aerodynamic accuracy, the results from the present simulations and from the reference LES [26,54] simulations are compared. The inflow wind speed is 5.33 m/s, and turbulence intensity is 9.4% which are selected for fair comparison with some LES results.

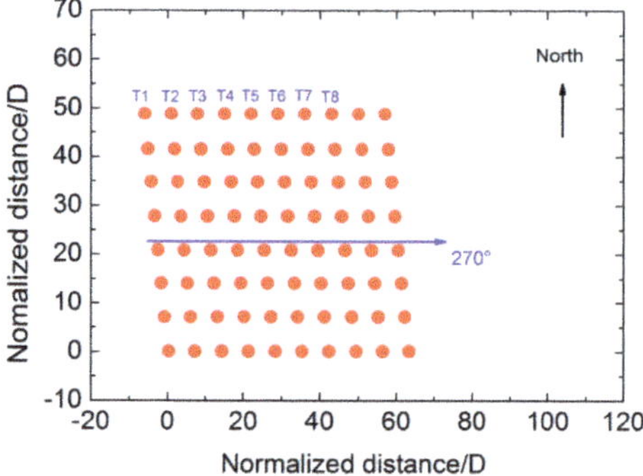

Figure 8. Layout of Horns Rev I wind farm.

The mesh topology used for this study is the same as in the previous simulations. The input airfoil data are now replaced to represent the Horns Rev wind turbines. The wind turbines under investigation are depicted in Figure 8 along the 270° directions. The wake profiles are shown in Figure 9 at downstream positions of 3D, 5D, 7D and 10D behind the first wind turbine where LES data are also available. General good agreements with LES are clearly seen from the results except very near wake. The wake expansion is well captured as well as the maximum velocity deficit. For the 7D and 10D cases, a slight over prediction of velocity deficit is seen from the current simulation. This indicates that the wake recovery at far downstream is slower using the present approach.

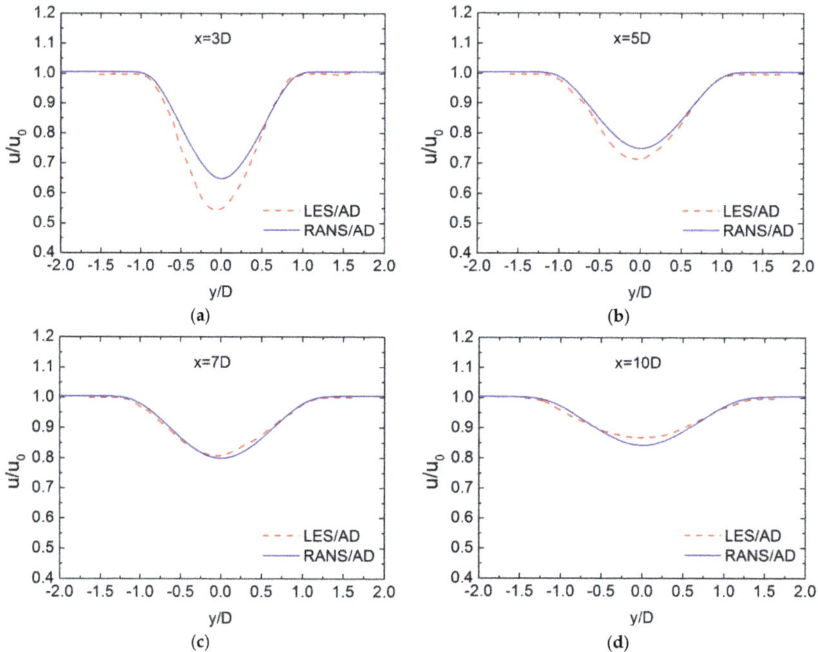

Figure 9. Comparisons between the LES and RANS lateral profiles normalized wake velocity for single wind turbine at down-stream positions of 3D, 5D, 7D and 10D. (**a**) 3D; (**b**) 5D; (**c**) 7D; (**d**) 10D.

The CFD simulation of the full column was carried out by extending the computational domain with a total number of 40 blocks. The resulted stream-wise velocity contours are shown in Figure 10 where the flow around wind turbine cluster can be observed.

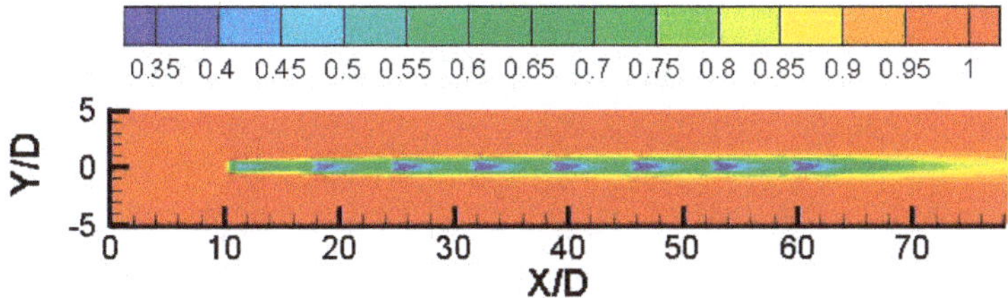

Figure 10. Contours of the stream-wise velocity for eight wind turbines.

In Figure 11, information about the power and thrust of the present wind turbines are also shown. Good agreements are seen from the power and thrust curves, which proves the model accuracy in a wide range of operational conditions.

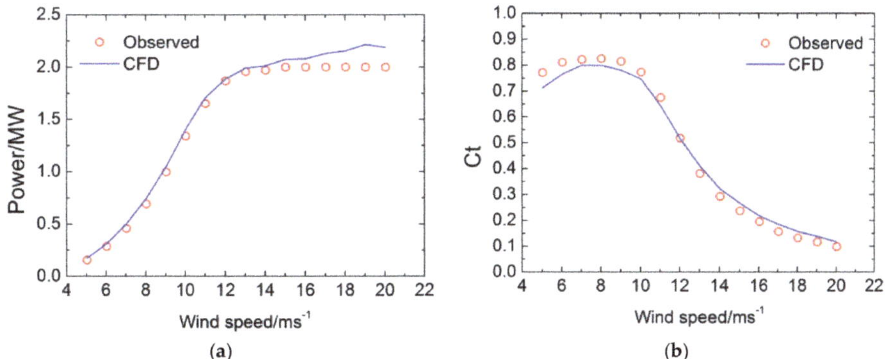

Figure 11. Comparisons between the observed data and RANS numerical for Cp, Ct, Power, and normalized power for 2 MW V80 wind turbines. (**a**) Power; (**b**) Ct.

3.4. Rotor Optimizations

3.4.1. Optimization Setup

This section describes the optimization problem used in present rotor aerodynamic design. With the optimization objectives of AEP and ATP, and the design variables of chord length, twist angle, relative thickness and rotational speed, the NSGA-II optimization method [55] is used to optimize the wind turbine rotor. In the NSGA-II optimization algorithm, the binary coding, crossover mutation and gene mutation are built to avoid premature phenomena during the optimization process. To better understand the optimization process, a sketch of the optimization diagram is given in Figure 12. The design variables are the blade geometrical data as well as the rotational speed. The "Bspline" curve is applied to reduce the number of control points. The CFD solver is embedded in the optimization loop which is a time-consuming part of the whole design process. The number of necessary iterations is determined by the NSGA-II optimization routine. Each evolution contains 30–50 populations and each population takes about 60 min.

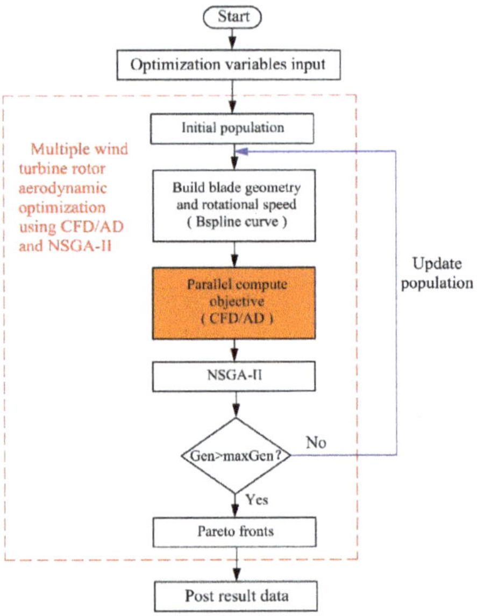

Figure 12. Wind turbine rotor aerodynamic optimization diagram.

The design objective is to maximize the annual energy production (AEP) and minimize the rotor thrust force (ATP, annual thrust production), thus the design objectives are to minimize both $1/\text{AEP}$ and ATP

$$\text{subject to } X \in R^n, \tag{20}$$

$$x_k^L \leq x_k \leq x_k^U, \; k = 1, \ldots, n \tag{21}$$

The vector X contains a total of n variables that are real numbers, R. The design variables, $X = [x_1, x_2, \ldots, x_n]$, are the control points that define the chord, twist angle and relative thickness as a function of blade span. The upper U and lower L notations, $x_k^L \leq x_k \leq x_k^U$, are the upper and lower limits of the control points. The 3rd-order Bsplines [56] are used to parameterize the chord, twist angle and relative thickness distributions, see Figure 13. The design is performed from a position at a radius of 0.337R to the tip of the blade. Thus, three control points for the chord, and two control points each for the twist angle and relative thickness distributions at the outboard of the blade are optimized. The 3rd-order Bspline curve is defined by

$$q_i(u) = \sum_{i=0}^{n} N_{i,p}(u) P_i \qquad u_a < u < u_b, \; i = 1, 2, \cdots, n$$

$$N_{i,0}(u) = \begin{cases} 1 & \text{if } u_i \leq u \leq u_{i+1} \\ 0 & \text{otherwise} \end{cases}$$

$$N_{i,p}(u) = \frac{u - u_i}{u_{i+p} - u_i} N_{i,p-1}(u) + \frac{u_{i+p+1} - u}{u_{i+p+1} - u_{i+1}} N_{i+1,p-1}(u) \tag{22}$$

$$u_0 = u_1 = \cdots = u_p = 0$$

$$u_{i+p} = u_{i+p-1} + |p_i - p_{i-1}|, \; i = 1, 2, \cdots, n-1$$

$$u_{n+p} = u_{n+p+1} = \cdots = u_{n+p+3} = 1$$

where $q_i(u)$ is a vector-valued function of the independent variable, u. $\{P_i\}$ are the control points, and the $\{N_{i,p}(u)\}$ are the 3rd degree Bspline basis functions. The ith Bspline basis function of p-degree (order $p+1$). u_i is the knot vector.

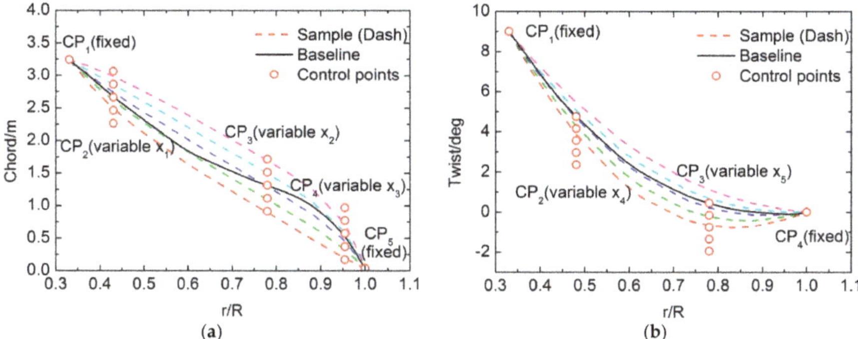

Figure 13. Distribution of the control points and the sample line of blade geometry. (a) Chord; (b) Twist angle.

To calculate the AEP and ATP, it is necessary to combine the power curve with the probability density of wind. The Weibull distribution function defining the probability density can be written in the following form

$$f(V_i < V < V_{i+1}) = \exp\left(-\left(\frac{V_i}{A}\right)^k\right) - \exp\left(-\left(\frac{V_{i+1}}{A}\right)^k\right) \tag{23}$$

where A is the scaling parameter, k is the shape factor and V_i is the wind speed distribution. If the wind turbine operates about 8760 h per year, its AEP and ATP can be evaluated as

$$AEP = \sum_{i=1}^{n-1} \frac{1}{2}(P(V_{i+1}) + P(V_i)) \times f(V_i < V < V_{i+1}) \times 8760 \tag{24}$$

$$ATP = \sum_{i=1}^{n-1} \frac{1}{2}(T(V_{i+1}) + T(V_i)) \times f(V_i < V < V_{i+1}) \times 8760 \tag{25}$$

where $P(V_i)$ and $T(V_i)$ are the power and thrust force at the wind speed of V_i.

3.4.2. Optimization Studies

The wind turbine rotor type of the Horns Rev I wind farm is chosen for the aerodynamic optimization. The selected wind turbines are located at the first and second rows (T1, T2 as depicted in Figure 8). It is worth noting that with the present method, it is possible to optimize the whole column of 8 wind turbines together. First, a larger number of wind turbines involve more design variables and thus a larger computational demand; second, it is the first two rows of wind turbines that have the major aerodynamic differences, as widely observed from measurements in several wind farms. According to the different optimization objectives, four optimization cases are carried out during the blade aerodynamic design.

Case 1: Baseline optimization for a single rotor. Using the optimization objectives of AEP and ATP, optimize the upstream wind turbine (the 1st row of rotors).

Case 2: Optimization of the 2nd row of rotors considering the wake effect from the 1st row of wind turbines. Using the optimization objectives of AEP and ATP, perform optimization for the 2nd wind turbine optimization.

Case 3: Further optimization of the rotational speed of the 1st row of wind turbines. The optimization objectives are the AEP and ATP for the two rows of wind turbines. The blade shape of the two types of rotors is obtained from Case 1 and Case 2, respectively, but the rotational speed of the 1st row of rotors will be optimized further.

Case 4: Based on Case 3, further optimization is performed for the two rows of wind turbines. Using the optimization objectives of both AEP and ATP, the two rows of wind turbines are optimized together.

To summarize these optimization cases: Case 1 follows a classical blade optimization routine; Case 2 only focuses on a downwind rotor, its blade shape is optimized; Case 3 optimizes the rotational speed of the first row of rotors to seek the global maximum energy production for two rows of rotors; Case 4 optimizes both the rotational speed of the first row of rotors and the blade geometry of the second row of rotors. By performing the 4 steps analysis, it expected that after the Case 4 study a more realistic optimization can be achieved for such a typical multiple wind turbine system.

The optimization Pareto fronts are shown in Figure 14 with the resulted Pareto front that measures how the solutions are distributed. The four subfigures (a–d) represents the above mentioned four cases. In the figures, the areas marked with *A*, *B* and *C* indicate the regions of feasible optimal solutions. The trend of area *A* suggests the solutions with high AEP, whereas low thrust is more focused in area *C*. As a multi-objective method, balance is then needed to weight between the high AEP and low ATP. In Figure 14a, Case 1 represents the standard optimization of a single rotor (the 1st row of rotors) without any wake effect. The high AEP and high ATP region is marked as area *A*, whereas area *B* finds good balance between low ATP and high AEP. In Figure 14b, the second row of wind turbine is optimized. Due to the wake effect from the first row of wind turbine, both the ATP and AEP are reduced comparing with Case 1 results. In Figure 14c, a high AEP area can hardly be achieved. The reason is that in Case 3 the AEP from the upstream rotor is reduced to increase the energy output for the downwind rotor. Such that, it is necessary to further optimize the downwind rotor. In Figure 14d, the optimization is continued such that the two rows of wind turbines are fully optimized with satisfactory results obtained.

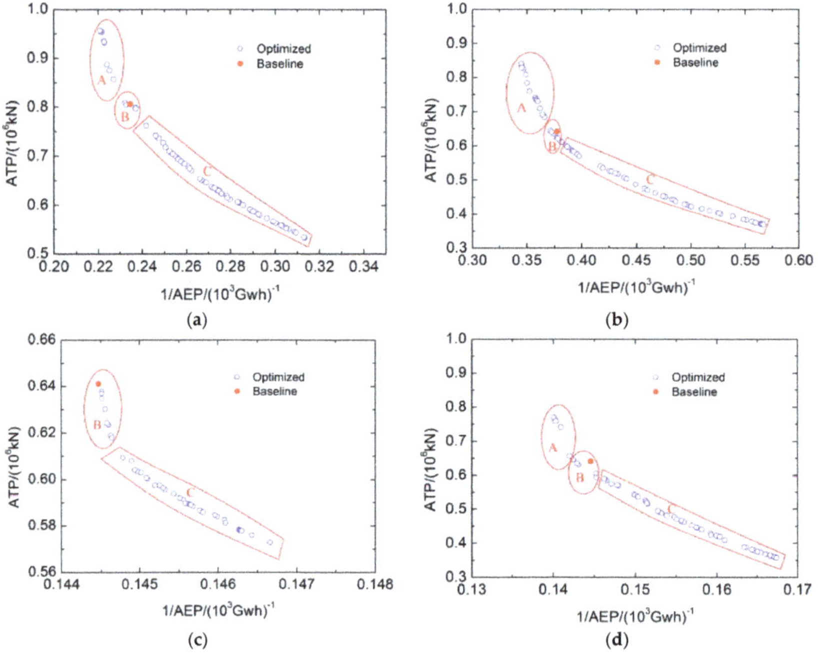

Figure 14. Pareto fronts of AEP and ATP in the 4 different cases. (**a**) Case 1; (**b**) Case 2; (**c**) Case 3; (**d**) Case 4.

Some key numbers are shown in Tables 2 and 3 after each optimization. For the Case 1 design, the upstream rotor is first optimized. As shown in Figure 14a, a trade-off between thrust and power can be found in area *B*. The final selected data is listed in Table 2 where the same ATP is maintained but the AEP is increased by 1.14%. For the Case 2 design, the downwind rotor is optimized. After optimization, the rotor in the wake has an increased power production of 1.78%, and there is also a small increase of thrust of 0.34%. Using the optimized solutions from Case 1 and Case 2, further investigation is carried out in Case 3. The rotational speed of the upstream wind turbine is considered to be a new design variable that may affect the total energy product. In Table 3, by comparing the AEP of each wind turbine with its reference value, it is seen that the 1st wind turbine has a negative change in AEP of -0.81% and the second wind turbine has a positive change in AEP of 1.2%. The overall gain of AEP is -0.04% because of the contribution of higher wind resource from the 1st wind turbine. The total thrust change of the two wind turbines is about 1.69%. It is seen that loads can be decreased with a certain amount by decreasing the front rotor speed, but it is not possible to gain more power from Case 3 study. A study in Case 4 is therefore performed to seek more energy from the 2nd rotor. The individual changes from each wind turbine and the overall changes are listed in Table 3. The reduction of the 1st rotor speed leads to a change of its AEP with -3.46%. The 2nd rotor is further optimized with a relatively large amount of AEP increase up to 9.71%. The total amount of AEP in the two-wind-turbine system is increased with 1.6%. The thrust is largely reduced for the 1st wind turbine under the new design condition at a relatively low TSR.

Table 2. Optimization results of Case 1 and Case 2.

Case	Case 1 (1st Rotor)		Case 2 (2nd Rotor)	
Objective	AEP	ATP	AEP	ATP
Reference	4.2665	0.8058	2.6559	0.6410
Optimized	4.3153	0.8058	2.7031	0.6432
Relative	1.14%	0	1.78%	0.34%

Case 1: Single wind turbine, consider both AEP&ATP; Case 2: Two wind turbines, consider both AEP and ATP of Downstream wind turbine.

Table 3. Optimization results of Case 3 and Case 4.

Case		Case 3 (1st Rotor)		Case 4 (Two Rotors)	
Objective		AEP	ATP	AEP	ATP
	1st	4.2653	0.8058	4.2653	0.8058
Reference	2rd	2.6559	0.6410	2.6559	0.6410
	Overall	6.9212	1.4468	6.9212	1.4468
	1st	4.2307	0.7919	4.1177	0.7522
Optimized	2rd	2.6875	0.6304	2.9139	0.6432
	Overall	6.9182	1.4224	7.0317	1.3955
	1st	-0.81%	-1.72%	-3.46%	-6.65%
Relative	2nd	1.2%	-1.65%	9.71%	0.34%
	Overall	-0.04%	-1.69%	1.6%	-3.55%

Case 3: Upstream WT rotation speed; consider AEP of overall WT, and ATP of Downstream WT; Case4: Upstream WT rotation speed and blade geometry; consider AEP of overall WT, and ATP of Downstream WT.

The final aerodynamic layouts of the optimized blades are gathered in Figure 15. The chord, twist and relative thickness distributions are given in Figure 15a–c, respectively. Based on the multi-objective optimizations, the chord length of the upstream wind turbine (Case 1) and downwind wind turbine (Case 2) deviates from the baseline chord distribution. The twist distribution of the two optimized blades is rather similar. The optimized rotor speed is shown in Figure 15d. With limited lower and upper bounds, the Case 4 optimization suggests a much lower rotational speed as compared with Case 3. As a result, the upwind rotor lost a certain amount of power which is gained by the downwind wind turbine. It is worth noting that the present study is only limited at aerodynamic optimization. Since the 1st wind turbine also has a decrease of thrust of 6.65%, there is a big potential to further reduce the structure weight.

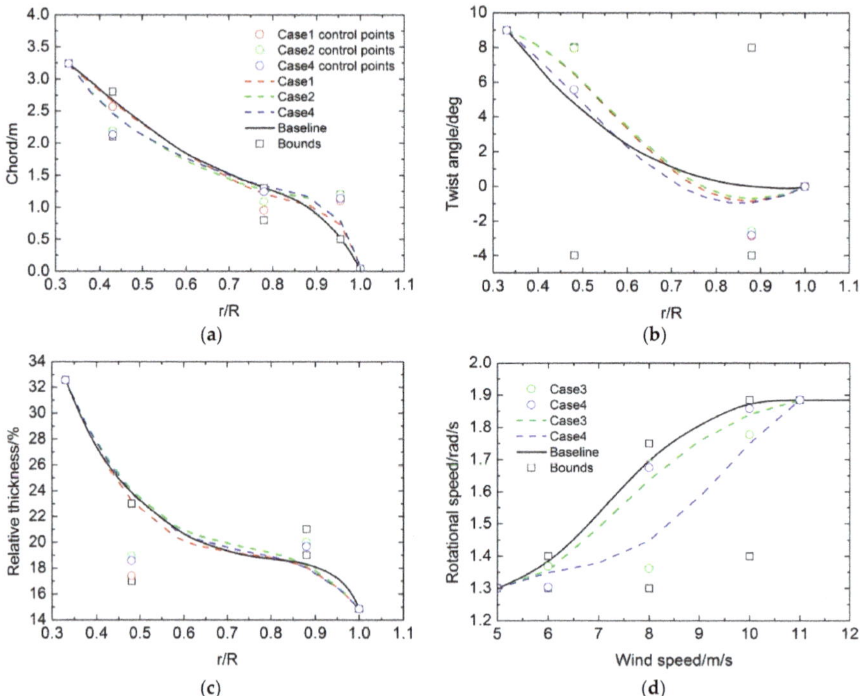

Figure 15. The comparisons of optimization results of wind turbine ((**a**–**c**) belong to Case 1, Case 2, Case 4; (**d**) belongs to Case 3 and Case 4). (**a**) Chord; (**b**) Twist angle; (**c**) Relative thickness; (**d**) Rotational speed.

4. Conclusions

This paper presented detailed numerical methodologies and convincing results for wind turbine rotor aerodynamic optimizations. To simulate the aerodynamic forces on the wind turbine rotor, a 2D actuator disc method is implemented in the incompressible flow solver favored by the axisymmetric boundary condition defined at the center line. The combined flow solver is more accurate and powerful than existing engineering models such as blade element momentum theory. On the other hand, it is order of magnitude faster than the 3D flow solver. However, the 2D and 3D solver implemented with same AD method should give same results. In addition to the AD flow simulations, two turbulent flow cases were investigated to proof the advantage of the modified k-ω turbulence model. With the fast CFD solver, the rotor optimization was carried out for single wind turbine with uniform flow and two wind turbines combined with wake interaction. The multi-objective optimization method NSGA-II is applied in the design loop and the two design objectives of high AEP and low ATP are required. The four optimization cases aimed at single rotor optimization without wake effect, single rotor optimization with wake effect, multi-wind turbine optimization with constant rotor speed and multi-wind turbine optimization with variable rotor speed. In general, single rotor optimizations are faster and they provide clear optimization trend. For the multi-wind turbine optimization cases, it was found that high aerodynamic performance is achieved when the upstream and downstream wind turbine both change their blade shape and rotational speed. These results indicate that traditional rotor optimization methods might not be sufficient due to the complex wake effect.

On a single PC with a developed parallel computation tool, the estimated time is 30 h for optimizing a single rotor, and for the coupled two-rotor optimizations, the simulation time is almost doubled. Future work will be carried out for more complicated design cases which may include the whole column of wind turbines such as the Horns Rev wind farm case. In such an optimization, the wind turbine spacing will be included as an important design parameter, to this end, the method can be applied for simple layout optimization.

Author Contributions: W.Z. conceived the rotor optimization method and wrote most part of the paper; J.C. carried out most of the programming part and the simulations; W.S., J.N.S. and T.W. put a lot of effort on guidance the overall structure of the paper as well as results and discussions.

Funding: This work was funded by the Jiangsu Province Natural Science Foundation (BK20160476), the National Science Foundation (11672261), and the National Basic Research Program of China (973 Program) (No. 2014CB046200), and the National Science Foundation (51761165022), and the Yangzhou University Science and Technology Innovation Training Foundation (2016CXJ039) and (2017CXJ041).

Acknowledgments: The corresponding author wish to express special acknowledgement to the National Nature Science Foundation under grant number 11672261, the Jiangsu Province Natural Science Foundation (BK20160476).

Conflicts of Interest: The authors declare no conflict of interest.

References

1. Glauert, H. Airplane propellers. In *Aerodynamic Theory*; Springer: Berlin/Heidelberg, Germany, 1935; pp. 169–360.
2. Hansen, M.O.L. *Aerodynamics of Wind Turbines*, 2nd ed.; Earthscan, Routledge: Abingdon, UK, 2010; pp. 45–62.
3. Sørensen, J.N. *General Momentum Theory*; Springer: Berlin/Heidelberg, Germany, 2016; pp. 43–58.
4. Sørensen, J.N.; Mikkelsen, R. On the Validity of the Blade Element Momentum Method. In Proceedings of the European Wind Energy Conference, Copenhagen, Denmark, 2–6 July 2001.
5. Ghadirian, A.; Dehghan, M.; Torabi, F. Considering induction factor using BEM method in wind farm layout optimization. *J. Wind Eng. Ind. Aerodyn.* **2014**, *129*, 31–39. [CrossRef]
6. Vaza, P.; Pinho, T. An extension of BEM method applied to horizontal axis wind turbine design. *Renew. Energy* **2011**, *36*, 1734–1740. [CrossRef]
7. Leishman, J.G.; Bhagwat, M.J.; Bagai, A. Free-vortex filament methods for the analysis of helicopter wakes. *J. Aircr.* **2003**, *39*, 759–775. [CrossRef]
8. Bareiss, R.; Wagner, S. The Free Wake/Hybrid Wake Code RVOLM—A Tool for Aerodynamic Analysis of Wind Turbines. In Proceedings of the European Community Wind Energy Conference, Lubeck-Travemunde, Germany, 8–12 March 1993; pp. 424–427.
9. Voutsinas, S.G.; Belessis, M.; Huberson, S. Dynamic Inflow Effects and Vortex Particle Methods. In Proceedings of the European Community Wind Energy Conference, Lubeck-Travemunde, Germany, 8–12 March 1993; pp. 428–431.
10. Ramos, G.N.; Hejlesen, M.M.; Sørensen, J.N.; Walther, J.H. Hybird vortex simulations using a three-dimensional viscous-inviscid panel method. *Wind Energy* **2017**, 1871–1889. [CrossRef]
11. Shen, X.; Chen, J.G.; Zhu, X.C.; Liu, P.Y.; Du, Z.H. Multi-objective optimization of wind turbine blades using lifting surface method. *Energy* **2015**, *90*, 1111–1121. [CrossRef]
12. Cao, J.F.; Wang, T.G.; Long, H.; Ke, S.T.; Xu, B.F. Dynamic loads and wake prediction for large wind turbines based on free wake method. *Trans. Nanjing Univ. Aeronaut. Astronaut.* **2015**, *32*, 240–249.
13. Zhu, W.J.; Shen, W.Z.; Sørensen, J.N. Integrated airfoil and blade design method for large wind turbines. *Renew. Energy* **2014**, *70*, 172–183. [CrossRef]
14. Sørensen, J.N.; Shen, W.Z.; Munduate, X. Analysis of the Wake States by a Full-Field Actuator Disc Model. *Wind Energy* **1998**, *1*, 73–88. [CrossRef]
15. Shen, W.Z.; Zhu, W.J.; Sørensen, J.N. Validation of the actuator line/Navier Stokes technique using mexico easurements. In *The Science of Making Torque from Wind*; Wiley: Hoboken, NJ, USA, 2010; pp. 227–234.
16. Shen, W.Z.; Zhang, J.H.; Sørensen, J.N. The Actuator Surface Model: A New Navier-Stokes Based Model for Rotor Computations. *J. Sol. Energy Eng.* **2009**, *13*, 011002. [CrossRef]

17. Bak, C. *Research in Aeroelasticity EFP-2006: Key Parameters in Aerodynamic Rotor Design*; RISØ-R-1611(EN); RISØ National Laboratory: Roskilde, Denmark, 2007.

18. Wang, X.D.; Shen, W.Z.; Zhu, W.J.; Sørensen, J.N.; Chen, J. Shape Optimization of Wind Turbine Blades. *Wind Energy* **2009**, *12*, 781–803.

19. Gupta, S.; Leishman, J.G. Comparison of momentum and vortex methods for the aerodynamic analysis of wind turbines. In Proceedings of the 43rd AIAA Aerospace Sciences Meeting and Exhibit, Reno, NV, USA, 10–13 January 2005; p. 594.

20. Vermeer, L.J.; Sørensen, J.N.; Crespo, A. Wind turbine wake aerodynamics. *Prog. Aerosp. Sci.* **2003**, *39*, 467–510. [CrossRef]

21. Sanderse, B.; Koren, B. Review of computational fluid dynamics for wind turbine wake aerodynamics. *Wind Energy* **2011**, *14*, 799–819. [CrossRef]

22. Göçmen, T.; Lann, P.; Réthoré, P.E.; Diaz, A.P.; Larsen, G.C.; Ott, S. Wind turbine wake models developed at the technical university of Denmark: A review. *Renew. Sustain. Energy Rev.* **2016**, *60*, 752–769. [CrossRef]

23. Jensen, N.O. *A Note on Wind Generator Interaction*; Report M-2411; Risø National Laboratory: Roskilde, Denmark, 1983.

24. Larsen, G.C. *A Simple Stationary Semi-Analytical Wake Model*; Technical Report Risø; Technical University of Denmark: Roskilde, Denmark, 2009.

25. Larsen, G.C.; Madsen, H.A.; Bingöl, F.; Mann, J. *Dynamic Wake Meandering Modeling*; Report R-1607(EN); Risø National Laboratory: Roskilde, Denmark, 2007.

26. Tian, L.L.; Zhu, W.J.; Shen, W.Z.; Zhao, N.; Shen, Z. Development and validation of a new two-dimensional wake model for wind turbine wakes. *J. Wind Eng. Ind. Aerodyn.* **2015**, *137*, 90–99. [CrossRef]

27. Tian, L.L.; Zhu, W.J.; Shen, W.Z.; Song, Y.L.; Zhao, N. Prediction of multi-wake problems using an improved Jensen wake model. *Renew. Energy* **2017**, *102*, 457–469. [CrossRef]

28. Wang, T.G. Unsteady Aerodynamic Modelling of Horizontal Axis Wind Turbine Performance. Ph.D. Thesis, University of Glasgow, Scotland, UK, 1999.

29. Ott, S.; Berg, J.; Nielsen, M. *Linearized CFD Models for Wakes*; Report R-1772(EN); Risø National Laboratory: Roskilde, Denmark, 2011.

30. Ainslie, J.F. Calculating the Flowfield in the Wake of Turbines. *J. Wind Eng. Ind. Aerodyn.* **1988**, *27*, 213–224. [CrossRef]

31. Sørensen, N.N.; Bechmann, A.; Pierre, E.R.; Frederik, Z. Near wake Reynolds-averaged Navier–Stokes predictions of the wake behind the MEXICO rotor in axial and yawed flow conditions. *Wind Energy* **2014**, *17*, 75–86. [CrossRef]

32. The, J.; Yu, H. A critical review on the simulations of wind turbine aerodynamics focusing on hybrid RANS-LES methods. *Energy* **2017**, *138*, 257–289. [CrossRef]

33. Allah, V.A.; Mayam, M.H.S. Large eddy simulation of flow around a single and two in-line horizontal-axis wind turbines. *Energy* **2017**, *121*, 533–544. [CrossRef]

34. Abdulqadir, S.A.; Iacovides, H.; Nasser, A. The physical modelling and aerodynamics of turbulent flows around horizontal axis wind turbines. *Energy* **2017**, *119*, 767–799. [CrossRef]

35. Troldborg, N.; Sørensen, J.N.; Mikkelsen, R.F.; Sørensen, N.N. A simple atmospheric boundary layer model applied to large eddy simulations of wind turbine wakes. *Wind Energy* **2014**, *17*, 657–669. [CrossRef]

36. Sørensen, J.N.; Shen, W.Z. Numerical modeling of wind turbine wakes. *J. Fluids Eng.* **2002**, *124*, 393–399. [CrossRef]

37. Shen, W.Z.; Zhu, W.J.; Sørensen, J.N. Actuator line/Navier-Stokes computations for the MEXICO rotor: comparison with detailed measurements. *Wind Energy* **2012**, *15*, 811–825. [CrossRef]

38. Deiterding, R.; Wood, S.L. A dynamical adaptive lattic Boltzmann method for predicting wake phenomena in fully coupled wind engineering problems. In Proceedings of the VI International Conference on Computational Methods for Coupled Problems in Science and Engineering Coupled Problems, Venice, Italy, 18–20 May 2015.

39. Angelidis, D.; Sotiropoulos, F. An immersed boundary-adaptive mesh refinement solver (IB-AMR) for high fidelity fully resolved wind turbine simulations. In Proceedings of the 68th Annual Meeting of the APS Division of Fluid Dynamics, Boston, MA, USA, 22–24 November 2015.

40. Zhu, W.J.; Behrens, T.; Shen, W.Z.; Sørensen, J.N. Hybrid immersed boundary method for airfoils with a trailing-edge flap. *AIAA J.* **2013**, *51*, 30–41. [CrossRef]

41. Menter, F.R. Two-Equation Eddy-Viscosity Turbulence Models for Engineering Applications. *AIAA J.* **1994**, *32*, 1598–1605. [CrossRef]

42. Wilcox, D.C. *Turbulence Modelling for CFD*, 1st ed.; DCW Industries: La Cañada Flintridge, CA, USA, 1993.

43. Prospathopoulos, J.M.; Politis, E.S. Evaluation of the effects of turbulence model enhancements on wind turbine wake predictions. *Wind Energy* **2011**, *14*, 285–300. [CrossRef]

44. Richards, P.; Hoxey, R. Appropriate boundary conditions for computational wind engineering models using the k-ϵ turbulence model. *J. Wind Eng. Ind. Aerodyn.* **1993**, *46*, 145–153. [CrossRef]

45. Spalart, P.R.; Rumsey, C.L. Effective In-flow Conditions for Turbulence Models in Aerodynamic Calculations. *AIAA J.* **2007**, *45*, 2544–2553. [CrossRef]

46. Tian, L.L.; Zhu, W.J.; Shen, W.Z.; Sørensen, J.N.; Zhao, N. Investigation of modified AD/RANS models for wind turbine wake predictions in large wind farm. *J. Phys. Conf. Ser.* **2014**, *524*, 012151. [CrossRef]

47. Shen, W.Z.; Mikkelsen, R.; Sørensen, J.N. Tip loss correction for Actuator/Navier-Stokes computations. *J. Sol. Energy Eng.* **2005**, *127*, 209–213. [CrossRef]

48. Michelsen, J.A. *Basis3D—A Platform for Development of Multiblock PDE Solvers*; Technical Report AFM; Technical University of Denmark: Kongens Lyngby, Denmark, 1992.

49. Sørensen, N.N. *General Purpose Flow Solver Applied Over Hills*; RISØ-R-827-(EN); Risø National Laboratory: Roskilde, Denmark, 1995.

50. Shen, W.Z.; Michelsen, J.A.; Sørensen, J.N. An improved Rhie-Chow interpolation for unsteady flow computations. *AIAA J.* **2001**, *39*, 2406–2409. [CrossRef]

51. Shen, W.Z.; Michelsen, J.A.; Sørensen, N.N.; Sørensen, J.N. An improved SIMPLEC method on Collocated grids for steady and unsteady flow computations. *Numer. Heat Transf.* **2003**, *43*, 221–239. [CrossRef]

52. Taylor, G.L. *Wake Measurements on the Nibe Wind Turbines in Denmark*; Contractor Report ETSU WN 5020; National Power-Technology and Environment Center: Leatherhead, UK, 1990.

53. Hansen, K.S.; Barthelmie, R.J.; Jensen, L.E. The impact of turbulence intensity and atmospheric stability on power deficits due to wind turbine wakes at Horns Rev wind farm. *Wind Energy* **2011**, *15*, 183–196. [CrossRef]

54. Wu, Y.T.; Fernando, P.A. Modeling turbine wakes and power losses within a wind farm using LES: An application to the Horns Rev offshore wind farm. *Renew. Energy* **2015**, *75*, 945–955. [CrossRef]

55. Deb, K.; Agrawal, S.; Pratap, A. A fast elitist nondominated sorting genetic algorithm for multi-objective optimization: NSGA-II. In Proceedings of the Parallel Problem Solving from Nature VI Conference, Paris, France, 18–20 September 2000; pp. 849–858.

56. Piegl, L.; Tiller, W. *The NURBS Book*, 2nd ed.; Springer: New York, NY, USA, 1997; ISBN 3-540-61545-8.

Article

Condition Monitoring of Wind Turbine Blades Using Active and Passive Thermography

Hadi Sanati, David Wood * and Qiao Sun

Department of Mechanical and Manufacturing Engineering, University of Calgary, 2500 University Drive N.W., Calgary, AB T2N 1N4, Canada; hadi.sanati@ucalgary.ca (H.S.); qsun@ucalgary.ca (Q.S.)
* Correspondence: dhwood@ucalgary.ca; Tel.: +1-(403)-220-3637

Received: 22 August 2018; Accepted: 16 October 2018; Published: 22 October 2018

Abstract: The failure of wind turbine blades is a major concern in the wind power industry due to the resulting high cost. It is, therefore, crucial to develop methods to monitor the integrity of wind turbine blades. Different methods are available to detect subsurface damage but most require close proximity between the sensor and the blade. Thermography, as a non-contact method, may avoid this problem. Both passive and active pulsed and step heating and cooling thermography techniques were investigated for different purposes. A section of a severely damaged blade and a small "plate" cut from the undamaged laminate section of the blade with holes of varying diameter and depth drilled from the rear to provide "known" defects were monitored. The raw thermal images captured by both active and passive thermography demonstrated that image processing was required to improve the quality of the thermal data. Different image processing algorithms were used to increase the thermal contrasts of subsurface defects in thermal images obtained by active thermography. A method called "Step Phase and Amplitude Thermography", which applies a transform-based algorithm to step heating and cooling data was used. This method was also applied, for the first time, to the passive thermography results. The outcomes of the image processing on both active and passive thermography indicated that the techniques employed could considerably increase the quality of the images and the visibility of internal defects. The signal-to-noise ratio of raw and processed images was calculated to quantitatively show that image processing methods considerably improve the ratios.

Keywords: thermography; wind turbine blades; defects; image processing; condition monitoring

1. Introduction

The most crucial components of wind turbines, the blades, are susceptible to different types of damage during their operation. The failure of one blade may damage nearby blades and wind turbines, increasing the total damage cost. Most blades consist of two halves made of a fiberglass composite and shear webs, which are glued together with strong adhesive materials [1]. The main function of the shear webs is to increase the strength of the structure. These bonded zones are potential sites for damage initiation and propagation [2]. Different surface and subsurface defects, including delamination, cracks, air inclusion, fiber-matrix debonding, and others, may be introduced to the blade during manufacturing or operation [3]. Harsh environmental conditions and airborne particles such as hail, snow, rain, ice, and dirt, expose wind turbine blades to more potential harm. Defective blades are rarely replaced because of the high cost of manufacturing. To prevent failure, blades need to be continuously monitored through Non-Destructive Testing (NDT) methods [4]. Different NDT techniques such as Ultrasonic Testing (UT), Acoustic Emission (AE), Fiber Bragg Grating (FBG) strain sensors, Vibration Analysis, and Tap Tests have been employed to inspect the integrity of wind turbine blades [5–9]. Conventional NDT techniques generally require close proximity between the sensor and the blade [10]. Since access to a blade is difficult and requires an industrial climber or crane, which can be dangerous and/or time-consuming, the practical implementation of conventional methods sometimes requires blade removal. Developing new NDT techniques that are capable of detecting faults in the blades from larger distances is essential.

Infrared (IR) thermography is a non-contact, long-distance NDT technique that can inspect extensive areas quickly by capturing thermal images of the object's surface. In general, defective areas alter the temperature distributions on the surface that are measured by IR cameras. Thermographic inspection is typically divided into two categories: active and passive. In active thermography, different heating sources such as flash and halogen lamps are employed for heating the object making the technique less usual for operating wind turbines. It is used here largely to allow comparison with passive thermography which utilizes solar radiation [2] to heat a blade (usually around sunrise) or to cool it at sunset. This method has been widely used to detect subsurface defects of different materials including metals [11], composites [12,13], and concrete [14].

Different studies have used thermography to detect faults in wind turbine blades. Meinlischmidt and Aderhold [2] employed passive thermography to detect internal structural features and subsurface defects such as poor bonding and delamination. Beattie and Rumsey [15] employed thermography to inspect blades during fatigue tests of a 13.1 m blade made from wood-epoxy-composite and a 4.25 m fiberglass blade. This experiment identified the root region of the blade as a defective area. Shi-bin [16] employed infrared thermal wave testing to detect subsurface faults such as foreign matter and air inclusions at various depths of a blade section. Galleguillos [17] conducted a new experiment to inspect an installed wind turbine blade. They mounted an IR camera on an unmanned aerial vehicle (UAV) and captured thermal images of installed blades while the blades were stationary, demonstrating the capability for fast data acquisition and inspection with this setup. Other research evaluated the suitability of different weather conditions for revealing the internal features of a blade section with thermal imaging [18,19]. Doroshtnasir [20] proposed a new passive thermography technique that can inspect operating blades from the ground. This experiment developed a new image processing technique to improve the thermal contrast quality by removing the effect of disruptive factors such as environmental reflections. Active thermography has been used by different researchers to quantitatively evaluate the presence of defects in different materials including composite and metallic samples. Lahiri [21] employed phase information obtained from the processing of active thermography data to determine quantitative information associated with flat-bottomed holes (FBHs) embedded in different materials including glass fiber reinforced polymer, high-density rubber and low-density rubber. Shin [22] used the Pulsed Phase Thermography (PPT) method to inspect the subsurface fatigue damage in adhesively bonded joints between fiber reinforcement polymer components. Maierhofer [23] compared the phase values obtained from pulsed and lock-in thermography applied on steel and Carbon Fiber Reinforced Plastic (CFRP) materials. They also compared the spatial resolution calculated from data captured through flash and lock-in thermography at different frequencies. In another study, Almond [24] used long pulsed thermography to detect the FBHs of different sizes and depths created in different materials including aluminum alloy, mild steel and stainless steel, and a CFRP composite plate.

Different image processing methods have been developed to improve the contrast of subsurface defects in images. Maldague and Marinetti [25] proposed PPT, which has the advantages of both pulsed and lock-in thermography. Lock-in thermography can detect deeper defects by continuously heating the surface using a periodic heat source [26] such as a modulated halogen lamp [27] but it may take a long time to detect a fault while pulsed thermography is fast. PPT uses a transform-based algorithm such as Fast Fourier Transform (FFT) to convert time domain data to frequency. Shin [24] employed this method to detect the initiation and propagation of defects in adhesively bonded joints under fatigue loading. This method has been widely used by different researchers to quantitatively and qualitatively evaluate the subsurface damages in different materials. Pawar [28], for example, inspected barely visible impact damage from low-velocity impacts. For this purpose, he first calibrated the defect depth with a blind frequency, the limited frequency at which the subsurface defect at a certain depth is visible in the recorded thermal data, for carbon epoxy laminate, and then applied the findings of depth and the blind frequency relationship to the specimen with barely visible impact damage on it. Castanedo [29] proposed an interactive methodology in a PPT experiment that connected acquisition parameters such as time and frequency resolution and storage capacity to each other in order to inspect defects at different depths with a single test. He used a combination of phase contrast and blind frequency and applied his proposed interactive methodology to a CFRP specimen with artificial defects at different depths to quantify the depth of the defects. Thermographic Signal Reconstruction [30] increases the quality of the thermal signatures associated with internal defects based on the known behavior of simple forms of the heat conduction equation. It improves the signal-to-noise ratio (SNR), while at the same time reducing image blurring and increasing the sensitivity [31]. This method was recently applied to the thermal image sequence of step heating thermography and gave reliable results [32]. Matched Filters (MF) have also been proposed to improve image contrast of subsurface defects by increasing the contrast of defective areas and reducing the signals from sound areas [31,33,34].

The present investigation developed passive and active thermographic inspection of wind turbine blades. During the active experiment, both pulsed and step heating thermography were employed and their results compared with each other (images were obtained and processed in some cases after the heating had finished and so strictly should be called "step cooling". We will use the term "step heating" in a purely generic manner to cover both cases). Several image processing techniques were applied to the raw thermal images to increase the contrast associated with internal defects. Once the most appropriate technique was determined, a passive thermography experiment was designed and the image processing technique was applied to the thermograms and the maps of surface temperature, to improve their quality. Passive thermography was also performed at different times of the day to assess the most favorable times for the best results. This paper is arranged as follows: Section 1 reviews the literature regarding NDT techniques and thermal imaging methods. The experimental procedures and materials are outlined in Section 2. Section 3 provides the theory of quantitative evaluation. Section 4 contains the results and discussion regarding experimental thermography for the inspection of subsurface defects. Finally, Section 5 provides a summary and conclusion.

2. Experimental Procedure

2.1. Materials

All samples used in this experiment came from a 50 m long wind turbine blade made of fiberglass composite obtained after it had been damaged in transit to a wind farm. The blade was never installed or operated. Figure 1a shows the 3 m long blade section with significant surface damage that was used for the passive thermography experiments. The laminate thickness in the damaged section was 14 mm. The chord length was approximately 1 m. The yellow/orange regions in Figure 1a are the exposed sandwich core on the rear section of the suction surface where there is very little laminate. Some patches of glue on the suction side resulted in different effusivity than the background and therefore generated spots with different brightnesses on the thermograms. The upper side of the blade section contained a crack which was visible to the naked eye.

A "defect plate" with dimensions of 170 mm × 195 mm × 8 mm was cut from the laminate skin of another section of the blade closer to the tip where the laminate skin was thinner. Flat-bottomed holes with different diameters and depths were drilled from the rear to produce a range of known "defects". The holes had diameters ranging from 4 mm to 20 mm with depths between 0.5 mm and 3 mm. Figure 1c is a schematic of the plate that illustrates the geometry and pattern of the defects. No holes penetrated the outer surface of the plate which corresponded to the outer surface of a blade and all thermograms were of the outer surface. The defect plate was attached to the surface of the damaged blade section during passive thermography experimentation. It was also the only blade material tested with active thermography. The holes were used to evaluate the minimum defect that can be detected using passive and active thermography.

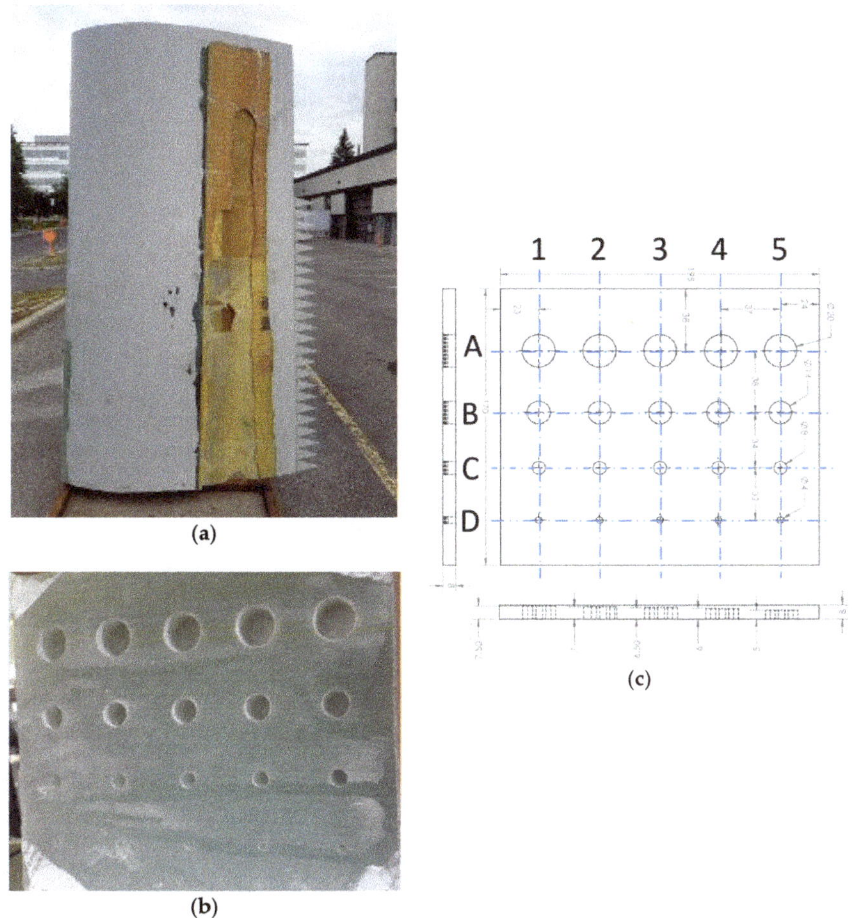

Figure 1. (**a**) The damaged wind turbine blade. (**b,c**) The defect plate with flat-bottomed holes. All holes were drilled from the rear and did not penetrate the outer surface.

2.2. Passive Thermography

In the passive thermography experiment, the suction side of the blade was monitored outdoors during a sunny day from morning until the afternoon. The blade's position was not changed relative to the IR camera during this time. This experiment, whose setup is depicted in Figure 2, sought to determine the most favorable conditions to reveal the most defects and to evaluate the fault detection capability of thermography when the blade is heated by the sun.

The experiment was conducted on a sunny day in July 2017 from 9:00 a.m. to 7:30 p.m. Sunrise and sunset on this day were 5:53 a.m. and 9:30 p.m., respectively. During the experiment, the sky was clear, the humidity was almost 36%, and the temperature varied between 16 °C and 26 °C. A T1030Sc IR camera made by FLIR Systems was located 4 m from the blade section and equipped with a 21.2 mm lens, resulting in a spatial resolution of 4 mm per pixel. ResearchIR, a software package developed by FLIR Systems that provides high-speed data recording and image analyzing capabilities was used to record thermograms at a frequency of 1 Hz.

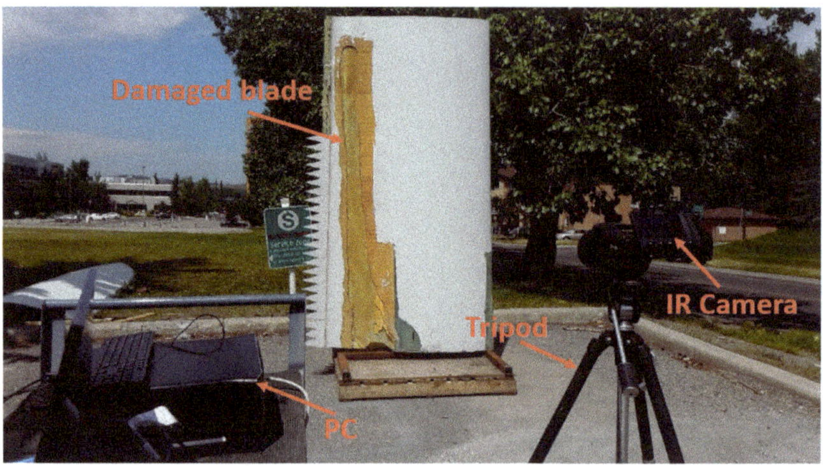

Figure 2. Passive thermography experiment.

2.3. Pulsed and Step Heating Thermography

These techniques were used only on the defect plate mounted 1.5 m from the IR camera. Despite being mounted near windows, the defect plate was always in a shadow during the experiments. During pulsed thermography, the defect plate was heated by a 2400 W flash lamp and thermal images were recorded at a frequency of 15 Hz immediately after flashing the sample. To heat the surface uniformly, the flash lamp was around 0.3 m from the object with the angle of around 15° with respect to the normal of the defect plate. The pulsed thermography experiment is depicted in Figure 3. The spatial resolution of thermal images obtained by active thermography was almost 0.5 mm per pixel.

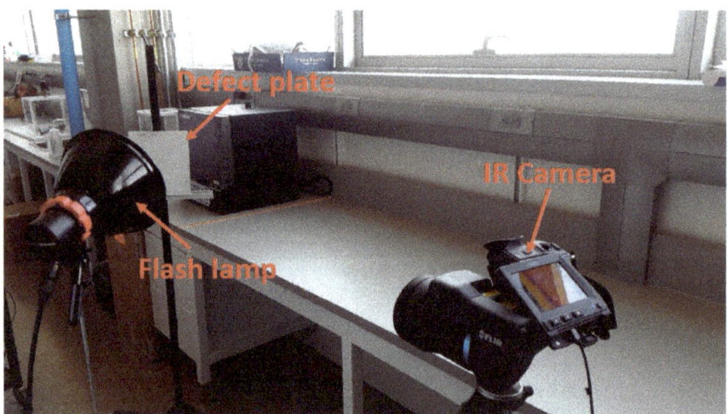

Figure 3. Pulsed thermography experiment.

The step heating thermography experiment, shown in Figure 4, used two 500 W halogen lamps to continuously heat the surface. The sample was heated between 10 and 75 s and the thermal evolution on the surface of the specimen was recorded at a frequency of 15 Hz. Once heating was finished, thermal decay was recorded with the same frequency. The room temperature was 23 ± 2 °C during the experiment. The thermal contrast associated with the defects could be observed after a few seconds of heating.

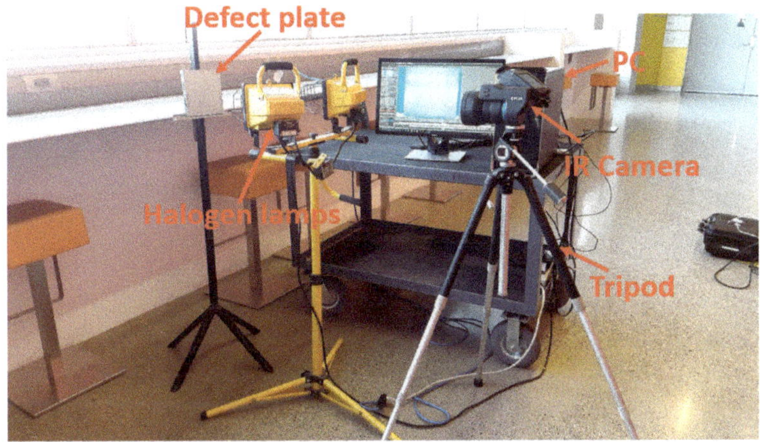

Figure 4. Step heating thermography experiment.

3. MF Algorithms

MF is a method to improve image contrast of subsurface defects by increasing the contrast of defective areas and reducing the signals from sound areas [31,33,34]. Different types of MF algorithms have been developed, all of which are based on the assumption that [31,34]:

$$T_{\text{obs}} = T_{\text{refl}} = T_{\text{ideal}} \tag{1}$$

at any time. T_{obs} is the temperature recorded by the IR camera, T_{refl} is the temperature response reflected from the defective area, and T_{ideal} is the ideal temperature response of the sound area. Equation (1) can be represented in vector form:

$$X = S + W \tag{2}$$

where these vectors collect all recordings over time and $X = T_{\text{obs}}$, $S = T_{\text{refl}}$, and $W = T_{\text{ideal}}$. Equation (2) is multiplied by a vector q that maximizes the visibility of the reflected temperature from the defective area and minimizes the response of non-defective areas. The vector q can, therefore, be calculated by [31,34]:

$$\max_{q}\|q^{T}S\|_{2} \text{ subject to } \min_{q}\|qW\|_{2} \tag{3}$$

where q^{T} is the transpose of q. Different methods for finding q result in the different types of MF algorithms. Each MF algorithm considers a certain q vector to increase the contrast of defective areas and decrease the signals from sound areas. The q vector for the spectral angle map (SAM) is

$$SAM = \frac{S^{T}X_{ij}}{\sqrt{S^{T}S}\sqrt{X_{ij}^{T}X_{ij}}} \tag{4}$$

where i and j imply that the calculation is repeated for all pixels of all thermal images to provide a single correlation image and S^T. The adaptive coherence estimator (ACE) uses as q:

$$ACE = \frac{S^T C^{-1} X_{ij}}{\sqrt{S^T C^{-1} S} \sqrt{X_{ij}^T C^{-1} X_{ij}}} \tag{5}$$

where C is covariance matrix of the ideal temperature vector defined as:

$$C = \frac{1}{m} \sum_{i=1}^{m} WW^T. \tag{6}$$

The t- and F-statistics use the different q vectors to improve the contrast associated with defective areas:

$$F_{stat} = \frac{\left(S^T R^{-1} x_{ij}\right)^2}{X_{ij}^T R^{-1} X_{ij} - \rho^2 \left(S^T R^{-1} X_{ij}\right)^2} \rho^2 (d - 1) \tag{7}$$

and

$$t_{stat} = \frac{S^T R^{-1} x_{ij}}{\sqrt{X_{ij}^T R^{-1} X_{ij} - \rho^2 \left(S^T R^{-1} X_{ij}\right)^2}} \rho \sqrt{d - 1} \tag{8}$$

where $\rho = \left(S^T R^{-1} S\right)^{-1/2}$.

4. Quantitative Evaluation

The SNR of the images is a measure of the quality of data. The traditional definition of SNR is the ratio of the average magnitude of a signal to the magnitude of the background noise, determined as described in Section 5. If a distorted signal, $y(n)$, is considered as:

$$y(n) = x(n) + u(n), \tag{9}$$

where $x(n)$ is a signal and $u(n)$ is the background noise, the SNR can be written for N samples as [35]:

$$SNR = \frac{\sum_{n=0}^{N-1} |[x(n)]|^2}{N\sigma^2}, \tag{10}$$

where $|[x(n)]|$ is the magnitude of a signal and σ^2 is the variance of the background noise. Since most signals have wide dynamic ranges, the SNR is usually given in decibels (dB), [36]:

$$SNR = 10 \log\left(\frac{\Delta^2}{\sigma^2}\right), \tag{11}$$

where Δ is the peak value of the signal. Equation (11) can be simplified as:

$$SNR = 20 \log\left(\frac{\Delta}{\sigma}\right). \tag{12}$$

5. Results and Discussion

The results of both the passive and active thermography are presented in this section. All temperatures were measured using the IR camera. The different image processing algorithms, employed to increase the quality of the thermal images, are described and their results are discussed in detail.

5.1. Active Thermography

The temperature distribution profiles depicted in Figure 5b are plotted along with the rows identified in Figure 5a. The most visible contrasts of the defects revealed, as expected, that the deeper the defect, in terms of the distance from the bottom of the hole to the surface, the less detectable it was. Moreover, the size of the defect was important. The raw thermograms do not show smaller defects located deep within the plate. Defects with a diameter of 4 mm were barely detected. It can also be seen that the defects in the middle part of the plate are clearer than those towards the plate edges. This was primarily caused by non-uniform heating where the middle part of the sample received more thermal energy than the boundaries. The temperature distribution profiles during the step heating thermography for all pixels along the lines shown in Figure 5a are illustrated in Figure 5c. Figure 5c was recorded after heating for 75 s. The signals generated by the defects located in the last row, which have the smallest diameters, were not strong enough to be easily detected. Contrary to pulse thermography, Figure 5c shows that the surface of the object has been uniformly heated during the step heating thermography, which increased the efficiency of this method in detecting internal defects.

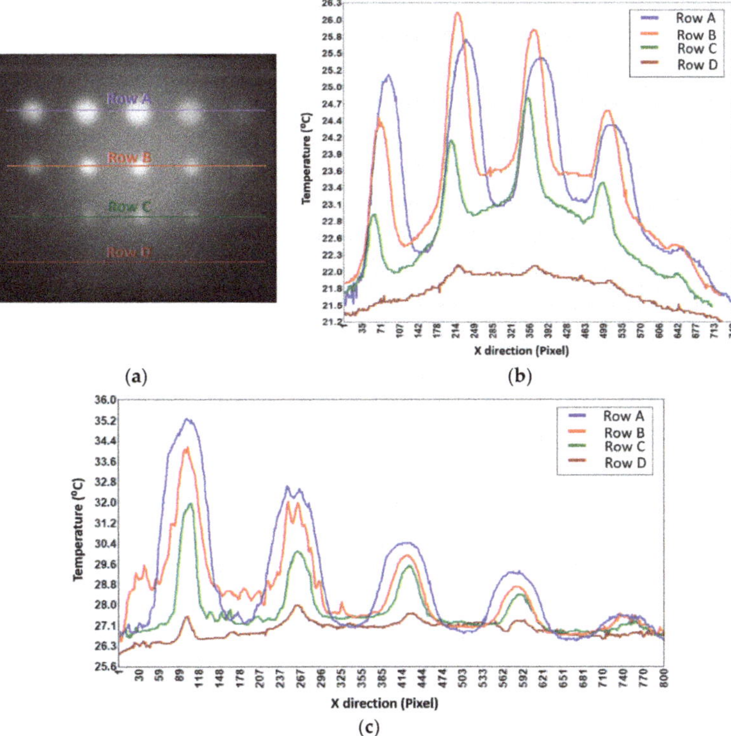

Figure 5. (a) Positions of temperature profiles. Temperature profiles using (b) pulsed and (c) step heating thermography. The rows are identified in (a).

Temperatures near the defects were measured and analyzed to better evaluate the effects of the defect's depth and size on the temperature variation. The temperatures measured at the locations of the flat-bottomed holes with various depths are illustrated in Figure 6. The sample was heated by two halogen lamps for 75 s, during which time data were recorded. It can be concluded from this figure that an increase in a defect's depth reduces its temperature variation, which makes detection more difficult because there is not a significant change in the surrounding temperature.

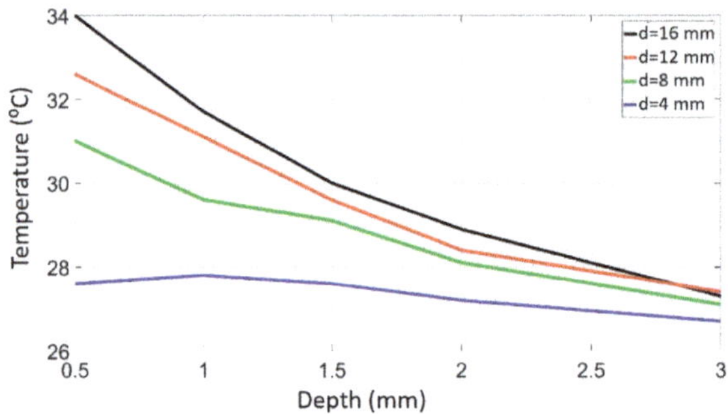

Figure 6. Effect of depth and size of the defect on the temperature distribution in a sample heated up by a halogen lamp for 75 s.

By increasing the size of the defect, the slope of the temperature variation also increased, which implied that the effect of the depth variation on the temperature was more significant for larger defects.

The SNRs for the defect plate profiles shown in Figure 5a were determined to provide a quantitative evaluation of the detection. The signal values of the defects were measured first. The difference between the signal values of the defects and the background signals was identified as the peak value signal for the SNR calculation. Then, by calculating the standard deviation of the background noise and using Equation (12), the SNR of each defect was calculated. By comparing the visual results obtained from image processing and the corresponded SNR tables, defects with a SNR of more than almost three times the background noise was considered as a defect that can be identified by this method. The SNRs, related to raw thermal data captured by flash and step heating (heating and cooling periods) thermography, are listed in Tables 1–3. Defect position names used in these tables are defined in Figure 1c. It can be observed from these tables that bigger defects closer to the surface generated high SNRs. It can also be concluded from the results presented in these tables that raw thermograms of flash thermography can provide more details than step heating thermography. It should be noted that the SNRs of raw thermograms are compared later with values of the processed thermal images to show the strength of each of the image processing algorithms.

Table 1. Signal to Noise Ratio (SNR) (dB) related to raw data captured by step heating thermography (heating period).

	1	2	3	4	5
A	18.4	18.2	17.8	14.4	3
B	18.1	18	17.7	12.9	6
C	17.9	17.5	17	11	NA
D	12.3	12	8.6	NA	NA

Table 2. SNR related to raw data captured by step heating thermography (cooling period).

	1	2	3	4	5
A	25.7	24.9	22.8	19.4	12.3
B	24.4	20	20.4	17.5	11.9
C	7.4	10.2	15.2	11.5	5.9
D	NA	2.58	2.5	NA	NA

Table 3. SNR related to raw data captured by flash thermography.

	1	2	3	4	5
A	16.3	18.5	18.5	13.7	3.4
B	10.2	15.8	17	11.9	2.9
C	2.5	4.2	13.1	4.9	2.6
D	NA	2.1	2.3	2.1	NA

5.2. Image Processing and Quantitative Evaluation

The results for the fourth row in Figure 5 show that some subsurface defects cannot generate visible contrasts. This suggests the need for image processing to improve the resolution. Different algorithms, including MF, and a combination of PPT as a transform-based technique were employed to increase the SNR and the visibility of the defects. We will use the term "Step Phase and Amplitude Thermography" (SPAT) for the analysis of thermograms after heating and successfully used for passive thermal data processing.

5.2.1. Matched Filters (MF)

Applying MF to the step heating thermograms improved the quality of the results, as shown in Figure 7. All four versions of MF increased the visibility and contrast of the defects. Three of these filters, including the SAM, the ACE, and the F-statistic, showed very promising results where all defects could be detected. However, the t-statistic highly improved the quality of the raw thermal data and compared to other MF methods provided less details regarding subsurface defects.

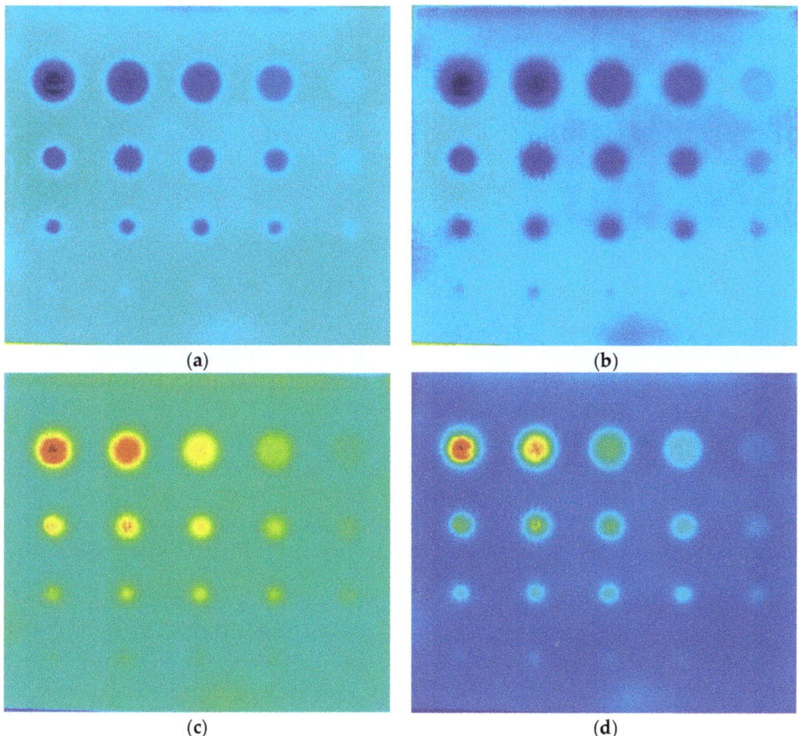

(a)

(b)

(c)

(d)

Figure 7. Four matched filters including (**a**) Spectral Angle Map (SAM), (**b**) Adaptive Coherence Estimator (ACE), (**c**) t-statistic and (**d**) F-statistic when the specimen was being step heated.

By using the above results, it can be seen that the diameter-to-depth ratio of the minimum detectable defect for each of SAM, ACE and F-statistic was 1.33, while this value for t-statistic was 2. To quantitatively evaluate the MF results, the SNRs for defective areas were determined. Figure 8a illustrates the signals of the F-statistic results obtained from the step heating data. The background noise around A2 is also shown in Figure 8b. These results were averaged to get the representative noise. It should be noted that all of the processed images have been normalized by dividing the intensity of each pixel over the difference between the maximum and minimum intensity available on the image.

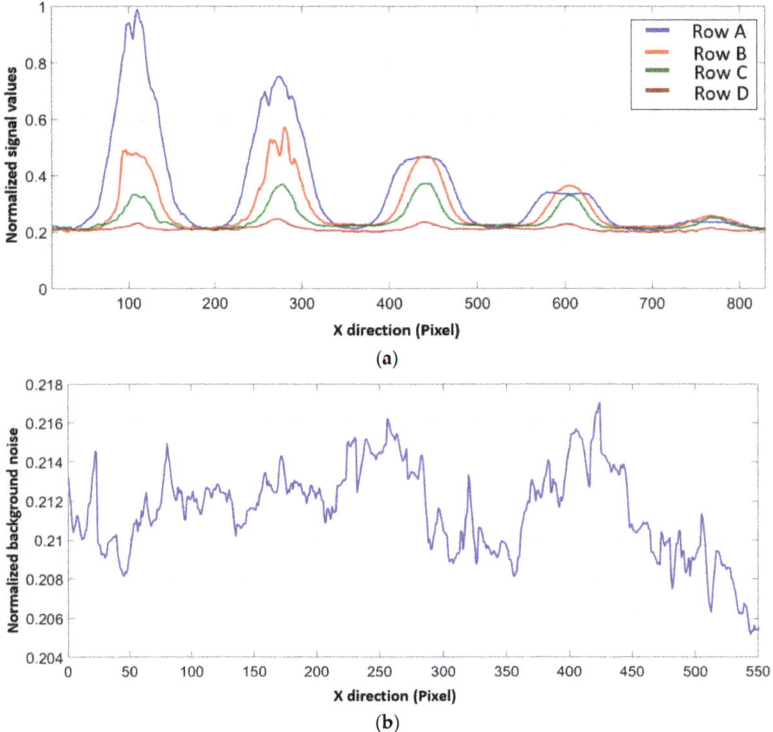

Figure 8. (**a**) Normalized signal values along the defect rows and (**b**) background noise around A2.

The SNRs are presented in Tables 4–7. The percentage improvement in the SNRs compared to the raw values is shown in parenthesis for each defect. These tables show that higher SNR values were obtained when MFs were applied to the step heating data. Therefore, the application of MFs to step heating data improved the resolution of the internal defects.

Tables 4–7 also reveal that larger defects closer to the surface generated higher SNR values. There were some exceptions to this general conclusion, however, primarily due to the non-uniformity of the heating.

Table 4. SNR (dB) for SAM on step heating thermal data.

	1	2	3	4	5
A	36.8 (99%)	35.3 (94%)	31.8 (78%)	26.9 (87%)	15.1 (396%)
B	33.5 (84%)	33.7 (87%)	31.2 (76%)	27.9 (116%)	17 (185%)
C	30.7 (72%)	29.3 (67%)	29.1 (70%)	26.3 (139%)	18.8
D	16.1 (31%)	18.7 (55%)	18.6 (116%)	13.9	5.5

Table 5. SNR (dB) for ACE on step heating thermal data.

	1	2	3	4	5
A	33.2 (79%)	31.9 (76%)	29.1 (63%)	24.3 (68%)	11.6 (282%)
B	30.6 (68%)	29.9 (66%)	28.5 (60%)	25.4 (97%)	15.3 (185%)
C	26.4 (48%)	27.1 (54%)	26.5 (55%)	24 (118%)	16.7
D	14.7 (19%)	16.9 (40%)	16.9 (96%)	11.7	8.5

Table 6. SNR (dB) for the *t*-statistic on step heating thermal data.

	1	2	3	4	5
A	44.3 (139%)	41.6 (129%)	34.9 (95%)	28.5 (90%)	15.7 (416%)
B	35.6 (95%)	37 (105%)	34.5 (95%)	38.8 (201%)	22.7 (280%)
C	28.2 (58%)	30.2 (72%)	30.3 (77%)	27.6 (151%)	19.4
D	13.5 (10%)	18.4 (53%)	16 (85%)	11.3	6.6

Table 7. SNR (dB) for the *F*-statistic on step heating thermal data.

	1	2	3	4	5
A	40.2 (117%)	37.6 (107%)	32.7 (83%)	27 (87%)	13.8 (354%)
B	33 (82%)	33.8 (88%)	32.2 (82%)	28.3 (120%)	17.8 (198%)
C	27.2 (52%)	28.9 (65%)	28.6 (68%)	26.1 (137%)	18.5
D	13.5 (10%)	17.2 (44%)	16.5 (92%)	11.6	2.7

5.2.2. Transform-Based Techniques: Step Phase and Amplitude Thermography

Pulsed thermograms were converted to a frequency domain using the FFT and the phase images were analyzed. Since the surface was excited uniformly during step heating, both phase and amplitude images of the transformed step heating thermograms were evaluated. Figure 9a,b shows a comparison of the raw pulsed thermograms and the normalized phase contrast obtained using the FFT. PPT significantly increased the contrast so that all defects, except D5, were detected. The same normalization method as the one used for the MF method was applied to the results obtained using the transformed-based techniques.

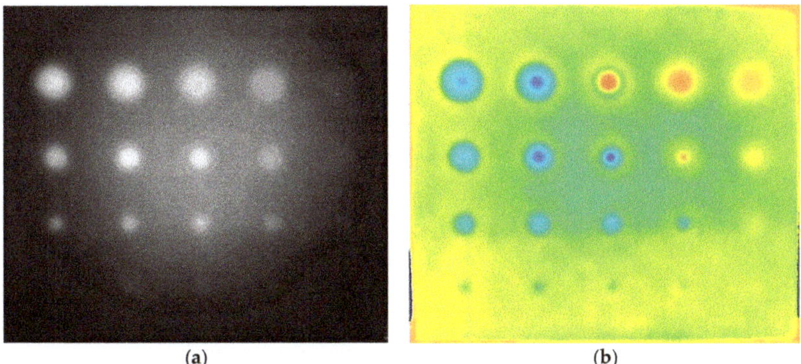

Figure 9. (a) Raw thermogram and (b) phase image (acquisition time = 53.2 s) obtained from the thermal image sequence recorded during cooling after flashing the surface.

The SNR values for the phase images depicted in Figure 9 are presented in Table 8, which shows that larger defects that were closer to the surface had higher SNRs. The first column of defects is an exception to this observation, mainly due to non-uniform heating of the surface.

Table 8. SNR (dB) for Pulsed Phase Thermography (PPT) data.

	1	2	3	4	5
A	25.5 (56%)	26.8 (45%)	24.9 (35%)	23.2 (69%)	20.7 (511%)
B	23.1 (127%)	26.5 (67%)	26.9 (58%)	23.4 (96%)	19.9 (591%)
C	19.6 (677%)	20.8 (397%)	19.8 (52%)	14.2 (190%)	12.9 (401%)
D	13.9	16.4	12.8	7 (234%)	NA

Figure 10a,b shows the phase and amplitude images at the minimum frequency where the FFT was applied to the step heating thermograms captured during cooling after heating for 75 s.

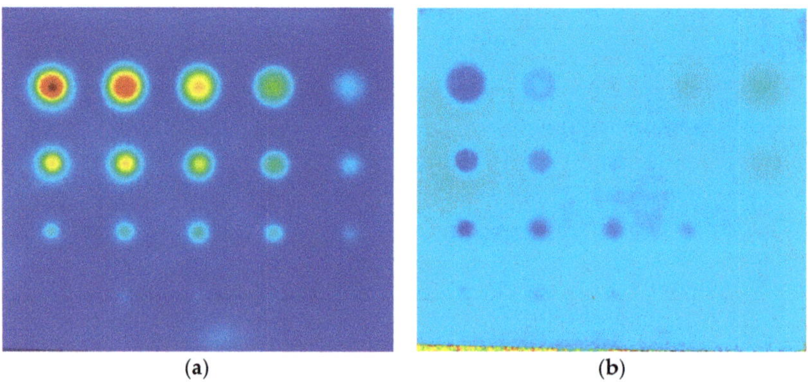

(a)   (b)

Figure 10. (a) Amplitude image and (b) phase image of the thermograms captured during cooling after 75 s of heating.

The diameter-to-depth ratio of the minimum detectable defect in each of phase images of PPT and SPAT was 2, while the amplitude of SPAT provided a better value of 1.33. It can be concluded from these results that the application of the FFT transform to step heating thermography was more effective than it was for flash thermography. This was especially true for the amplitude data, where clearer results with higher contrasts were achieved. Phase images revealed that defects with a better contrast compared to the amplitude images, in some cases. This demonstrated that a reliable inspection could be achieved by evaluating both results. Figure 11 shows the results of an FFT transform applied to thermograms captured during heating for 75 s. The phase image captured all defects—assumed to be indicated by the signals that were more than three times the background level—demonstrating that phase images extracted from the heating data provided more visibility compared to the data obtained from cooling. The amplitude images obtained during cooling were less noisy and contained more detail, which allowed the shape of defects to be determined.

Figure 12a presents a further analysis of the normalized amplitude contrast distribution of the thermograms captured during cooling after 75 s of heating. The curves in this figure are the amplitude variation along the lines shown in Figure 5a. A significant change occurred between the sound and defective areas in the amplitudes, leading to a sharp boundary around the defects. The normalized phase contrast distribution of thermograms obtained after heating for 75 s is depicted in Figure 12b. Bigger defects closer to the surface generated higher phase contrasts and were subsequently more visible in the phase image. By analyzing these plots, all defects, except D5, on the defect plate were detected.

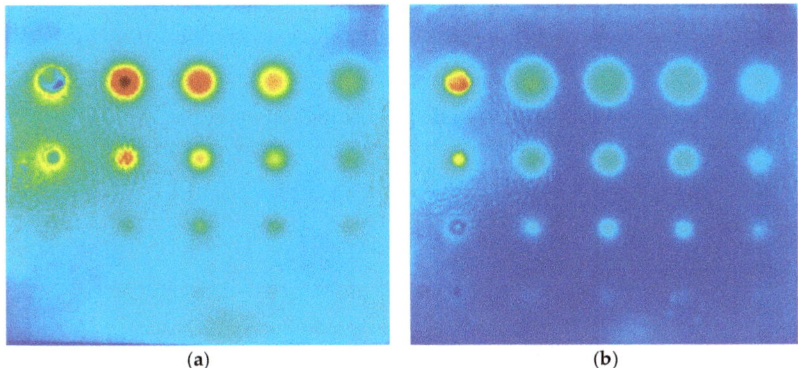

Figure 11. (a) Amplitude image and (b) phase image of thermograms captured during heating for 75 s.

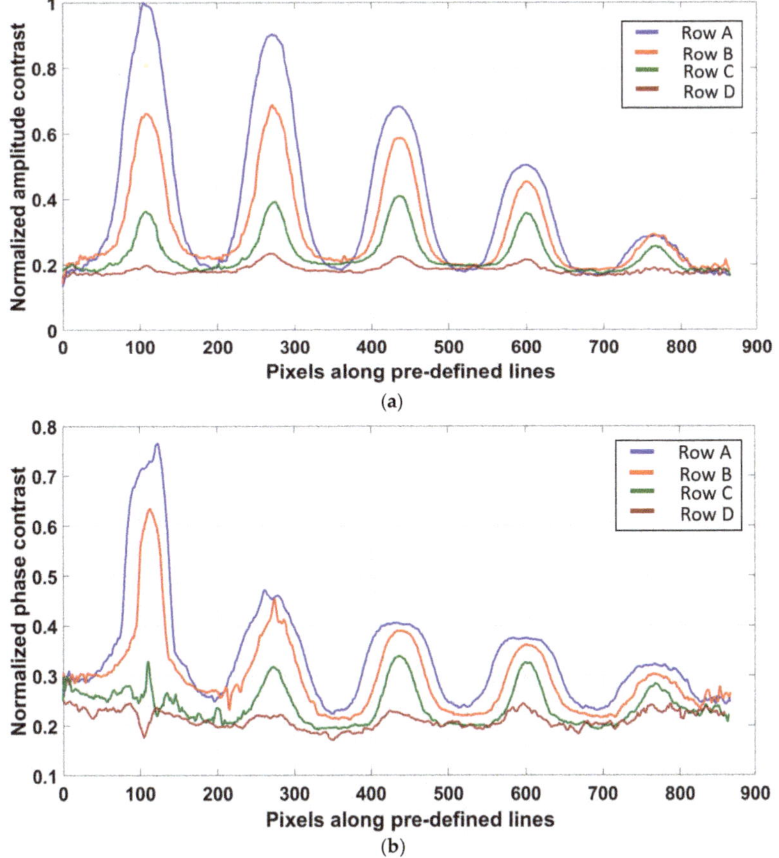

Figure 12. (a) Normalized amplitude value and (b) normalized phase value distributions of the defects where the thermograms were obtained during cooling and heating, respectively.

The SNR values for the phase and amplitude images captured by the application of FFT on the step heating data are listed in Tables 9–11. These results demonstrate that amplitude images extracted from thermal data captured during cooling had higher SNR values and revealed more details of subsurface defects. The SNRs of both the phase and amplitude images show that larger defects closer to the surface provided higher values.

Table 9. SNR (dB) of amplitude image captured during cooling after 75 s heating.

	1	2	3	4	5
A	46.2 (80%)	45.1 (81%)	42.2 (85%)	38.4 (98%)	29.7 (141%)
B	41 (68%)	41.3 (106%)	39.6 (94%)	36.7 (109%)	28.7 (140%)
C	33.1 (349%)	33.7 (231%)	34.6 (128%)	33.4 (190%)	25.9 (338%)
D	14.9	23.5 (809%)	21.2 (763%)	17	13.6

Table 10. SNR (dB) of phase image captured during 75 s heating.

	1	2	3	4	5
A	38.5 (183%)	30.9 (144%)	29.2 (127%)	27.6 (180%)	23.6 (774%)
B	36 (302%)	32 (160%)	29.2 (132%)	27.5 (208%)	23.2 (896%)
C	21.5 (1211%)	26.3 (707%)	27.8 (164%)	26.6 (580%)	22.9 (907%)
D	19	13.3 (1044%)	18.3 (827%)	17.7 (710%)	12.7

Table 11. SNR (dB) of amplitude image captured during 75 s heating.

	1	2	3	4	5
A	33.6 (106%)	41.9 (127%)	40.8 (120%)	38.3 (179%)	31.6 (834%)
B	26.9 (163%)	38.5 (142%)	37.6 (121%)	34.7 (191%)	29.2 (915%)
C	20.3 (706%)	26.2 (527%)	30.1 (130%)	28.7 (484%)	24.2 (842%)
D	8	13.1 (537%)	12.7 (455%)	10.8 (416%)	9.4

By comparing the results presented in Tables 8–11, it can be concluded from the results presented in Tables 9–11 that the application of the FFT transform to step heating thermography was more effective than its application to flash thermography. This was especially true for the amplitude data, where higher SNRs were achieved. The amplitude images extracted from step heating thermography were an effective means of revealing subsurface damage, as they detected even small and deep defects.

By comparing the results presented in Tables 1–9 related to the SNRs of raw and processed thermal data, it is concluded that the image processing algorithms significantly increased the visibility of subsurface defects especially those of smaller sizes, which are embedded in deeper areas.

5.3. Passive Thermography

Passive thermal imaging of the damaged blade section was conducted at different times of the day. Typical results are shown in Figure 13. The results demonstrated that passive thermography was capable of capturing cracks, delamination, and internal features of the blade section.

Early morning experiments provided a visible contrast of the defects on the defect plate attached to the damaged blade section primarily due to the considerable temperature change on the blade during this period [2,18,19]. All defects in the plate, except the smallest 4 mm ones, were detected during this period. Less useful information about the defects was obtained when the experiment was performed around noon. None of the defects were visible during the evening (around 6 pm), which was mainly due to the balanced temperature on the blade surface after several hours of heating.

Thermal contrasts associated with dirt (glue on the surface), within the blue box in Figure 13, were most pronounced at noon. These contrasts faded during the afternoon. The blade's internal features such as shear webs were detected as cold regions during the morning and at noon but became hot signatures during the evening after several hours of heating.

Cracks and delamination on the suction side were apparent near the blade's trailing edge. At noon, with peak sunlight, cracks and delamination were detected. Delamination in the upper area, identified by the green box in Figure 13, were not detected clearly during the morning. The evening thermograms did not provide any information regarding cracks and delamination, so noon was the best time for crack and delamination monitoring.

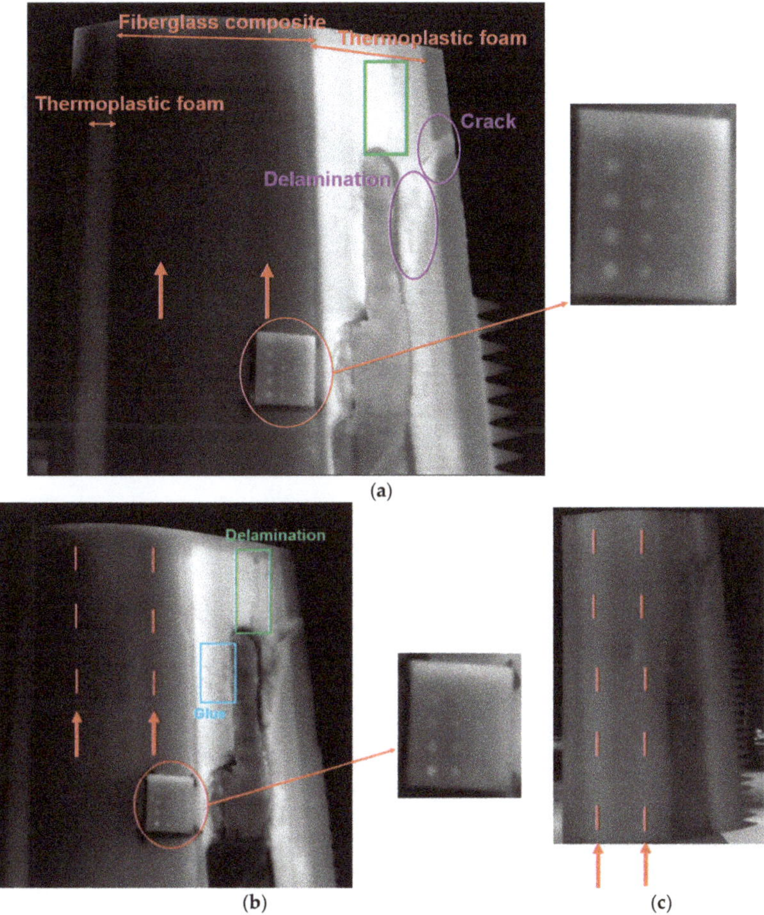

Figure 13. (a) Thermographic results of the experiment around 9 a.m., (b) noon and (c) 6 p.m. (sunrise and sunset were around 5.53 a.m. and 9.30 p.m., respectively). The vertical arrows and dashed lines indicate the shear webs.

The transform-based technique, discussed in Section 5.2.2, SPAT, was employed for the first time to increase the quality of the passive thermography results. The phase images captured using this method were not sensitive to non-uniform heating. The FFT was applied to passive thermograms obtained at morning. The results are shown in Figure 14. Part (b) illustrates that the amplitude images considerably increased the quality and visibility of the visualized subsurface defects, as the visibility of cracks, delamination, and a large portion of the flat-bottomed hole defects was improved. Phase images have noticeably increased the contrast of shear webs signatures, marked by red arrows in Figure 14a.

Cracks and delamination are marked by white boxes in Figure 14b. The thermoplastic foam near the leading edge was detected in the amplitude results and is highlighted by the red box in Figure 14b. This method not only increased the quality of the thermal images and improved the detectability of thermography but also eliminated the false indications associated with environmental reflections, dirt, and dust on the surface.

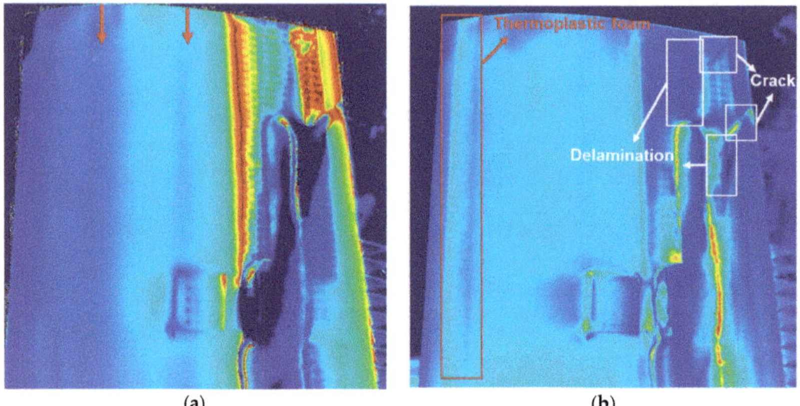

(a)   (b)

Figure 14. (a) Phase images of the passive thermograms captured during the morning at a frequency of 0.00184 Hz and (b) amplitude image of the passive thermograms recorded during the morning at a frequency of 0.0165 Hz.

6. Conclusions

The blades, the most critical components of wind turbines, are susceptible to failure due to initiation and propagation of subsurface damage in a number of forms. This study investigated thermography techniques for monitoring the condition of wind turbine blades. Active thermography, using pulsed and step heating, was conducted on a specially-constructed defect plate to evaluate this method's potential for detecting subsurface defects. The 170×195 mm plate was cut from the laminate skin of a wind turbine blade. Flat-bottomed holes were drilled from the inside to produce "defects" with known diameters and penetrations. The results demonstrated that active thermography is a powerful method for the monitoring and fault detection of wind turbine blades but that the signals generated by some small defects could not be detected.

Passive thermography was conducted on a damaged blade section and attached defect plate. This experiment was conducted at different times of day to determine the most favorable time of a day for maximum defect detection. The results showed that early morning and noon were best for detecting certain types of defects. Cracks, delamination, and surface dirt generated the most visible signatures around noon, while defects such as flat-bottomed holes in the defect plate were more visible in the morning as the sun heated the targets. This conclusion applies to the day time experiments as there were not any overnight tests.

The raw thermograms obtained by both passive and active thermography could not reveal the small defects located deep in the laminate skin. Different image processing algorithms including Matched Filters, Thermal Signal Reconstruction, and Pulsed Phase Thermography were used to increase the quality of active thermograms. "Step Phase and Amplitude Thermography", was used to analyze the step heating data. This technique gave the best detection of defects as measured by the diameter-to-depth ratio of 1.33. All the image processing algorithms improved the contrast in the active thermograms but some drawbacks can be considered for the Matched Filters method. The Matched Filters method is not fully automated and requires manual selection of points in a sound area of the damaged blade, which is time-consuming and affects the quality of results. In order to quantitatively evaluate the results, the signal-to-noise ratios of the raw and processed images were calculated. Higher ratios can be obtained when image processing algorithms are applied to the raw thermal data. As an obvious observation, larger defects closer to the surface generated higher ratios.

Step Phase and Amplitude Thermography, as a successful method for improving active thermography results, was applied to passive thermal data. The quality of passive thermograms was increased as a result. This method could also eliminate the false indications associated with environmental reflections and dirt on the surface. Nevertheless, it was not possible to resolve the smallest defects of 4 mm diameter whatever the depth. The minimum diameter-to-depth ratio for these defects was thus 1.25.

Author Contributions: The overall conception of this project was by D.W. and Q.S. who obtained the damaged wind turbine blade. The specific aims and implementation of them was done by H.S. as were all the experiments. All three authors contributed to writing the paper.

Funding: Hadi Sanati and David Wood acknowledge the Schulich endowment to the Faculty of Engineering, University of Calgary for providing funding for this work.

Conflicts of Interest: The authors declare no conflict of interest.

References

1. Jüngert, A. Damage Detection in wind turbine blades using two different acoustic techniques. In Proceedings of the 7th fib PhD Symposium in Civil Engineering, Stuttgart, Germany, 10–13 September 2008.

2. Meinlschmidt, P.; Aderhold, J. Thermographic inspection of rotor blades. In Proceedings of the 9th European Conference on NDT, Berlin, Germany, 25–29 September 2006.

3. Tao, N.; Zeng, Z.; Feng, L.; Li, X.; Li, Y.; Zhang, C. The application of pulsed thermography in the inspection of wind turbine blades. In *International Symposium on Photoelectronic Detection and Imaging 2011: Terahertz Wave Technologies and Applications*; SPIE: Washington, DC, USA, 2011; p. 819319.

4. Habali, S.M.; Saleh, I.A. Local design, testing and manufacturing of small mixed airfoil wind turbine blades of glass fiber reinforced plastics: Part II: Manufacturing of the blade and rotor. *Energy Convers. Manag.* **2000**, *41*, 281–298. [CrossRef]

5. Lading, L.; Mcgugan, M.; Sendrup, P.; Rheinländer, J.; Rusborg, J. *Fundamentals for Remote Structural Health Monitoring of Wind Turbine Blades—A Preproject Annex B—Sensors and Non-Destructive Testing Methods for Damage Detection in Wind Turbine Blades*; Risø National Laboratory: Roskilde, Denmark, 2002.

6. Rumsey, M.; Paquette, J. Structural health monitoring of wind turbine blades. *Proc. SPIE* **2008**, *6933*, 69330E.

7. Schroeder, K.; Ecke, W.; Apitz, J.; Lembke, E. A fibre Bragg grating sensor system monitors operational load in a wind turbine rotor blade. *Meas. Sci. Technol.* **2006**, *17*, 1167.

8. Liu, W.; Tang, B.; Jian, Y. Status and problems of wind turbine structural health monitoring techniques in China. *Renew. Energy* **2010**, *35*, 1414–1418. [CrossRef]

9. Larsen, G.; Hansen, M.; Baumgart, A.; Carlén, I. *Modal Analysis of Wind Turbine Blades*; Risø National Laboratory: Roskilde, Denmark, 2002.

10. Borum, K.; McGugan, M. Condition monitoring of wind turbine blades. In Proceedings of the 27th Riso International Symposium on Materials Science: Polymer Composite Materials for Wind Power Turbines, Roskilde, Denmark, 4–7 September 2006.

11. Li, T.; Almond, D.P.; Rees, D.A.S. Crack imaging by scanning laser-line thermography and laser-spot thermography. *Meas. Sci. Technol. Mar.* **2011**, *22*, 35701.

12. Pawar, S.S.; Peters, K. Through-the-thickness identification of impact damage in composite laminates through pulsed phase thermography. *Meas. Sci. Technol.* **2013**, *24*, 115601. [CrossRef]

13. Maierhofer, C.; Myrach, P.; Reischel, M.; Steinfurth, H.; Röllig, M.; Kunert, M. Characterizing damage in CFRP structures using flash thermography in reflection and transmission configurations. *Compos. Part B Eng.* **2014**, *57*, 35–46.

14. Maierhofer, C.; Arndt, R.; Röllig, M.; Rieck, C.; Walther, A.; Scheel, H.; Hillemeier, B. Application of impulse-thermography for non-destructive assessment of concrete structures. *Cem. Concr. Compos.* **2006**, *28*, 393–401. [CrossRef]

15. Beattie, A.; Rumsey, M. Non-destructive evaluation of wind turbine blades using an infrared camera. In Proceedings of the 37th Aerospace Sciences Meeting and Exhibit, Reno, NV, USA, 11–14 January 1999.

16. bin Zhao, S.; Zhang, C.; Wu, N. Infrared thermal wave nondestructive testing for rotor blades in wind turbine generators non-destructive evaluation and damage monitoring. In *International Symposium on Photoelectronic Detection and Imaging 2009: Advances in Infrared Imaging and Applications*; SPIE: Washington, DC, USA, 2009.

17. Galleguillos, C.; Zorrilla, A.; Jimenez, A.; Diaz, L.; Montiano, Á.L.; Barroso, M.; Viguria, A.; Lasagni, F. Thermographic non-destructive inspection of wind turbine blades using unmanned aerial systems. *Plast. Rubber Compos.* **2015**, *44*, 98–103. [CrossRef]

18. Worzewski, T.; Krankenhagen, R.; Doroshtnasir, M. Thermographic inspection of a wind turbine rotor blade segment utilizing natural conditions as excitation source, Part I: Solar excitation for detecting deep structures in GFRP. *Infrared Phys. Technol.* **2016**, *76*, 756–766. [CrossRef]

19. Worzewski, T.; Krankenhagen, R. Thermographic inspection of wind turbine rotor blade segment utilizing natural conditions as excitation source, Part II: The effect of climatic conditions on thermographic inspections–A long term outdoor experiment. *Infrared Phys. Technol.* **2016**, *76*, 767–776. [CrossRef]

20. Doroshtnasir, M.; Worzewski, T.; Krankenhagen, R.; Röllig, M. On-site inspection of potential defects in wind turbine rotor blades with thermography. *Wind Energy* **2016**, *19*, 1407–1422. [CrossRef]

21. Lahiri, B.B.; Bagavathiappan, S.; Reshmi, P.R.; Philip, J.; Jayakumar, T.; Raj, B. Quantification of defects in composites and rubber materials using active thermography. *Infrared Phys. Technol.* **2012**, *55*, 191–199. [CrossRef]

22. Shin, P.H.; Webb, S.C.; Peters, K.J. Pulsed phase thermography imaging of fatigue-loaded composite adhesively bonded joints. *NDT E Int.* **2016**, *79*, 7–16. [CrossRef]

23. Maierhofer, C.; Röllig, M.; Krankenhagen, R.; Myrach, P. Comparison of quantitative defect characterization using pulse-phase and lock-in thermography. *Appl. Opt.* **2016**, *55*, D76–D86. [CrossRef] [PubMed]

24. Almond, D.P.; Angioni, S.L.; Pickering, S.G. Long pulse excitation thermographic non-destructive evaluation. *NDT E Int.* **2017**, *87*, 7–14. [CrossRef]

25. Maldague, X.; Marinetti, S. Pulse phase infrared thermography. *J. Appl. Phys.* **1996**, *79*, 2694–2698. [CrossRef]

26. Chatterjee, K.; Tuli, S.; Pickering, S.; Almond, D. A comparison of the pulsed, lock-in and frequency modulated thermography nondestructive evaluation techniques. *NDT E Int.* **2011**, *44*, 655–667. [CrossRef]

27. Montanini, M. Quantitative determination of subsurface defects in a reference specimen made of Plexiglas by means of lock-in and pulse phase infrared thermography. *Infrared Phys. Technol.* **2010**, *53*, 363–371. [CrossRef]

28. Pawar, S.S. Identification of Impact Damage in Composite Laminates through Integrated Pulsed Phase Thermography and Embedded Thermal Sensors. Ph.D. Thesis, North Carolina State University, Raleigh, NC, USA, 2013.

29. Ibarra, C.C. Quantitative Subsurface Defect Evaluation by Pulsed Phase Thermography: Depth Retrieval with the Phase. Ph.D. Thesis, Laval University, Quebec, QC, Canada, 2005.

30. Shepard, S.M.; Lhota, J.R.; Rubadeux, B.A.; Wang, D.; Ahmed, T. Reconstruction and enhancement of active thermographic image sequences. *Opt. Eng.* **2003**, *42*, 1337. [CrossRef]

31. Larsen, C. Document Flash Thermography. Master's Thesis, Utah State University, Logan, UT, USA, 2011.

32. Roche, J.M.; Balageas, D.L. Common tools for quantitative pulse and step-heating thermography—Part II: Experimental investigation. *Quant. Infrared Thermogr. J.* **2015**, *12*, 1–23. [CrossRef]

33. Foy, B.R. *Overview of Target Detection Algorithms for Hyperspectral Data; Rep. to NNSA, Rep. No. LU-UR-09-00593*; Los Alamos Natl. Lab.: Los Alamos, NM, USA, 2009.

34. Kretzmann, J. Evaluating the industrial application of non-destructive inspection of composites using transient thermography. Master's Thesis, Stellenbosch Stellenbosch University, Stellenbosch, South Africa, 2016.

35. Xia, X.G. A Quantitative Analysis of SNR in the Short-Time Fourier Transform Domain for Multicomponent Signals. *IEEE Trans. Signal Process.* **1998**, *46*, 200–203.

36. Jain, J.; Jain, A. Displacement Measurement and Its Application in Interframe Image Coding. *IEEE Trans. Commun.* **1981**, *29*, 1799–1808. [CrossRef]

Article

CFD Simulations of Flows in a Wind Farm in Complex Terrain and Comparisons to Measurements

Matias Sessarego [1], **Wen Zhong Shen** [1,*], **Maarten Paul van der Laan** [2],
Kurt Schaldemose Hansen [1] and **Wei Jun Zhu** [3]

[1] Department of Wind Energy, Technical University of Denmark, Lyngby Campus, 2800 Kgs. Lyngby, Denmark; matse@dtu.dk (M.S.); kuhan@dtu.dk (K.S.H.)
[2] Department of Wind Energy, Technical University of Denmark, Risø Campus, 4000 Roskilde, Denmark; plaa@dtu.dk
[3] School of Hydraulic, Energy and Power Engineering, Yangzhou University, Yangzhou 225127, China; wjzh@dtu.dk
* Correspondence: wzsh@dtu.dk; Tel.: +45-45-25-4317

Received: 11 April 2018; Accepted: 9 May 2018; Published: 15 May 2018

Abstract: This article describes Computational Fluid Dynamics (CFD) simulations of flows in a wind farm in complex terrain in Shaanxi, China and the comparisons of the computational results with utility scale field measurements. The CFD simulations performed in the study are using either a Reynolds-Averaged Navier–Stokes (RANS) or Large-Eddy Simulation (LES) solver. The RANS method together with an Actuator Disc (AD) approach is employed to predict the performance of the 25 wind turbines in the farm, while the LES and Actuator Line (AL) technique is used to obtain a detailed description of the flow field around a specific wind turbine #14 near two met masts. The AD-RANS simulation results are compared with the mean values of power obtained from field measurements. Furthermore, the AL-LES results are compared with the mean values of power, rotor speed, and wind speed measured from the wind turbine and its nearby two masts. Results from the simulations indicate that both AD-RANS and AL-LES methods can reasonably predict the performance of the wind farm and wind turbine #14, respectively, in complex terrain in Shaanxi. The mean percent difference obtained for power in the AD-RANS simulations was approximately 20%. Percent differences obtained for power and rotor RPM in the AL-LES varied between 0.08% and 11.6%. The mean percent differences in the AL-LES for power and rotor RPM are approximately 7% and 1%, respectively.

Keywords: computational fluid dynamics; wind farm; complex terrain; SCADA; met mast measurements

1. Introduction

Onshore wind is one of the most competitive sources of new generation capacity [1]. According to the International Renewable Energy Agency (IRENA), the global weighted average cost of electricity was USD 0.06/kWh for onshore wind power in 2017, which falls within the range of generation costs for fossil-based electricity [1]. However, the performance and lifetime of onshore wind turbines are strongly influenced by the site-specific wind resource and terrain environment. Wind flow and wind farm modeling software aims to predict and simulate the characteristics of the wind resource as well as the performance of a proposed or existing wind farm, e.g., to estimate its annual energy production. The current study focuses on high-fidelity numerical approaches, i.e., Computational Fluid Dynamics (CFD), for estimating the performance of wind turbines and wind farms in complex terrain.

A large variety of CFD methodologies have been employed to predict the flow characteristics of wind farms in offshore [2], flat [3] and complex [4–9] terrain environments. The most commonly employed high-fidelity CFD methods involve either the Reynolds-Averaged Navier–Stokes (RANS)

and Large-Eddy Simulation (LES) techniques to determine the wind flow over a farm. LES solvers are approximately three orders more computationally expensive than RANS solvers, since the large scale turbulence is resolved on a grid and only the smaller scale turbulence is modeled, e.g., using a sub-grid scale (SGS) model [3]. RANS solvers are generally performed in a steady-state manner where all scales of turbulence are modeled [3]. Since the turbulence length scale on a wind turbine rotor surface is too small in relation to the size of the wind farm, the full three-dimensional geometry of the rotor is commonly simplified using the Actuator Disc (AD) or Actuator Line (AL) techniques [10].

Makridis and Chick [4] investigated wind turbine wakes and neutral atmospheric wind flow over a moderately complex terrain using the CFD software Fluent (version 12.0, ANSYS, Inc., Canonsburg, PA, USA). RANS was used to model the wind flow and an AD implementation of Fluent was used to model the wind turbine rotor effects. Comparisons were made to measurements and WAsP (Wind Atlas Analysis and Application Program) simulations. Castellani et al. [9] performed AD-RANS analyses of a sub-cluster of four wind turbines from a wind farm sited in Italy in complex terrain. The dominant 270 degrees wind flow direction over the four wind turbines was studied using simulation results and yaw measurements. Schulz [5,6] performed Delayed-Detached-Eddy-Simulations (DES) of a 2.4 MW wind turbine in complex terrain including a turbulent atmospheric boundary layer. The wind turbine was modeled by meshing the tower, nacelle, hub, joints and blades. Tabib [7] performed LES and RANS simulations of the onshore Bessaker wind farm. Tabib [8] then analyzed wake-terrain interactions by comparing two LES simulations of a multi-wind-turbine layout with annual power data at the actual wind farm. Tabib used the AL technique to model the wind turbine.

In the current article, both AD-RANS and AL-LES techniques are employed to compute the flow field and performance of a utility scale wind farm located in Jianbian, Shaanxi, China, consisting of twenty-five 2 MW wind turbines (WTs) in complex terrain. Measurements from a supervisory control and data acquisition (SCADA) system containing sensors from the twenty-five WTs and measurements from two met masts in the farm are used to compare with the CFD simulations. The main contribution of the present work is the better understanding of the level of accuracy and capabilities of AD-RANS and AL-LES solvers when predicting the wind farm performance in complex terrain environments. The results of this study can be used by other researchers, who do not have access to measurement data, to compare their own wind farm CFD simulation results in similar complex terrain environments. The comparison will enable researchers to quantify how close their wind farm CFD simulation results might be if field measurements were available. A description of how the mesh, wind turbine model and wind farm layout are prepared, and the input parameters used in the CFD simulations can be used as reference.

The background regarding the CFD solvers, wind farm and methodology for creating the terrain grids will be described first in Section 2. Then, results from both the AD-RANS and AL-LES simulations compared with field measurements will be shown in Section 3. Finally, conclusions are given in Section 4.

2. Methodology

The RANS and LES simulations are performed using the in-house developed flow solver EllipSys3D [11,12] with the AD and AL methodologies as described in [13,14], respectively, in neutral atmospheric conditions. EllipSys3D solves the Navier–Stokes equations in general curvilinear coordinates using a block structured finite-volume approach [10]. Both the AD and AL include controllers for the rotational speed and pitch. Section 2.1 contains information about the wind turbine model used in the AD and AL techniques. Sections 2.2 and 2.3 describe the AD-RANS and AL-LES implementations in more detail, respectively. Section 2.4 summarizes the flow cases used in the present study.

2.1. Wind Turbine Model

The twenty-five 2 MW wind turbines in the wind farm are designed and manufactured in China by China Shipbuilding Industry Corporation (CSIC) Haizhuang (HZ) Windpower Equipment Co., Limited (Chongqing, China). Table 1 contains the basic information about the wind turbine. A computer-aided design (CAD) model of the blade was provided by the blade manufacturer (confidential), which was used to extract six airfoil profiles needed for the AD and AL models. The airfoil profiles were used as input to the viscous-inviscid airfoil flow solver, Q^3UIC [15], to generate the polar data containing the lift and drag coefficients as a function of the angle of attack.

The power and thrust as a function of wind speed were provided by the manufacturer, but not the rotor RPM (revolutions per minute) and pitch schedule. Therefore, the AD and AL models were developed iteratively to match as closely as possible to the manufacturer's data as well as the wind turbine measurement data for the RPM and pitch. Figure 1a,b depict the electrical power and rotor thrust as a function of the wind speed from the manufacturer's data and the developed model used in the AD and AL techniques. Figure 1b shows that the developed wind turbine model over-predicts the rotor thrust between 3 and 10 m/s compared to the manufacturer's data. However, an estimate of rotor thrust based on strain-gage measurements showed deviations from the manufacturer's data as well [16], and is in close agreement to the wind turbine model. Nevertheless, rotor thrust discrepancies will influence the wake deficit comparisons shown in Section 3.2.

Table 1. China Shipbuilding Industry Corporation (CSIC) Haizhuang (HZ) wind turbine specifications.

Name	CSIC HZ93-2.0MW
Diameter	93 m
Power rating	2.0 MW
Hub height	67 m
Power control	Variable speed, collective blade pitch control

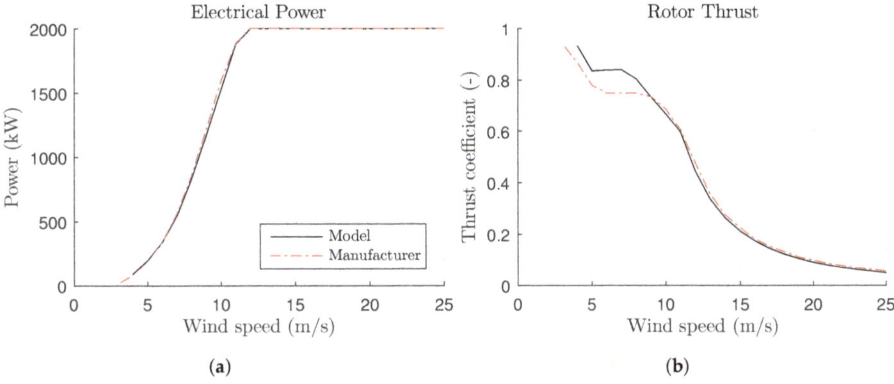

Figure 1. Comparison of the manufacturer's data and the developed model for the (**a**) electrical power and (**b**) rotor thrust as a function of the wind speed.

2.2. Actuator Disc Reynolds-Averaged Navier–Stokes Simulation

The k-ϵ-f_P eddy-viscosity model [3,17] is used to model the turbulence in RANS. A log-law is used with a surface roughness of $z_0 = 0.1$ m over the entire domain to simulate the site vegetation. Ideally, z_0 should be based on measurement data, e.g., by considering the inflow profile according to measurements in the wind farm, however a rough estimate for z_0 was used instead. Figure 2a depicts the terrain and wind turbine setup for the AD-RANS where the wind turbine rotor diameter,

D, is 93 m. In the AD-RANS setup, the mesh is cylindrical in shape with a radius of 28 km and a height of approximately 25 km. The 7.8 km^2 terrain grid is situated at the center of the mesh at WT10, since WT10 is approximately located at the center of the wind farm. The cylindrical mesh allows variable wind directions without changing the grid. Inflow/outflow boundary conditions are applied on the outer-boundary of the mesh according to the wind direction. The velocities at the top and inlet boundaries are prescribed. The mesh contains 280 blocks with each block having 48^3 grid points. The terrain grid is 10 × 10 blocks. There are approximately 6 (horizontal) × 16 (vertical) grid cells on each AD.

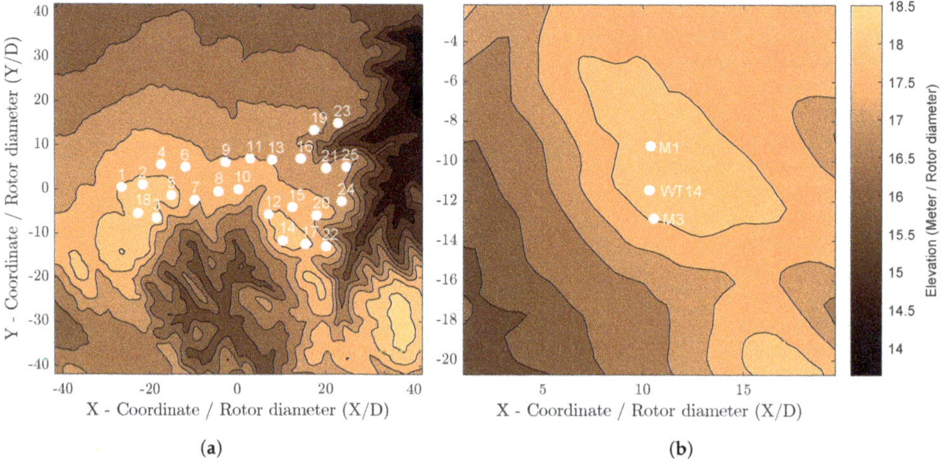

Figure 2. Computational layouts of the (**a**) Shaanxi wind farm with twenty-five wind turbines and of (**b**) wind turbine 14 with met masts M1 and M3. The wind turbines are labeled from 1 to 25.

Given an amount of computational resources to work with, since the amount of computational resources is limited, the computational mesh is designed such that there is more grid points at the center to capture terrain effects and the effect of each AD on the flow field. At the center, the velocities gradients due to complex terrain and the wind turbine presence is greatest. Further away from the center and towards the cylindrical boundary of the mesh, a much coarser grid is used over a long distance to minimize the computational cost and at the same time reduce the effect of the boundary conditions on the flow field at the center. The computational grid gave slightly less than the recommended amount of grid points on each AD, which based on experience should be a minimum of 8 horizontal and 8 vertical grid cells [17]. The effect of a refined mesh, where the terrain grid is 24 × 10 blocks, is shown in Section 3.1.

Sensors are placed at hub height for all twenty-five wind turbines as well as on the met masts in the CFD setup. The (a) outer and (b) terrain computational grids for the AD-RANS computations are shown in Figure 3. Each AD-RANS computation is performed using one-hundred-and-forty 2.8 GHz processors on a Linux cluster, where each processor is allocated for two blocks out of the 280 blocks. Each AD-RANS computation requires approximately 3 to 4 h for completion.

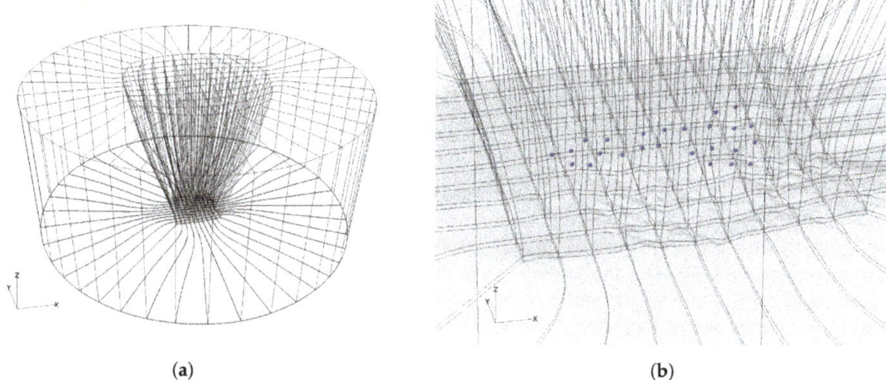

(a)   (b)

Figure 3. (**a**) outer and (**b**) terrain computational grids used in the actuator disc Reynolds-averaged Navier-Stokes (AD-RANS) computations. The circles in (**b**) are the actuator discs.

2.3. Actuator Line Large-Eddy Simulation

Figure 2b depicts the terrain and wind turbine setup. The AL-LES provides much more detailed information about the wind flow than the AD-RANS, but at the expense of a much higher computational cost. Therefore, in contrast to the AD-RANS simulations of the entire Shaanxi wind farm, the AL-LES is applied to a much smaller area, specifically around WT14. The smaller area contains WT14 and the two met masts M1 and M3. The area around WT14 was selected because it is near the two met masts and the results from the simulations can be compared with measurements from the met masts. The terrain is centered at WT14 and is $1728\,m^2$ in size. The mesh is also cylindrical in shape with a radius of 6.5 km, a height of approximately 3.2 km and contains a total of 1404 blocks. The terrain grid is 18 × 18 blocks. Each block contains 48^3 grid points as well. There is approximately 47 (horizontal) × 39 (vertical) grid cells on the rotor swept area, which is similar to the number of grid cells used in Refs. [10,18]. The boundary conditions are the same as in Section 2.2. The (a) outer and (b) terrain computational grids for the AL-LES computations are shown in Figure 4. Each AL-LES computation is performed using seven-hundred-and-to 2.8 GHz processors on a Linux cluster, where each processor is allocated for two blocks out of the 1404 blocks.

To attain a flow field over the vicinity of WT14, the flow field in AL-LES was scaled using the sensor at 70 m height on the upstream met mast, since it has almost the same height as the wind turbine hub (67 m). For the 4 and 7 m/s at 180 degrees, the upstream and downstream met masts are M3 and M1, respectively. For the 4 and 7 m/s at 0 degree, the upstream and downstream met masts are M1 and M3, respectively. In addition, WT14, M1, and M3 are located at elevation levels of 1695, 1692, and 1678 m, respectively. To simulate the wind shear, a prescribed profile is imposed at the first time-step in the AL-LES using a shear exponent of 0.3, roughly estimated from the met-mast measurements.

Two sets of simulations are carried out: inflow with and without synthetic turbulence. The inflow simulations without synthetic turbulence are performed to scale the flow field more quickly, as mentioned above, since the computational cost is much less than the inflow simulations with synthetic turbulence. The AL-LES computations without and with synthetic inflow turbulence require approximately two and five days for completion, respectively. Simulations without synthetic inflow turbulence are performed until steady-state values for power, rotor RPM, and wind speed are achieved (i.e., approximately 34 s real time). Simulations with synthetic inflow turbulence are performed to obtain 600 s (10 min) of real time, excluding initial transients.

Synthetic inflow turbulence is simulated by prescribing a plane, $372 \times 372\,m^2$ in size, 2.5 rotor diameters upstream of the wind turbine. Turbulence boxes are generated as a pre-processing step using the Mann model [19,20] and slices from the box are gradually fed into the plane during the simulation. In an attempt to represent the high turbulence magnitudes observed in the measurements, a high surface roughness of 0.4 was selected to generate the turbulence boxes. The turbulence is induced in a square 2.5 rotor diameters upstream of the wind turbine because it is scaled, using the surface roughness in the Mann model, in an attempt to match the turbulence intensity seen in the measurements in that specific area of the domain. Of course, a much larger square can be imposed upstream of the wind turbine with a constant surface roughness to cover the entire domain. However, this would imply that the turbulence intensity is uniform across the entire domain, which is likely not the case. It is assumed that creating a plane larger than $372 \times 372\,\text{m}^2$ would not influence the results.

In addition, the magnitude of the turbulence released at the plane was multiplied by a scaling factor of 1.5. The scaling factor does not have a physical meaning and is used when comparing with measurements where the source of the additional turbulence observed in the measurements cannot be captured in the LES simulation alone. The time-averaged solutions from the simulations without and with synthetic inflow turbulence are compared to the mean values from SCADA.

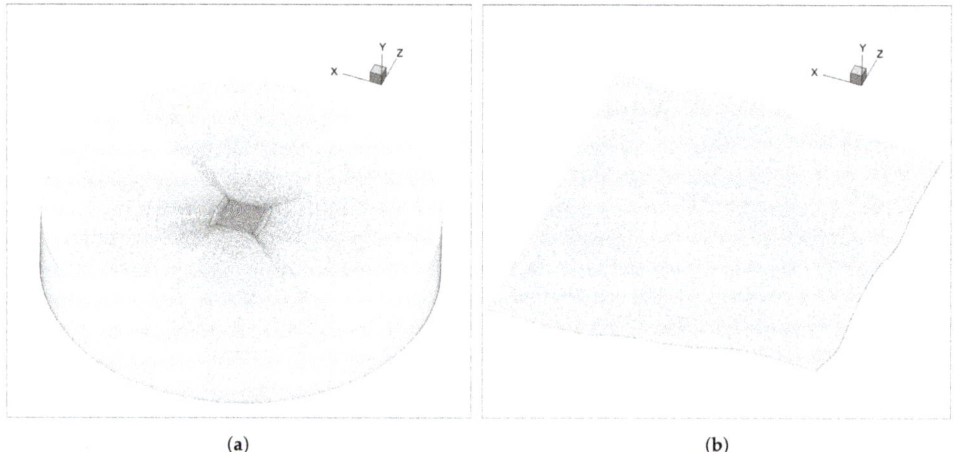

(a)   (b)

Figure 4. (**a**) outer and (**b**) terrain computational grids used in the AL-LES computations.

2.4. Flow Cases

Based on twelve months of met mast and wind turbine measurements, nine wind farm flow cases (FC1) have been identified for comparing the AD-RANS simulations. The AD-RANS flow cases are shown in Table 2. The wind turbine furthest upstream and that is the least influenced by the wakes of nearby wind turbines is chosen as the wind speed reference in Table 2. The met masts were not used as references for the wind speed because they are influenced by the wakes of nearby wind turbines for the 300° and 330° wind directions. It was not possible to derive the wind direction from the wind turbine SCADA, thus the wind turbine could not be used as the wind direction reference. The met masts are used to determine the reference wind direction. Due to the large computational expense, only four cases (FC2) are investigated for the AL-LES simulations. The AL-LES flow cases are shown in Table 3. The met masts were used as reference for the wind speed and direction in FC2. The ±5° and ±1 m/s cases in Tables 2 and 3 represent the bin intervals used to collect the SCADA measurements. Met mast and SCADA measurements used in the comparisons are filtered and represent neutral atmospheric conditions. Section 3 contains the AD-RANS and AL-LES simulation results for the FC1 and FC2 flow cases, respectively.

Table 2. Inflow conditions for the actuator disc Reynolds-averaged Navier-Stokes (AD-RANS) flow cases (FC1) where h = height and WT = wind turbine.

FC1	Sector	Wind Speed, h = 67 m	Reference Wind Speed	Reference Wind Direction
1.1	$180 \pm 5°$	$4 \pm 1\,\mathrm{m/s}$	WT14, h = 67 m	M3, h = 70 m
1.2	$300 \pm 5°$	$4 \pm 1\,\mathrm{m/s}$	WT11, h = 67 m	M1, h = 70 m
1.3	$330 \pm 5°$	$4 \pm 1\,\mathrm{m/s}$	WT11, h = 67 m	M1, h = 70 m
1.4	$180 \pm 5°$	$7 \pm 1\,\mathrm{m/s}$	WT14, h = 67 m	M3, h = 70 m
1.5	$300 \pm 5°$	$7 \pm 1\,\mathrm{m/s}$	WT11, h = 67 m	M1, h = 70 m
1.6	$330 \pm 5°$	$7 \pm 1\,\mathrm{m/s}$	WT11, h = 67 m	M1, h = 70 m
1.7	$180 \pm 5°$	$10 \pm 1\,\mathrm{m/s}$	WT14, h = 67 m	M3, h = 70 m
1.8	$300 \pm 5°$	$10 \pm 1\,\mathrm{m/s}$	WT11, h = 67 m	M1, h = 70 m
1.9	$330 \pm 5°$	$10 \pm 1\,\mathrm{m/s}$	WT11, h = 67 m	M1, h = 70 m

Table 3. Inflow conditions for the actuator line large-eddy simulation (AL-LES) flow cases (FC2) where h = height.

FC2	Sector	Wind Speed, h = 67 m	Reference Wind Speed	Reference Wind Direction
2.1	$180 \pm 5°$	$4 \pm 1\,\mathrm{m/s}$	M3, h = 70 m	M3, h = 70 m
2.2	$180 \pm 5°$	$7 \pm 1\,\mathrm{m/s}$	M3, h = 70 m	M3, h = 70 m
2.3	$0 \pm 5°$	$4 \pm 1\,\mathrm{m/s}$	M1, h = 70 m	M1, h = 70 m
2.4	$0 \pm 5°$	$7 \pm 1\,\mathrm{m/s}$	M1, h = 70 m	M1, h = 70 m

3. Results

The current section summarizes the results obtained from the AD-RANS and AL-LES simulations and their comparisons to the SCADA measurements. The results are divided into two subsections, actuator disc Reynolds-averaged Navier–Stokes and actuator line large-eddy simulation.

3.1. Actuator Disc Reynolds-Averaged Navier–Stokes Results

Prior to simulating the entire wind farm using AD-RANS, the inputs for the far upstream wind direction and speed should be modified to include the effect of the turning angle and speed-up. The wind direction and speed throughout the complex terrain will not be equal to the far upstream wind direction and speed where the terrain is assumed to be flat (see e.g., Figure 3). The flow will change direction and gain or lose speed when encountering, e.g., hills, valleys, and cliffs. For an accurate comparison with the measurements, the computed wind direction and speed at the reference met mast and wind turbine must take into account the effect of the turning angle and speed-up. The turning angle and speed-up are obtained from a pre-computation without wind turbines at the reference met mast and wind turbine locations, respectively. The turning angle and speed-up are calculated using the reference wind speed and direction as shown in Table 2. Table 4 summarizes the turning angles and speed-ups obtained for FC1. In the pre-computation without wind turbines, e.g., FC 1.1 in Table 4, the far upstream wind direction and speed were set to 180 degrees and $4\,\mathrm{m/s}$, respectively. The wind direction and speed computed at M3 and WT14 was $(180 + 4.28)$ degrees and $(4 \times 1.25)\,\mathrm{m/s}$, respectively. The far upstream wind direction and speed are then set to $(180 - 4.38)$ degrees and $(4/1.25)\,\mathrm{m/s}$ to obtain approximately 180 degrees and 4 m/s at M3 and WT14, respectively.

Table 4. Turning angles and speed-ups obtained for FC1.

FC1	1.1	1.2	1.3	1.4	1.5	1.6	1.7	1.8	1.9
Turning angle (degrees)	4.28	−4.78	1.29	4.28	−4.76	1.29	4.28	−4.75	1.28
Speed-up (-)	1.25	1.12	1.13	1.25	1.12	1.13	1.25	1.12	1.13

Figure 5 contains the results for the flow cases defined in Table 2, which consist of the AD-RANS results and the comparisons with the measurement data from SCADA. The notation *wd* and *ws* shown in the title for each figure denotes wind direction and wind speed, respectively. The wind direction corresponds to the measurements on met masts M1 and M3, while the wind speed corresponds to the measurements on the reference wind turbine (ref. WT). The figure contains the predicted, labeled as AD# and AD# $\pm$ 5° where # is the wind direction in degrees, and measured, labeled as SCADA, electrical power output for all 25 WTs. Each WT is labeled between 1 and 25, refer to Figure 2a. The AD# results represent a single run with the wind direction, #, which is also specified in the sub-figure titles. The AD# $\pm$ 5° results represent the average of five runs: -5, -2, 0, $+2$, and $+5$ degrees relative to *wd*. For example, AD# for FC 1.1 is a single run at 180 degrees (# = 180°), and AD# $\pm$ 5° is the average of 5 runs: $180 - 5$, $180 - 2$, 180, $180 + 2$, and $180 + 5$ degrees. The SCADA measurements represent averaged values in ± 5 degree bins. Although difficult to see, the standard error (SE) in Figure 5 is equal to SCADA $\pm$ 5° but includes the error bars (may require zoom-in).

In Figure 5, AD# $\pm$ 5° should more closely represent the SCADA measurements than AD#, since the SCADA measurements are grouped and averaged in ± 5 degree bins. The SCADA results cannot be determined exactly at 180, 300 or 330 degrees in Figure 5. Nevertheless, the AD-RANS results in Figure 5 show that AD# $\pm$ 5° is almost identical to AD#. In general, there is an agreement between the AD-RANS and the SCADA. However, there are significant differences in, for example, FC 1.1 for all the WTs and WT16 in FC 1.7. A possible explanation for the former case is that for the south wind at a small wind speed of 4 m/s, the wind distribution over the terrain may be non-uniform. Large variations in wind speed is captured in the SCADA but not in the RANS. The latter discrepancy is due to the delayed wake recovery caused by low dissipation in the employed turbulence model. The percent difference for the latter discrepancy is 57.4% and 66.4% based on AD# $\pm$ 5° and AD#, respectively. The latter discrepancy is the same as observed in WT16 for FC 1.4 as well. Figure 6 shows the wind speed magnitude (a) for the entire wind farm and (b) in the vicinity of WT16 for FC 1.7. In Figure 6b, WT16 is in the wake of WT17 and possibly WT20 as well, despite being more than 10 diameters away. The higher power output for WT16 in the SCADA indicates that WT16 may not be influenced as much by the WT17 and WT20 wakes.

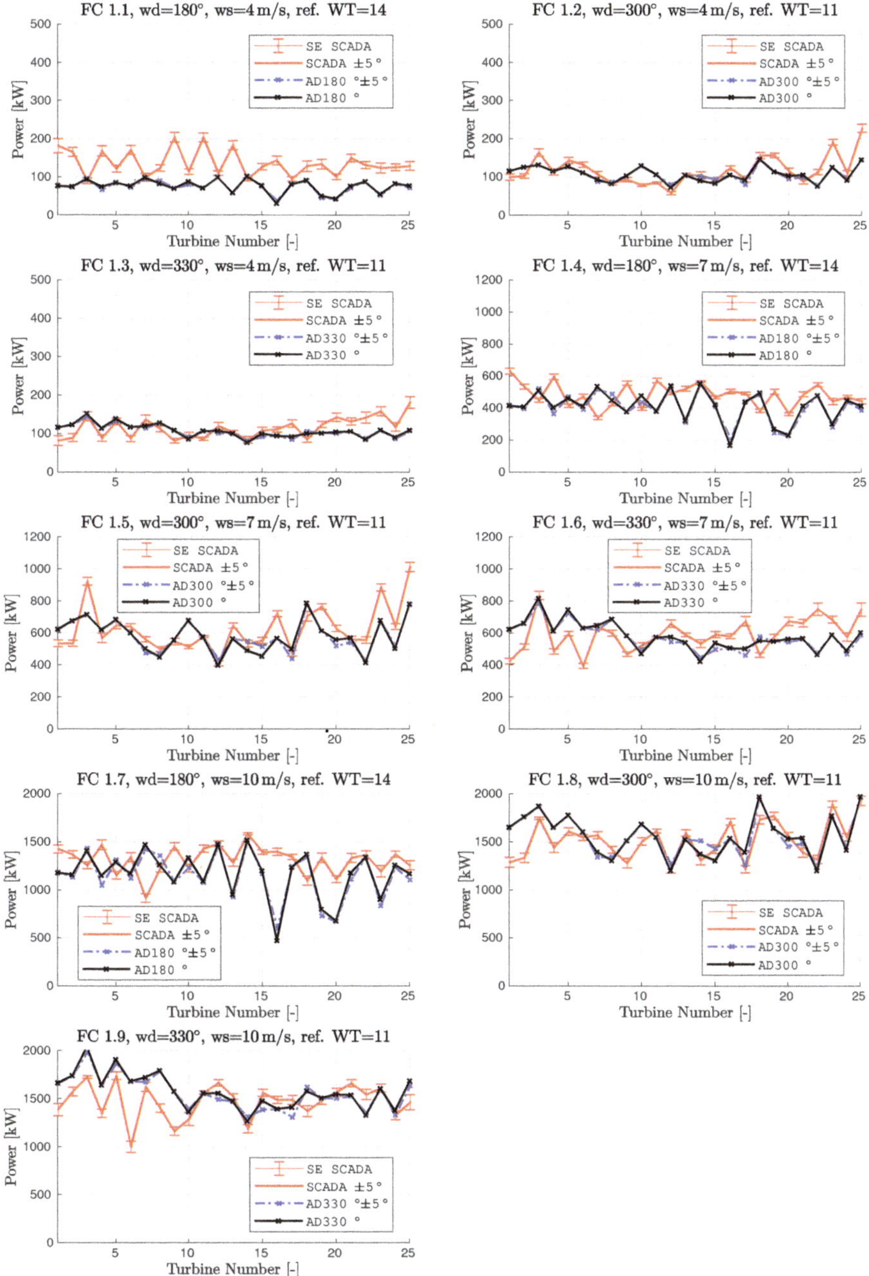

Figure 5. Predicted power from AD-RANS (AD#) for wind turbines 1 to 25 compared to supervisory control and data acquisition (SCADA) for 4, 7 and 10 m/s wind speeds (ws) at wind directions (wd) of 180, 300 and 330 degrees. The wind speed, ws, is referenced from the reference wind turbine (ref. WT). SE is the standard error and # is the wind direction in degrees.

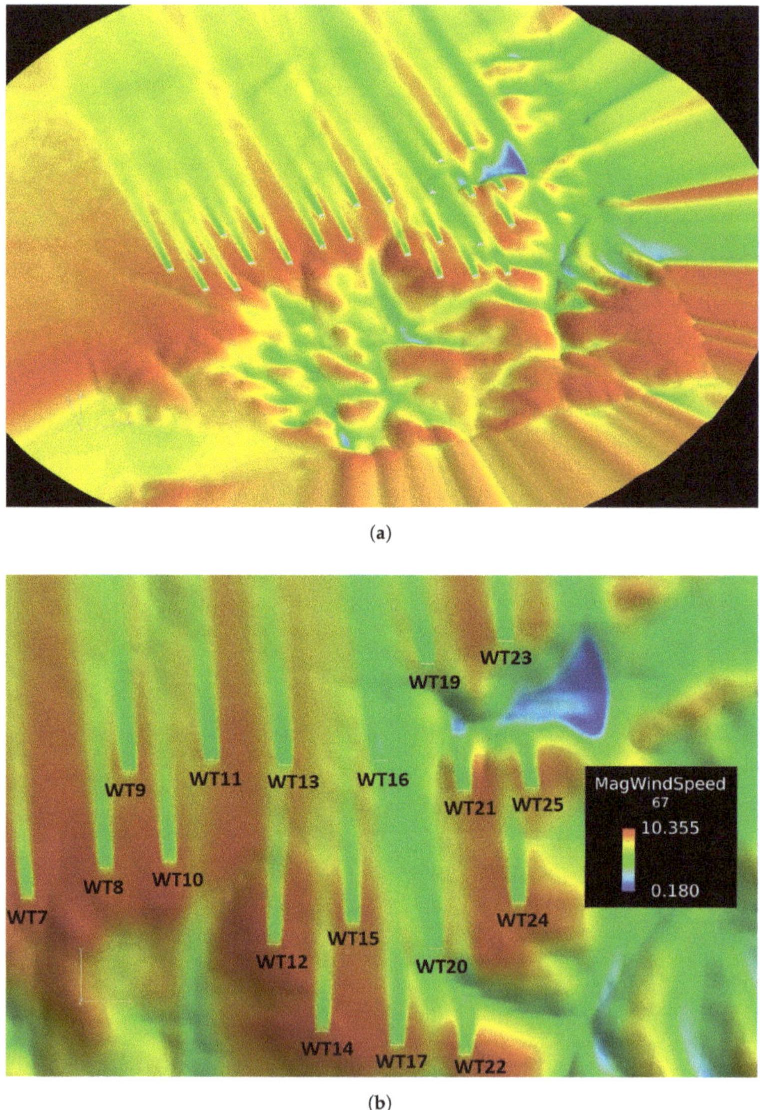

Figure 6. Wind speed magnitude (**a**) for the entire wind farm and (**b**) in the vicinity of wind turbine 16 for FC 1.7.

The predicted power from AD-RANS for two terrain mesh grids: 10×10 (AD Base) and 24×10 blocks (AD Fine), compared to SCADA for a wind direction of 180 degrees at (a) 4 m/s, (b) 7 m/s and (c) 10 m/s is shown in Figure 7. The mesh with 24×10 blocks is shown in Figure 7d. In comparison to the terrain mesh with 10×10 blocks, the refined mesh has a greater concentration of points on the left (west) and right (east) sides to capture the terrain effects in greater detail. The non-uniform mesh results in between 7 and 15 horizontal grid points per AD. Figure 7a–c shows larger differences in predicted power between AD Fine and AD Base for larger wind speeds.

A possible explanation is that the refined mesh includes more terrain detail, and the effects of terrain are pronounced at higher wind speeds. Nevertheless, refining the mesh did not result in significant improvements in comparison to SCADA measurements.

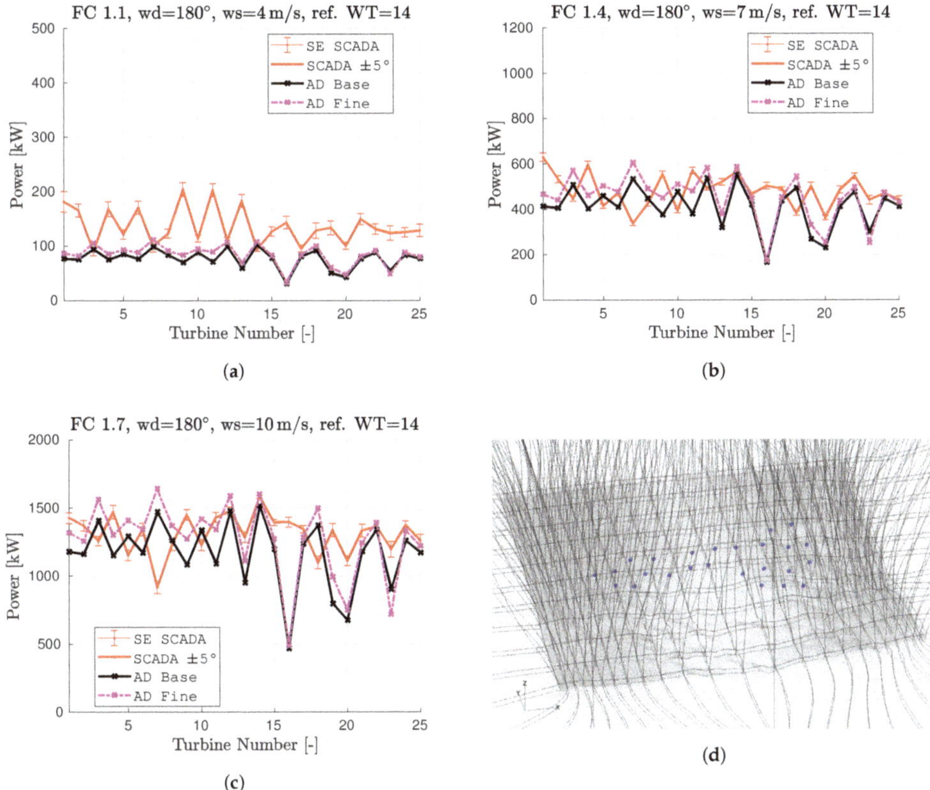

Figure 7. Predicted power from AD-RANS for two terrain mesh grids: 10×10 (AD Base) and 24×10 blocks (AD Fine), compared to SCADA for a wind direction of 180 degrees at (**a**) $4\,\text{m/s}$, (**b**) $7\,\text{m/s}$ and (**c**) $10\,\text{m/s}$. The mesh (**d**) with 24×10 blocks.

In summary, the comparison of the RANS simulations with the measurements is a challenging task. The wind directions and wind speeds are always changing in the wind farm, and this cannot be measured precisely. In other words, the reference wind directions and wind speeds are difficult to determine, which may lead to a high measurement uncertainty when the data is processed into bins. The significant changes in power levels between 0 and 360 degrees is seen in the power measurements. Percent differences obtained for power in the AD-RANS varied between 0.1% (FC 1.3, WT3) and 73.4% (FC 1.1, WT16) based on AD# $\pm\,5°$. Based on AD#, minimum and maximum values are 0.1% (FC 1.7, WT12) and 78.4% (FC 1.1, WT16), respectively. The mean percent difference for power from all twenty-five wind turbines varied between 9.6% (FC 1.8) and 42.1% (FC 1.1) based on AD# $\pm\,5°$, or 10.1% (FC 1.8) and 40.9% (FC 1.1) based on AD#. The mean percent difference for power from all flow cases is 20.5% and 20.0% for AD# $\pm\,5°$ and AD#, respectively. Therefore, averaging RANS simulations in the ± 5 degree range may reduce extreme percent differences slightly (e.g., 78.4% down to 73.4% for FC 1.1, WT16), but does not reduce the mean percent difference.

3.2. Actuator Line Large-Eddy Simulation Results

Figure 8 depicts the iso-surface of vorticity for the AL-LES simulation without synthetic inflow turbulence of WT14 at the 7 m/s wind speed and 180 degrees wind direction, i.e., south to north. The comparison of AL-LES results with measurement data are shown in Table 5 and Figures 9–11. In Table 5, the mean power (kW), rotor speed (RPM), and the horizontal streamwise velocity at 30, 50, and 70-m heights on met masts M1 and M3 for the four flow cases are shown. The table includes the results from the AL-LES in black, the mean measurement values in blue, and the percent difference between the AL-LES and the measurements in purple. The standard deviations (STDs) from the measurements for a 1-year period are also included in the table.

Table 5. Summarized values from AL-LES without synthetic inflow turbulence in black, measurements in blue, and percent differences in purple for the FC2 flow cases. The flow field is scaled using the horizontal streamwise velocity (streamwise vel.) measured at the upstream met mast at 70 m (in bold). STD is the standard deviation of the measurements for a 1-year period.

AL-LES (Black), Measured (Blue), % diff. (Purple)	180°, 4 m/s	STD	180°, 7 m/s	STD	0°, 4 m/s	STD	0°, 7 m/s	STD
	142.0		714.3		123.1		685.1	
Power (kW)	156.6	81.2	709.0	177.4	117.8	65.8	775.0	234.3
	−9.32%		0.75%		4.50%		−11.60%	
	8.860		14.720		8.860		14.520	
Rotor speed (RPM)	8.868	3.253	14.680	0.587	8.940	1.742	14.646	0.613
	−0.08%		0.27%		−0.89%		−0.86%	
Streamwise vel. (m/s)	**4.149**		**7.027**		1.560		3.066	
@ 70 m	**4.143**	0.659	**7.036**	0.528	2.263	0.894	3.862	0.824
M3	**0.15%**		**−0.12%**		−31.06%		−20.61%	
Streamwise vel. (m/s)	3.891		6.590		2.308		4.302	
@ 50 m	3.867	0.681	6.669	0.523	2.025	0.927	3.643	0.887
M3	0.62%		−1.19%		13.98%		18.10%	
Streamwise vel. (m/s)	3.547		6.009		3.217		4.845	
@ 30 m	3.719	0.766	6.294	0.621	2.463	0.705	5.143	0.761
M3	−4.62%		−4.54%		30.57%		−5.80%	
Streamwise vel. (m/s)	1.902		3.294		**4.068**		**7.155**	
@ 70 m	2.671	0.712	3.707	0.695	**4.050**	0.406	**7.114**	0.620
M1	−28.81%		−11.13%		**0.46%**		**0.58%**	
Streamwise vel. (m/s)	2.554		4.457		3.741		6.582	
@ 50 m	2.614	0.738	3.933	0.923	3.888	0.501	6.814	0.658
M1	−2.30%		13.30%		−3.76%		−3.42%	
Streamwise vel. (m/s)	3.554		6.042		3.306		5.818	
@ 30 m	3.090	0.767	5.104	0.830	3.588	0.704	6.343	0.728
M1	15.01%		18.38%		−7.85%		−8.28%	

Larger percent differences were obtained for the horizontal streamwise velocities, particularly when comparing AL-LES and the measurements for the met mast located downstream of WT14. For example, the percent difference in the horizontal streamwise velocity at 70 m height on met mast M1 for the 180 degrees and 4 m/s flow case is 28.81%. The reason for the large difference is because M1 is located downstream of WT14 as shown in Figure 2b. The complexity and stochastic nature of the turbulence from the wake of WT14 introduces larger variations and uncertainty in the flow field. There is a reasonable agreement between the AL-LES and the measurements for the power and rotor speed on WT14, the largest percent difference being 11.60% in power for the 0 degree and 7 m/s flow case. The smallest percent difference is 0.08% in rotor speed for the 180 degrees and 4 m/s flow case. The mean percent differences from the four flow cases for power and rotor speed are 6.54% and 0.53%, respectively.

Figures 9–11 are the results in Table 5 displayed in a bar-chart format. The results for the horizontal streamwise velocity comparisons are grouped by upstream (Figure 9) and downstream (Figure 10) locations. In Figures 9 and 10, bars are paired to represent comparisons between AL-LES and measurements for each one of the sensor locations: 70 m, 50 m and 30 m heights on the met mast. The horizontal streamwise velocity comparisons for the upstream mast in Figure 9 are in better agreement than the comparisons for the downstream mast shown in Figure 10 for two reasons. First, as mentioned in Section 2.3, the flow field in the AL-LES was scaled with the measurements using the sensor at 70 m height on the upstream met mast. Second, as mentioned previously for Table 5, larger differences occur for the downstream mast because of the complexity and stochastic nature of the turbulence generated from the wake of WT14. The wake of WT14 introduces larger variations and uncertainty in the flow field. In Figure 11, the largest difference between the AL-LES and SCADA appears for the power in the 0° and 7 m/s flow case. All the other comparisons in Figure 11 generally agree.

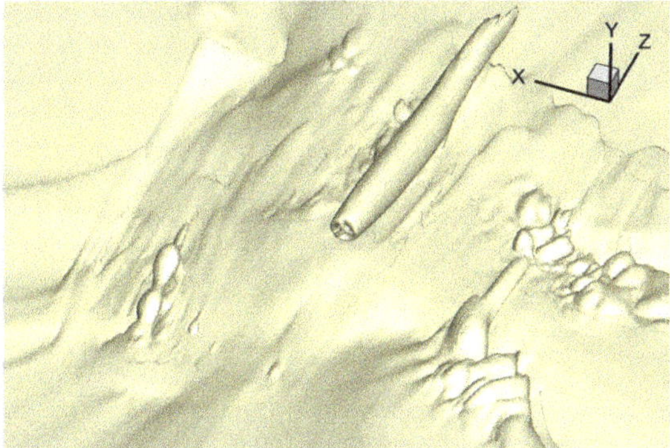

Figure 8. Iso-surface of vorticity for the actuator line large-eddy simulation of wind turbine 14 without synthetic inflow turbulence for 7 m/s wind speed and 180 degrees wind direction (south to north). The domain shown is the same as in Figure 4b.

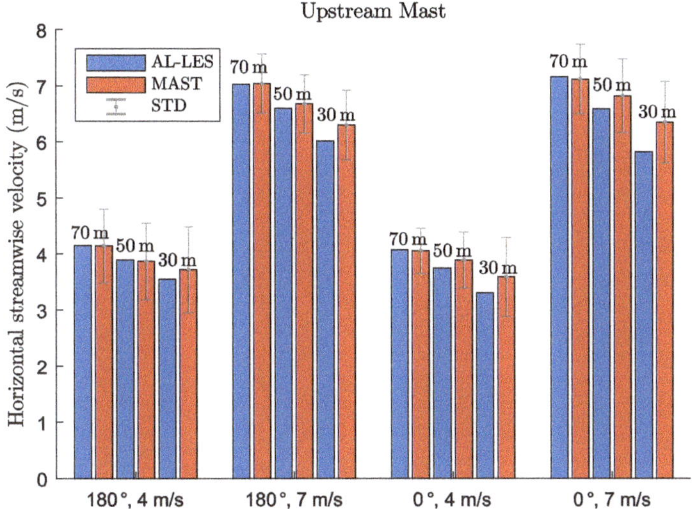

Figure 9. Comparison of AL-LES without synthetic inflow turbulence and met mast data for the horizontal streamwise velocity at 30-, 50- and 70-m heights on masts M1 and M3 when M1 and M3 are upstream of wind turbine 14. STD is the standard deviation of the measurements for a 1-year period.

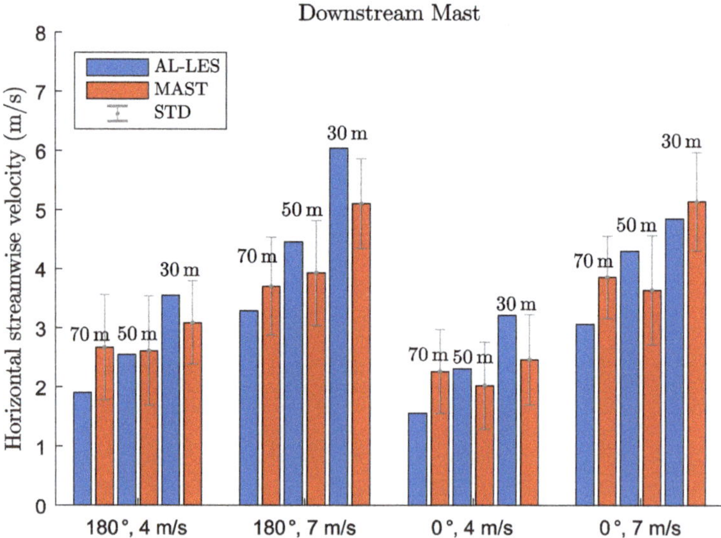

Figure 10. Comparison of AL-LES without synthetic inflow turbulence and met mast data for the horizontal streamwise velocity at 30-, 50- and 70-m heights on masts M1 and M3 when M1 and M3 are downstream of wind turbine 14. STD is the standard deviation of the measurements for a 1-year period.

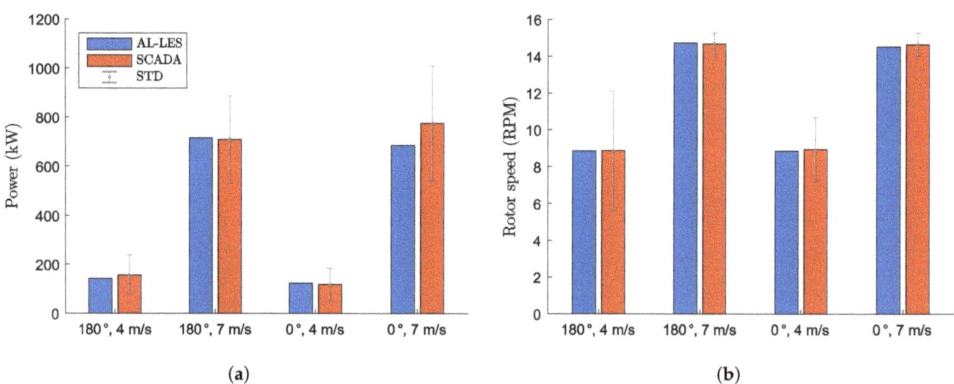

Figure 11. Comparison of AL-LES without synthetic inflow turbulence and SCADA for (**a**) power and (**b**) rotor speed of wind turbine 14 for the four FC2 flow cases. STD is the standard deviation of the measurements or a 1-year period.

Figure 12 depicts the iso-surface of vorticity for the AL-LES simulation of WT14 with synthetic inflow turbulence at the 7 m/s wind speed and 180 degrees wind direction. In Figure 12, the synthetic turbulent inflow input is seen by the rectangular plane, $372 \times 372\,\text{m}^2$ in size, 2.5 rotor diameters upstream of the wind turbine. Since slices from the Mann turbulence box were gradually fed into the plane during the unsteady simulation, a rectangular box is seen encompassing the wind turbine and its wake. The comparison of AL-LES results with measurement data is shown in Table 6.

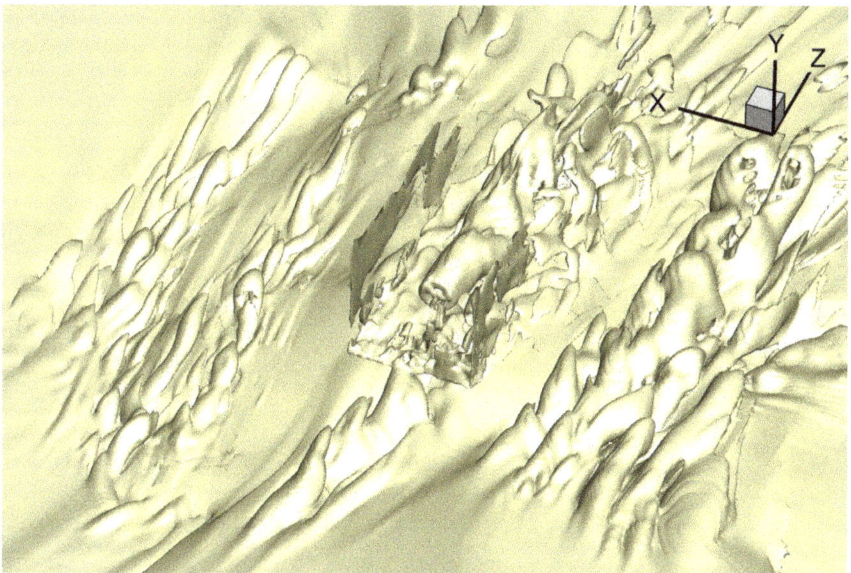

Figure 12. Iso-surface of vorticity for the actuator line large-eddy simulation of wind turbine 14 with synthetic inflow turbulence for 7 m/s wind speed and 180 degrees wind direction (south to north). The domain shown is the same as in Figure 4b. The synthetic inflow turbulence input is seen by the rectangular plane 2.5 rotor diameters upstream of the wind turbine.

The flow field in the AL-LES simulation with synthetic inflow turbulence was scaled using the sensor at 70 m height on the upstream met mast based on the simulations without synthetic inflow turbulence. The scaling is more difficult and not as accurate as in the case without synthetic inflow turbulence because of the large variation in wind speed in the AL-LES simulation with synthetic inflow turbulence. For example, for the 7 m/s wind speed and 0 degree wind direction case at 70 m on M1, the percent difference between the measurement and the AL-LES is 1.81%. For the case without synthetic inflow turbulence in Table 5, the percent difference is only 0.58%. The percent differences obtained for the horizontal streamwise velocity at 30-, 50- and 70-m heights on masts M1 and M3 are similar to the case without synthetic inflow turbulence. The percent differences obtained for the power and RPM are also similar. The percent differences obtained for power and RPM vary between 0.44% and 11.42%. The mean percent differences from the four flow cases for power and rotor speed are 6.77% and 1.47%, respectively. Figures 13–15 are the results in Table 6 displayed in a bar-chart format identical to Figures 9–11. All comparisons in Figures 13–15 are very similar to Figures 9–11.

The turbulence intensities based on the horizontal streamwise velocity in Table 6 are shown in bar-chart format in Figures 16 and 17. It was not possible to attain very good agreement for the turbulence intensities. The highest percent difference is 55.23% at 30 m on M1 for the 4 m/s and 180 degrees case. The turbulence intensities obtained from the AL-LES largely depended on the chosen input parameters for the surface roughness (=0.4) when generating the Mann turbulence box and the turbulence scaling factor (=1.5), see Section 2.3.

Table 6. Summarized values from AL-LES with synthetic inflow turbulence in black, measurements in blue, and percent differences in purple for the FC2 flow cases. The flow field is scaled using the horizontal streamwise velocity (streamwise vel.) measured at the upstream met mast at 70 m (in bold). STD is the standard deviation of the measurements for a 1-year period.

AL-LES (Black), Measured (Blue), % diff. (Purple)	180°, 4 m/s	STD	180°, 7 m/s	STD	0°, 4 m/s	STD	0°, 7 m/s	STD
Power (kW)	141.6		730.3		114.2		686.5	
	156.6	81.2	709.0	177.4	117.8	65.8	775.0	234.3
	−9.59%		3.01%		−3.05%		−11.42%	
Rotor speed (RPM)	8.988		14.523		8.899		14.207	
	8.868	3.253	14.684	0.587	8.938	1.742	14.646	0.613
	1.36%		−1.10%		−0.44%		−3.00%	
Streamwise vel. (m/s)	**4.152**		**7.156**		1.623		3.538	
@ 70 m	**4.143**	0.659	**7.036**	0.528	2.263	0.894	3.862	0.824
M3	**0.23%**		**1.70%**		−28.28%		−8.38%	
Streamwise vel. (m/s)	3.858		6.648		1.970		4.151	
@ 50 m	3.867	0.681	6.669	0.523	2.025	0.927	3.643	0.887
M3	−0.22%		−0.33%		−2.71%		13.94%	
Streamwise vel. (m/s)	3.644		6.291		2.914		5.484	
@ 30 m	3.719	0.766	6.294	0.621	2.463	0.705	5.143	0.761
M3	−2.01%		−0.06%		18.34%		6.63%	
Streamwise vel. (m/s)	2.095		3.934		**4.031**		**7.243**	
@ 70 m	2.671	0.712	3.707	0.695	**4.050**	0.406	**7.114**	0.620
M1	−21.59%		6.12%		**−0.47%**		**1.81%**	
Streamwise vel. (m/s)	2.558		4.705		3.695		6.655	
@ 50 m	2.614	0.738	3.933	0.923	3.888	0.501	6.814	0.658
M1	−2.17%		19.61%		−4.95%		−2.33%	
Streamwise vel. (m/s)	3.459		6.033		3.202		5.805	
@ 30 m	3.090	0.767	5.104	0.830	3.588	0.704	6.343	0.728
M1	11.94%		18.20%		−10.75%		−8.48%	
Turbulence intensity (-)	0.123		0.123		0.206		0.234	
@ 70 m	0.108	0.046	0.095	0.028	0.264	0.122	0.229	0.063
M3	13.92%		29.60%		−22.21%		2.16%	
Turbulence intensity (-)	0.131		0.138		0.212		0.221	
@ 50 m	0.131	0.062	0.105	0.030	0.278	0.126	0.255	0.069
M3	0.00%		31.60%		−23.52%		−13.40%	
Turbulence intensity (-)	0.064		0.084		0.185		0.185	
@ 30 m	0.134	0.068	0.115	0.025	0.271	0.096	0.242	0.041
M3	−52.06%		−27.08%		−31.72%		−23.53%	
Turbulence intensity (-)	0.145		0.161		0.105		0.097	
@ 70 m	0.258	0.083	0.231	0.056	0.111	0.076	0.110	0.026
M1	−43.99%		−30.17%		−5.17%		−11.82%	
Turbulence intensity (-)	0.206		0.188		0.107		0.107	
@ 50 m	0.274	0.092	0.262	0.054	0.128	0.080	0.115	0.028
M1	−24.75%		−28.07%		−16.01%		−6.59%	
Turbulence intensity (-)	0.104		0.098		0.144		0.161	
@ 30 m	0.232	0.076	0.215	0.038	0.153	0.094	0.132	0.029
M1	−55.23%		−54.23%		−5.51%		21.74%	

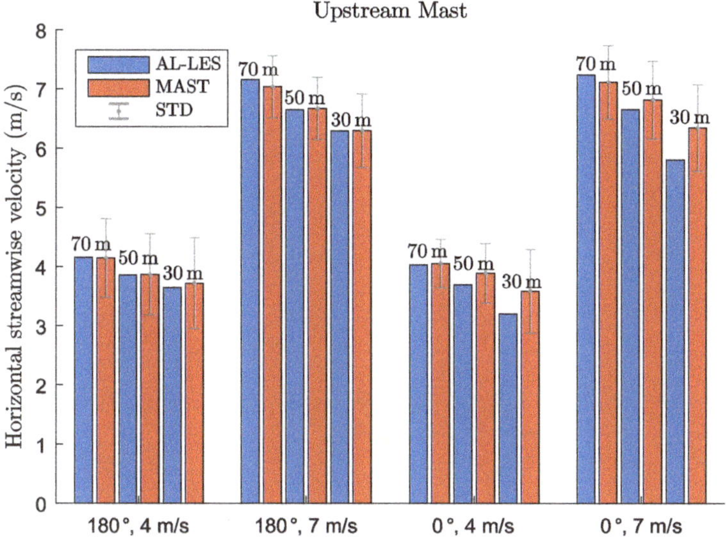

Figure 13. Comparison of mean AL-LES results with synthetic inflow turbulence and met mast data for the horizontal streamwise velocity at 30-, 50- and 70-m heights on masts M1 and M3 when M1 and M3 are upstream of wind turbine 14. STD is the standard deviation of the measurements for a 1-year period.

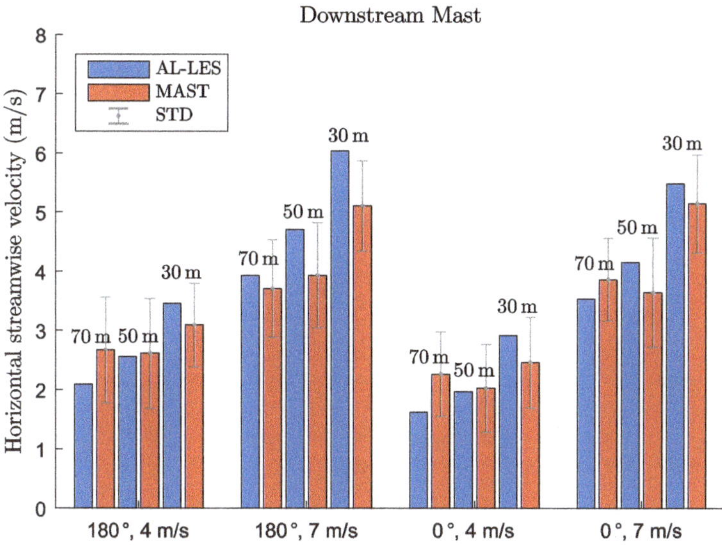

Figure 14. Comparison of mean AL-LES results with synthetic inflow turbulence and met mast data for the horizontal streamwise velocity at 30-, 50- and 70-m heights on masts M1 and M3 when M1 and M3 are downstream of wind turbine 14. STD is the standard deviation of the measurements for a 1-year period.

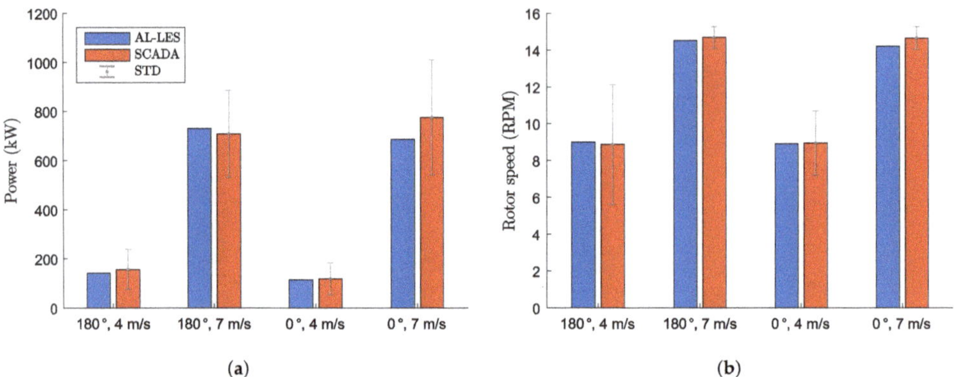

Figure 15. Comparison of mean AL-LES results with synthetic inflow turbulence and SCADA for (**a**) power and (**b**) rotor speed of wind turbine 14 for the four FC2 flow cases. STD is the standard deviation of the measurements for a 1-year period.

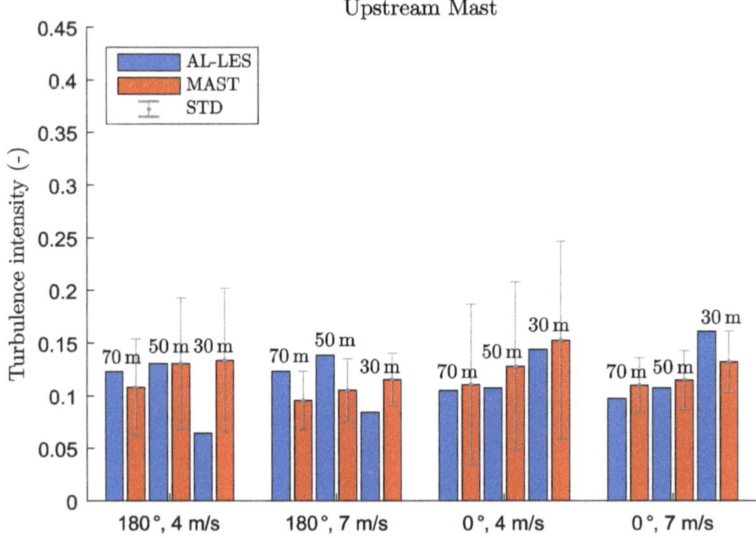

Figure 16. Comparison of mean AL-LES results with synthetic inflow turbulence and met mast data for turbulence intensities at 30-, 50- and 70-m heights on masts M1 and M3 when M1 and M3 are upstream of wind turbine 14. The turbulence intensities are based on the horizontal streamwise velocity and STD is the standard deviation of the measurements for a 1-year period.

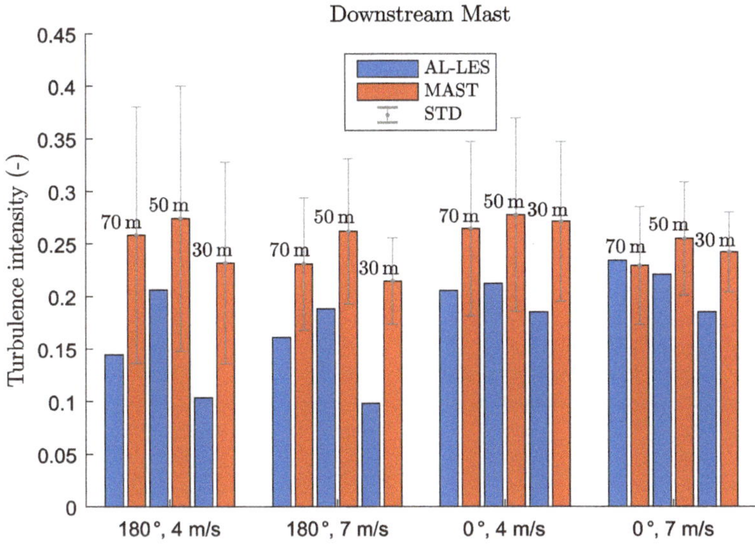

Figure 17. Comparison of mean AL-LES results with synthetic inflow turbulence and met mast data for turbulence intensities at 30-, 50- and 70-m heights on masts M1 and M3 when M1 and M3 are downstream of wind turbine 14. The turbulence intensities are based on the horizontal streamwise velocity and STD is the standard deviation of the measurements for a 1-year period.

4. Conclusions

This article described Computational Fluid Dynamics (CFD) simulations of flows in a wind farm in complex terrain in Shaanxi, China and the comparisons of the computational results with utility scale field measurements. The CFD simulations in the study were performed using Reynolds-Averaged Navier–Stokes (RANS) and Large-Eddy Simulation (LES) solvers. The RANS method together with an Actuator Disc (AD) approach was employed to predict the performance of the 25 wind turbines in the farm, while the LES method with an Actuator Line (AL) technique was used to obtain a detailed description of the flow field around one specific wind turbine near two met masts. The AD-RANS simulation results were compared with the mean values of power obtained from field measurements. Furthermore, the AL-LES results were compared with the mean values of power, rotor speed, and streamwise velocity measured from the wind turbine and its nearby two masts.

Results from both the AD-RANS and AL-LES simulations indicated that both methods can reasonably predict the flow patterns and performance of the wind farm and wind turbine #14, respectively, in complex terrain in Shaanxi. The mean percent difference obtained for power in the AD-RANS simulations was approximately 20%. Percent differences obtained for power and rotor RPM in the AL-LES varied between 0.08% and 11.6%. The mean percent differences in the AL-LES for power and rotor RPM are approximately 7% and 1%, respectively. Careful consideration of the measurement data as well as the computational setup, such as the meshing of terrain and flow domain, wind turbine model and wind farm layout, and the flow input parameters is needed to obtain reasonable or improved results. Uncertainty quantification and statistical models should be employed. For example, the uncertainty in the measurement data can be incorporated into the CFD solver to provide more reliable outputs of power, rotor rotational speed, wind velocities, etc.

Author Contributions: W.Z.S. and K.S.H. conceived and designed the project; M.S. performed the computations and analyzed the results under the guidance of M.P.v.d.L. and W.J.Z.; M.S., W.Z.S. and M.P.v.d.L. wrote the paper.

Funding: This research was supported by the Sino-Danish cooperative project "Wind farm layout optimization in complex terrain" funded by the Danish Energy Agency (EUDP J.nr. 64013-0405) and the Ministry of Science and Technology of China (NO. 2014DFG62530).

Acknowledgments: The authors wish to give special thanks to the Danish and Chinese partners (DTU Wind Energy, EMD International A/S, North West Survey and Design Institute of Hydro China Consultant Corporation, and HoHai University) for their active collaborations.

Conflicts of Interest: The authors declare no conflict of interest.

Abbreviations

The following abbreviations are used in this manuscript:

AD	actuator disc
AL	actuator line
CFD	computational fluid dynamics
CSIC	China Shipbuilding Industry Corporation
DES	delayed-detached-eddy simulation
FC1/FC2	flow case 1, flow case 2
HZ	Haizhuang
LES	large-eddy simulation
MW	megawatt
RANS	Reynolds-Averaged Navier–Stokes
RPM	revolutions per minute
SCADA	supervisory control and data acquisition
SE	standard error
SGS	sub-grid scale
STD	standard deviation
wd	wind direction
ws	wind speed
WT(s)	wind turbine(s)

References

1. International Renewable Energy Agency. *Renewable Power Generation Costs in 2017*; Technical Report; IRENA: Abu Dhabi, UAE, 2018; ISBN 978-92-9260-040-2.
2. Storey, R.C.; Norris, S.E.; Stol, K.A.; Cater, J.E. Large eddy simulation of dynamically controlled wind turbines in an offshore environment. *Wind Energy* **2013**, *16*, 845–864. [CrossRef]
3. Van der Laan, M.P.; Sørensen, N.N.; Réthoré, P.E.; Mann, J.; Kelly, M.C.; Troldborg, N. The k-ε-f_P model applied to double wind turbine wakes using different actuator disk force methods. *Wind Energy* **2015**, *18*, 2223–2240. [CrossRef]
4. Makridis, A.; Chick, J. Validation of a CFD model of wind turbine wakes with terrain effects. *J. Wind Eng. Ind. Aerodyn.* **2013**, *123*, 12–29. [CrossRef]
5. Schulz, C.; Klein, L.; Weihing, P.; Lutz, T.; Krämer, E. CFD Studies on Wind Turbines in Complex Terrain under Atmospheric Inflow Conditions. *J. Phys. Conf. Ser.* **2014**, *524*, 012134. [CrossRef]
6. Schulz, C.; Klein, L.; Weihing, P.; Lutz, T. Investigations into the Interaction of a Wind Turbine with Atmospheric Turbulence in Complex Terrain. *J. Phys. Conf. Ser.* **2016**, *753*, 032016. [CrossRef]
7. Tabib, M.; Rasheed, A.; Kvamsdal, T. LES and RANS simulation of onshore Bessaker wind farm: Analysing terrain and wake effects on wind farm performance. *J. Phys. Conf. Ser.* **2015**, *625*, 012032. [CrossRef]
8. Tabib, M.; Rasheed, A.; Fuchs, F. Analyzing complex wake-terrain interactions and its implications on wind-farm performance. *J. Phys. Conf. Ser.* **2016**, *753*, 032063. [CrossRef]
9. Castellani, F.; Astolfi, D.; Mana, M.; Piccioni, E.; Becchetti, M.; Terzi, L. Investigation of terrain and wake effects on the performance of wind farms in complex terrain using numerical and experimental data. *Wind Energy* **2017**, *20*, 1277–1289. [CrossRef]

10. Sørensen, J.N.; Mikkelsen, R.F.; Henningson, D.S.; Ivanell, S.; Sarmast, S.; Andersen, S.J. Simulation of wind turbine wakes using the actuator line technique. *Philos. Trans. Ser. A Math. Phys. Eng. Sci.* **2015**, *373*, 20140071. [CrossRef]
11. Sørensen, N.N. General Purpose Flow Solver Applied to Flow over Hills. Ph.D. Thesis, Technical University of Denmark, Roskilde, Denmark, 1995.
12. Michelsen, J.A. *Basis3D—A Platform for Development of Multiblock PDE Solvers*; Technical Report; Technical University of Denmark: Roskilde, Denmark, 1992.
13. Réthoré, P.E.; van der Laan, P.; Troldborg, N.; Zahle, F.; Sørensen, N.N. Verification and validation of an actuator disc model. *Wind Energy* **2014**, *17*, 919–937. [CrossRef]
14. Sørensen, J.N.; Shen, W.Z. Numerical Modeling of Wind Turbine Wakes. *J. Fluids Eng.* **2002**, *124*, 393–399. [CrossRef]
15. Ramos-García, N.; Sørensen, J.N.; Shen, W.Z. A strong viscous-inviscid interaction model for rotating airfoils. *Wind Energy* **2014**, *17*, 1957–1984. [CrossRef]
16. Han, X.; Liu, D.; Xu, C.; Shen, W.Z. Atmospheric stability and topography effects on wind turbine performance and wake properties in complex terrain. *Renew. Energy* **2018**, *126*, 640–651. [CrossRef]
17. Van der Laan, M.P.; Sørensen, N.N.; Réthoré, P.E.; Mann, J.; Kelly, M.C.; Troldborg, N.; Schepers, J.G.; Machefaux, E. An improved k-ε model applied to a wind turbine wake in atmospheric turbulence. *Wind Energy* **2015**, *18*, 889–907. [CrossRef]
18. Andersen, S.J.; Sørensen, J.N.; Mikkelsen, R. Simulation of the inherent turbulence and wake interaction inside an infinitely long row of wind turbines. *J. Turbul.* **2013**, *14*, 1–24, [CrossRef]
19. Mann, J. The spatial structure of neutral atmospheric surface-layer turbulence. *J. Fluid Mech.* **1994**, *273*, 141–168. [CrossRef]
20. Mann, J. Wind field simulation. *Probab. Eng. Mech.* **1998**, *13*, 269–282. [CrossRef]

 *applied sciences*

Article

An Optimization Framework for Wind Farm Design in Complex Terrain

Ju Feng [1], Wen Zhong Shen [1,*] and Ye Li [2]

[1]   Department of Wind Energy, Technical University of Denmark, 2800 Kgs. Lyngby, Denmark; jufen@dtu.dk
[2]   School of Naval Architecture, Ocean and Civil Engineering, Shanghai Jiao Tong University, Shanghai 201100, China; ye.li@sjtu.edu.cn
*   Correspondence: wzsh@dtu.dk

Received: 28 August 2018; Accepted: 22 October 2018; Published: 25 October 2018

Featured Application: Analysis and optimization of wind farm design in complex terrain.

Abstract: Designing wind farms in complex terrain is an important task, especially for countries with a large portion of complex terrain territory. To tackle this task, an optimization framework is developed in this study, which combines the solution from a wind resource assessment tool, an engineering wake model adapted for complex terrain, and an advanced wind farm layout optimization algorithm. Various realistic constraints are modelled and considered, such as the inclusive and exclusive boundaries, minimal distances between turbines, and specific requirements on wind resource and terrain conditions. The default objective function in this framework is the total net annual energy production (AEP) of the wind farm, and the Random Search algorithm is employed to solve the optimization problem. A new algorithm called Heuristic Fill is also developed in this study to find good initial layouts for optimizing wind farms in complex terrain. The ability of the framework is demonstrated in a case study based on a real wind farm with 25 turbines in complex terrain. Results show that the framework can find a better design, with 2.70% higher net AEP than the original design, while keeping the occupied area and minimal distance between turbines at the same level. Comparison with two popular algorithms (Particle Swarm Optimization and Genetic Algorithm) also shows the superiority of the Random Search algorithm.

Keywords: wind farm; layout optimization; design; random search; complex terrain

---

1. Introduction

In the past two decades, the world has witnessed a remarkable growth of wind energy development. According to the latest statistics from the Global Wind Energy Council (GWEC), the global cumulative installed wind capacity has increased from 23.9 GW in 2001 to 439.1 GW in 2017, representing an average annual increase by 20.4% [1]. Although offshore wind is now attracting a lot of interests and has grown rapidly in recent years, especially in northern Europe, today it represents only 4.3% of the global installed capacity of wind energy [2]. The dominance of onshore wind is mainly due to its longer development history, lower cost, and easier deployment.

As a result of the rapid growth of onshore wind, a lot of onshore wind farms have been built worldwide. Since many suitable sites in flat terrain have already been developed with wind farms, more and more wind farms are going to be built in complex terrain, especially for countries featured by a large percentage of topography covered with mountains, such as China.

Compared with a wind farm on flat terrain, a wind farm built on complex terrain benefits from the possible richer wind resource at certain locations (brought by the speed-up effect due to the terrain topography change), but it is also more likely to be exposed to more complex flow conditions, higher fatigue loads, more expensive installation, operation and maintenance costs, and other disadvantages [3].

Designing wind farms in complex terrain is not a trivial task, mainly due to the complex interactions of the boundary layer flow with the complex terrain and wind turbine wakes, and also due to the multi-disciplinary nature of the wind farm design problem. This problem typically involves different design and engineering tasks, which may come from technical, logistical, environmental, economical, legal, and/or even social considerations.

For most industrial practitioners, the design of wind farms conventionally concerns mainly the micro-siting of wind turbines based on consideration of wind resource, i.e., determining the exact position for each turbine at a selected wind farm site [4]. Typically, a wind farm developer/designer uses a wind resource assessment tool, such as WAsP [5] (the industry-standard PC software for wind resource assessment, siting, and energy yield calculations for wind turbines and wind farms), to assess the wind resource of the wind farm site based on wind measurement data during a period [6]. Then, the final design of the wind farm, mainly the layout of turbines, is determined for maximizing the annual energy production (AEP) while considering certain constraints, such as the proximity of wind turbines. This final layout is usually obtained by either manual adjustments or using some optimization techniques [7]. After finding the turbine layout, other design tasks related to foundations, access roads, electrical system, and other components or systems are handled separately, often by specialized engineering consulting companies.

The general problem of modelling the wind flow over complex terrain has been an important research field for a very long time. Seventy years ago, Queney published a review of theoretical models of inviscid flow over hills and mountains [8]. More recently, Wood gave a historical review of studies (until 2000) on wind flow over complex terrain [9]. Similarly, in the wind energy community, the problem of wind resource assessment in complex terrain was investigated by many researchers over a long period [10].

In contrast, wind farm design optimization has only become a hot research area quite recently and the majority of the published studies in this field deal with wind farms in flat terrain or offshore [11]. Few studies investigated the wind farm design optimization problems in complex terrain. Song et al. [12] proposed a bionic method for optimizing the turbine layout in complex terrain. In this study the objective was to maximize the power output, and the wake flow was simulated by using the virtual particle wake model. They also developed a new greedy algorithm for this problem in a later study [13]. Feng and Shen [14] considered the layout optimization problem for a wind farm on a 2D Gaussian hill, using computational fluid dynamics (CFD) simulations for the background flow field and an adapted Jensen wake model for the wake effect. They used the random search algorithm [15,16] to optimize the layout for maximizing the total power. Kuo et al. [17] proposed an algorithm that couples CFD with mixed-integer programming (MIP) to optimize the layouts in complex terrain. In this study, the wind farm domain was discretized into cells to use the MIP method.

All the studies mentioned above aimed to maximize either the total power [12–14] or the total kinetic energy [17] of wind farm, with constraints on wind farm boundary and minimal distance requirements.

In this paper, we present an optimization framework for wind farm design in complex terrain, which was developed in the Sino-Danish research cooperation project, FarmOpt [18]. This framework combines the state-of-the-art flow field solver, a fast engineering wake model, and an advanced optimization algorithm to solve the design optimization problem of wind farms in complex terrain. Various realistic objectives, constraints, and requirements that one might encounter in real life wind farm developments can be considered in this framework.

We will first introduce the modelling methodology in Section 2. Then, the optimization framework is presented in Section 3. Results of a real wind farm case study will be described and discussed in Section 4. Section 5 compares the performance of the Random Search (RS) algorithm with two popular algorithms, i.e., Particle Swarm Optimization (PSO) and the Genetic Algorithm (GA). Finally, conclusions are given in Section 6.

2. Wind Farm Modelling

A reliable estimation of AEP is crucial for wind farm design, since AEP represents the amount of electricity that a given wind farm can generate in a year, which in turn determines the annual income that the wind farm's owner can obtain. This makes AEP either the objective function or a critical component in the objective function for any wind farm design optimization problem. For example, the levelized cost of energy (LCOE), which is a popular choice for evaluating wind farm designs [19], can be calculated by considering AEP, together with capital expenditure (CAPEX), operational expenditure (OPEX), and some financial parameters, such as the interest rate. In this section, we present the aspects of wind farm modelling related to the AEP estimation.

2.1. Site Condition

Due to terrain effects, i.e., the effect of topographic changes on the flow filed over a complex terrain [9], different locations at a complex terrain site will experience different wind conditions. Using onsite wind measurements and standard wind resource assessment tools, such as WAsP [5] and WindPRO [20], one could get the essential information on location-specific site conditions, i.e., wind resource and terrain effects, for any given inflow wind direction sector.

For flat or moderately complex terrain, linear models, such as the conventional WAsP [6], provide reasonable results. For more complex terrain, nonlinear models, such as WAsP CFD [21], are needed. In the current framework, we assume that the location-specific values of wind resource and terrain effects have been obtained by using a wind resource assessment tool.

Considering a rectangle area covering the interested area for a wind farm design, we first discretize the area into a number of grids, defined by a range of x coordinates $[x_1, x_2, \ldots, x_{N_x}]$ and a range of y coordinates $\left[y_1, y_2, \ldots, y_{N_y}\right]$. For a certain number of wind direction sectors defined by a range of far field inflow wind directions $\left[\theta_1^\infty, \theta_2^\infty, \ldots, \theta_{N_\theta}^\infty\right]$, we can then obtain the relevant sector wise values of wind resource and terrain effects variables, such as Weibull-A, Weibull-k, frequency, speed up factor, terrain induced turning angle, and mean wind speed, at a given height above the ground. Note that Weibull-A is the scale parameter and Weibull-k is the shape parameter of Weibull distribution, which is the most commonly used distribution for wind speed [4]. Values of variables that do not depend on inflow wind direction, such as elevation and overall mean wind speed, are also obtained.

Using linear interpolation, we can then easily estimate the values of these variables for any arbitrary position (x, y) in this area, under any far field inflow wind direction, θ^∞, if wind direction dependent variables are concerned. The variables important for estimating AEP are listed in Table 1.

Table 1. Site condition variables important for AEP (annual energy production) estimation.

Variable	Unit	Notation
Elevation	m	$z(x, y)$
Weibull-A	m/s	$A(x, y, \theta^\infty)$
Weibull-k	-	$k(x, y, \theta^\infty)$
Frequency	-/°	$freq(x, y, \theta^\infty)$
Speed up	-	$sp(x, y, \theta^\infty)$
Turning angle	°	$tu(x, y, \theta^\infty)$

In Table 1, speed up is defined as the ratio between the local wind speed, v, at location (x, y) and the far field wind speed, v^∞, at the same height above the ground for a given far field inflow wind direction, θ^∞, while turning angle, tu, denotes the difference between the local wind direction, θ, and the far field inflow wind direction, θ^∞. Thus, they can be computed as:

$$sp(x, y, \theta^\infty) = v(x, y, \theta^\infty)/v^\infty(\theta^\infty), \tag{1}$$

$$tu(x, y, \theta^\infty) = \theta(x, y, \theta^\infty) - \theta^\infty. \tag{2}$$

As a demonstration, four interpolated site condition variables are shown in Figure 1 for a real wind farm site in complex terrain. Note that the results on wind resource and terrain effects are imported from WAsP CFD simulations and the wind farm on this site will be used in Sections 4 and 5 as a test case.

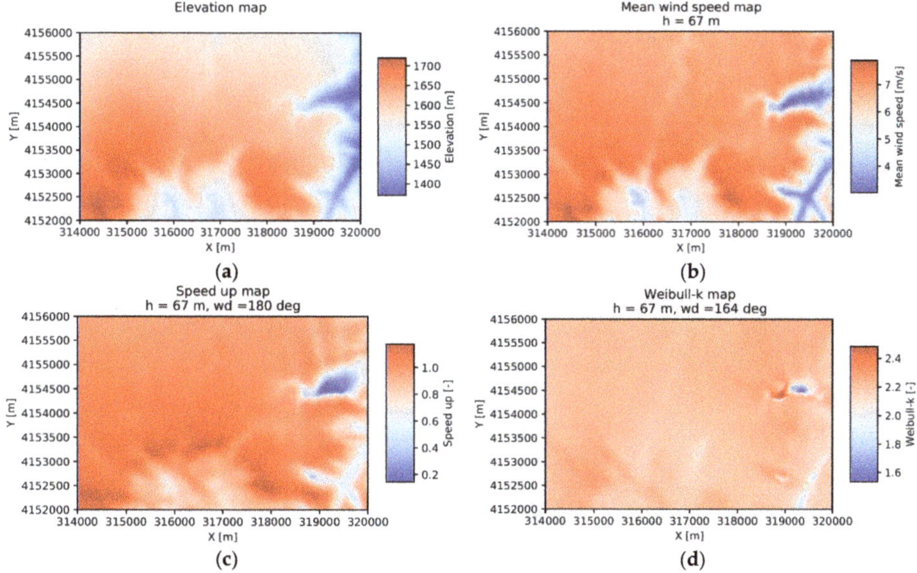

Figure 1. Interpolated site condition variables for a wind farm site in complex terrain: (**a**) elevation; (**b**) mean wind speed at 67 m height above the ground; (**c**) speed up at 67 m height under the far field inflow wind direction of 180°; (**d**) Weibull-k at 67 m height under the far field inflow wind direction of 164°.

2.2. Wake Model

Modelling the wake effects of wind turbines is essential for AEP estimation, but quite challenging for wind farms in complex terrain due to the complex interplay between terrain and wake flow [22]. While the CFD can model the wake flow field of a wind farm in complex terrain with a reasonable accuracy [23], its high computational cost makes it unsuitable for wind farm design optimization, since typical optimization algorithms require a large number of design evaluations.

Several studies have proposed a few fast wake models for wind farms in complex terrain: Song et al. proposed a virtual particle wake model [24] and applied it in wind farm layout optimizations [12,13]; Feng and Shen developed an adapted Jensen wake model in [14]; Kuo et al. [25] proposed a wake model by solving a simplified variation of the Navies-Stokes equations, which yields results with a reasonable accuracy and has a much less computational cost when compared with full CFD simulations.

In the current framework, the wake flow field of a wind farm in complex terrain is modelled by coupling the adapted Jensen wake model with the terrain flows obtained by CFD simulations, i.e., flow fields of the wind farm site without wind turbines, such as those imported from WAsP CFD simulations. The adapted Jensen wake model assumes the wake centerline of a wind turbine's wake follows the terrain at the same height above the ground along the local inflow wind direction, while its wake zone expands linearly and its wake deficit develops following the same rule as the original Jensen wake model [14]. The schematic of this wake model is shown in Figure 2.

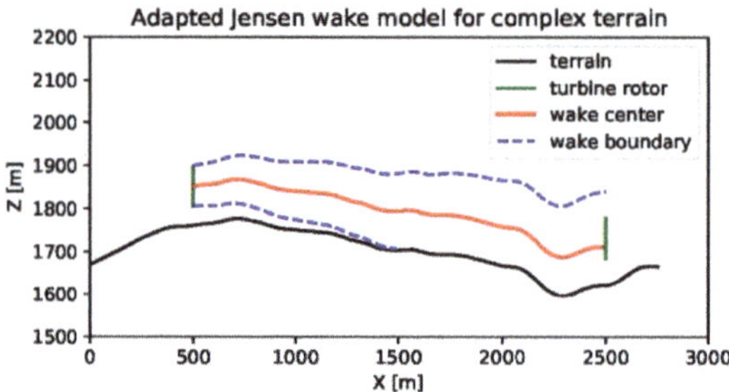

Figure 2. Schematic of the adapted Jensen wake model for a wind turbine in complex terrain.

Considering the wake effect between an upwind turbine, i, located at (x_i, y_i) and a downwind turbine, j, located at (x_j, y_j) for a given far field inflow condition: v^∞ and θ^∞, we can calculate the downwind distance, $d_{ij}^{down}(\theta^\infty)$, and the crosswind distance, $d_{ij}^{cross}(\theta^\infty)$, from the upwind turbine to the location of the downwind turbine. The downwind distance can be computed by integration following the wake centerline (the red line in Figure 2) based on the elevation, $z(x, y)$, and the local inflow wind direction, $\theta_i(\theta^\infty) = \theta^\infty + tu(x_i, y_i, \theta^\infty)$, while the crosswind distance can be easily obtained by considering the coordinates of these two turbines and the local inflow wind direction of the upwind turbine. Then, the local inflow wind speed at turbine i and the effective wake deficit turbine i caused on turbine j can be calculated by:

$$v_i(v^\infty, \theta^\infty) = sp(x_i, y_i, \theta^\infty) \cdot v^\infty, \tag{3}$$

$$\Delta v_{ij}(v^\infty, \theta^\infty) = \frac{A_{ij}^{ol}(\theta^\infty)}{A_j^{rotor}} \cdot v_i(v^\infty, \theta^\infty) \cdot \left(\frac{1 - \sqrt{1 - C_T(v_i(v^\infty, \theta^\infty))}}{1 + \alpha \cdot d_{ij}^{down}(\theta^\infty)/R_i} \right), \tag{4}$$

in which $A_{ij}^{ol}(\theta^\infty)$ is the overlapping area between the wake zone of turbine i and the rotor area of turbine j; $A_j^{rotor} = \pi R_j^2$ is the rotor area of turbine j; $C_T(\cdot)$ represents the thrust coefficient function of the wind turbine; α denotes the wake decay coefficient, and R_i represents the rotor radius of turbine i. Note that $A_{ij}^{ol}(\theta^\infty)$ can be calculated based on the wake zone radius, $R_{wake} = R_i + \alpha \cdot d_{ij}^{down}(\theta^\infty)$, the rotor radius, R_j, and the relevant crosswind distance, $d_{ij}^{cross}(\theta^\infty)$, using the method described in [14]. Additionally, $\alpha = 0.075$ is used in this study.

Denoting the set of indexes of all the turbines upwind of turbine j as I^{up}, and using the energy deficit balance assumption [14], we can then calculate the effective wind speed of turbine j under the far field inflow wind condition, v^∞ and θ^∞, as:

$$\bar{v}_j(v^\infty, \theta^\infty) = v_j(v^\infty, \theta^\infty) - \sqrt{\sum_{i \in I^{up}} \left(\Delta v_{ij}(v^\infty, \theta^\infty) \right)^2}. \tag{5}$$

A more detailed description of the wake model governed by Equations (3)–(5) is referred to in [14].

2.3. AEP Estimation

After the local wind speeds with and without wake effects of each wind turbine are obtained by using Equations (3)–(5), it is then easy to calculate the corresponding power outputs using the power curve of the wind turbine.

Note that the site condition described in Section 2.1 provides wind resource parameters at each turbine site for a given far field inflow wind direction, i.e.,: $A_i(\theta^\infty) = A(x_i, y_i, \theta^\infty)$, $k_i(\theta^\infty) = k(x_i, y_i, \theta^\infty)$ and $freq_i(\theta^\infty) = freq(x_i, y_i, \theta^\infty)$. Based on these parameters, the probability of each turbine site's local wind condition, i.e., $v_i(v^\infty, \theta^\infty)$ and $\theta_i(\theta^\infty)$, with respect to the far field inflow wind condition (v^∞ and θ^∞) can be estimated by using the joint distribution proposed in [26] as:

$$
\begin{aligned}
pdf_i(v^\infty, \theta^\infty) &= pdf(v_i(v^\infty, \theta^\infty), \theta_i(\theta^\infty)) \\
&= \frac{k_i(\theta^\infty)}{A_i(\theta^\infty)} \cdot \left(\frac{v_i(v^\infty, \theta^\infty)}{A_i(\theta^\infty)} \right)^{k_i(\theta^\infty)-1} \cdot \exp\left(-\left(\frac{v_i(v^\infty, \theta^\infty)}{A_i(\theta^\infty)} \right)^{k_i(\theta^\infty)} \right) \cdot freq_i(\theta^\infty)
\end{aligned}
\tag{6}
$$

Thus, for a wind farm composed of N_{wt} turbines, its gross and net AEP, i.e., AEP with and without wake effects, can be estimated as:

$$
\text{AEP}_{gross} = \sum_{i=1}^{N_{wt}} 8760 \cdot \eta_i \cdot \iint P(v_i(v^\infty, \theta^\infty)) \cdot pdf_i(v^\infty, \theta^\infty) dv^\infty \theta^\infty,
\tag{7}
$$

$$
\text{AEP}_{net} = \sum_{i=1}^{N_{wt}} 8760 \cdot \eta_i \cdot \iint P(\bar{v}_i(v^\infty, \theta^\infty)) \cdot pdf_i(v^\infty, \theta^\infty) dv^\infty \theta^\infty,
\tag{8}
$$

in which 8760 is the total number of hours in a year, η_i denotes the availability factor of turbine i, and $P(\cdot)$ represents the power curve of the wind turbine. The integrations in Equations (7) and (8) are done numerically after properly discretizing the interested ranges of v^∞ and θ^∞. In this study, the availability factors of all the turbines are assumed to be 1.0 and the discretization used in numerical integration is $\Delta v^\infty = 1$ m/s, and $\Delta\theta^\infty = 5°$.

3. Optimization Framework

3.1. Problem Formulation

For a wind farm with N_{wt} turbines, its design can be specified by the turbines' locations, i.e., $X = [x_1, x_2, \ldots, x_{N_{wt}}]$ and $Y = [y_1, y_2, \ldots, y_{N_{wt}}]$. Then, a general optimization problem of the wind farm design can be formulated as:

$$
\left.
\begin{aligned}
\min \quad & f_m(X, Y), & m = 1,2,\ldots,M, \\
\text{subject to :} \quad & g_k(X, Y) \geq 0, & k = 1,2,\ldots,K; \\
& X^{(L)} \leq X \leq X^{(U)}; \\
& Y^{(L)} \leq Y \leq Y^{(U)}.
\end{aligned}
\right\}
\tag{9}
$$

where f_m is the mth objective function, g_k is the kth inequality constraint function, and $X^{(L)}$, $X^{(U)}$, $Y^{(L)}$, and $Y^{(U)}$ denote the lower and upper bounds.

Common objective functions for wind farm design optimization include AEP, LCOE, profit, noise emission, and so on. Typical constraints that can be modelled as inequality constraint functions include wind farm boundary, exclusive zones, and wind turbine proximity. In the current framework, the default objective function is AEP and the considered constraints are summarized in the next subsection.

3.2. Constraints

Wind farm design is subject to various constraints, which may come from technical, logistical, environmental, economical, legal, and/or even social considerations. In the current framework, three types of constraints are considered: (1) inclusive and exclusive boundaries; (2) minimal distance requirements; and (3) bounds on certain site condition variables.

Inclusive and exclusive boundaries denote the feasible and infeasible area for placing turbines, which may come from limitations and requirements on leased land, existing roads and properties, soil conditions, and so on. As a general method, polygons can be used to model inclusive and exclusive zones defined by the inclusive and exclusive boundaries. Thus, for a wind farm with N_{inc} inclusive and N_{exc} exclusive boundaries, the individual inclusive and exclusive zones and the overall feasible zone can be modeled as:

$$S^i_{inc} = \left\{ (x,\, y) \,\middle|\, a^i_k x + b^i_k y \leq c^i_k, \quad k = 1, 2, \ldots, m^i \right\}, \quad i = 1, 2, \ldots, N_{inc} \tag{10}$$

$$S^j_{exc} = \left\{ (x,\, y) \,\middle|\, a^j_l x + b^j_l y \leq c^j_l, \quad l = 1, 2, \ldots, n^j \right\}, \quad j = 1, 2, \ldots, N_{exc} \tag{11}$$

$$S_{\text{feasible}} = \left\{ (x,\, y) \,\middle|\, (x,\, y) \in \cup_{i=1}^{N_{inc}} S^i_{inc}, \quad (x,\, y) \notin \cup_{j=1}^{N_{exc}} S^j_{exc} \right\} \tag{12}$$

where m^i and n^j are the polygon edge numbers of the ith inclusive and jth exclusive boundaries, respectively.

Thus, the constraints on inclusive and exclusive boundaries of the wind farm design can be written as:

$$(x_i,\, y_i) \in S_{\text{feasible}}, \quad \text{for } i = 1, 2, \ldots, N_{\text{wt}}. \tag{13}$$

The second type of constraint is also typical in the engineering practice, which requires the distance between any two turbines larger than a minimal value. This is because a shorter distance between turbines will give a larger wake loss and a higher turbulence intensity, which results in a higher level of fatigue loads and maintenance cost, and even a shorter lifetime of turbines. Also, a minimal distance is required to make sure two turbines' blades never contact each other and one turbine never falls on the other turbine. In the current framework, a fixed minimal distance requirement, $Dist_{\min}$, is assumed for any two turbines. Thus, the constraints on the minimal distance requirement are governed by:

$$\sqrt{(x_i - x_j)^2 + (y_i - y_j)^2} - Dist_{\min} \geq 0 \quad \text{for } i, j = 1, 2, \ldots, N_{\text{wt}} \text{ and } i \neq j. \tag{14}$$

The third type of constraints exposes requirements on certain site condition variables to rule out certain unfavorable sites, such as those with too low a mean wind speed, too high a turbulence intensity, or too rugged terrain. When applied properly, this type of constraint can help the optimizer to focus on the more favorable design space and speed up the convergence process, as found in [18].

To measure the degree of the terrain ruggedness, a terrain ruggedness index (TRI) is introduced. This index was first proposed by Riley et al. [27] as a quantitative measure of topographic heterogeneity. A dimensionless version of this index is developed in the current framework, which can be calculated based on the elevation data of grids as defined in Section 2.1. Given the grid sizes along the x and y directions are Δx and Δy, the TRI at grid $(x_i,\, y_j)$ can be computed based on its elevation and the elevation values of the eight surrounding grids as:

$$
\begin{aligned}
\text{TRI}(x_i,\, y_j) = \; &\frac{1}{8}\left\{ \frac{\left[z(x_i,\, y_j) - z(x_{i+1},\, y_j)\right]^2}{\Delta x^2} + \frac{\left[z(x_i,\, y_j) - z(x_{i-1},\, y_j)\right]^2}{\Delta x^2} \right. \\[2mm]
&+ \frac{\left[z(x_i,\, y_j) - z(x_i,\, y_{j+1})\right]^2}{\Delta y^2} + \frac{\left[z(x_i,\, y_j) - z(x_i,\, y_{j-1})\right]^2}{\Delta y^2} + \frac{\left[z(x_i,\, y_j) - z(x_{i+1},\, y_{j+1})\right]^2}{\Delta x^2 + \Delta y^2} \\[2mm]
&\left. + \frac{\left[z(x_i,\, y_j) - z(x_{i+1},\, y_{j-1})\right]^2}{\Delta x^2 + \Delta y^2} + \frac{\left[z(x_i,\, y_j) - z(x_{i-1},\, y_{j+1})\right]^2}{\Delta x^2 + \Delta y^2} + \frac{\left[z(x_i,\, y_j) - z(x_{i-1},\, y_{j-1})\right]^2}{\Delta x^2 + \Delta y^2} \right\}^{1/2}
\end{aligned}
\tag{15}
$$

For the wind farm site in Figure 1a, the TRI map calculated in Equation (15) is shown in Figure 3a. Note that the grid sizes are chosen as $\Delta x = \Delta y = 25$ m in this study according to the default setting of WAsP, which makes the total number of grids in the rectangle area shown in Figure 1 as 38,801. In the current framework, the constraints on TRI and mean wind speed, v^{mean}, are considered. Note that the mean wind speed values on the grids, $v^{mean}(x_i,\, y_j)$, can be imported from wind resource assessment tools or calculated based on the sector wise wind resource parameters, i.e., Weibull-A, Weibull-k, and frequency. For an arbitrary site, linear interpolation is used to find the value of TRI and v^{mean}. Thus, constraints on TRI and mean wind speed can be written as:

$$
\text{TRI}_{max} - \text{TRI}(x_i,\, y_i) \geq 0, \quad \text{for } i = 1, 2, \ldots, N_{\text{wt}},
\tag{16}
$$

$$
v^{mean}(x_i,\, y_i) - v^{mean}_{min} \geq 0, \quad \text{for } i = 1, 2, \ldots, N_{\text{wt}}.
\tag{17}
$$

where TRI_{max} is the allowed maximal TRI and v^{mean}_{min} is the allowed minimal mean wind speed.

For the wind farm site in Figure 1, the feasibility of each grid, $(x_i,\, y_j)$, can then be estimated as $F(x_i,\, y_j)$ based on the first and third types of constraints, i.e., constraints defined in Equations (13), (16) and (17). Note that $F(x_i,\, y_j)$, with a value of 1 (true) or 0 (false), means whether the given site, $(x_i,\, y_j)$, is a feasible site to install a turbine, i.e., a site satisfies the constraints on inclusive/exclusive boundaries and bounds of certain site condition variables. The map of $F(x_i,\, y_j)$ is shown in Figure 3b. Note that the black area in this figure denotes the feasible area. Compared with the feasible area defined by Equations (13) and (14), the bounds on TRI and v^{mean} help to reduce the feasible grids from 12,150 to 9218, representing a 24.13% reduction in the feasible design space. This reduction can help certain optimization algorithms, such as RS, to limit the search space and converge faster.

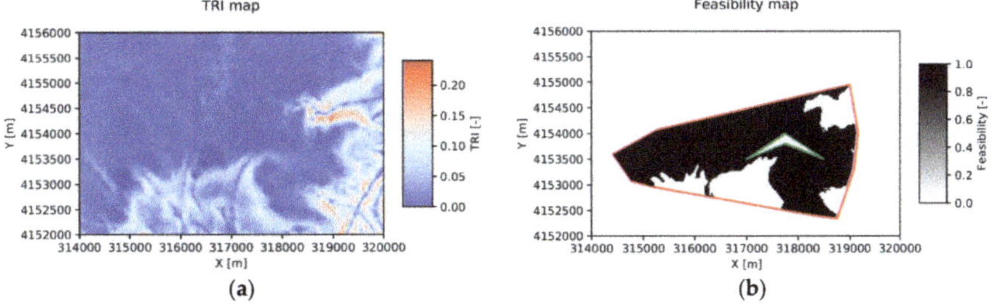

Figure 3. Constraint modelling: (**a**) TRI (terrain ruggedness index) map; (**b**) feasibility map with one inclusive (red line) and one exclusive (green line) boundary, and bounds on TRI ($\text{TRI}_{max} = 0.05$) and v^{mean} ($v^{mean}_{min} = 6.5$ m/s).

3.3. Optimization Algorithm

To solve the optimization problem, the Random Search (RS) algorithm is used in the current framework. This algorithm was first proposed by Feng and Shen in [15] and improved in [16]. Recently, they also extended the algorithm to cover multiple objective cases [28] and overall design cases with multiple types of turbines [19], and to maximize the robustness of wind farm power production under wind condition uncertainties [29]. It has shown a superior performance when compared to other popular algorithms, such as GA (genetic algorithm) [16], NSGA-II (Non-dominated Sorting Genetic Algorithm II) [28], and mixed-discrete PSO (Particle Swarm Optimization) [19].

RS is a single solution search method. At each step, a new feasible design, i.e., a design satisfies all the constraints and bounds, is generated by slightly changing the current design, i.e., randomly choosing a turbine and moving it to a random position inside the feasible area. This new design is then compared with the current design with respect to the objective function. If the new design is better, it becomes the current design. If not, the current design remains unchanged. This step is iteratively repeated until a given stop condition is met. Normally, the stop condition can be set as a maximal number of steps (equivalent to the number of objective function evaluations). The procedure of this algorithm is shown as a flowchart in Figure 4.

There are two parameters controlling the RS algorithm: maximal step size, M_{max}, which controls how far away a chosen turbine can be possibly moved from its current position; and maximal number of evaluations, E_{max}, which defines when to terminate the optimization process. M_{max} should be set large enough to allow the algorithm to explore the design space more thoroughly and escape local minima more likely. When no trials and experimentations are first applied, M_{max} could be set in the order of the maximal distance between any two points in the feasible wind farm area and E_{max} should be in the range of 1000 to 10,000.

Due to the randomness involved in the optimization process, meta-heuristics, such as RS, exhibits a stochastic nature, meaning different runs typically obtain different results. This also means that the quality of the initial solution(s) can have profound influences on the final optimization results and the convergence speed [30]. Saavedra-Moreno et al. [31] tackled this problem by seeding an evolutionary algorithm with initial solutions obtained by a greedy heuristic algorithm, and applied this method in wind farm layout optimization.

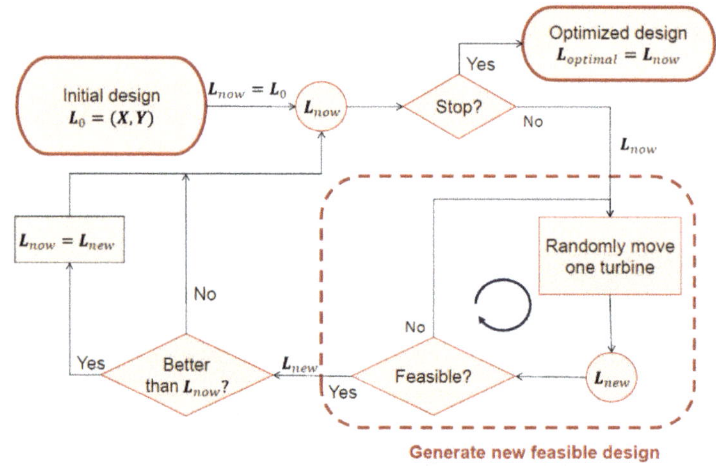

Figure 4. Flowchart of the Random Search algorithm.

In this study, a simple algorithm called Heuristic Fill is proposed to find a good initial design, based solely on site conditions and constraints. As described in Section 3.2, the feasibility of each grid, (x_i, y_j), can be calculated as $F(x_i, y_j)$ based on constraints defined in Equations (13), (16), and (17). The mean wind speed of each grid, $v^{mean}(x_i, y_j)$, can also be obtained from site conditions. The algorithm then tries to fill the feasible grids with N_{wt} wind turbines one by one according to the corresponding mean wind speed from high to low, while respecting the minimal distance constraints in Equation (14). This algorithm can be summarized in the pseudo code shown in Algorithm 1.

Algorithm 1. Pseudo code of the Heuristic Fill algorithm.

(1)   Find the set of feasible grid points:

$$G_{feasible} = \left\{ (x_i, y_j) \left| F(x_i, y_j) = 1, \text{ for } i = 1, 2, \ldots, N_x, j = 1, 2, \ldots, N_y \right. \right\};$$

(2)   Sort the feasible grid points according to mean wind speed, i.e., find a list of feasible grid points as

$$\left[(x_{j1}, y_{j1}), (x_{j2}, y_{j2}), \ldots, (x_{jm}, y_{jm}), \right], \text{ where } (x_{jk}, y_{jk}) \in G_{feasible} \text{ with } k = 1, 2, \ldots, m \text{ and}$$

$$v^{mean}(x_{jk}, y_{jk}) \geq vv^{mean}(x_{jk+1}, y_{jk+1}) \text{ with } k = 1, 2, \ldots, m - 1;$$

(3)   Heuristically fill with N_{wt} turbine sites:

      (a)   Set $w = 1, g = 1$;

      (b)   **While** $1 < w \leq N_{wt}$ and $\sqrt{(x_i - x_{ig})^2 + \left(y_i - y_{jg}\right)^2} < Dist_{min}$ holds for $i = 1, \ldots, w - 1$:

           $g = g + 1$

           End While

           Set $x_w = x_{ig}, y_w = y_{jg}, g = g + 1, w = w + 1$;

      (c)   Set the initial layout as $X = [x_w | w = 1, 2, \ldots, N_{wt}], Y = [y_w | w = 1, 2, \ldots, N_{wt}]$.

3.4. Framework

Putting things together, we can describe the framework in the architecture diagram shown in Figure 5. Note that WAsP (or another wind resource assessment code) needs to be used only once to get the relevant results for defining the site conditions, since the wake modelling and AEP estimation, as well as the modelling of constraints and optimization process, are all implemented by the modules inside the framework as introduced in the previous subsections. Thus, this framework can work as a stand-alone tool for wind farm design optimization. If objective functions other than AEP and more constraints are considered, the relevant modules can also be extended accordingly, while the optimization algorithm and the architecture remain unchanged.

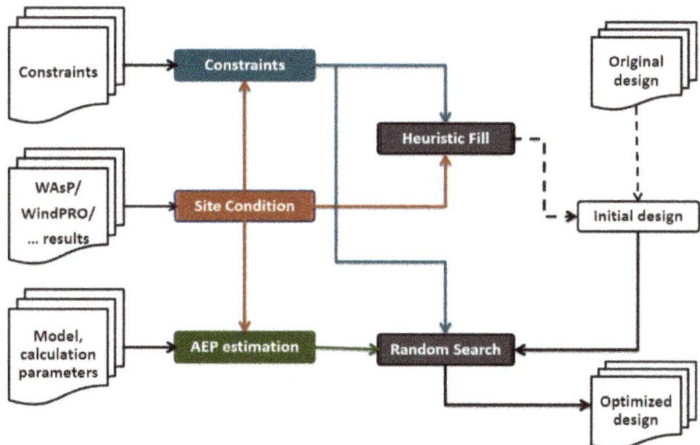

Figure 5. Architecture of the framework.

4. Case Study and Results

As a demonstration, a case study for an anonymous wind farm in complex terrain using the proposed framework is presented here. This wind farm is located in Northwest China, and has a total capacity of 50 MW with 25 turbines. This wind farm has been studied in [23,32] to investigate the wake flow simulation in complex terrain using CFD, and the preliminary results of the layout optimization study of this wind farm have been presented in [18]. Some of this wind farm's site conditions have already been shown in Figure 1. Its original layout is shown in Figure 6 and the turbine characteristics are shown in Figure 7. This type of turbine is variable speed, pitch regulated and has a rotor diameter of $D = 93$ m and a hub height of $H = 67$ m. The cut-in, rated, and cut-out wind speeds are 3 m/s, 12 m/s, and 25 m/s, respectively.

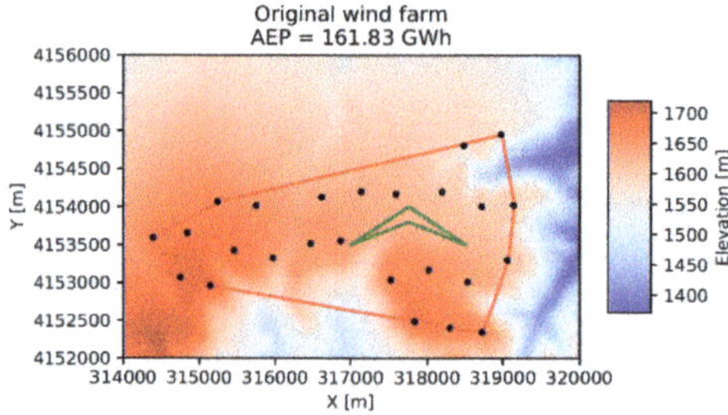

Figure 6. Original design of the anonymous wind farm in complex terrain.

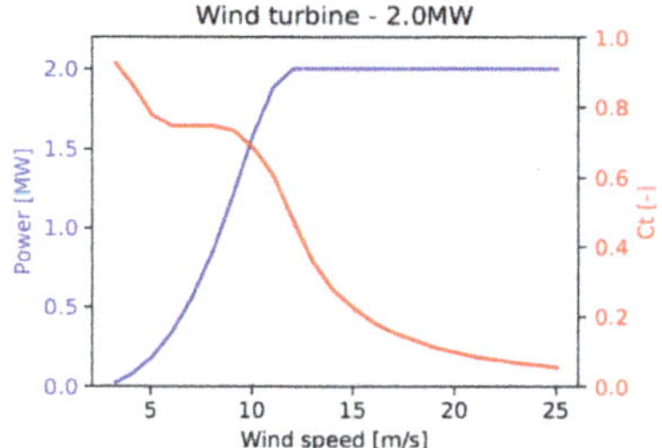

Figure 7. Turbine characteristics.

Note that constraints on inclusive and exclusive boundaries and bounds on parameters that are considered in this case study are shown in Figure 3. Additionally, the minimal distance requirement used in Equation (14) is chosen as the minimal distance between the turbines in the original layout, which is $D_{min} = 403.8$ m. The inclusive boundary shown with red line in Figure 6 is set as the minimal convex polygon that covers all the turbines, and the exclusive boundary (the green line) is arbitrarily set to test the modelling capacity of the exclusive boundary of the framework. The choice of the inclusive boundary is to make sure the optimized wind farm occupies the same or a smaller area and the required cost of electrical cables and access roads that connect all turbines stays at the same level, so that the comparison of AEP with the original wind farm makes sense.

To find a good feasible initial design, the Heuristic Fill algorithm is applied in this wind farm, which obtains an initial design with AEP $= 164.61$ GWh, as shown in Figure 8.

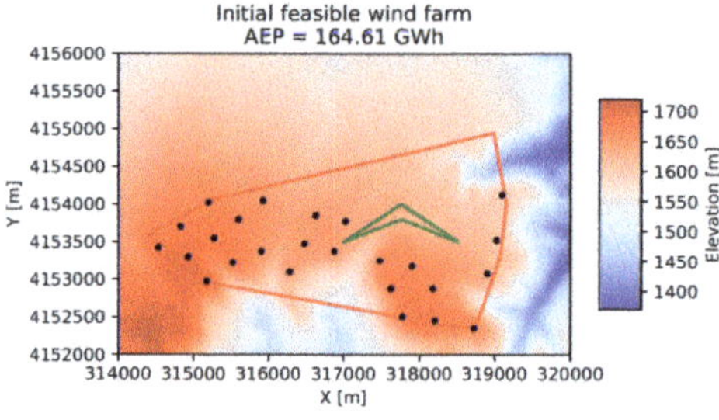

Figure 8. Initial design obtained by the Heuristic Fill algorithm.

Compared to the original wind farm, the design found by the Heuristic Fill algorithm obtains an impressive 1.72% increase of AEP (from 161.83 GWh to 164.61 GWh). Note that this increase is achieved by a deterministic filling process without any searching process, thus can be done in several seconds.

To show the performance of the proposed framework, multiple runs of Random Search with different settings of the optimization algorithm are carried out, i.e., optimization runs with different numbers of total evaluation number, E_{max}, and maximal step size, M_{max}. For each of these parameter combinations, 10 optimization runs are done to obtain the statistics. Cases starting from the original design (as shown in Figure 6) and the initial design found by the Heuristic Fill algorithm (as shown in Figure 8) are both tried. The results are summarized in Table 2. Note that the percentages of AEP that increase in the optimized wind farm design compared to the original wind farm (161.83 GWh) are marked with underlines inside the parentheses, and the CPU time shown here is the mean value per run for 10 runs by a Python 3.6 implementation of the framework on a laptop with an Intel® i5-2520M CPU @2.50 GHz.

Table 2. Performance of the framework in multiple runs with different settings.

Opt. Setting		AEP [GWh] (<u>Increase [%]</u>)				CPU Time [s]
E_{max} [-]	M_{max} [m]	Initial	Opt. min	Opt. mean	Opt. max	Mean
1000	5000	161.83	164.62 (<u>1.72</u>)	165.03 (<u>1.98</u>)	165.67 (<u>2.37</u>)	2137.5
1000	50	164.61	165.18 (<u>2.07</u>)	165.27 (<u>2.12</u>)	165.30 (<u>2.14</u>)	1824.6
1000	500	164.61	165.17 (<u>2.06</u>)	165.28 (<u>2.13</u>)	165.41 (<u>2.21</u>)	1923.7
1000	5000	164.61	165.17 (<u>2.06</u>)	165.45 (<u>2.24</u>)	165.79 (<u>2.45</u>)	2272.3
5000	5000	164.61	165.76 (<u>2.42</u>)	166.02 (<u>2.58</u>)	166.17 (<u>2.68</u>)	12,042.6
10,000	5000	164.61	165.95 (<u>2.55</u>)	166.11 (<u>2.64</u>)	**166.21 (<u>2.70</u>)**	24,613.0

As shown in Table 2, allowing the optimization process to run longer, i.e., with a larger number of evaluations, generally obtains better results (see the last three rows in Table 2). The other advantage for an optimization process is starting from a better initial design, as the results from the initial design obtained by the Heuristic Fill algorithm are much better than those from the original design using the same optimization setting (see the first and fourth rows of the results in Table 2).

The maximal step size also has an effect on the results, as it controls the degree of change that can be made to the current best design in each step. As discussed in Section 3.3, this parameter should be set large enough to explore the design space more thoroughly and thus find a better solution after a sufficient large number of evaluations. This is supported by the comparison of optimization runs with the same E_{max}, but different M_{max}, as shown in the second to fourth rows of Table 2.

The best optimized design found by all the optimization runs summarized in Table 2 is under the optimization setting: $E_{max} = 10,000$, $M_{max} = 5000$ m, with the optimized AEP as 166.21 GWh, representing a 2.70% increase to the original wind farm design. This design is found by the Random Search algorithm starting from the initial design obtained by the Heuristic Fill algorithm, and shown in Figure 9.

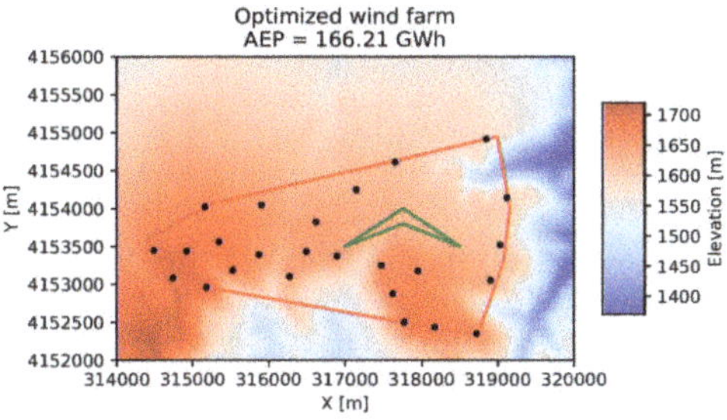

Figure 9. Best optimized design found by the framework with $E_{max} = 10,000$, $M_{max} = 5000$ m.

To better compare the performance of the original, initial, and optimized designs as shown in Figures 6, 8 and 9, their gross and net AEP values are listed in Table 3.

Table 3. AEP values comparison of the original, initial, and optimized wind farms.

Wind Farm	Net AEP [GWh]	Gross AEP [GWh]	Wake Loss [%]
Original (Figure 6)	161.83	169.75	4.67
Initial (Figure 8)	164.61	175.42	6.16
Optimized (Figure 9)	166.21	174.91	4.97

As Table 3 shows, the initial design found by the Heuristic Fill algorithm has both higher gross AEP and higher net AEP than the original design. Although the wake loss is higher for the initial design than the original one (6.16% versus 4.67%), the larger increase (3.34%) of gross AEP thanks to the higher mean wind speed of the sites found by the Heuristic Fill algorithm compensates the higher wake loss and still yields a substantial increase (1.72%) of net AEP. Note that the higher wake loss of the initial design can also be seen from the layout shown in Figure 8, as the turbines are placed much closer to each other than the original layout shown in Figure 6.

Since the optimized design in Figure 9 is obtained by the optimization run starting from the initial design in Figure 8, we can also compare these two designs. Apparently, the optimization algorithm makes a small sacrifice in gross AEP, i.e., allowing the gross AEP to decrease by 0.28% from 175.42 GWh to 174.91 GWh, while decreasing the wake loss percentage from 6.16% to 4.97%. The combined effect allows the optimized design to gain a 0.97% increase of net AEP to the initial design (from 164.61 GWh to 166.21 GWh), representing a 2.70% increase of net AEP when compared to the original design (from 161.83 GWh to 166.21 GWh).

Examining the layout of the initial design in Figure 8 and the layout of the optimized design in Figure 9, we can see that the turbines are scattered away from each other while most of the turbines are still located at high elevation sites. This shows that the optimization algorithm tries to lower wake effects while not sacrifice too much on the gross AEP, as suggested by the net and gross AEP comparison. To show the performance of the framework in multiple runs, the evolution histories of the 10 runs under the optimization setting: $E_{max} = 10,000$, $M_{max} = 5000$ m are shown in Figure 10, which demonstrate the random nature of the RS algorithm.

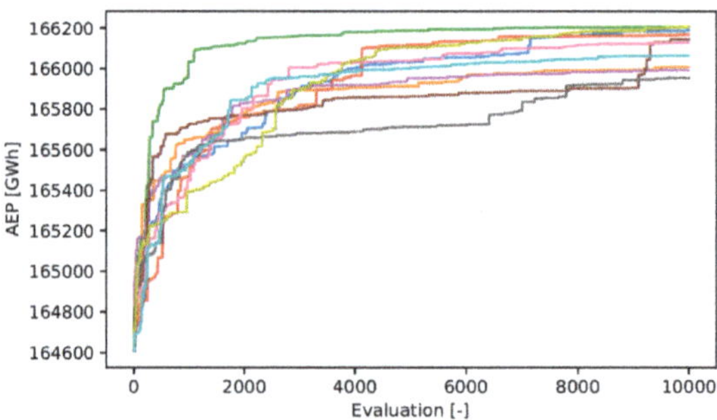

Figure 10. Evolution histories of the 10 Random Search runs under optimization setting: $E_{max} = 10,000$, $M_{max} = 5000$ m.

As Figure 10 shows, different runs of the optimization process yield different results. Thus, if it is possible, it is always advised to have multiple runs (10 or more) of the optimization process and use the best result. Also, it is shown for these 10 runs that the largest portion of the AEP increase is achieved in the first 2000 evaluations.

5. Comparison with Other Algorithms

To better show the effectiveness of the RS algorithm for wind farm design in complex terrain, we compare its performance with two widely used algorithms: Particle Swarm Optimization (PSO) and Genetic Algorithm (GA) [33].

PSO is a population based meta-heuristic algorithm inspired by bird flocking, which maintains a population of solutions (called particles) and moves each particle in the design space according to the global best solution in the population and the local best solution found by each particle. In this study, the version of PSO used by Chowdhury et al. in [34] is implemented in Python with the algorithm parameters set as recommended in Table 4 of [34]. Details of this algorithm are referred to in the paper [34].

GA is also a nature inspired population based meta-heuristic algorithm for optimization problems. Various versions of GA have been applied to solve the wind farm layout optimization problem, most of which are binary coded [33]. As a binary-coded GA makes restrictions on where to place turbines and thus limits the design space, we choose to implement a real-coded GA (RCGA) recently proposed by Chuang et al. [35], which has no such restrictions. This version of RCGA was developed for constrained optimization based on three specially designed evolutionary operators, i.e., ranking selection, direction-based crossover, and dynamic random mutation, and showed a better performance than traditional versions of RCGA for a variety of benchmark constrained optimization problems [35]. A Python version of this algorithm is implemented in this study using the parameters recommended by the authors of [35]. Details of this algorithm can be found in [35].

For the sake of simplicity, we consider the design optimization problem for the same wind farm as in the case study with only bounds on the design variables, i.e., x and y coordinates of wind turbines. No constraints on wind resource, terrain parameters, and minimal distances are considered. The bounds are set as the boundary of the rectangle area shown in Figure 6, i.e., $314,000 \text{ m} \leq x \leq 320,000 \text{ m}$ and $4,152,000 \text{ m} \leq y \leq 4,156,000 \text{ m}$. Following the recommendations in [34,35], the size of the population is set as 250 for both PSO and RCGA.

Two scenarios for setting the initial solution(s) are tested. In the random scenario, 250 initial layouts are generated randomly by putting 25 turbines randomly inside the bounds. These initial layouts are used in PSO and RCGA as the initial population, while the worst performing initial layout, i.e., the layout with lowest AEP, is used as the initial solution in RS.

As the initial solution(s) has influences on the optimization results obtained by meta-heuristics, there has been an effort on seeding good initial solution(s) for meta-heuristics in the field of wind farm layout optimization [31]. We address this issue in the heuristic fill scenario. In this scenario, an initial layout is first obtained by using the Heuristic Fill algorithm described in Section 3 with the minimal distance requirement set as four times of the rotor diameter. Then, a group of 124 layouts are generated by running RS 124 times with $E_{\max} = 100$ and $M_{\max} = 5000 \text{ m}$ from the initial layout found by Heuristic Fill. Thus, these 124 initial layouts have a higher or the same AEP as the one found by the Heuristic Fill, as they are improved better ones found by RS. After this process, we have 125 good initial layouts, which are then seeded in the initial populations of PSO and GA, while the remaining halves of the initial populations are filled with random layouts. In this way, the initial populations for both PSO and GA have some good quality initial solutions and some random initial solutions, thus they have a good degree of randomness, which is essential for the diversification. For RS, the worst performing layout among the 125 good initial layouts is used as the initial solution.

The performance of these three algorithms with a different number of evaluations ($E_{\max}$) for both scenarios are summarized in Table 4.

Table 4. Performance comparison of three algorithms: PSO (particle swarm optimization), RCGA (real-coded genetic algorithm) and RS (random search).

Initial Scenario	Algorithm	AEP [GWh]			E_{max} [-]
		Worst Initial	Best Initial	Optimized	
Random	PSO	119.12	152.44	160.76	5000
	RCGA	119.12	152.44	156.68	5000
	RS	119.12	119.12	173.58	5000
	PSO	119.12	152.44	161.66	50,000
	RCGA	119.12	152.44	165.43	50,000
	RS	119.12	119.12	**174.26**	10,000
Heuristic fill	PSO	170.27	171.18	171.74	10,000
	RCGA	170.27	171.18	171.96	10,000
	RS	170.27	170.27	174.10	10,000
	PSO	170.27	171.18	171.84	50,000
	RCGA	170.27	171.18	172.58	50,000
	RS	170.27	170.27	173.80	5000

Note: 'Worst/best initial' represents the worst/best performing initial solution among the 250 random initial layouts for the random scenario and the 125 seeded good initial layouts for the heuristic fill scenario. These two values are the same for RS as it maintains only one solution. 'Optimized' denotes the global best solution found by PSO/RCGA.

It is worth to note that in both scenarios, RS starts from the worst initial solution among the solutions seeded in the initial population for PSO and GA, thus with a disadvantage in terms of the quality of the initial solution(s). Despite this disadvantage, RS shows a much better performance than PSO and RCGA. As Table 4 shows, RS with 10,000 evaluations obtains the best optimized layout in both scenarios. Compared with PSO and RCGA, RS with only 5000 evaluations largely outperforms PSO and RCGA with 50,000 evaluations (equivalent to 200 generations) in both scenarios, which clearly demonstrates the effectiveness of RS. To better visualize the difference between algorithms, the evolution histories of the best solutions in the different runs of PSO, RCGA, and RS are shown in Figure 11.

Figure 11 shows that: (1) starting from good initial solutions makes PSO and RCGA converge to much better optimized solutions, while its benefit for RS is not so profound; (2) RCGA obtains better results than PSO after a sufficient number of evaluations (20,000); and (3) RS converges faster to a better solution than PSO and RCGA in both scenarios. Based on the results described above, it is natural to conclude that RS used in the current framework performs much better than PSO and RCGA for this case.

The best optimized design, which is found by RS with 10,000 evaluations in the random scenario, is shown in Figure 12.

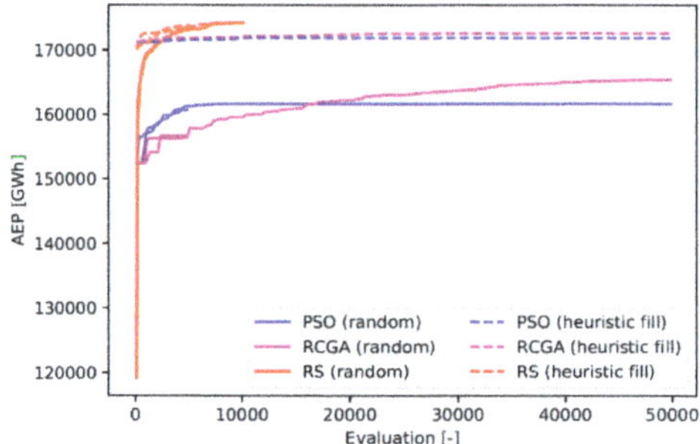

Figure 11. Evolution histories of the best solutions in different runs of PSO (particle swarm optimization), RCGA (real-coded genetic algorithm), and RS (random search).

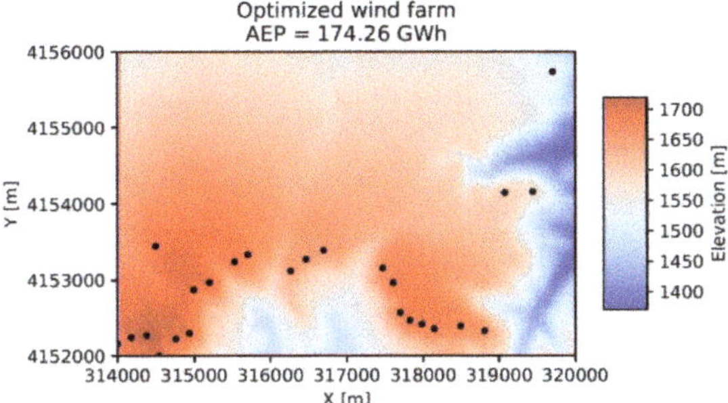

Figure 12. Best optimized design for the wind farm without any constraints (found by Random Search with $E_{\mathrm{max}} = 10,000$ and $M_{\mathrm{max}} = 5000$ m in the random scenario)

Without the limitation of inclusive and exclusive boundaries, the turbines in the optimized design in Figure 12 are spread out in the whole rectangle area and mostly occupying high elevation sites. Note that these sites also have high mean wind speeds as shown in Figure 1b. To place more turbines on the limited high elevation sites, some of the turbines are placed quite close (down to 170.44 m), which could cause problems on fatigue loads for certain turbines. Although the optimized wind farm has a much higher AEP (174.26 GWh), representing a 7.68% increase over the original wind farm, it should be interpreted with caution, as in one aspect, it occupies a much larger area and thus requires more investments on electrical cables and access roads, and in the other aspect, it places several turbines too close, thus it will have difficulty satisfying the requirements on fatigue loads.

6. Conclusions

Wind farm design in complex terrain is a crucial yet challenging task. In this work, we present an overall design optimization framework for tackling this task. This framework combines results from state-of-the-art wind resource assessment tools, such as WAsP CFD, a fast engineering wake model, and an advanced optimization algorithm, to solve the wind farm design optimization problem in complex terrain under various realistic constraints and requirements. This makes the framework a valuable tool that can be used by designers/developers who need to optimize wind farm designs in complex terrain under a real-life scenario.

While the framework in its default setting considers maximizing the total net AEP, its modular architecture makes it easy to be extended to consider other objective function(s). More constraints, requirements, and other optimization algorithms can also be easily implemented and added.

Other contributions of this work include: a terrain ruggedness index (TRI) to characterize the terrain feature and model constraints on the terrain ruggedness; and a fast and simple algorithm called Heuristic Fill to find good initial designs for wind farms in complex terrain, by using a deterministic process based on wind resource considerations and constraints.

The case study of a real wind farm in complex terrain demonstrates the effectiveness of the framework and the performance of proposed algorithms. An impressive increase of net AEP (up to 2.70%) is achieved by using the framework, while respecting realistic constraints on inclusive and exclusive boundaries, minimal mean wind speed, and maximal TRI, and minimal distance constraints between turbines. In comparison with two widely used algorithms (PSO and real-coded GA) for the same wind farm without any constraints, the Random Search algorithm also shows a much better performance in terms of convergence speed and optimization results. This of course demonstrates the advantages of the developed framework and proves the usefulness for wind farm designers/developers.

Although the wind farm design considered in the current framework concerns only the turbine layout, other design variables, such as number, hub height(s), and type(s) of turbines, foundations, access roads, and electrical systems, also play important roles for an overall design optimization of a wind farm. Our future works will investigate these problems and extend the framework in this direction.

Author Contributions: J.F. and W.Z.S. conceived and designed the study; J.F. wrote the code and performed the computation; J.F., W.Z.S. and Y.L. analyzed the results and wrote the paper.

Funding: This research was supported by the Sino-Danish cooperative project "Wind farm layout optimization in complex terrain" funded by the Danish Energy Agency (EUDP J.nr. 64013-0405) and the Ministry of Science and Technology of China (No. 2014DFG62530).

Acknowledgments: The authors wish to give special thanks to the Danish and Chinese partners (DTU Wind Energy, EMD International A/S, North West Survey and Design Institute of Hydro China Consultant Corporation, and HoHai University) for their active collaborations.

References

1. GWEC: Global Cumulative Installed Capacity 2001–2017. Available online: http://gwec.net/global-figures/graphs/ (accessed on 7 July 2018).
2. GWEC: Global Cumulative and Annual Offshore Wind Capacity End 2017. Available online: http://gwec.net/global-figures/graphs/ (accessed on 7 July 2018).
3. Alfredsson, P.H.; Segalini, A. Wind farms in complex terrains: An introduction. *Phil. Trans. R. Soc. A* **2017**, *375*, 20160096. [CrossRef] [PubMed]
4. Manwell, J.; McGowan, J.; Rogers, A. *Wind Energy Explained: Theory, Design, and Application*, 2nd ed.; John Wiely & Sons: Hoboken, NJ, USA, 2009.
5. DTU Wind Energy: WAsP. Available online: http://www.wasp.dk/wasp (accessed on 7 July 2018).
6. Mortensen, N.G. *Wind Resource Assessment Using the WAsP Software*; DTU Wind Energy E-0135; Technical University of Denmark: Roskilde, Denmark, 2016.
7. Burton, T.; Jenkins, N.; Sharpe, D.; Bossanyi, E. *Wind Energy Handbook*, 2nd ed.; John Wiely & Sons: Hoboken, NJ, USA, 2011.
8. Queney, P. The problem of airflow over mountains: A summary of theoretical studies. *Bull. Am. Meteorol. Soc.* **1948**, *29*, 16–26. [CrossRef]
9. Wood, N. Wind flow over complex terrain: A historical perspective and the prospect for large-eddy modelling. *Bound.-Lay. Meteorol.* **2000**, *96*, 11–32. [CrossRef]
10. Landberg, L.; Myllerup, L.; Rathmann, O.; Petersen, E.L.; Jørgensen, B.H.; Badger, J.; Mortensen, N.G. Wind resource estimation—An overview. *Wind Energ.* **2003**, *6*, 261–271. [CrossRef]
11. González, J.S.; Payán, M.B.; Santos, J.M.R.; González-Longatt, F. A review and recent developments in the optimal wind-turbine micro-siting problem. *Renew. Sust. Energ. Rev.* **2014**, *30*, 133–144. [CrossRef]
12. Song, M.X.; Chen, K.; He, Z.Y.; Zhang, X. Bionic optimization for micro-siting of wind farm on complex terrain. *Renew. Energ.* **2013**, *50*, 551–557. [CrossRef]
13. Song, M.X.; Chen, K.; He, Z.Y.; Zhang, X. Optimization of wind farm micro-siting for complex terrain using greedy algorithm. *Energy* **2014**, *67*, 454–459. [CrossRef]
14. Feng, J.; Shen, W.Z. Wind farm layout optimization in complex terrain: A preliminary study on a Gaussian hill. *J. Phys. Conf. Ser.* **2014**, *524*, 012146. [CrossRef]
15. Feng, J.; Shen, W.Z. Optimization of wind farm layout: A refinement method by random search. In Proceedings of the 2013 International Conference on aerodynamics of Offshore Wind Energy Systems and wakes (ICOWES 2013), Lyngby, Denmark, 17–19 June 2013.
16. Feng, J.; Shen, W.Z. Solving the wind farm layout optimization problem using random search algorithm. *Renew. Energ.* **2015**, *78*, 182–192. [CrossRef]
17. Kuo, J.Y.; Romero, D.A.; Beck, J.C.; Amon, C.H. Wind farm layout optimization on complex terrains—Integrating a CFD wake model with mixed-integer programming. *Appl. Energ.* **2016**, *178*, 404–414. [CrossRef]

18. Feng, J.; Shen, W.Z.; Hansen, K.S.; Vignaroli, A.; Bechmann, A.; Zhu, W.J.; Larsen, G.C.; Ott, S.; Nielsen, M.; Jogararu, M.M.; et al. Wind farm design in complex terrain: The FarmOpt methodology. In Proceedings of the China Wind Power 2017, Beijing, China, 17–19 October 2017.

19. Feng, J.; Shen, W.Z. Design optimization of offshore wind farms with multiple types of wind turbines. *Appl. Energ.* **2017**, *205*, 1283–1297. [CrossRef]

20. EMD: WindPRO. Available online: https://www.emd.dk/windpro/ (accessed on 7 July 2018).

21. DTU Wind Energy: WAsP CFD. Available online: http://www.wasp.dk/waspcfd (accessed on 7 July 2018).

22. Politis, E.S.; Prospathopoulos, J.; Cabezon, D.; Hansen, K.S.; Chaviaropoulos, P.K.; Barthelmie, R.J. Modeling wake effects in large wind farms in complex terrain: The problem, the methods and the issues. *Wind Energ.* **2012**, *15*, 161–182. [CrossRef]

23. Sessarego, M.; Shen, W.Z.; van der Laan, M.P.; Hansen, K.S.; Zhu, W.J. CFD Simulations of flows in a wind farm in complex terrain and comparisons to measurements. *Appl. Sci.* **2018**, *8*, 788. [CrossRef]

24. Song, M.X.; Chen, K.; He, Z.Y.; Zhang, X. Wake flow model of wind turbine using particle simulation. *Renew. Energ.* **2012**, *41*, 185–190. [CrossRef]

25. Kuo, J.; Rehman, D.; Romero, D.A.; Amon, C.H. A novel wake model for wind farm design on complex terrains. *J. Wind Eng. Ind. Aerod.* **2018**, *174*, 94–102. [CrossRef]

26. Feng, J.; Shen, W.Z. Modelling wind for wind farm layout optimization using joint distribution of wind speed and wind direction. *Energies* **2015**, *8*, 3075–3092. [CrossRef]

27. Riley, S.J.; DeGloria, S.D.; Elliot, R. A terrain ruggedness index that quantifies topographic heterogeneity. *Intermt. J. Sci.* **1999**, *5*, 23–27.

28. Feng, J.; Shen, W.Z.; Xu, C. Multi-objective random search algorithm for simultaneously optimizing wind farm layout and number of turbines. *J. Phys. Conf. Ser.* **2016**, *753*, 032011. [CrossRef]

29. Feng, J.; Shen, W.Z. Wind farm power production in the changing wind: Robustness quantification and layout optimization. *Energ. Convers. Manag.* **2017**, *148*, 905–914. [CrossRef]

30. Blum, C.; Roli, A. Metaheuristics in combinatorial optimization: Overview and conceptual comparison. *ACM Comput. Surv.* **2003**, *35*, 268–308. [CrossRef]

31. Saavedra-Moreno, B.; Salcedo-Sanz, S.; Paniagua-Tineo, A.; Prieto, L.; Portilla-Figueras, A. Seeding evolutionary algorithms with heuristics for optimal wind turbines positioning in wind farms. *Renew. Energ.* **2011**, *36*, 2838–2844. [CrossRef]

32. Han, X.; Liu, D.; Xu, C.; Shen, W.Z. Atmospheric stability and topography effects on wind turbine performance and wake properties in complex terrain. *Renew. Energ.* **2018**, *126*, 640–651. [CrossRef]

33. Khan, S.A.; Rehman, S. Iterative non-deterministic algorithms in on-shore wind farm design: A brief survey. *Renew. Sust. Energ. Rev.* **2013**, *19*, 370–384. [CrossRef]

34. Chowdhury, S.; Zhang, J.; Messac, A.; Castillo, L. Unrestricted wind farm layout optimization (UWFLO): Investigating key factors influencing the maximum power generation. *Renew. Energ.* **2012**, *38*, 16–30. [CrossRef]

35. Chuang, Y.C.; Chen, C.T.; Hwang, C. A simple and efficient real-coded genetic algorithm for constrained optimization. *Appl. Soft Comput.* **2016**, *38*, 87–105. [CrossRef]

Article

Development of an Efficient Numerical Method for Wind Turbine Flow, Sound Generation, and Propagation under Multi-Wake Conditions

Zhenye Sun [1], Wei Jun Zhu [1,2,*], Wen Zhong Shen [2], Emre Barlas [3], Jens Nørkær Sørensen [2], Jiufa Cao [1] and Hua Yang [1]

[1] School of Hydraulic Energy and Power Engineering, Yangzhou University, Yangzhou 225127, China; zhenye_sun@yzu.edu.cn (Z.S.); jfcao@yzu.edu.cn (J.C.); yanghua@yzu.edu.cn (H.Y.)
[2] Technical University of Denmark, 2800 Lyngby, Denmark; wzsh@dtu.dk (W.Z.S.); jnso@dtu.dk (J.N.S.)
[3] Goldwind Energy ApS, 8382 Hinnerup, Denmark; eba@goldwinddenmark.com
* Correspondence: wjzhu@yzu.edu.cn; Tel.: +86-139-2160-2334

Received: 26 November 2018; Accepted: 20 December 2018; Published: 28 December 2018

Abstract: The propagation of aerodynamic noise from multi-wind turbines is studied. An efficient hybrid method is developed to jointly predict the aerodynamic and aeroacoustics performances of wind turbines, such as blade loading, rotor power, rotor aerodynamic noise sources, and propagation of noise. This numerical method combined the simulations of wind turbine flow, noise source and its propagation which is solved for long propagation path and under complex flow environment. The results from computational fluid dynamics (CFD) calculations not only provide wind turbine power and thrust information, but also provide detailed wake flow. The wake flow is computed with a 2D actuator disc (AD) method that is based on the axisymmetric flow assumption. The relative inflow velocity and angle of attack (AOA) of each blade element form input data to the noise source model. The noise source is also the initial condition for the wave equation that solves long distance noise propagation in frequency domain. Simulations were conducted under different atmospheric conditions which showed that wake flow is an important part that has to be included in wind turbine noise propagation.

Keywords: wind turbine noise source; wind turbine noise propagation; wind turbine wake

1. Introduction

The generation and propagation of wind turbine aerodynamic noise exhibit several special characteristics as compared to some other industrial noise problems. Considering the generation of wind turbine aerodynamic noise, the rotor aerodynamic noise level depends on airfoil profiles at cross sections, blade shape (the spanwise distribution of chord, twist, and airfoils), rotor size, rotational speed, yaw angle, tilt angle, pitch setting, the angular position of blades, wind speed profile, inflow turbulence level, density, and viscosity of air. From the perspective of aerodynamic noise propagation (ANPropagation), aerodynamic noise source (ANSource) of large wind turbines is often located at high altitude which leads to more significant problem for ANPropagation over long range. Intrinsically, the study of wind turbine noise is a multi-disciplinary subject which should consider wind farm aerodynamic noise generation (ANGeneration), ANPropagation, and energy production. However, ANGeneration from a single airfoil has a similar basis in the nature of aerodynamics and aeroacoustics as that from a large wind farm. Sophisticated numerical methods are needed to understand the physical mechanisms of flow induced noise. Computational aero-acoustics (CAA) methods were widely applied favored by the high-performance computing technology. However, advanced large eddy simulation (LES)/CAA method still requires heavy computational effort.

Consequently, conventional CAA methods are neither efficient to model wind farm noise source, nor to predict long range ANPropagation. In the following paragraphs, feasible aerodynamic and aero-acoustic modeling techniques are briefly reviewed and the numerical methods employed in the current study are introduced.

As a prerequisite for wind turbine noise source and propagation modeling, a proper aerodynamic tool should be considered. There are many methods developed for wind turbine aerodynamic simulations. The classic blade element momentum (BEM) method [1,2] is a commonly used semi-engineering technique that simulates the aerodynamics of a single wind turbine. With a BEM model, one can apply engineering approaches to compute the noise source level of a wind turbine. When the prediction accuracy increases, the model complexity as well as the computational cost also increases. It is known that the CFD based methods predict the rotor aerodynamic forces and the wake flow behind the rotor more accurately. Considering long-distance ANPropagation of wind turbines, CFD methods become time consuming because the wake flow should be simulated in the large range whose size is the same as that of the ANPropagation. In the present study, the velocity field is the input data for the ANPropagation solver. Instead of conventional CFD methods, other sophisticated methods are developed for wind turbine applications, such as the AD/AL/AS (actuator disc /actuator line /actuator surface) methods, these methods have good trade-off between model accuracy and computational efficiency. The actuator disc method [3,4] is applied in this study which has relatively high computational efficiency. The rotor is assumed as a permeable disc. The aerodynamic loadings of the rotor are treated as external volume force terms of the momentum equations of Navier–Stokes (NS) equations. The rotor disc of a horizontal axis wind turbine (HAWT) has an inherent axisymmetric characteristic. In this study, to reduce computational time the 3D-AD method is simplified to 2D-AD. The boundary condition of rotor center is axisymmetric, such that the obtained flowfield with 2D-NS equations will be exactly identical to that of the 3D case. Since the axisymmetric assumption is made, it has to be mentioned that the solution only holds correct for axisymmetric uniform inflow conditions and rotor under no yaw and tilt manipulations.

Dealing with the modeling of ANSource, the CAA methods, based on high fidelity techniques [5–10], can model the physical details of flow induced noise mechanisms. Nevertheless, the atmospheric acoustic scientists mainly focus on sound propagation problems with larger scales or sizes [11–13]. It is true that the governing equations of most CAA based methods can solve both wind turbine ANGeneration and ANPropagation problems, but the heavy simulation cost often limits the computational domain into small size. It is clear that more efficient methods with less computational effort are needed. To satisfy this requirement, the flow passing wind turbine, the ANSource and the ANPropagation through long path are naturally combined together in this study. The engineering ANSource model for wind turbine [14–16] is built into the above mentioned AD flow solver. Such that the relative inflow velocity and AOA of each blade section are obtained from the AD-CFD simulation and then fed into the noise source model of wind turbine.

Afterwards, the relative sound pressure level (RSPL) from wind turbines is simulated over large distances from the turbine positions to the desired receivers. The wind turbine ANSource are the key inputs to the propagation solver [11,12]. The wind turbine wake flow, computed by the AD-CFD method, is considered as the medium of ANPropagation. The complex flow environment is found as an important factor for long-path wind turbine ANPropagation. In the previous study of Lee [17] and Heimann [18], the influences of atmospheric conditions on ANPropagation were investigated. Both studies showed that there is large noise variation under different wind profile and atmospheric stability conditions. More recently, this parabolic equation (PE) method is used by Barlas et al. [19,20] for wind turbine noise applications. It was clearly shown that the wind turbine ANPropagation has different refraction characteristics under various wake conditions. The present work focuses on an integrated method for modeling of flow, ANSource and ANPropagation, and the detailed phenomenon of ANPropagation through multi-wind turbine wakes are studied.

This paper is organized following the sequences: In Section 2, the numerical method for wind turbine flow simulation, ANGeneration and ANPropagation modeling are described; Section 3 provides results of various validation cases first and more results from the integrated flow-noise model afterwards; conclusions are provided in the last section.

2. Numerical Methods

The related numerical methods applied in the study are described separately in this section: (1) The RANS/AD model for wind turbine flow simulation; (2) The BPM (named after Brooks TF, Pope DS, Marcolini MA [14]) wind turbine ANGeneration model; (3) The wind turbine ANPropagation model. The relations between the solvers are shown in the schematic diagram below in Figure 1. It is seen that the flowfield data are the input to both the wind turbine noise source and propagation models.

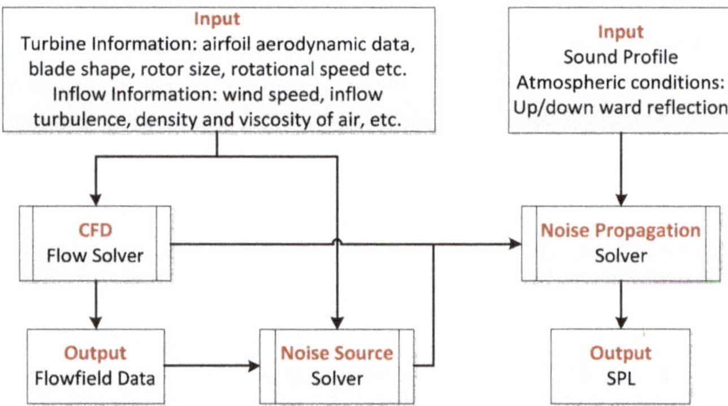

Figure 1. Schematic diagram of relations between the developed solvers.

2.1. RANS/AD Method

To simulate the flow passes wind turbine rotor, the general equations are modified by adding the aerodynamic loadings of the rotor as external volume force terms in the momentum equations of the NS equations

$$\frac{\partial u_i}{\partial x_i} = 0 \quad (i = 1, 2) \tag{1}$$

$$\frac{\partial u_i}{\partial t} + u_j \frac{\partial u_i}{\partial x_j} = -\frac{1}{\rho}\frac{\partial P}{\partial x_i} + \frac{\partial}{\partial x_j}\left(v\frac{\partial u_i}{\partial x_j} - \overline{u'_i u'_j}\right) + f_{ext} \tag{2}$$

$$-\rho\overline{u'_i u'_j} = \mu_t\left(\frac{\partial u_i}{\partial x_j} + \frac{\partial u_j}{\partial x_i}\right) - \frac{2}{3}\rho k\delta_{ij} \tag{3}$$

where x_i is the displacement in the inertial coordinate system and u_i is the velocity components. t, ρ, and P represent time, air density and pressure respectively. v is the kinematic viscosity term, f_{ext} is the added source term. With the assumption based on the axisymmetric property of the HAWT, 2D-RANS/AD method is utilized such that the computational effort can be greatly decreased.

The rotor is assumed as permeable disc such that the mass equation remains unchanged. To avoid numerical instability the volume force in Equation (2) should be smeared along the flow direction using 1D Gaussian function. For the elements which are away from the disc with a normal distance of d, the smeared force f' is computed by convolution computation

$$f\prime = f_{AD} \otimes \eta^{1D} \tag{4}$$

$$\eta^{1D}(d) = 1/\left(\varepsilon\sqrt{\pi}\right)\exp\left[-(d/\varepsilon)^2\right] \tag{5}$$

where the free parameter $\varepsilon = 3\Delta z$ is three times larger than the axial size of reference grid Δz. What is more, the body forces of rotor can be computed using blade element theory with the aerodynamic data of airfoil and the simulated information of flowfield.

The present 2D-AD method solves the ideal axisymmetric flowfield. The axial and tangential induced velocities are naturally combined with the NS equations, such that the inflow angle and the relative inflow velocity are determined from

$$\phi = \tan^{-1}\left(\frac{V_0 - W_z}{\Omega r + W_\theta}\right), \ V_{rel}^2 = (V_0 - W_z)^2 + (\Omega r + W_\theta)^2 \tag{6}$$

The local AOA is easily computed by $\alpha = \phi - \gamma$, where γ is the real twist angle (added by pitch) at local blade element. The lift and drag forces per unit span are computed using Equation (7) with the aerodynamic data of airfoil.

$$L = \frac{1}{2}\rho V_{rel}^2 cBC_l, \ D = \frac{1}{2}\rho V_{rel}^2 cBC_d \tag{7}$$

The blade forces are updated after each iteration using newly computed velocity field, then the external volume force terms of the NS equations apply renewed blade forces in the next iteration.

2.2. BPM Model for Noise Source or Generation

The original BPM noise model was developed by Brooks et al. [14] which is able to predict airfoil self-noise. The model was based on scaling the wind tunnel experimental data using the NACA 0012 airfoil with chord lengths varies from 2.5 cm to 61 cm. Here, the BPM model was integrated with the BEM method [15] in order to deal with the complicated flow for wind turbine. The required boundary layer thickness data in the BPM model is obtained from the Xfoil computation so that this data varies with the airfoils rather than fixed with NACA 0012. A general expression of the BPM model is shown in Equation (8)

$$SPL = 10log_{10}\left(\frac{\delta M^n lD}{r^2}\right) + G_1(St) + G_2(Re) + G_3(\delta) + C \tag{8}$$

In the equation, it is shown that rotor aerodynamic noise is mainly a function of parameter δ which represents boundary layer thickness. The boundary layer thickness depends on airfoil shape, Reynolds number, AOA, turbulence level etc. l is the spanwise length of an airfoil segment. G_1, G_2, and G_3 are related to the Strouhal number St, Mach number M, Reynolds number Re and boundary layer thickness parameter δ. Detailed descriptions of noise mechanisms can be found in references [14,15]. ANSource level of rotor generally depends on the blade shape and operational conditions, such as rotational speed, wind speed, yaw, tilt, etc. Instead of combining the engineering BEM model, the BPM noise model is directly integrated into the NS equations with the AD model. The BPM model is a passive subroutine that can be activated anytime with the inflow velocity and AOA provided from AD simulations. The trailing edge boundary layer thickness is interpolated from the existing database prepared with Xfoil through a wide range of Reynolds numbers and AOAs. To take clean and rough blade surface conditions into account, the boundary layer thickness database is created under different surface roughness cases, such that wind turbine noise under free transition and fully turbulent flows can be modelled. A fully-turbulent boundary layer thickness database is utilized to calculate the noise of rotor whose blade surface is rough, and free transition database is applied for a rotor with a clean surface.

2.3. Noise Propagation Method

Sound speed in an ambient flow is written as $c_{eff} = c + v_x$ where c is the constant sound speed at a given air temperature, v_x is the wind speed component through the propagation path. The wind speed components may come from field measurements, empirical expressions or numerical simulations.

For an incompressible wind flow, the air density is assumed as a constant, the wave equation in a moving atmosphere reads [12]

$$\left[\nabla^2 + k^2(1+\epsilon) - \frac{2i}{\omega}\frac{\partial v_i}{\partial x_j}\frac{\partial^2}{\partial x_i \partial x_j} + \frac{2ik}{c_0}v \cdot \nabla\right]P'(r) = 0 \tag{9}$$

where $k = \omega/c_0$, ω is the radian frequency of sound wave, c_0 is the reference sound speed, $P'(r)$ denotes the monochromatic field of sound pressure and $\varepsilon = (c_0/c)^2 - 1$. If the effect from ambient flow is neglected, this equation can be further reduced to Helmholtz equation. The ANSource of a wind turbine is placed on the left boundary of the computational domain depicted in Figure 2. A classical approach to represent a monopole ANSource is to distribute the source along the vertical direction which is also called starting field. In this study, the rotor noise source is located at the hub height as sketched at the left side of the boundary. Based on the initial source strength, the equation or the sound pressure amplitude is computed along the x-direction, such that solution is renewed from $P(x)$ to $P(x + \Delta x)$. The boundary condition applied on the ground surface is calculated from the impedance of the ground. The empirical impedance model for absorbing materials developed by Delany and Bazely [21] is used in this study,

$$Z = 1 + 0.0511\left(\frac{\sigma}{f}\right)^{0.75} + i\,0.0768\left(\frac{\sigma}{f}\right)^{0.73} \tag{10}$$

where Z is the normalized acoustic impedance, σ represents the flow resistivity and f is the frequency. The similar approach can be applied on the top surface. A constant of the normalized acoustic impedance $Z = 1$ is implemented on the top grid, such that the vertical waves through the top surface is vanished without any reflection. However, to sufficiently eliminate wave reflections, an absorbing layer is defined on top of the grid. As shown in the figure, the grey region sketched on top of the domain is the absorbing layer. In the study, the width of such an absorbing layer is defined as 50λ. This implies that for lower frequency sound propagation, a larger absorbing layer is needed. The sound pressure along the vertical line is updated from step n to $n + 1$ until the receiver position is reached. The solution procedure starts with the separation of the forward and backward propagation waves. Here, only the forward wave propagates from source towards receiver is considered (in the positive x-axis), which is in the downwind direction. Detailed mathematical manipulations of the equation are found in [12,13].

According to the methodologies described in Sections 2.1 and 2.2, the noise spectrum of a given wind turbine is the logarithmic sum of sound pressure level (SPL) from all the blade elements such that

$$L_p(f) = 10log_{10}\left(\sum_i^n 10^{0.1SPL_{total}^i}\right) \tag{11}$$

where n is the number of the blade elements and SPL_{total}^i is the SPL of the ith blade element, which accounts for all the noise mechanisms. The SPL of wind turbine ANSource $L_p(f)$ is the initial condition for calculating long-path ANPropagation. The sound power level $L_w(f)$, which is a representation of wind turbine total noise strength, it does not change with propagation range. The SPL at the receiver is

$$L_p(f) = L_w(f) - 10log_{10}4\pi D^2 - \alpha D + \Delta L \tag{12}$$

where the first term on the right side is the sound power level, the second term $-10\log_{10} 4\pi D^2$ is the geometric attenuation term that corresponds to the sound intensity loss when the sound wave sphere is expanding. The third term shows the attenuation due to air absorption, which is proportional to the absorption coefficient α and the propagation length D. ΔL is the relative SPL differences caused by factors such as ground reflection/absorption (flat/irregular terrain), atmospheric refraction, wind and turbulence and sound barriers, etc. It does not include the effects of geometric attenuation and air absorption. The ΔL varies along the propagation path with these factors, which is beneficial to clearly manifest the effects of propagation. Most of the discussions and results in Section 3 are dealing with this relative pressure level which is abbreviated as RSPL. The values of α depend on sound frequency, air temperature, pressure and humidity. In the following studies, unless specified separately, a relative humidity of 70% and temperature of 10 °C are assumed.

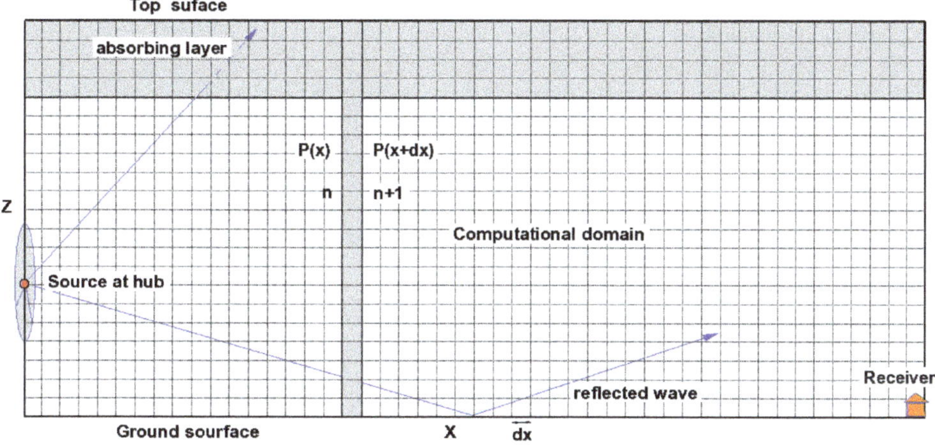

Figure 2. Demonstration of grids in the plane of propagation utilized in the simulations.

The wave equation is solved based on each frequency. The ANSource and wake information of given wind turbines from BPM/AD computations are combined into the ANPropagation solver. For each frequency, the corresponding ANSource, known as the starting field, is individually applied as the initial condition to the computational domain. Detailed descriptions about starting field formulation can be found in publication from Salomons [11]. The horizontal size of the simulation range is determined by the location of the receiver. However, the size of the vertical domain is flexible such that the height of domain can vary with the simulated frequency. The maximum vertical size of the grid should be less than $\Delta x = \Delta z = \lambda/8$ so that a sound wave can be resolved with at least eight grid points. As a result, the domain height is inversely proportional to frequencies for a fixed number of vertical grids. Besides, a larger CFD domain should be used in the simulation such that the flow field can fully cover the noise propagation domain.

3. Results and Discussions

A few validations about ANPropagation cases are shown first. Under various flow conditions, the flow passing wind turbine rotor, ANGeneration and ANPropagation problems are discussed in detail. The RSPL ΔL over a long distance is computed for each frequency. After the validations, the wind turbine ANPropagation across multi-wakes is shown. In all the cases, the temperature at the ground level is 15 °C, such that the equivalent sound speed is 340 m/s.

3.1. Validation of Long-Range Sound Propagations without Ambient Flow

This validation case considers the idealized long-range sound propagation without any moving medium or temperature gradient effects. The terrain contains a flat and acoustically soft surface with a total length of 200 m. The flow resistivity of a grassland are usually in range from 100 kPa · s · m^{-2} to 300 kPa · s · m^{-2}. The absorbing ground surfaces in the current simulation has a flow resistivity of 250 kPa · s · m^{-2}. The ANSource locates at $x = 0$ m, $z = 0.5$ m and the receiver at $x = 200$ m, $z = 1.5$ m. The current results are compared with the data presented in reference [22] where either measured data or data obtained from other models are available. The sound transmission losses over 200 m are shown in Figure 3 where the selected frequencies in the 1/3-octave band are gathered. At very low frequencies, i.e., f = 20 Hz and f = 40 Hz, the RSPL ΔL is increased with the propagation distance. At higher frequencies, ΔL gradually decreases along the propagation distance. The ΔL values at 200 m are compared in Figure 4 which shows that there is a large propagation loss between 400 Hz and 1000 Hz. The present simulation agrees with the reference data and the simulation obtained from the Nord2000 commercial model [23].

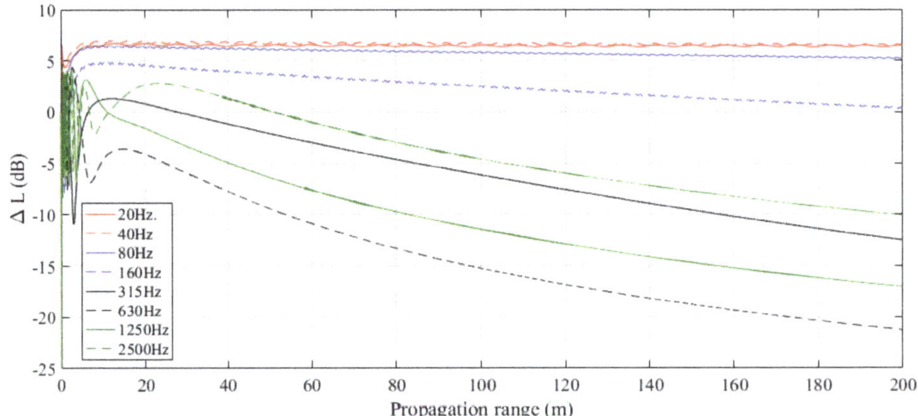

Figure 3. Sound transmission loss over 200 m at various frequencies (soft ground).

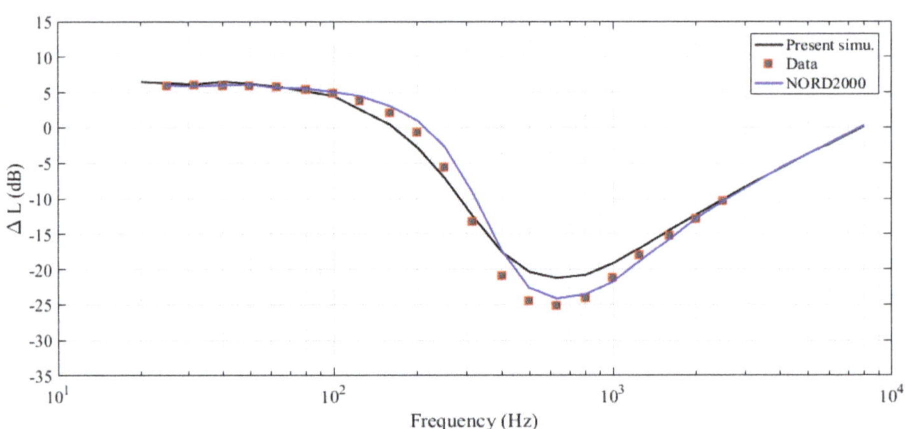

Figure 4. Comparisons of the RSPL in 1/3-octave bands.

3.2. Validation of Long-Range Sound Propagations with a Positive Temperature Gradient

In this case, the terrain distance, source, and receiver locations remain the same, but the ground is an acoustically hard surface, meaning that the flow resistivity is close to infinity. The temperature gradient was measured as 0.0846 K/m. If a linear temperature variation is assumed, the temperature and sound speed profiles are easily computed using the following relations

$$T = \frac{\partial T}{\partial z} \cdot z + T_0 \tag{13}$$

$$c = c_0 \cdot \sqrt{T/T_0} \tag{14}$$

Such a temperature gradient leads to a downward sound propagation. As a result, the RSPL along the propagation direction will probably increase. In Figure 5, it is seen that there is a general increase of ΔL along the propagation path except near the source locations ($x < 40$ m). The ΔL values are gathered at $x = 200$ m distance and are depicted in Figure 6. It is found that the ΔL values are positive between 20 Hz and 5 kHz. The current prediction fits well with the reference data.

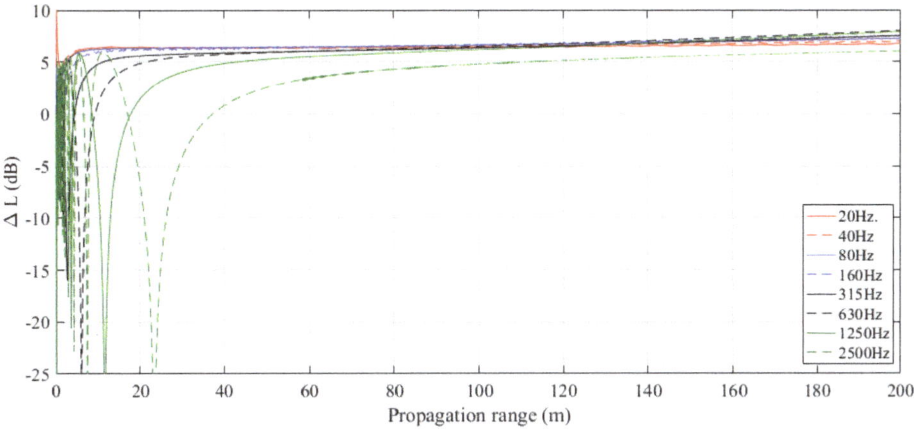

Figure 5. Sound transmission loss over 200 m at various frequencies (rigid ground and downward refraction).

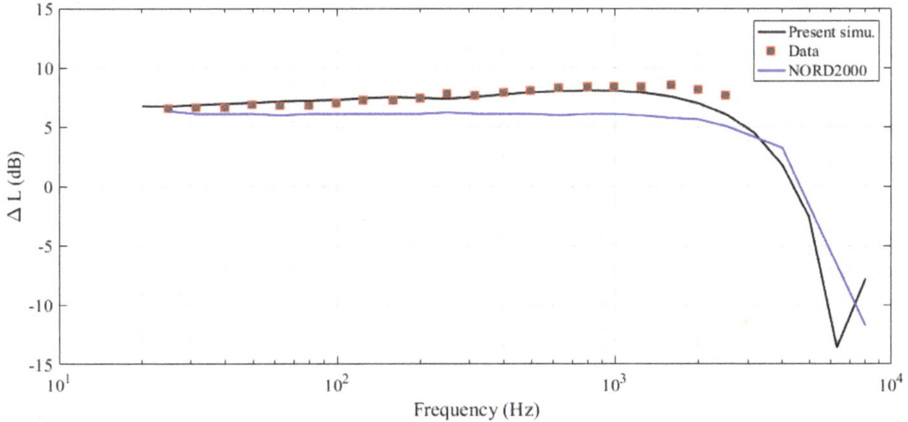

Figure 6. Comparisons of the RSPL in 1/3-octave bands at a distance of 200 m.

3.3. Validation of Long-Range Sound Propagations with Upward Refraction under Wind Condition

Similar as shown in the previous test case, the ground acoustic impedance is also infinity. However, an upward refraction case is considered which is caused by a moving atmosphere. The wind speed recorded at 10 m height above the ground is -4.165 m/s, and the surface roughness length is 0.1 m. In a moving medium, the sound speed can be computed using the equation below

$$c_{eff}(z) = c_0 + b \ln\left(\frac{z}{z_0} + 1\right) \tag{15}$$

where c_{eff} is the profile of effective sound speed along the vertical direction, $c_0 = 340$ m/s is the sound speed and $z_0 = 0.1$ m is the roughness height. Knowing the wind speed at 10 m height and the roughness length, it is easy to show that $b \approx -1$. Such a sound speed profile represents a slightly upward atmospheric refraction. In Figure 7, except at very low frequencies, the RSPL decays very fast along the propagation path. Figure 8 shows a very good agreement with the reference data which indicates the capability of the current method to handle noise propagations through moving media.

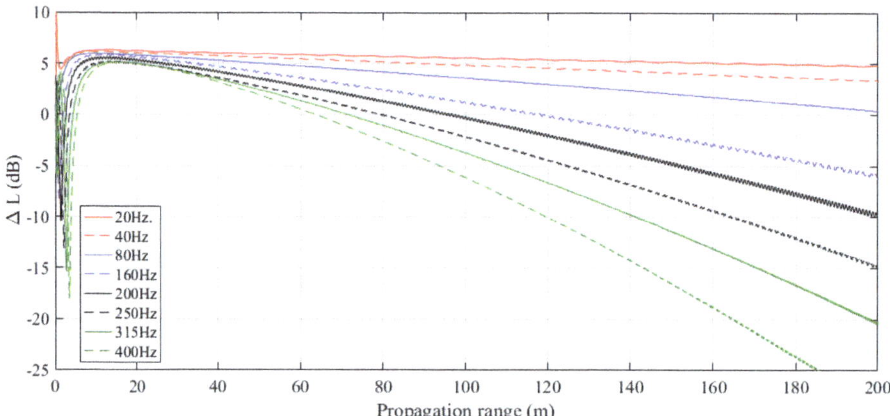

Figure 7. Sound transmission loss over 200 m at various frequencies (rigid ground and upward refraction).

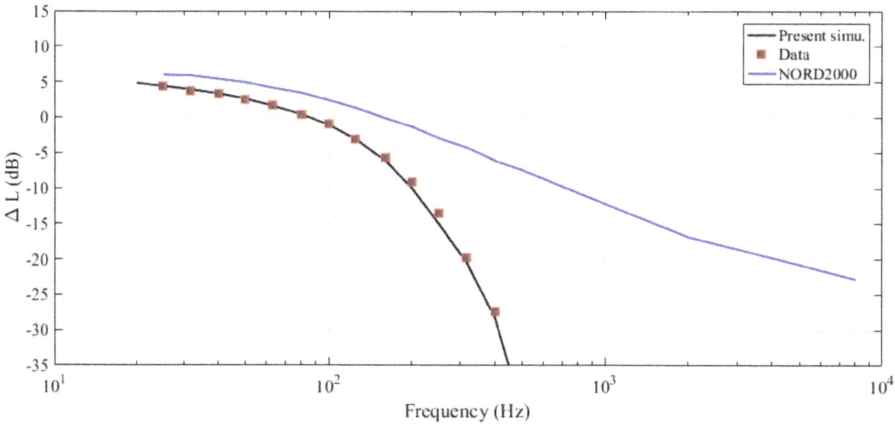

Figure 8. Comparisons of the RSPL in 1/3-octave bands.

3.4. Coupled Flow and Noise Simulations for a Wind Turbine Array

3.4.1. Wake Modeling of a Wind Turbine Cluster

Wind turbine flow are simulated for a Vestas wind turbine called NM80. The AOA and the relative velocity at all blade elements from RANS/AD simulations are inputted to the BPM model. The AD solver is much more efficient because it represents the rotor by volume force term rather than the dense body-fitted-mesh around blade surface. The flow computations are carried out with the in-house developed flow solver [24–26] which is an incompressible flow solver.

The grid for AD simulation and a typical flowfield with multi-wake structures are depicted in Figure 9. In Figure 9a, two wind turbines, depicted with thick white lines, are used for demonstration. The inflow boundary is applied on the left and top domain edge (numbered as 1 in Figure 9), the axisymmetric boundary on bottom edge (numbered as 2) and outflow boundary on the right edge (numbered as 3). The computational domain consists of the several blocks whose number is the same as that of wind turbines to be investigated. Each block contains grid points of 64 × 64. Using this approach, multi-wake flowfield can be efficiently modelled. A typical horizontal velocity contour plot is shown in Figure 9b.

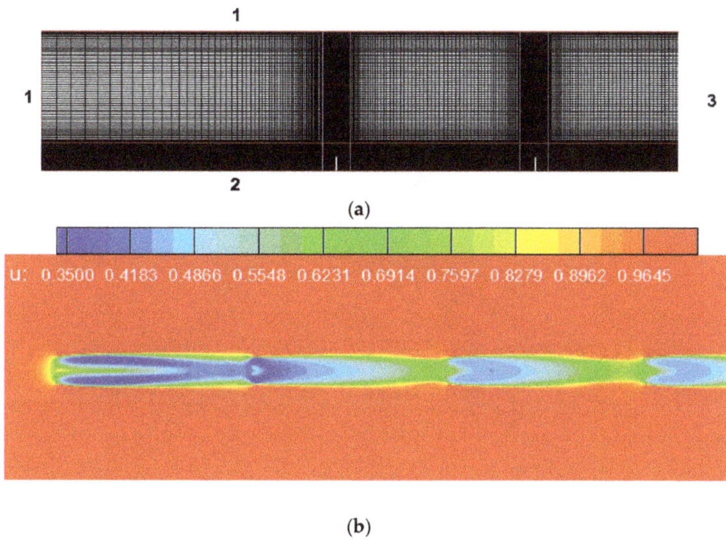

Figure 9. (**a**) Mesh topology; (**b**) stream-wise flow with multi-wakes.

To show the model accuracy, in Figure 10, some results from the current simulations using RANS turbulence model are compared with the LES results under same flow conditions [27,28]. On the modeling of wind turbine wake, many studies can be found [29–31]. In the present study, it is not of our primary interest to go into more details on this topic. Some more details of the AD modeling approach and cross comparisons can be found from our previous works [32–34].

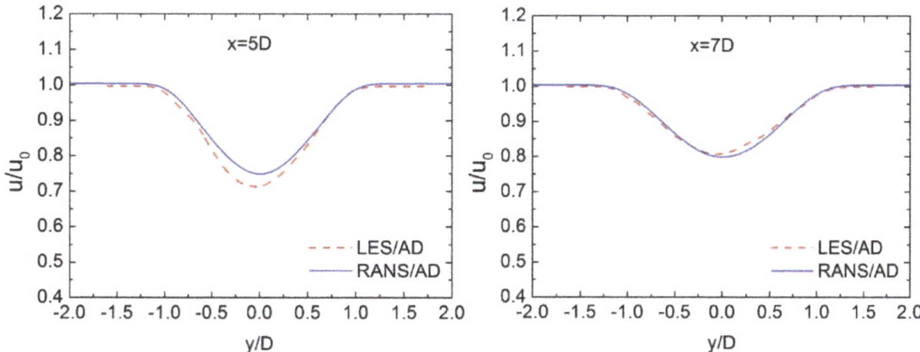

Figure 10. Comparisons of the normalized wake velocity from LES and RANS simulations, for down-stream locations of 5D and 7D after a single wind turbine.

3.4.2. Modeling of the Wind Turbine ANSource

As is shown in the above AD simulation, every wind turbine has different inflow conditions because some wind turbines are in the wake of another, which leads to different ANSources.

The simulation were conducted for a free-stream wind velocity of 8 m/s. For a receiver at 120 m far away, the detailed noise spectra solely from the first upstream wind turbine are shown in Figure 11. The noise spectra for other downstream wind turbines will be shown in the later sections. As shown in the figure, the total SPL spectrum is name as SPL-TOTAL which is logarithmically summed from several noise mechanisms. These noise mechanisms include: noise from blade tip (SPL-TIP); from trailing edge bluntness (SPL-TEBLUNT); from the shedding of laminar boundary layer vortex (SPL-LBLVS); from flow separation (SPL-SEPARATION); from turbulent boundary layer trailing edge (SPL-TBLTE); from turbulent inflow (SPL-INFLOW) [14–16]. Later, the total noise spectrum will be applied into the noise propagation computations.

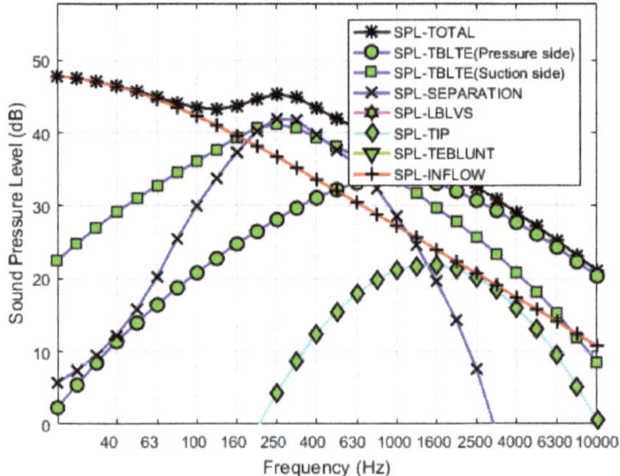

Figure 11. Aerodynamic noise source spectra of a single wind turbine.

3.4.3. Sound Profiles Inside wind Turbine Wakes

Although the flow and the noise source are modelled without wind and temperature gradients, in reality the effective sound speed profile will be affected by wind and temperature profiles. It was shown that the combinations of wind speed and temperature gradient forms different types of sound speed profiles [35]. Three idealized initial sound profiles are considered in the following sound propagation simulations. In Figure 12a, a uniform sound speed at temperature of 15 °C is assumed at the left boundary. The wind turbines are separated with a normalized distance of seven rotor diameters. For example, the second wind turbine is located at $x = 560$ m and the third wind turbine is located at $x = 1120$ m. Inside the computational domain, the sound speed is superimposed with the wind turbine wake field. Similarly, Figure 12b,c represent downward and upward sound speed contours. These sound profiles merge with the wind turbine multi-wakes and show distinct characteristics. The sound profiles have significant effects on wind turbine noise propagation as shown in the next subsection.

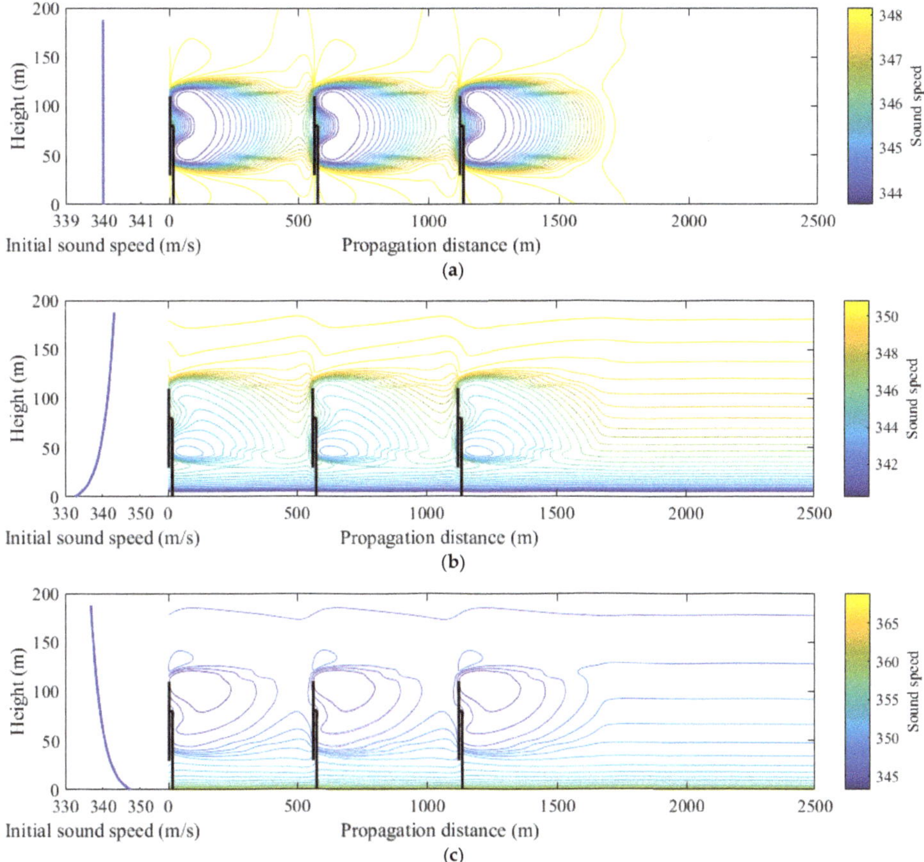

Figure 12. Initial sound speed profiles at upstream and effective sound speed contours in the wake flow. (**a**) Uniform initial sound profile transmitted into wake; (**b**) upward refraction sound profile transmitted into wake; (**c**) downward refraction sound profile transmitted into wake.

3.4.4. Wind Turbine ANPropagation across Multi-Wake

As shown in Figure 13, the RSPL or ΔL is simulated along 1500 m propagation distance, at $f = 300$ Hz and with the noise spectrum in Figure 11. The first wind turbine locates at $x = 0$ m, the second at 560 m and the third at 1120 m. The distance between each wind turbine are relative large. The wind turbine ANSource are placed at a height of 70 m. The ANSources from the second and third wind turbines, depicted as white lines, are not considered so as to focus on the propagation phenomenon through the wakes. In Figure 13, RSPL results are shown under several conditions: (1) Case 1, ANPropagation under homogeneous atmospheric condition, no wake effect; (2) Case 2, ANPropagation under homogeneous atmospheric condition, considering multi-wakes; (3) Case 3, ANPropagation under atmospheric condition of down-ward refraction, considering multi-wakes; (4) Case 4, ANPropagation under atmospheric condition of upward refraction, considering multi-wakes. It is shown that the wind turbine ANPropagation is clearly affected by the complicated flow environment. For the case without any wake effect shown in Figure 13a, the ANPropagation is mainly influenced by the reflection and absorption of ground. As is shown in Figure 13b, the contours of RSPL is more complicated than that of Figure 13a which distinctly shows the influences of wake. For atmospheric condition of downward refracting depicted in Figure 13c, the 'noise pollution' band also shows a downward reflecting behavior. Additionally, a low-noise zone, shown in blue, can be found near $x = 1000$ m. In Figure 13d, a much quieter situation is observed under an upward refraction condition. In such a case, the quiet zone is located between wind turbine #2 and #3. It should be emphasized that uniform inflow boundary condition is applied in the flowfield simulations of Figure 13c,d. However, the un-uniform or down/up-ward atmospheric conditions, which means atmospheric altitude-dependent variation of the sound speed, are used in the ANPropagation computations.

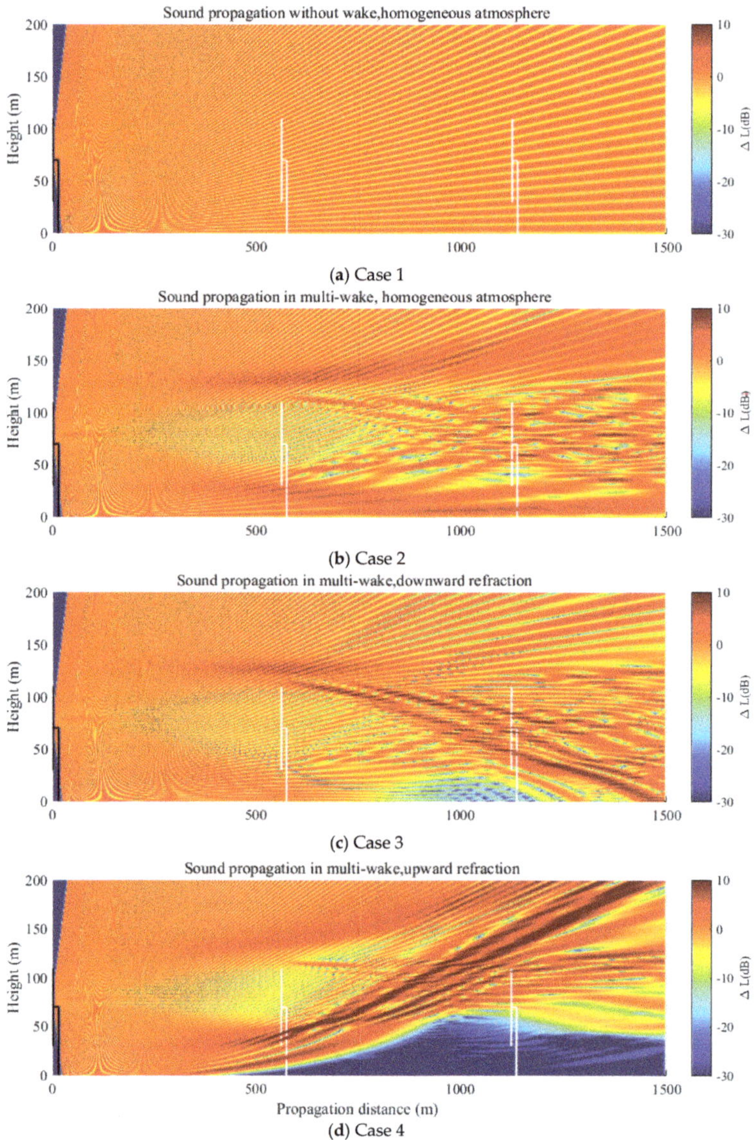

Figure 13. ANPropagation from the first ANSource: (**a**) Case 1, homogeneous atmospheric condition, no wake effect; (**b**) Case 2, homogeneous atmospheric condition with multi-wakes; (**c**) Case 3, downward refraction and multi-wakes; (**d**) Case 4, upward refraction and multi-wakes.

Manifesting the noise received by humans, the RSPL or ΔL at $h = 2$ m is extracted from Figure 13 and shown in Figure 14.The sound attenuation ΔL of Case 1 starts to vary linearly with the propagation distance at around $x = 400$ m, because the reflection effect becomes weaker after that. In contrast, the ΔL of Case 2, 3, and 4 vary obviously even after a long distance. In Cases 2 and 3, ΔL fluctuates over distance due to the wake effect and the strong downward refraction respectively. Whereas in Case 4, a more significant RSPL loss is shown which is caused by the upward refraction. It should be mentioned that the ΔL values might be quite different when the observer height changes. From this comparison, it can be concluded that ANPropagation can be greatly influenced by the environmental conditions. What is more, the change of SPL in the near downstream is much smaller than that at far downstream ($x > 400$ m), which shows the significance of long distance wind turbine noise predictions under complex atmospheric conditions. The near field ($x < 400$ m) simulations and measurements only belong to ANSource study which is not identical to the long-path ANPropagation.

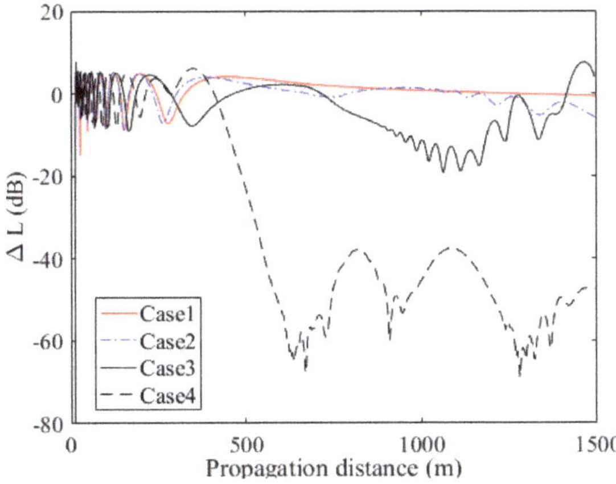

Figure 14. The RSPL at 2 m height extracted from four cases of Figure 13.

Result shown in Figure 14 represents the ΔL solely due to propagation which is computed at a fixed frequency of $f = 300$ Hz. Repeating the calculation for many frequencies and extracting the data at a receiver 1500 m away and 2 m height, a full spectrum of ΔL will be obtained. Using Equation (12) and considering all the effects (including ΔL), the SPL at receiver's location under Case 1 and Case 3 is gathered and shown in Figure 15. Besides, the spectrum of ANSource and at receiver's location (without ΔL) are also depicted in Figure 15. In this work, the ANSource is considered in an 1/3-octave band from 20 Hz to 5000 Hz. Since a minimal eight grid points are needed to resolve a sound wave, the grid number for high-frequency simulations increases sharply if the same domain size is adopted. For instance, there is a 100-fold increase in the grid number for frequency of 3000 Hz compared with that of 300 Hz. To reduce the computational effort, frequencies above 5000 Hz are not considered here. Another reason is that the noise reduction due to air absorption, $-\alpha D$ in Equation (12), is much larger than the other factors at large frequency ($f > 5000$ Hz). The SPL at the source location is shown with black line with asterisk and labeled as 'Source: Lp' in Figure 15. The black lines with stars represent the SPL at receiver (2 m height at $x = 1500$ m) without considering the propagation loss ΔL in Equation (12). The blue line (Case 1) and red line (Case 3) consider all the factors, including geometric attenuation, air absorption and ΔL, in the long-path ANPropagation simulations. The big deviation between the two black lines demonstrates that the geometrical attenuation ($10 \log_{10} 4\pi D^2$) contributes most to the ANPropagation loss. The even larger reduction of noise, when $f > 1000$ Hz, means that air absorption has a larger effect at high frequencies. The differences between blue/red lines and the black line with stars implies the value of ΔL. The value of blue and red lines are larger than that of black line with stars, which means positive ΔL near low frequency (f = 30 Hz). However, Case 1 and Case 3 show opposite ΔL around f = 200 Hz which implies the influences of wake and downward refraction.

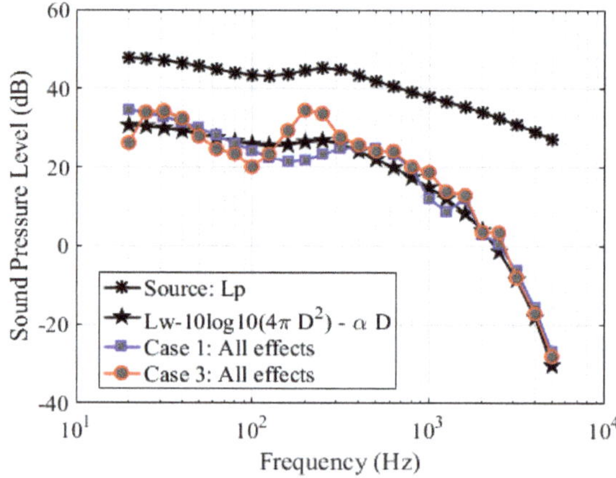

Figure 15. SPL of wind turbine #1 (Source: Lp) received at 1500 m distance (2 m height): with geometric attenuation and air absorption, Case 1 and Case 3 with all propagation effects.

Similar as the previous simulations, the following study focuses on the noise propagation area between 1120 m and 2500 m, and a higher frequency f = 1000 Hz is selected. In this region, the three individual wind turbine noise sources propagate and superimpose on each other. In Figure 16a, the results cover the entire domain behind the last wind turbine. The solution differs from Figure 13a in the sense that it is the RSPL from all the three wind turbines. In Figure 16b, the uniform sound profile at the inlet boundary is affected by the multi-wake (see Figure 12a), such that the resulted RSPL is enhanced around wake center. With the prescribed sound profile of Figure 12b, the RSPL is refracted downwards, as shown in Figure 16c. Similarly, using the sound profile given in Figure 12c, the resultant sound propagation is refracted upwards, see Figure 16d.

Figure 17 better illustrates the superposition of the sound pressure propagates from each wind turbine, which is a decomposition of Figure 16d. In the first figure, noise is generated from wind turbine #1 and propagates through the wake flow. The other two wind turbines are depicted with white lines, which imply that they are currently excluded for noise propagation. In the next simulation, shown in Figure 17b, ANPropagation from wind turbine #2 is individually considered, which excluded noise propagation from wind turbine #1 and #3. Similarly, wind turbine #3 is then considered for ANPropagation, which does not take account of the two upstream wind turbine noise propagations. The total RSPL is the summation of sound pressure following the logarithmic sumation rule, which is shown in Figure 17d.

Figure 18 shows the RSPL or ΔL extracted from a line at 2 m high. The overall sound pressure variation for different cases is smaller as compared to the previous results shown in Figure 14. This is due to the mixing of three individually simulated propagation field. Among all the test cases, Case 3 can be regarded as the worst-case scenario. For example, around the distance of x = 1800 m, there is an obvious increase of SPL due the strong downward refraction at this region, which is also observed in Figure 16c at x = 1800 m.

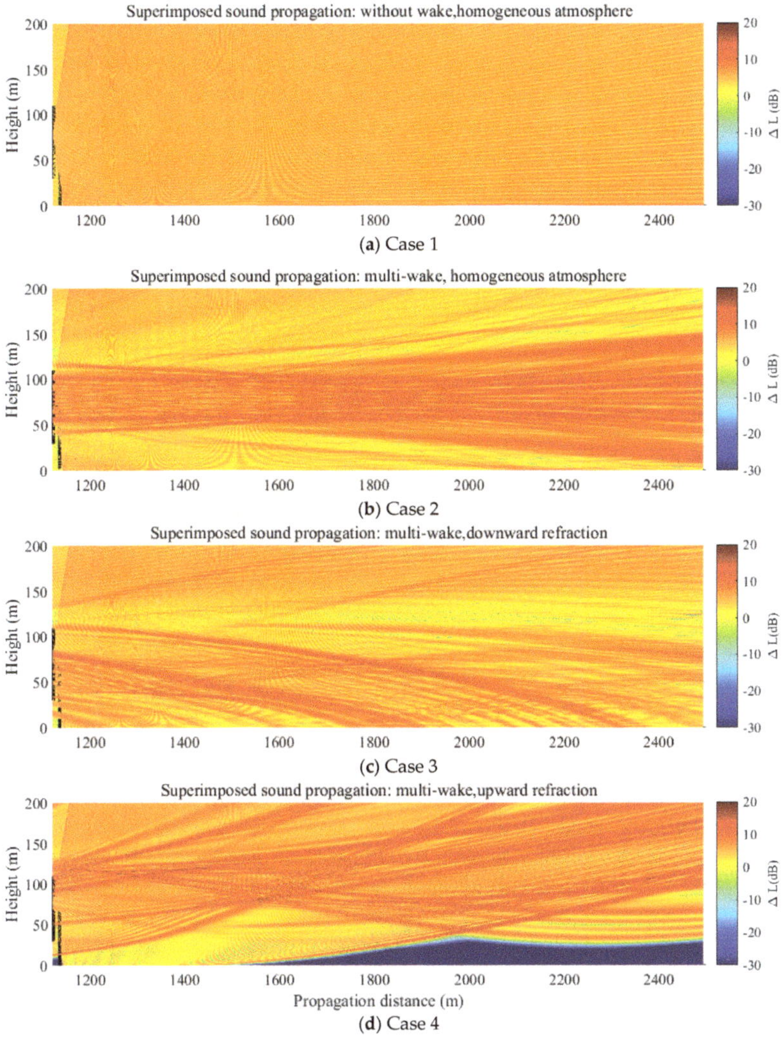

Figure 16. ANPropagation from all ANSources: (**a**) Case 1, homogeneous atmospheric condition, no wake effect; (**b**) Case 2, homogeneous atmospheric condition with multi-wakes; (**c**) Case 3, downward refraction and multi-wakes; (**d**) Case 4, upward refraction and multi-wakes.

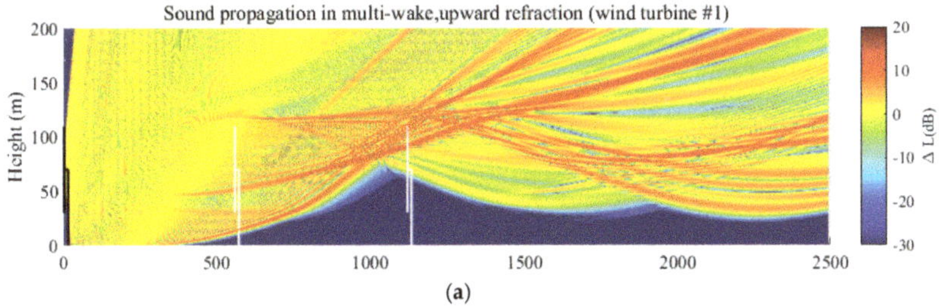

Figure 17. *Cont.*

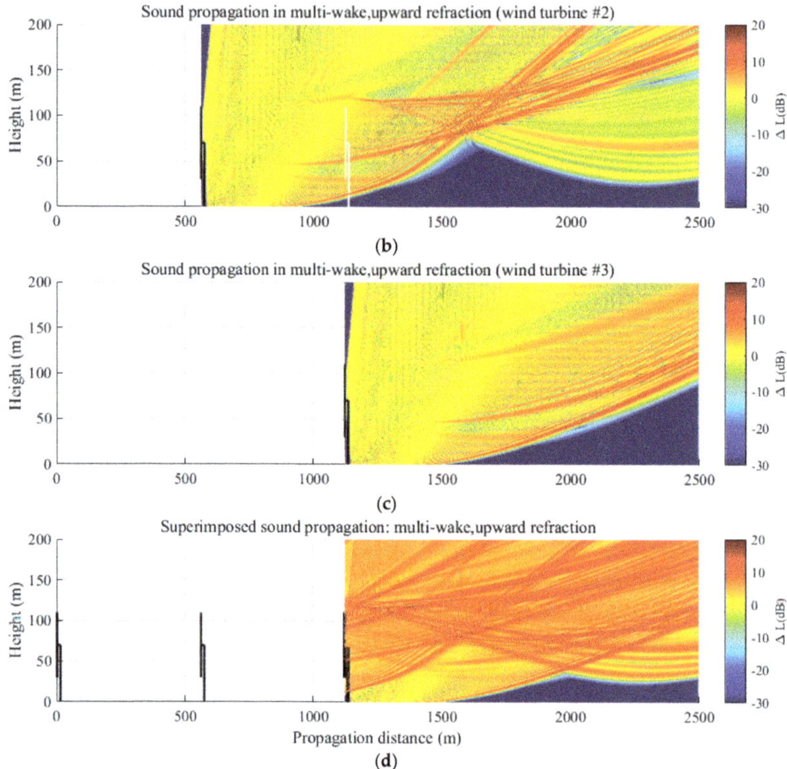

Figure 17. RSPL simulated at f = 1000 Hz and b = −3: (**a**) ANPropagation by wind turbine #1 downstream of the wakes of wind turbines # 1, 2, 3; (**b**) ANPropagation by wind turbine #2 downstream of the wakes of wind turbines #2, 3; (**c**) ANPropagation by wind turbine #3 downstream of the wakes of wind turbine #3; (**d**) superimposed RSPL from all wind turbines #1, 2, 3.

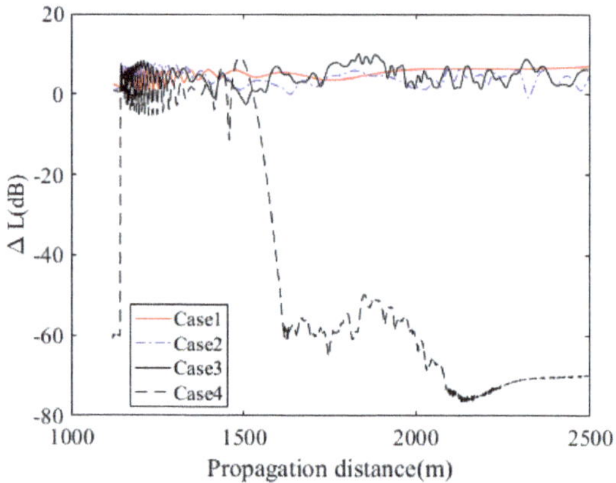

Figure 18. The RSPL at 2 m height extracted from four cases of Figure 16.

Similar to the procedures previously mentioned for Figure 15, Figure 19a–c show the noise spectra from wind turbines #1, 2, and 3, respectively. The difference is that the three wind turbines are individually considered for its noise generation and propagation at the given frequencies (20 Hz–

5000 Hz). Due to the presence of velocity deficit in the wake, the inflow conditions of the two downstream wind turbines are slightly different, and thus the ANSources of the three wind turbines are slightly different from each wind turbine. In each figure, the original ANSource is shown on top of the other noise spectra. Similar to the situations in Figure 15, geometrical attenuation plays an primary role in the large noise reduction for ANPropagation at $f < 1000$ Hz, and the noise absorption by air has a strong effect at $f > 1000$ Hz. As expected, Case 3 shows more distinct characteristics for the frequencies. In Figure 19, the receiver is fixed at $x = 2500$ m, $z = 2$ m, therefore the sound pressure is largely attenuated for wind turbine #1 due to the longest propagation distance. A similar behavior is seen for wind turbines #2 and #3, where the noise level is the highest for wind turbine #3 due to the shortest relative distance from source to receiver. Figure 19a–c also show that at different relative propagation distances 2500 m (#1), 1940 m (#2) and 1380 m (#3), the distribution of SPL maintains some similarity. The differences of these sound spectra come from the combined effect of wake flow and the propagation distance. Finally, the total noise received from the three wind turbines can be got through logarithmically adding the sound spectra from individual wind turbines. In Figure 19d, the sources Lp1, Lp2, and Lp3 represent the ANSource level of the three wind turbines. The major difference is seen at the middle frequency range, which is due to the different input velocity data obtained from flow simulation. Case 1 and Case 3 all show that the most significant influence comes from wind turbine #3, which is not a surprise. To highlight more detailed difference in the noise spectra, a few typical values are compared in Table 1. At $f = 20$ Hz, the summated SPL from three turbines (Case 3) is 37.92 dB which is 3 dB higher than that from wind turbine #3 (Case 3). This difference becomes smaller when frequency increases. This implies that the low frequency noise generated from the far upstream wind turbines still plays an important role. On the other hand, if a simple sound propagation algorithm is applied, such as shown in the last column of Table 1, a large error is observed in a frequency range below 1000 Hz. It indicates that the high frequency noise will be heavily attenuated over long propagation distance except in special cases such as sound propagation over water (hard surface), stronger downward refraction under special climate conditions.

Table 1. Comparisons of SPL at some selected frequencies

	Case 3 (Figure 19c)	**Case 3 (Figure 19d)**	$L_w - 10\log_{10} 4\pi D^2 - \alpha D$ **(Figure 19d)**
20 (Hz)	34.64 dB	37.92 dB	33.89 dB
100 (Hz)	12.61 dB	15.05 dB	28.53 dB
250 (Hz)	26.25 dB	28.33 dB	27.57 dB
1000 (Hz)	17.10 dB	17.99 dB	18.00 dB

Finally, it is interesting to predict the noise spectra at different receivers. The receiver height is again fixed at $z = 2$ m, and the horizontal distances measured from wind turbine #1 are: $x = [1620, 1820, 2020, 2220, 2420]$ m. The noise spectra are shown in Figure 20 and marked with different colors. The general trend is that the noise level decreases as observer distance increases. However, it does not always hold true at some distances and frequencies. For example, at $x = 1820$ m (22.75D), except at low frequencies, increase of noise level is observed as compared to the $x = 1620$ m case. The simulations show that the low frequency noise might be the major wind turbine noise source in the farfield. Since the spectra are obtained in farfield, the overall SPLs are quite low. According to the wind turbine noise regulations in many countries (such as 45 dB), the calculated noise evidently fulfill the limit. From the results, it is seen that SPLs over 1000 Hz is below 20 dB. For future studies, it is indicated that more focus should be paid on long-path ANPropagation at low frequencies.

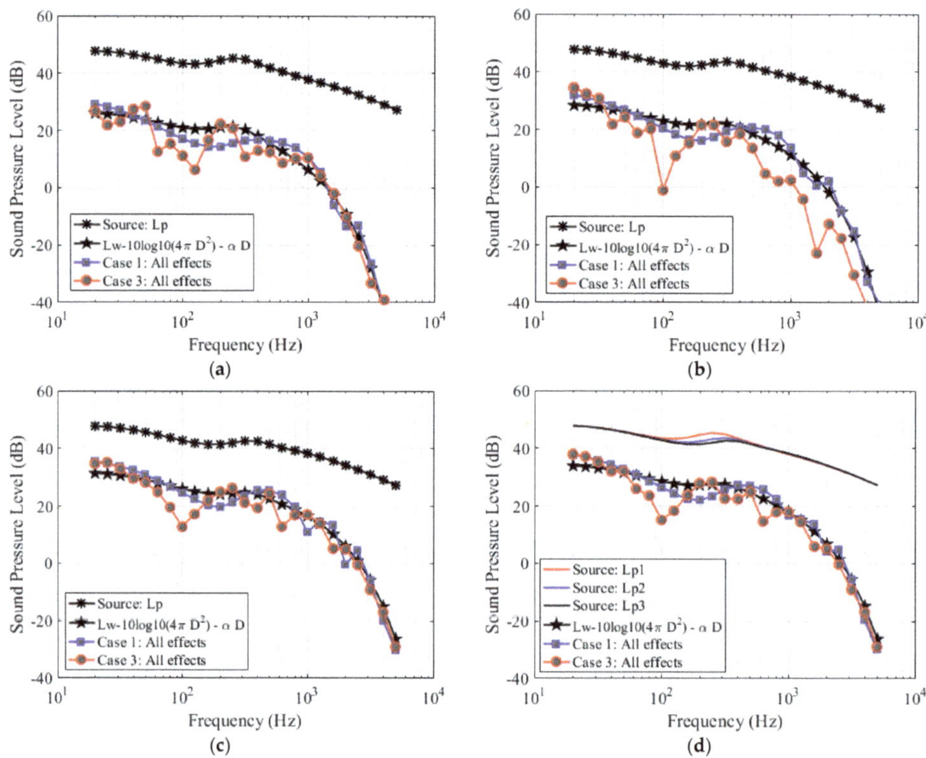

Figure 19. Receiver is fixed at $x = 2500$ m away from wind turbine #1, and the receiver height is 2 m. (**a**) SPL from wind turbine #1 (Source: Lp) to the receiver; (**b**) SPL from wind turbine #2 (Source: Lp) to the receiver; (**c**) SPL from wind turbine #3 (Source: Lp) to the receiver; (**d**) Total SPL from wind turbines #1, 2, 3 (Source: Lp1, Lp2, Lp3).

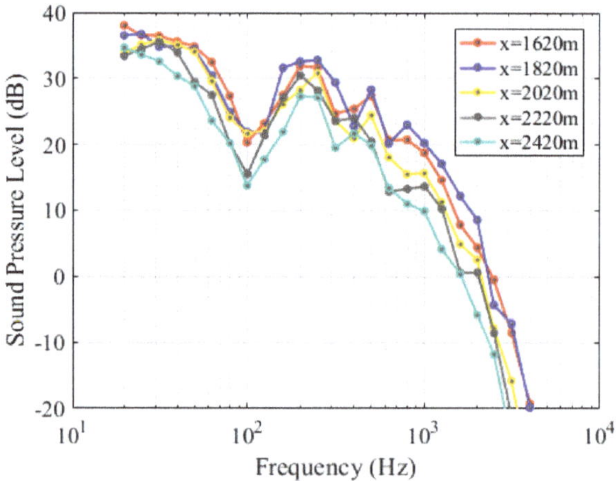

Figure 20. Total SPL from wind turbines #1, 2, 3 received at various distance.

4. Conclusions

Combing the flowfield aerodynamic simulation and the aeroacoustics computation of multi-wind turbines is the focus of this paper. To achieve this objective, ANSource modeling, flow simulation and ANPropagation are individually described and then integrated. The numerical computations are carried out using the in-house developed tool EllipSys. The ANSource modeling and ANPropagation modeling are implemented in the platform of EllipSys. The steady CFD simulations are performed to obtain the flowfield with a relatively efficient 2D-AD solver. The ANSource of wind turbine is calculated using the flowfield information from CFD computations. Then the ANPropagation is simulated which needs the data from both CFD simulations and ANSource computations. Results from different case studies show that a long-path ANPropagation are greatly influenced by different flow conditions, atmosphere conditions and wake condition. An overall trend is that the ANSource from wind turbines is broadband, but the received noise level at long travelling distances from multi-wind turbines is complicated. The noise spectra show that the low frequency noise from a wind turbine cluster might be an important issue to be further investigated. It has been shown in the results that at a distance over 1 km (about 12 rotor diameter), the high frequency components are largely attenuated, say at $f > 1000$ Hz. In a large wind farm with superimposition of noise sources from multi-wind turbines, the computational load will be heavy if high frequency noise is simulated. The ray theory, as compared in Sections 3.1–3.2, is a high-frequency approach therefore the low frequency solutions are not very accurate if strong refraction and diffraction are represented. The present numerical approach has same accuracy over all frequency range. However, to solve high frequency ANPropagation, higher grid density is required, thus increasing computational time. Fortunately, results presented in Figures 18 and 19 indicate that it is the low frequency noise components that contribute to the long range wind turbine ANPropagation. The receiver distances shown in Figure 19 ranges from 20 to 30 rotor diameters. It is clear that at such a large distance, attenuation of low frequency noise is still not significant.

Author Contributions: Conceptualization, W.J.Z. and Z.S.; Software, W.J.Z. and E.B.; Validation, Z.S. and W.J.Z.; Formal analysis, Z.S. and W.J.Z.; Resources, E.B., W.Z.S. and J.N.S.; Data curation, Z.S., W.J.Z., and J.C.; Writing—original draft preparation, Z.S. and W.J.Z.; Writing—Z.S., W.J.Z., W.Z.S., J.N.S., J.C., and H.Y.; Visualization, Z.S. and W.J.Z.; Supervision, W.J.Z.; Project administration, W.J.Z.; Funding acquisition, W.J.Z.

Funding: This research was funded by National Nature Science Foundation of China, grant number 11672261.

Acknowledgments: The authors wish to express acknowledgement to the National Nature Science Foundation of China under grant number 11672261.

Conflicts of Interest: The authors declare no conflict of interest.

References

1. Glauert, H. *Airplane Propellers. Aerodynamic Theory*; Springer: Berlin/Heidelberg, Germany, 1935; pp. 169–360.
2. Sun, Z.Y.; Chen, J.; Shen, W.Z.; Zhu, W.J. Improved blade element momentum theory for wind turbine aerodynamic computations. *Renew. Energy* **2016**, *96*, 824–831. [CrossRef]
3. Sørensen, J.N.; Shen, W.Z. Numerical Modeling of Wind Turbine Wakes. *J. Fluids Eng.* **2002**, *124*, 393. [CrossRef]
4. Sørensen, J.N.; Shen, W.Z.; Munduate, X. Analysis of wake states by a full-field actuator disc model. *Wind Energy* **1998**, *1*, 73–88. [CrossRef]
5. Lighthill, M.J. On Sound Generated Aerodynamically. I. General Theory. *Proc. R. Soc. Lond.* **1952**, *211*, 564–587. [CrossRef]
6. Curle, N. The Influence of Solid Boundaries upon Aerodynamic Sound. *Proc. R. Soc. Lond.* **1955**, *231*, 505–514. [CrossRef]
7. Williams, J.E.; Hall, L.H. Aerodynamic sound generation by turbulent flow in the vicinity of a scattering half plane. *J. Fluid Mech.* **2006**, *40*, 657–670. [CrossRef]
8. Hardin, J.C.; Pope, D.S. An acoustic/viscous splitting technique for computational aeroacoustics. *Theor. Comput. Fluid Dyn.* **1994**, *6*, 323–340. [CrossRef]

9.   Zhu, W.J.; Shen, W.Z.; Sørensen, J.N. High-order numerical simulations of flow-induced noise. *Int. J. Numer. Methods Fluids* **2011**, *66*, 17–37. [CrossRef]

10. Zhu, W.J.; Shen, W.Z.; Barlas, E.; Bertagnolio, F.; Sørensen, J.N. Wind turbine noise generation and propagation modeling at DTU Wind Energy: A review. *Renew. Sustain. Energy Rev.* **2018**, *88*, 133–150. [CrossRef]

11. Salomons, E.M. *Computational Atmospheric Acoustics*; Springer Science & Business Media: Dordrecht, The Netherlands, 2001; ISBN 978-1-4020-0390-5.

12. Ostashev, V.E.; Wilson, D.K. Acoustics in Moving Inhomogeneous Media. *J. Acoust. Soc. Am.* **1999**, *105*, 2067. [CrossRef]

13. Blancbenon, P.; Dallois, L.; Juvé, D. Long Range Sound Propagation in a Turbulent Atmosphere Within the Parabolic Approximation. *Acta Acust. United Acust.* **2001**, *87*, 659–669. [CrossRef]

14. Brooks, T.F.; Pope, D.S.; Marcolini, M.A. *Airfoil Self-Noise and Prediction*; National Aeronautics and Space Administration: Washington, DC, USA, 1989.

15. Zhu, W.J.; Heilskov, N.; Shen, W.Z.; Sørensen, J.N. Modeling of Aerodynamically Generated Noise From Wind Turbines. *J. Sol. Energy Eng.* **2005**, *127*, 517–528. [CrossRef]

16. Lau, A.S.H.; Kim, J.W.; Hurault, J.; Vronsky, T. A study on the prediction of aerofoil trailing-edge noise for wind-turbine applications. *Wind Energy* **2017**, *20*, 233–252. [CrossRef]

17. Lee, S.; Lee, D.; Honhoff, S. Prediction of far-field wind turbine noise propagation with parabolic equation. *J. Acoust. Soc. Am.* **2016**, *140*, 767–778. [CrossRef] [PubMed]

18. Heimann, D.; Käsler, Y.; Gross, G. The wake of a wind turbine and its influence on sound propagation. *Meteorol. Z.* **2011**, *20*, 449–460. [CrossRef]

19. Barlas, E.; Zhu, W.J.; Shen, W.Z.; Kelly, M.; Andersen, S.J. Effects of wind turbine wake on atmospheric sound propagation. *Appl. Acoust.* **2017**, *122*, 51–61. [CrossRef]

20. Barlas, E.; Zhu, W.J.; Shen, W.Z.; Dag, K.O.; Moriarty, P. Consistent modelling of wind turbine noise propagation from source to receiver. *J. Acoust. Soc. Am.* **2017**, *142*, 3297. [CrossRef]

21. Delany, M.E.; Bazley, E.N. Acoustical properties of fibrous absorbent materials. *J. Acoust. Soc. Am.* **2005**, *48*, 105–116. [CrossRef]

22. Jónsson, G.B.; Jacobsen, F. A Comparison of Two Engineering Models for Outdoor Sound Propagation: Harmonoise and Nord2000. *Acta Acust. United Acust.* **2008**, *94*, 282–289. [CrossRef]

23. Plovsing, B.; Kragh, J. *Nord2000-Comprehensive Outdoor Sound Propagation Model. Part 2: Propagation in an Atmosphere with Refraction*; Delta Acoustics & Vibration Report; DELTA: Copenhagen, Denmark, 2006.

24. Mikkelsen, R. Actuator Disc Methods Applied to Wind Turbines. Ph.D. Thesis, Technical University of Denmark, Copenhagen, Denmark, 2003.

25. Michelsen, J.A. *Basis3D-A Platform for Development of Multiblock PDE Solvers*; Technical Report; University of Denmark: Copenhagen, Denmark, 1992.

26. Sørensen, N. *General Purpose Flow Solver Applied over Hills*; Technical Report; Risø National Laboratory: Roskilde, Denmark, 1995.

27. Tian, L.L.; Zhu, W.J.; Shen, W.Z.; Zhao, N.; Shen, Z.W. Development and validation of a new two-dimensional wake model for wind turbine wakes. *J. Wind Eng. Ind. Aerodyn.* **2015**, *137*, 90–99. [CrossRef]

28. Wu, Y.T.; Porté-Agel, F. Modeling turbine wakes and power losses within a wind farm using LES: An application to the Horns Rev offshore wind farm. *Renew. Energy* **2015**, *75*, 945–955. [CrossRef]

29. Sarmast, S.; Segalini, A.; Mikkelsen, R.F.; Ivanell, S. Comparison of the near-wake between actuator-line simulations and a simplified vortex model of a horizontal-axis wind turbine. *Wind Energy* **2016**, *19*, 471–481. [CrossRef]

30. Troldborg, N.; Sørensen, J.N.; Mikkelsen, R. Numerical simulations of wake characteristics of a wind turbine in uniform inflow. *Wind Energy* **2010**, *13*, 86–99. [CrossRef]

31. Troldborg, N.; Zahle, F.; Réthoré, P.E.; Sørensen, N.N. Comparison of wind turbine wake properties in non-sheared inflow predicted by different computational fluid dynamics rotor models. *Wind Energy* **2015**, *18*, 1239–1250. [CrossRef]

32. Tian, L.L.; Zhu, W.J.; Shen, W.Z.; Sørensen, J.N.; Zhao, N. Investigation of modified AD/RANS models for wind turbine wake predictions in large wind farm. *J. Phys.* **2014**. [CrossRef]

33. Tian, L.L.; Zhu, W.J.; Shen, W.Z.; Song, Y.L.; Zhao, N. Prediction of multi-wake problems using an improved Jensen wake model. *Renew. Energy* **2017**, *102*, 457–469. [CrossRef]

34. Zhu, W.J.; Shen, W.Z.; Sørensen, J.N.; Yang, H. Verification of a novel innovative blade root design for wind turbines using a hybrid numerical method. *Energy* **2017**, *141*, 1661–1670. [CrossRef]

35. Wilson, D.K.; Pettit, C.L.; Ostashev, V.E. Sound propagation in the atmospheric boundary layer. *Acoust. Today* **2015**, *11*, 44–53. [CrossRef]

 *applied sciences*

Article

Design and Testing of a LUT Airfoil for Straight-Bladed Vertical Axis Wind Turbines

Shoutu Li [1,2,3]**, Ye Li** [4,5,6,7,]*****, Congxin Yang** [1,2,3]**, Xuyao Zhang** [1,2,3]**, Qing Wang** [1,2,3]**, Deshun Li** [1,2,3]**, Wei Zhong** [8] **and Tongguang Wang** [8]

[1] School of Energy and Power Engineering, Lanzhou University of Technology, Lanzhou 730050, China; lishoutu@lut.edu.cn (S.L.); ycxwind@163.com (C.Y.); zxy0932@163.com (X.Z.); wangqing_lut@foxmail.com (Q.W.); lideshun_8510@sina.com (D.L.)

[2] Gansu Provincial Technology Centre for Wind Turbines, Lanzhou 730050, China

[3] Key Laboratory of Fluid machinery and Systems, Lanzhou 730050, China

[4] School of Naval Architecture, Ocean and Civil Engineering, Shanghai Jiao Tong University, Shanghai 200240, China

[5] State Key Laboratory of Ocean Engineering, School of Naval Architecture, Ocean and Civil Engineering, Shanghai Jiao Tong University, Shanghai 200240, China

[6] Collaborative Innovation Center for Advanced Ship and Deep-Sea Exploration, Shanghai Jiao Tong University, Shanghai 200240, China

[7] Key Laboratory of Hydrodynamics (Ministry of Education), Shanghai Jiao Tong University, Shanghai 200240, China

[8] Jiangsu Key Laboratory of Hi-Tech Research for Wind Turbine Design, Nanjing University of Aeronautics and Astronautics, Nanjing 210016, China; zhongwei@nuaa.edu.cn (W.Z.); tgwang@nuaa.edu.cn (T.W.)

***** Correspondence: ye.li@sjtu.edu.cn; Tel.: +86-021-3420-8250

Received: 29 September 2018; Accepted: 13 November 2018; Published: 16 November 2018

Abstract: The airfoil plays an important role in improving the performance of wind turbines. However, there is less research dedicated to the airfoils for Vertical Axis Wind Turbines (VAWTs) compared to the research on Horizontal Axis Wind Turbines (HAWTs). With the objective of maximizing the aerodynamic performance of the airfoil by optimizing its geometrical parameters and by considering the law of motion of VAWTs, a new airfoil, designated the LUT airfoil (Lanzhou University of Technology), was designed for lift-driven VAWTs by employing the sequential quadratic programming optimization method. Afterwards, the pressure on the surface of the airfoil and the flow velocity were measured in steady conditions by employing wind tunnel experiments and particle image velocimetry technology. Then, the distribution of the pressure coefficient and aerodynamic loads were analyzed for the LUT airfoil under free transition. The results show that the LUT airfoil has a moderate thickness (20.77%) and moderate camber (1.11%). Moreover, compared to the airfoils commonly used for VAWTs, the LUT airfoil, with a wide drag bucket and gentle stall performance, achieves a higher maximum lift coefficient and lift–drag ratios at the Reynolds numbers 3×10^5 and 5×10^5.

Keywords: airfoil design; aerodynamic; wind tunnel experiment; VAWTs (Vertical axis wind turbines)

---

1. Introduction

In an urban or offshore environment, Vertical Axis Wind Turbines (VAWTs), especially the lift-driven type, have unique advantages due to their lower cost, lower noise, simpler structure, and ability to actively accept wind energy from different directions [1–3]. Moreover, with the development of distributed wind energy, VAWTs function more efficiently in certain areas, particularly areas with large fluctuations in wind direction [4,5]. However, compared to Horizontal Axis Wind Turbines (HAWTs), for a long time, VAWTs have experienced slow development and received very little financial support. One reason is the lack of completed theoretical research on VAWTs. For example, although the multiple flow tube theory, which is based on blade element momentum (BEM), is applied to the VAWT design, the results are still far from satisfactory because of the lack of adequate theoretical correction [6]. Another possible reason is that HAWTs have thousands of special airfoils, such as the FFA (FLYGTEKNISKA FORSOKSANSTALTEN) series, DU (Delft University) series, RISØ series, and S series. However, little research has been performed on the airfoils for VAWTs, resulting in the lack of good blades for VAWTs, which further affects VAWT development. Meanwhile, the airfoil plays an essential role in the design of a wind turbine and greatly affects the wind turbine's performance. Therefore, the above factors have led to not only the slow development of VAWTs but also to a heavily debated topic, i.e., the types of airfoil to use in a VAWT [7].

Many studies that have been conducted on the aerodynamic performance of VAWTs have been carried out with a focus on commonly used aircraft airfoils. In particular, in previous research, the symmetrical airfoils of the NACA 00XX series, such as NACA 0012, NACA 0015, and NACA 0018, are widely applied to VAWTs because they help to improve the power coefficient of VAWTs [8]; nevertheless, symmetrical airfoils are unable to properly self-start in VAWTs [9]. Other studies [10,11] have pointed out that VAWTs with a symmetrical airfoil result in poor performance at low Reynolds numbers (R_e), because the self-start difficulty is worsened under such conditions. Studying the effects of the airfoil on the aerodynamic performance of VAWTs, Mohamed et al. [12] focused on 25 kinds of airfoils, and the results indicate that the cambered airfoil LS(1)-0413 increases the power coefficient compared to NACA 0018. In addition, the cambered airfoil NACA 63-415 has a wider operating range. The thickness and camber of the airfoil influence VAWTs according to studies of the NACA-4 digit modified airfoil family or the NACA four-series airfoils [13–15]. However, when the VAWTs work at a higher tip-speed ratio, the aerodynamic performance of the VAWTs decreases with the constant increase in both the camber and thickness of the airfoil [11,13]. Therefore, based on these previous results, it is necessary to consider a moderate camber and moderate thickness in the design of an airfoil for use in a VAWT.

Additionally, according to previous research, some specially purposed airfoils are developed for VAWTs. Achieving great success, the Sandia National Laboratory and Natural Resources Canada carried out a huge volume of work on airfoils for VAWTs through wind tunnel and field experiments [16,17]. The studies of the Sandia National Laboratory show that the natural laminar flow airfoil is suitable for VAWTs. As a result, the Somers airfoils are designed based on their research results; however, they are applied to the curved-bladed Darrieus turbines primarily [7]. Meanwhile, Islam performed meaningful work on VAWTs [9,18,19] by demonstrating the proper camber and thickness most suitable for VAWTs. Moreover, the MI-VAWT1 airfoil is designed to solve the problems of the overall cost, the aerodynamic performance, and blade strength. At the same time, the aerodynamic performance of the MI-VAWT1 airfoil is superior to that of NACA 0015. In addition, the Technical University of Denmark (DTU) has been dedicated to the research of airfoils for VAWTs. Its studies [20,21] depict some very thick airfoils that were designed by considering the structural stiffness of the blade and the aerodynamic characteristics. For instance, the AIR001 airfoil (max. thickness of 25%) can improve the aerodynamic performance in the case of a larger tip-speed ratio, while the DU12W262 airfoil (max. thickness of 26.2%) can increase blade strength. Other research [9,22,23] has explored specially purposed airfoils for VAWTs, such as the DU06-W-200, EN0005, WUP1615, and "TWT series" airfoils from Tokai University. In brief, there are different design aims for different special-purpose airfoils.

As mentioned above, a great deal of investigation has been done to understand the aerodynamic performance of VAWTs and the characteristics of the airfoil. However, compared to HAWTs, the research on special-purpose VAWT airfoils, which affect the aerodynamic performance of VAWTs, is limited. In addition, there exists room for improvement of many aspects, such as the maximum lift coefficient, maximum lift–drag coefficient, wide drag bucket, and stall performance, in the aerodynamic performance of the existing airfoil. Additionally, it is very important that the airfoil has good performance at a low R_e due to the operating environment of VAWTs. Therefore, the objective of this paper is to design an airfoil—named the LUT airfoil, a reference to Lanzhou University of Technology—that shows good aerodynamic characteristics for VAWTs and to investigate the characteristics of the airfoil at low R_e values by wind tunnel experiments. This paper is divided into three main parts: illustrating the design basis and optimization method in Section 2, describing the experimental equipment and procedures in Section 3, and reporting the profile and the aerodynamic performance of the LUT model in Section 4.

2. Basis and Method of Design for the LUT Airfoil

2.1. Design Basis

Figure 1 shows the structure and operation of an H-type VAWT (H-VAWT). The rotational axis of H-VAWTs is always perpendicular to the incoming airflow V_{in}, which results in a constantly changing local angle of attack and a readily occurring dynamic stall [24], especially with a rotor at low speed. Moreover, the VAWTs typically operate in low-R_e environments. Nevertheless, reasonable design of special airfoil for VAWTs is one of the effective methods to solve the above problems. Therefore, the characteristics of an airfoil at a low R_e and a wide range of angles of attack (α) were also considered factors in the design of the LUT airfoil.

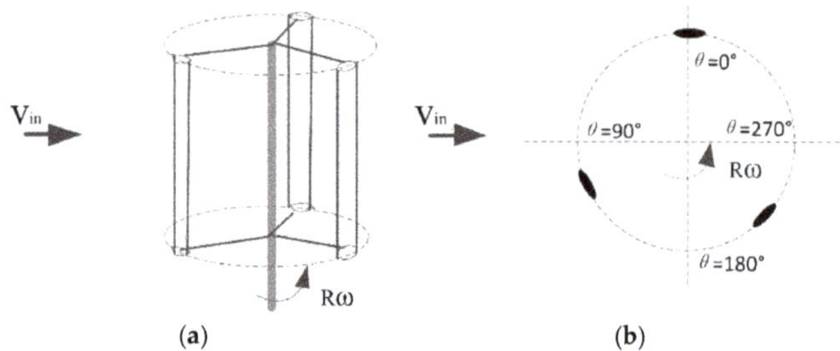

Figure 1. Schematic of an H-type Vertical Axis Wind Turbines (H-VAWT): (**a**) 3D schematic; and (**b**) 2D cross-section. The rotor of the H-VAWT turns counterclockwise; the azimuth angle θ is defined to help describe the active conditions of an H-VAWT. R is the radius of the rotor in units of mm. ω is the rotation angular velocity of the rotor in units of rad/s.

2.2. LUT Airfoil Optimization and Design Method

Optimization algorithm and numerical simulation are widely used in design of the airfoil for VAWTs. Therefore, in this study, the sequential quadratic programming (SQP) method [25,26] was adopted to design the LUT airfoil, because it can properly solve the problem of constrained optimization, maintain the approximate second-order drop speed, and effectively handle constraint conditions. The optimization process is detailed in the following.

The optimization problem can essentially be boiled down to a nonlinear programming problem, as follows:

$$\begin{cases} Min & f(X) \\ s.t. & c_i(X)=0 \quad i=1,2,\cdots,m_e \\ & c_i(X)\geq 0 \quad i=m_e+1,\cdots,m \end{cases} \tag{1}$$

where $f(X)$ is the objective function, $c_i(X)$ is the constraint condition, m_e is the number of constraint conditions of the equation, and m is the number of total constraint conditions.

For the convenience of derivation, the Lagrange function of Equation (1) is:

$$L(X,\lambda) = f(X) - \sum_{i=1}^{m_e} \lambda_i \cdot c_i(X) \tag{2}$$

The gradient vector of this function of x and the Hesse matrix are denoted, respectively, by:

$$\nabla_x L(X, \lambda) = \nabla f(X) - \sum_{i=1}^{m_e} \lambda_i \nabla c_i(X) \tag{3}$$

$$\nabla_x^2 L(X, \lambda) = \nabla^2 f(X) - \sum_{i=1}^{m_e} \lambda_i \nabla^2 c_i(X) \tag{4}$$

For $\lambda = (\lambda_1, \lambda_2, \cdots, \lambda_{m_i})$, the gradient vector and Hesse matrix of the Lagrange function of X and λ, corresponding to $\nabla L(X, \lambda)$ and $\nabla^2 L(X, \lambda)$, respectively, are defined as follows:

$$\nabla L(X, \lambda) = \begin{bmatrix} \nabla f(X) - \nabla c(X)\lambda \\ -c(X) \end{bmatrix} \tag{5}$$

$$\nabla^2 L(X, \lambda) = \begin{bmatrix} \nabla_x^2 L(X, \lambda) & -\nabla c(X) \\ -\nabla c(X)^T & 0 \end{bmatrix} \tag{6}$$

where $\nabla c(X)$ is the Jacobian matrix for the vector function at X, that is, the matrix of $n \times m$ with the partial derivative $\frac{\partial c_j(X)}{\partial x_i}$ of the (i, j) element. The first-order Taylor of the gradient vector is expanded to:

$$\nabla L(X, \lambda) = \nabla L\left(X^k, \lambda^k\right) + \nabla^2 L\left(X^k, \lambda^k\right) \begin{bmatrix} X - X^k \\ \lambda - \lambda^k \end{bmatrix} \tag{7}$$

where k represents the current iteration. It is a necessary condition that the gradient is equal to zero for the optimal solution of the Lagrange function, that is, $\nabla L(X, \lambda) = 0$. Substituting Equations (5) and (6) into the first-order Taylor expansion of the gradient vector produces:

$$\begin{bmatrix} \nabla_x^2 L\left(X^k, \lambda^k\right) & -\nabla c\left(X^k\right) \\ -\nabla c\left(X^k\right)^T & 0 \end{bmatrix} \begin{bmatrix} X - X^k \\ \lambda - \lambda^k \end{bmatrix} = \begin{bmatrix} -\nabla f\left(X^k\right) + \nabla c\left(X^k\right)\lambda^k \\ c\left(X^k\right) \end{bmatrix} \tag{8}$$

By defining $d = X - X^k$ and replacing $\nabla_x^2 L\left(X^k, \lambda^k\right)$ with the Hessen matrix B^k, we can obtain:

$$\begin{bmatrix} B^k & -\nabla c\left(X^k\right) \\ -\nabla c\left(X^k\right)^T & 0 \end{bmatrix} \begin{bmatrix} d \\ \lambda \end{bmatrix} = \begin{bmatrix} -\nabla f\left(X^k\right) \\ c\left(X^k\right) \end{bmatrix} \tag{9}$$

This equation is the exact K-T condition of the quadratic programming problem:

$$\begin{cases} Min & \frac{1}{2}d^T B^k d + \nabla f\left(X^k\right)^T d \\ s.t. & \nabla c\left(X^k\right)^T d + c\left(X^k\right) = 0 \end{cases} \tag{10}$$

where the vector d represents the search direction, and B^k is the approximation of the Hessen matrix. In this way, a common constraint problem is transformed into a quadratic programming problem. Moreover, on this basis, it can be naturally extended to inequality constraints:

$$\begin{cases} Min & \frac{1}{2}d^T B^k d + \nabla f\left(X^k\right)^T d \\ & \nabla c\left(X^k\right)^T d + c\left(X^k\right) = 0 \\ s.t. & \nabla c\left(X^k\right)^T d + c\left(X^k\right) \geq 0 \end{cases} \tag{11}$$

The above SQP method is only locally convergent in theory. To converge to the optimal solution, a one-dimensional search is required to ensure its overall convergence. The search direction d is obtained by Equation (11), which is the downward direction of many penalty functions. For example:

$$F_\sigma(X) = f(X) + \sigma\left[\sum_{i=1}^{m_e}|c_i(X)| + \sum_{i=m_e+1}^{m}|min\{c_i(X),0\}|\right] \tag{12}$$

where σ is the penalty factor that satisfies $\sigma > \max\{|\lambda_i|\, i = 1,2,\cdots,m\}$. A one-dimensional search based on $F_\sigma'\left(X^k,d\right) < 0$ is carried out along with the d direction, and then the step length t^k that satisfies $F_\sigma\left(X^k + t^k d\right) < F_\sigma\left(X^k\right)$ is obtained. Meanwhile, the next iteration point can be obtained:

$$X^{k+1} = X^k + t^k d^k \tag{13}$$

The flowchart describing the optimization of the airfoil's aerodynamic shape based on the SQP method is shown in Appendix A.

Additionally, the aerodynamic performance of the LUT airfoil is obtained under free transition by employing ANSYS Fluent code during the process of optimization; the numerical solution of the Reynolds-averaged Navier–Stokes (RANS) equation is used, and the Shear–Stress–Transport (*SST*) *k*-ω turbulence model is employed [27]. Pressure–velocity coupling is achieved by the SIMPLEC scheme, where the second-order upwind discretization scheme is utilized to calculate the pressure, and the second-order implicit formula is adopted to test temporal discretization. Figure 2a demonstrates the geometric scheme and the boundary conditions of the numerical simulation. The walls of the airfoil are set as non-slip walls, while the other walls are set as free-slip walls. The O-grids are positioned near the airfoil, and the first grid's spacing is less than 1×10^{-5} to ensure that the Y^+ is less than 1.

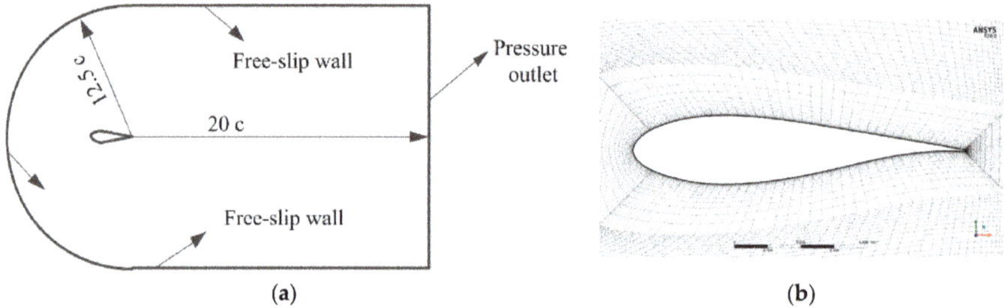

(a) (b)

Figure 2. Scheme of the geometry and boundary conditions for the numerical simulation: (a) The semicircular far field is set as a velocity inlet located at 12.5 c (c is chord length) from the trailing edge of the airfoil, while the outlet condition is the pressure outlet located at 20 c from the trailing edge of the airfoil. (b) The boundary layer consists of 40 cell layers: the height of first layer is 1×10^{-6}; the growth rate of the grid is 1.05 in the boundary layer.

The design results are introduced in the results and discussion section.

This study required a LUT airfoil with a good aerodynamic performance; meanwhile, to ensure the structural strength of the blade for the LUT airfoil, the corresponding airfoil thickness and bending constraints were considered. Therefore, the objective function and constraint conditions can be summarized as:

$$\begin{cases} Min & C_d/C_L \\ & \max(thickness) > \max(thickness0) \\ s.t. & \\ & \max(curve) > \max(curve0) \end{cases} \tag{14}$$

where the thickness0 and curve0 represent the optimized benchmark corresponding to the maximum thickness and the maximum camber of the NACA 0018 airfoil, respectively; the design Reynolds number is 5×10^5; and the design angle of attack is $\alpha = 9°$.

Figure 3 exhibits the results of optimization for $R_e = 5 \times 10^5$. In Figure 3a, the profile of the LUT airfoil was obtained through the SQP method. Figure 3b shows the checked result of the mesh independence in the numerical simulation, where the maximum lift–drag ratio (C_L/C_d) of the NACA 0018 airfoil was investigated at an angle of attack of $\alpha = 9.25°$ [28]. When the total number of cells was higher than 5.8×10^5, the maximum C_L/C_d value of NACA 0018 was converged. Therefore, in this study, more than 6×10^5 cells were employed in the numerical simulation method.

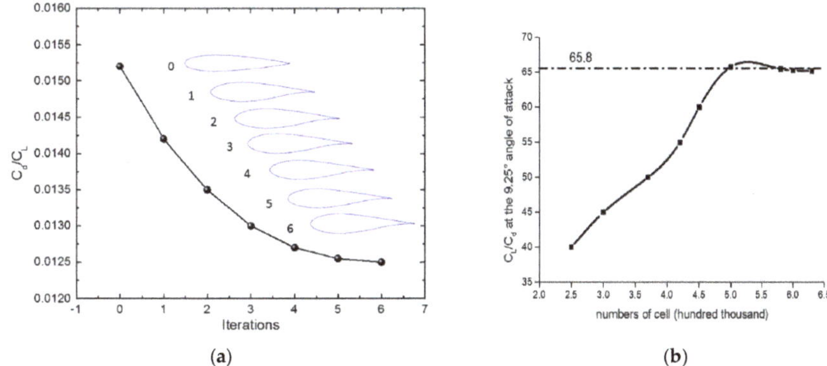

(a) (b)

Figure 3. Airfoil optimization and convergence process. The process of the optimization was from 0 to 6, where the 0 and 6 represent the NACA 0018 airfoil and Lanzhou University of Technology (LUT) airfoil, respectively. (**a**) The constraint conditions were the minimum C_d/C_L, thickness, and camber of the airfoil for the LUT airfoil. (**b**) The maximum lift–drag ratio of the NACA 0018 is 65.8 at an angle of attack of 9.25° [28]. In this Figure, the R_e is 5×10^5.

Figure 4 presents the profile of the LUT airfoil. The LUT airfoil has a moderately maximum camber of 1.11% that is positioned at 78.9% of the chord; a maximum thickness of 20.77% is located at 28.3% of the chord; and the leading-edge radius (r_0) is at 3.84% of the chord. Compared with the DU06-W-200 and NACA 0018 airfoils, the position of the maximum camber is close to the trailing edge for the LUT, which can form a trailing-edge loading to get effective lift [29,30]. Moreover, the moderately maximum thickness not only ensures the strength of the blade for the LUT airfoil, but also helps improve the performance of the VAWTs with the LUT airfoil at high tip speed ratio.

Meanwhile, in the design, we tried to maintain a constant curvature of the upper surface to avoid an increase in flow velocity on the upper surface. Thus, the maximum thickness of the LUT was achieved by adding the thickness of the lower surface, which helps to improve the resistance of the leading-edge roughness. Additionally, the LUT airfoil has a larger leading-edge radius, helping to improve the resistance of the leading-edge roughness. Theoretically, as a result, the LUT has a low sensitivity to roughness. The sensitivity to the roughness of the LUT airfoil will be investigated in a future wind tunnel experiment. However, compared with the DU06-W-200 and NACA 0018, the position of the maximum thickness is slightly closer to the leading edge.

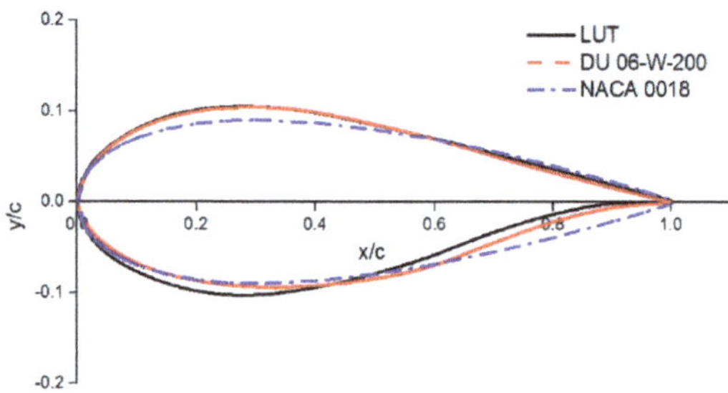

Figure 4. The profiles of the three airfoils. The profiles of the NACA 0018 and DU06-W-200 models come from the material specified on the website [28]. As shown in this Figure, the marks in black, red, and blue represent the profiles of the LUT, DU06-W-200, and NACA 0018 airfoils, respectively.

3. Experimental Equipment and Procedures

In this work, the aerodynamic performance of the LUT model was investigated by measuring the surface pressure of the LUT model at different values of R_e. A description of the main equipment used in the experiment is as follows.

3.1. Wind Tunnel

Wind tunnel tests were carried out in an open-circuit, closed-test-section, low-speed, and low-turbulence wind tunnel at the Northwestern Polytechnical University, as shown in Figure 5. The rectangular test section is 400 mm × 1000 mm, the length is 2800 mm, and the maximum wind speed is 70 m/s. The turbulence intensity is adjustable between 0.02% and 0.3% by changing the screen numbers. For the case of the empty test section in the stream direction with 12 layers of screens and a contraction ratio of 22.6, the turbulence intensity is 0.03%, 0.025%, and 0.02% at 12, 15, and 30 m/s freestream velocities, respectively [31], and the accuracy of the wind speed is less than ±3%.

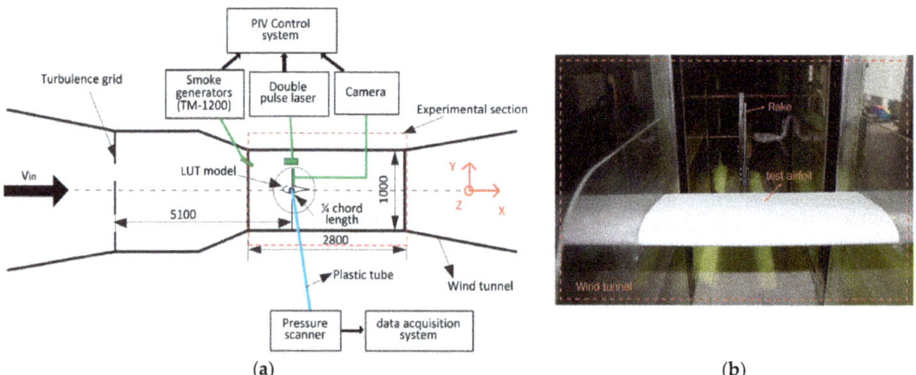

(a)   (b)

Figure 5. Experimental scheme of the wind tunnel and the testing of the LUT airfoil: (**a**) schematic of the wind tunnel and the experimental devices; and (**b**) experiment section. The cross-section of the wind tunnel is 400 mm × 1000 mm and the length is 2800 mm.

The coordinate system of the wind tunnel is defined by the x-, y-, and z-axes, which represent the freestream, perpendicular direction, and span of the airfoil, respectively.

3.2. Test LUT Airfoil Model

In this study, the test LUT airfoil was designed by the authors. The cross-section width of the LUT model is the same as that specified above, the chord of the LUT is 200 mm, and the span is 400 mm. As shown in Figure 6, to investigate the aerodynamic performance of the LUT airfoil under a static condition, 50 pressure taps with a diameter of 0.7 mm are distributed on the LUT airfoil model, where the distribution average density of the pressure taps is 0.7:0.19:0.4 from the leading edge to the trailing edge of the LUT model.

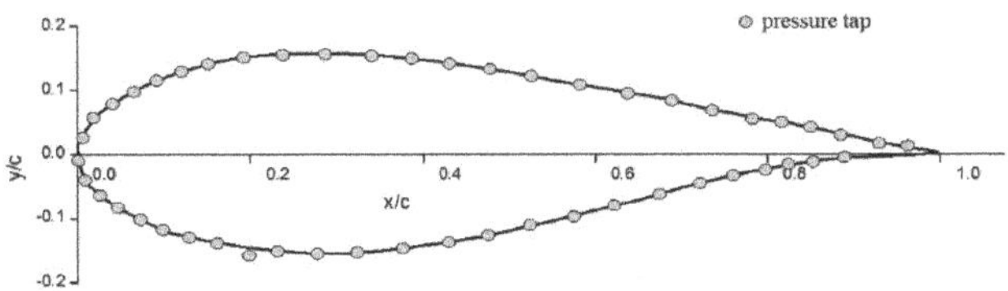

Figure 6. The profile and the distribution of the pressure taps for the test LUT airfoil. The total number of pressure taps is 50 on the surface of the airfoil, and the density of the pressure taps is closely distributed at the leading edge and the trailing edge of the test LUT airfoil.

3.3. Pressure Measurement and PIV (Particle Image Velocimetry) Devices

The pressure measurement devices include an electronic scanning measurement system (DSY104), pressure scanner (PSI9816), and angle of attack control mechanism. The accuracy of the pressure measurement and the PSI9816 pressure scanning are less than $\pm0.2\%$ FS (Full Scale) and $\pm0.05\%$ FS, respectively. The accuracy of the angle of attack control is $\pm2'$. The Particle Image Velocimetry (PIV) systems include a double pulse laser with a maximum operating frequency of 15 Hz, a camera resolution of 2048 $\times$ 2048, a field range of 312 mm $\times$ 312 mm, and an image resolution of 152.66 μm/pixel. The tracer particles were produced by large smoke generators (TM-1200).

The pressure on the LUT airfoil surface is given by the pressure coefficient $C_p = (P_i - P_\infty)/(0.5\rho V_\infty^2)$, where P_i is the pressure of the pressure tap i, P_∞ is the freestream static pressure, V_∞ is the freestream speed, and ρ is the air density. In this work, the drag coefficient was calculated by the wake investigation method. The wake rake had a height of 300 mm and was installed 0.9 c downstream from the test airfoil trailing edge. The calculation formula of the drag coefficient is:

$$c_d = \int_{wake\ rake\ height} c'_x d(y/c) \tag{15}$$

where y is the y-coordinate value of the total pressure tap for the wake rake, whose integral function is:

$$c'_x = 2\left(\frac{p_1}{p_\infty}\right)^{1/\gamma}\left(\frac{p_{01}}{p_0}\right)^{(\gamma-1)/\gamma}\left[\frac{1-(p_1/p_{01})^{(\gamma-1)/\gamma}}{1-(p_\infty/p_0)^{(\gamma-1)/\gamma}}\right]^{0.5} \cdot \left[\left(\frac{1-(p_\infty/p_{01})^{(\gamma-1)/\gamma}}{1-(p_\infty/p_0)^{(\gamma-1)/\gamma}}\right)^{0.5}\right] - c'_{x0} \tag{16}$$

where C'_x is the local drag coefficient of the total pressure tap for the wake rake, p_1 is the static pressure at the wake of the model, p_∞ is the freestream static pressure, p_{01} is the total pressure of the wake rake, p_0 is the freestream total pressure, C'_{x0} is the C'_x arithmetic mean at the wake of the airfoil model, and γ is the specific heat ratio, which typically takes the value $\gamma = 1.4$.

The lift coefficient C_L was calculated by the relationship of the coordinate system, the drag coefficient C_d, and the normal force coefficient C_n:

$$C_L = C_n \cdot cos\alpha - (C_d - C_n \cdot sin\alpha) \cdot tan\ \alpha \tag{17}$$

In the wind tunnel experiment, it was necessary to eliminate errors of measurement. Therefore, the pressure values of the pressure taps and the wake rake were referenced when the freestream velocity was 0 m/s and the angle of attack was 0°. Uncertainty analysis of the coefficients of both the lift and the drag was evaluated with the accuracy of the PSI9816 scanner.

4. Results and Discussion

4.1. Static LUT Airfoil Performance at Low Reynolds Number

In this section, the performance of the clean, static LUT airfoil is described for $R_e = 3 \times 10^5$ and 5×10^5 (these two low Reynolds numbers are suitable for the typical working environment of VAWTs), respectively.

4.1.1. Pressure Coefficient Distribution on the Surface for the LUT Airfoil

In Figure 7, the distribution of the pressure coefficients on the LUT airfoil surface is shown for some specific angles of attack at $R_e = 3 \times 10^5$ and at $R_e = 5 \times 10^5$. In this figure, the lines in black, red, and blue denoting the pressure coefficient (C_p) correspond to angles of attack of C_L/C_d-max, C_L-max, and post-stall, respectively.

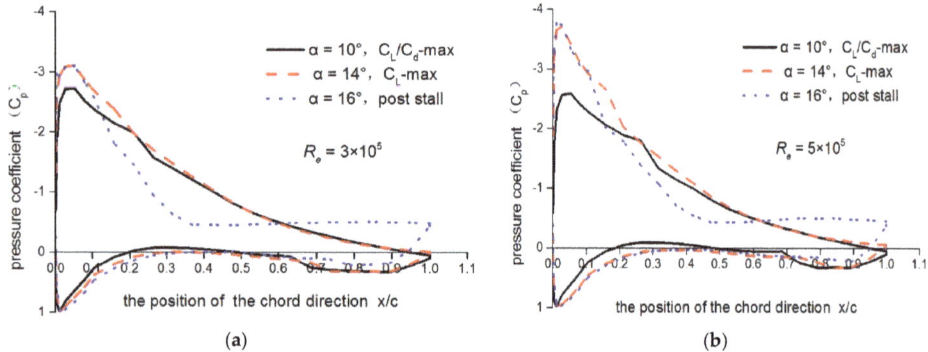

Figure 7. Pressure coefficient distribution of the LUT airfoil: (**a**) the case of $R_e = 3 \times 10^5$; and (**b**) the case of $R_e = 5 \times 10^5$. The data were measured when the surface of the LUT airfoil was clean. The lines in black, red, and blue show the pressure coefficients corresponding to angles of attack of C_L/C_d-max, C_L-max, and post-stall, respectively.

In the case of $R_e = 3 \times 10^5$, the LUT airfoil has a maximum lift coefficient of $C_{L\text{-max}} = 1.47$ at an angle of attack of $\alpha = 14°$ and a maximum lift–drag ratio of $C_L/C_{d\text{-max}} = 64.4$ at an angle of attack of $\alpha = 10°$. After an angle of attack of $\alpha = 14°$, the LUT airfoil starts to stall, and separation occurs on the suction surface at about 39% of the chord with an angle of attack of $\alpha = 16°$. Therefore, the pressure coefficient on the suction surface of the LUT airfoil keeps a constant value from about 39% of the chord to the trailing edge. In the case of $R_e = 5 \times 10^5$, the LUT airfoil has a maximum lift coefficient of $C_{L\text{-max}} = 1.5$ at an angle of attack of $\alpha = 14°$ and a maximum lift–drag ratio of $C_L/C_{d\text{-max}} = 79.8$ with an angle of attack of $\alpha = 10°$. After an angle of attack $\alpha = 14°$, the LUT airfoil starts to stall, and separation occurs on the suction surface at about 48% of the chord when the angle of attack is 16°. Moreover, after 48% of the chord, the pressure coefficient on the suction surface of the LUT airfoil keeps a constant value; that is, the LUT airfoil is fully stalled.

Comparing the cases of $R_e = 3 \times 10^5$ and $R_e = 5 \times 10^5$, the suction peak value increases and moves to the leading edge of the LUT airfoil with the increase in R_e. This indicates that the region of adverse pressure gradient on the suction surface of the LUT airfoil is extended when $R_e = 5 \times 10^5$. However, due to the effect of the turbulence, the flow separation is delayed [32]. Additionally, the pressure was not measured at large angles of attack due to the measuring range of the wake rake. In the future, the performance of the LUT needs to be studied at large angles of attack.

Figure 8 shows the changes in the pressure coefficient distribution with increasing angles of attack at different values of R_e for the LUT airfoil. In the case of $R_e = 3 \times 10^5$, the free transition position is located at 53% and 35% of the chord when the angle of attack is 0° and 8°, respectively. Until an angle of attack of $\alpha = 21°$, the pressure of the upper surface from the leading edge to the trailing edge remains constant due to the complete separation of the flows on the upper surface; that is, leading-edge separation is occurring. Combined with Figure 7, the transition position on the upper surface slightly develops in the range $8° < \alpha < 14°$ and is located at about 30% of the chord, indicating that the LUT is characterized by low drag, because the desirable pressure gradient exists at about 30% of the chord on the upper surface of the LUT airfoil [33], while the pressure coefficient on the lower surface is not obviously developed in the range $8° < \alpha < 14°$. Comparing the case of $R_e = 3 \times 10^5$ to the case of $R_e = 5 \times 10^5$, it is observed that the free transition position is more toward the leading edge with the increase in R_e. An angle of attack of $\alpha = 8°$, for example, results in a transition position located at about 29% of the chord when R_e is 5×10^5. In particular, in the case of $R_e = 5 \times 10^5$, the pressure on the lower surface significantly drops at an angle of attack of $\alpha = 21°$. This shows that obvious separation on the lower surface exists in the LUT airfoil at an angle of attack of $\alpha = 21°$.

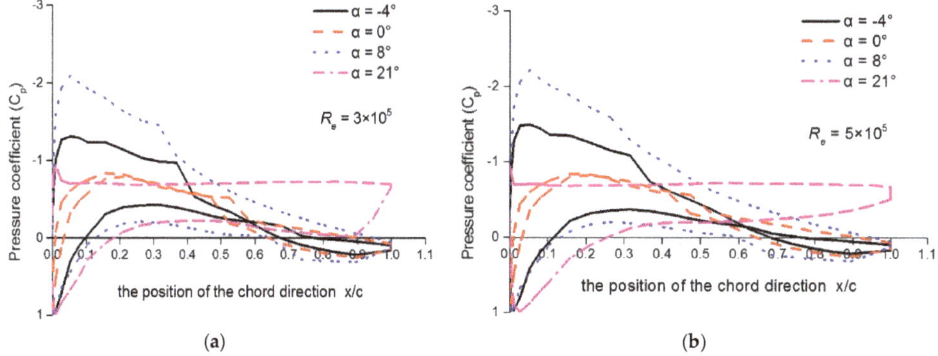

Figure 8. Pressure distribution for the LUT airfoil: (**a**) the case of $R_e = 3 \times 10^5$; and (**b**) the case of $R_e = 5 \times 10^5$. The data were measured when the surface of the LUT airfoil was clean. As shown in this figure, the lines in black, red, blue, and pink show the pressure coefficients corresponding to the angles $\alpha = -4°$, $\alpha = 0°$, $\alpha = 8°$, and $\alpha = 21°$, respectively.

The above discussions indicate that the distribution of the pressure coefficient of the LUT is better before a stall angle of $\alpha = 14°$ for both the case of $R_e = 3 \times 10^5$ and $R_e = 5 \times 10^5$; moreover, the free transition position at about 30% chord almost keeps constant before the stall, it shows that the pressure distribution on the upper surface is steady between $R_e = 3 \times 10^5$ and $R_e = 5 \times 10^5$ for the LUT airfoil. However, in the case of $R_e = 5 \times 10^5$, the pressure distribution on the lower surface is poor where the separation of the flows is obvious after a stall angle of $\alpha = 14°$.

4.1.2. Lift Coefficient of the LUT Airfoil

As shown in Figure 9, we investigated the relationship between the lift coefficient and the angles of attack at two Reynolds numbers, $R_e = 3 \times 10^5$ and $R_e = 5 \times 10^5$, obtained from Equation (17). In this Figure, the lines in black and red of the lift coefficient correspond to the results of the experiment and the numerical simulation (design results) for the LUT airfoil, respectively. In the case of $R_e = 3 \times 10^5$, the lift coefficient of the LUT airfoil linearly increases in the extending direction of the angles of attack from $\alpha = -9°$ to $\alpha = 12°$ (this region is called the linear region). At an angle of attack of $\alpha = 14°$, the lift coefficient reaches its maximum value at $C_{L\text{-max}} = 1.47$, and then the curve of the lift coefficient displays a gentle downward trend until an angle of attack of $\alpha = 21°$, a phenomenon called a stall. This indicates that the LUT airfoil has good stall performance.

In the case of $R_e = 5 \times 10^5$, the trend for the lift coefficient curve is the same as that of $R_e = 3 \times 10^5$ at the angles of attack from $\alpha = -9°$ to $\alpha = 12°$. However, the maximum lift coefficient $C_{\text{L-max}}$ is 1.5 at an angle of attack of $\alpha = 14°$. Comparing the results of the experiment to the results of the numerical simulation (design results), in the linearly increasing areas, the values of the lift coefficient are almost the same for the two values of R_e. However, after the stall angle of attack, the values of the lift coefficient for $R_e = 5 \times 10^5$ have a slight difference and are larger than the value of the lift coefficient for $R_e = 3 \times 10^5$. While the numerical simulation results are larger than those of the experiment, these differences are very small. Therefore, the results of the numerical simulation (design results) are in agreement with the results of the experiment reported in this paper.

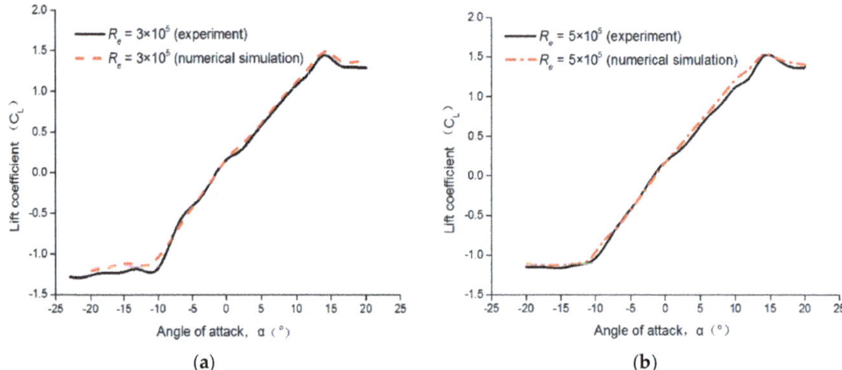

Figure 9. Lift coefficient for the LUT airfoil at different values of R_e: (**a**) the case of $R_e = 3 \times 10^5$; and (**b**) the case of $R_e = 5 \times 10^5$. The lines in black and red of the lift coefficient correspond to the experimental results and the numerical simulation results, respectively.

4.1.3. Drag Coefficient of the LUT Airfoil

The relationship between the drag coefficient and the angles of attack are displayed for the two Reynolds numbers, $R_e = 3 \times 10^5$ and $R_e = 5 \times 10^5$, in Figure 10, where the drag coefficient was obtained via Equations (15) and (16).

In Figure 10, the lines in black and red of the drag coefficient correspond to the experimental results and the numerical simulation results (design results), respectively. In the cases of $R_e = 3 \times 10^5$ and $R_e = 5 \times 10^5$, the values of the drag coefficient are smaller and remain almost constant in the range $-10° < \alpha < 10°$, because the most attached flow is on the surface of the LUT airfoil. The numerical simulation results agree with the experimental results in this region. However, the values of the drag coefficient increase rapidly when the angle of attack is greater than $\alpha = 13°$ or less than $\alpha = -10°$. The results of the experiment and the numerical simulation show a slight difference, but these errors are acceptable. Therefore, the results of the numerical simulation (design results) are in agreement with the results of the experiment reported in this paper.

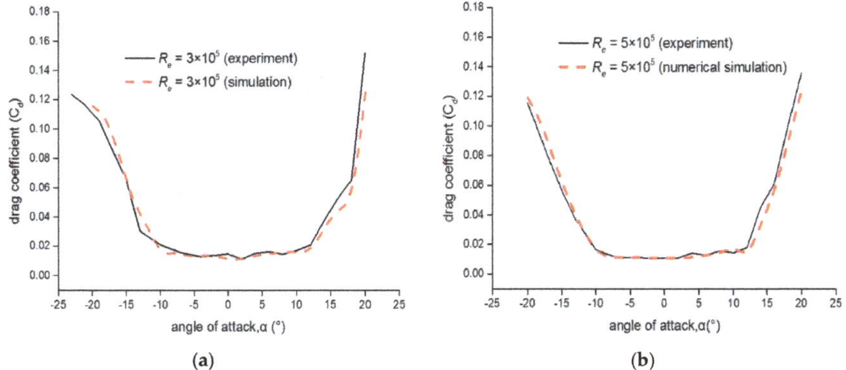

Figure 10. Drag coefficient for the LUT airfoil at different values of R_e: (**a**) the case of $R_e = 3 \times 10^5$; and (**b**) the case of $R_e = 5 \times 10^5$. The lines in black and red of the drag coefficient correspond to the experimental results and the numerical simulation results, respectively.

Figures 9 and 10 demonstrate that the results of the experiment agree with the results of the numerical simulation at low angles of attack. However, at high angles of attack where the flow separation occurs, there are some slight differences between the results of the experiment and the numerical simulation. On the one hand, the characteristics of the lift and drag of the LUT airfoil are obtained through integral surface pressure in this experiment, while the results of the numerical simulation with the *SST k-ω* turbulence model are acquired with solving the Reynolds-averaged Navier–Stokes equation; on the other hand, even though the current numerical methods have been greatly improved, for example, the predicted result of the *SST k-ω* turbulence model is accurate, but the *SST k-ω* turbulence model cannot completely solve the turbulence problem [34,35]. In addition, the wall of the LUT model is infinitely smooth in the numerical simulation, but in the current experiment, although the surface of the LUT model has very low roughness, the roughness of the surface is limited. Therefore, the results of the experiment and the numerical simulation have some differences in the region of the flow separation, while the results of the experiment and the numerical simulation are almost the same at low angles of attack.

Additionally, the maximum lift coefficient of the LUT airfoil has slow change with the increase in R_e from $R_e = 3 \times 10^5$ to $R_e = 5 \times 10^5$.

4.1.4. Lift–Drag Ratios of the LUT Airfoil

Figure 11a presents the lift coefficient versus the drag coefficient at different values of R_e for the LUT airfoil, where the lift coefficient and drag coefficient are described in and were obtained according to Sections 4.1.1 and 4.1.2. The lines in black and red of the lift–drag ratios correspond to $R_e = 3 \times 10^5$ and $R_e = 5 \times 10^5$, respectively. The results show that the LUT airfoil has the maximum lift–drag ratio $C_L/C_{d\text{-max}} = 64.4$ at an angle of attack of $\alpha = 10°$ when $R_e = 3 \times 10^5$, and it has the maximum lift–drag ratio $C_L/C_{d\text{-max}} = 79.8$ at an angle of attack of $\alpha = 10°$ when $R_e = 5 \times 10^5$. Moreover, combined with Figure 10, the lift–drag ratios linearly increase in the range $-10° < \alpha < 10°$. Therefore, the LUT airfoil presents a wider drag bucket, which is a desired characteristic for the airfoil of VAWTs due to the law of motion of the VAWTs [19]. Figure 11b shows the lift–drag ratios versus the lift coefficients, where the colors of the lines have the same meaning as in Figure 11a. The lift–drag ratios of the LUT increase almost linearly with the increase in the lift coefficient in the range $-1 \leq C_L \leq 1.25$.

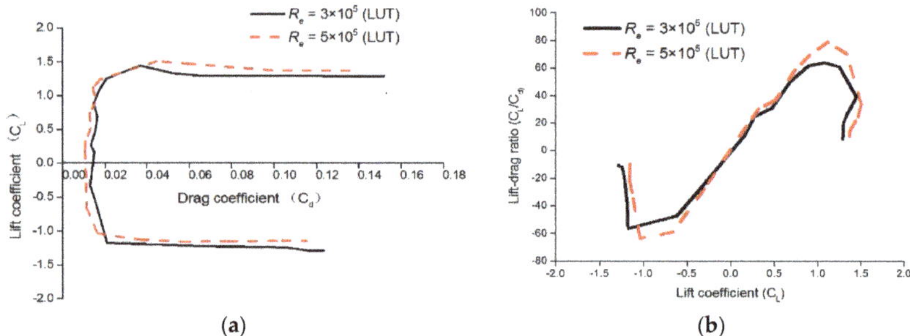

Figure 11. The characteristics of lift–drag ratios for the LUT airfoil at different values of R_e: (**a**) the relationship between the lift coefficient and the drag coefficient; and (**b**) the relationship between lift–drag ratios and the lift coefficient. The lines in black and red of the lift–drag ratios correspond to $R_e = 3 \times 10^5$ and $R_e = 5 \times 10^5$, respectively.

In Figure 12, the flow separation on the surface of the LUT is displayed at different angles of attack when the R_e is 3×10^5, measured using PIV technology. Figure 12 shows that the separation point gradually moves from the trailing edge to the leading edge with the increase in the angle of attack, which is consistent with the results in Figures 7 and 8. Moreover, the separation vortex becomes high due to the gradual flow separation. Especially at the angle of attack $\alpha = 22°$ in Figure 12g, the leading-edge separation results in the formation of the wake vortex and the detached vortex. At this moment, the pressure distribution on the upper surface of the LUT airfoil keeps a constant value, as shown in Figure 8a. Figure 12d–f displays the development of the flow separation on the upper surface of the LUT airfoil, which results in decreasing pressure on the upper surface of the airfoil, as shown in Figure 7a. However, the development of the vortices on the upper surface of the LUT airfoil are slow; combined with Figure 8a, it indicates that the LUT airfoil has good stall performance. Meanwhile, the changes in flow separation on the lower surface of the LUT airfoil are presented in Figure 12a–c when the angle of attack is a negative value. The pressure distribution on the LUT airfoil surface at a negative angle of attack is shown in Figure 8. However, due to the limited field range of the PIV here, it is difficult to observe the development of the wake vortices. Meanwhile, at a large negative angle of attack, the flow separation on the lower surface is obvious, and the separation point on the lower surface moves toward the leading edge with increasing angles of attack.

However, because the placement of the PIV equipment was only convenient for photographing the upper surface of the LUT model, the phenomenon of flow separation was not obviously surveyed. Consequently, we could only survey the vortices near the trailing edge of the LUT.

Table 1 shows the main geometrical and aerodynamic parameters of a part of the commonly used and special-purpose airfoils for straight-bladed VAWTs when R_e is 3×10^5 and 5×10^5. In addition, the MI-VAWT1 airfoil [9] is also a dedicated airfoil for VAWTs; it has a 21% thickness and 0.44% camber. Compared with the results for Table 1, the geometrical parameters of the LUT airfoil balance the strength of the blade and the aerodynamic performance for a VAWTs with the LUT airfoil. Especially, the LUT airfoil performs better in the maximum lift coefficient and the maximum lift–drag ratios at a low R_e, it is very important, because VAWTs typically operate at a low R_e [19]. However, the stall angle of attack of the LUT airfoil is smaller than that of the airfoil in Table 1.

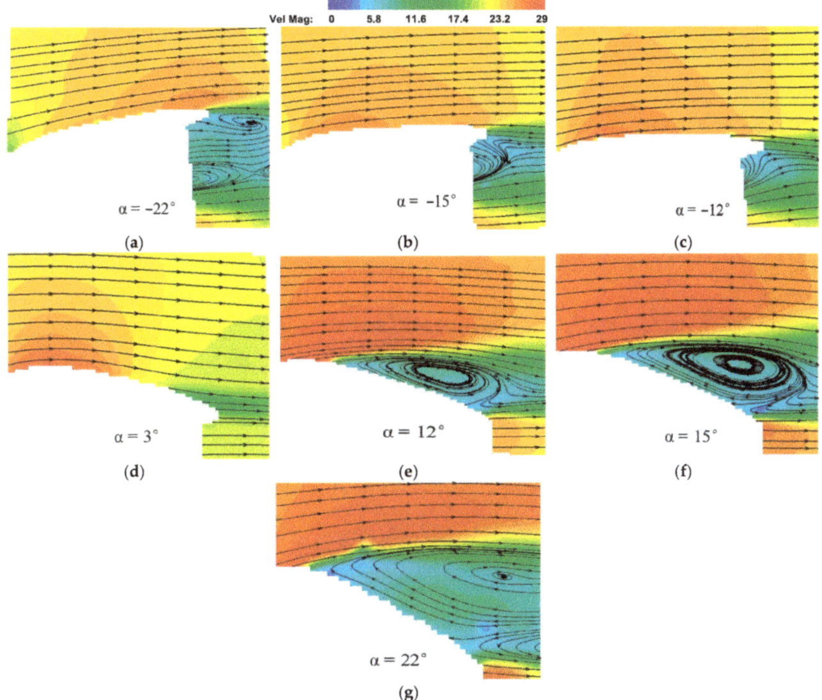

Figure 12. Contours of the velocity and stream lines at different angles of attack: (**a–g**) the contours of the velocity at different angles of attack. This figure shows the transient contours from the angle of attack $\alpha = -20°$ to the angle of attack $\alpha = 22°$ at $R_e = 3 \times 10^5$. However, only the varying velocity on the upper surface of the LUT and the larger vortex are captured in this figure due to the picture being limited by the ability of the Particle Image Velocimetry (PIV) equipment.

Table 1. Main geometrical parameters and aerodynamic parameters of a part of commonly used airfoils for straight-bladed VAWTs.

Airfoil	R_e (1×10^5)	Max. Thickness	(x/c) Max. Thickness	Max. Camber	(x/c) Max. Camber	Radius Leading Edge (r_0/c)	Max. C_L/C_d (α)	Max. C_L (α)
Du06-W-200 [28]	3	19.8%	31.1%	0.5%	84.6%	1.7442%	58.05 (9.75°)	1.4153 (16.75°)
	5						71.5 (9.5°)	1.4384 (18°)
NACA0018 [28]	3	18%	30%	0	0	3.1017%	57.09 (8.25°)	1.24 (18.75)
	5						65.8 (9.25°)	1.2624 (16.5°)
NACA0015 [28]	5	15%	30%	0	0	2.3742%	66.43 (7.5°)	1.2731 (16.75°)
DU12W262 [21]	-	26.2%	37%	-	-	-	-	-
AIR013 [20]	-	34.9%	-	2.29%	-	7.8%	-	-

Note: "-" indicates that the data of the corresponding locations are not presented in the literature.

5. Conclusions

In this study, we designed an asymmetrical airfoil for straight-bladed VAWTs based on the law of motion of the VAWTs by employing the SQP optimization method; we called this design the LUT airfoil. Meanwhile, the aerodynamic performance of the LUT airfoil was investigated at two Reynolds numbers, 3×10^5 and 5×10^5, under the conditions of a low-speed, low-turbulence wind tunnel at the Northwestern Polytechnical University, China; the flow field around the LUT airfoil was observed by PIV technology.

The LUT airfoil has a 20.77% moderate thickness and 1.11% moderate camber. Moreover, the position of the maximum camber is close to the trailing edge, which forms a trailing-edge loading to improve the efficient lift of the LUT airfoil. The maximum thickness of the LUT airfoil is achieved by restricting the thickness of the upper surface while adding the thickness of the lower surface. It is conducive to delay flow separation on the upper surface. Moreover, the moderate thickness helps to ensure the structural strength of the blade and to improve the performance of VAWTs with the LUT airfoil at high tip speed ratio.

Ultimately, the main objectives of achieving a higher maximum lift coefficient, higher maximum lift–drag coefficient, and gentle stall characteristics were achieved. Compared with the airfoil commonly used for VAWTs, the LUT airfoil displays a better aerodynamic performance at a low R_e. Additionally, the maximum lift coefficient of the LUT airfoil is steady with the increase in R_e between 3×10^5 and 5×10^5. The LUT airfoil retains the desirable characteristic of low drag over a wide range of angles of attack in the range $-10° < \alpha < 10°$, that is, the LUT airfoil has a wide drag bucket. The distribution of the pressure coefficient agrees with the results obtained using PIV. Theoretically, due to the larger 3.84% leading-edge radius and other geometric features, the LUT airfoil can reduce the sensitivity to roughness.

In future work, the LUT airfoil's sensitivity to roughness and noise level will be tested by wind tunnel experiments.

Author Contributions: Methodology, S.L., Y.L. and Q.W.; Writing—Original Draft Preparation, S.L. and X.Z.; and Writing—Review and Editing, Y.L., C.Y., D.L., W.Z. and T.W.

Funding: The authors acknowledge funding from the Natural Science Foundation of GANSU (grant No. 1508RJYA098), the National Natural Science Foundation of China (No. 51506088 and No. 51766009), and the National Basic Research Program of China (No. 2014CB046201)

Acknowledgments: The authors would like to acknowledge and thank the Natural Science Foundation of GANSU (grant No. 1508RJYA098); the National Natural Science Foundation of China (No. 51506088 and No. 51766009); the National Basic Research Program of China (No. 2014CB046201); the people who provided many good suggestions for this paper; and the Northwestern Polytechnical University and China Aerodynamics Research and Development Center for providing the experiment instruments and wind tunnel.

Nomenclature

H-VAWT	H-type Vertical Axis Wind Turbines
VAWT	Vertical Axis Wind Turbines
HAWT	Horizontal Axis Wind Turbines
LUT	The name of the newly designed airfoil (Lanzhou University of Technology)
DTU	Technical University of Denmark
SST	Shear Stress Transport
R_e	Reynolds number
α	angle of attack, °
θ	azimuthal angle, °
ω	rotational speed of the wind turbine, rad/s
V_{in}	freestream velocity, m/s
R	rotor radius, m
C	airfoil chord length, mm
SQP	Sequential Quadratic Programming
σ	penalty factor
C_L/C_d	lift–drag ratios
C_L	lift coefficient
C_d	drag coefficient
r_0	leading-edge radius
C_p	Pressure coefficient
PIV	Particle Image Velocimetry
FS	Full Scale

Appendix A

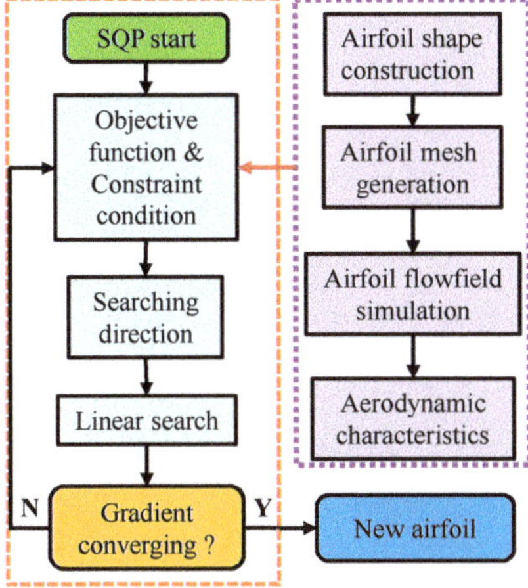

Figure A1. The optimization flow of the airfoil profile based on the SQP method.

References

1. Ho, A.; Mbistrova, A.; Corbetta, G. *The European Offshore Wind Industry-Key Trends and Statistics 2015*; Technical Report; European Wind Energy Association: Brussels, Belgium, February 2016.
2. Battisti, L.; Benini, E.; Brighenti, A.; Dell'Anna, S.; Raciti Castelli, M. Small wind turbine effectiveness in the urban environment. *Renew. Energy* **2018**, *129*, 102–113. [CrossRef]
3. Guo, Y.; Liu, L.; Gao, X.; Xu, W. Aerodynamics and Motion Performance of the H-Type Floating Vertical Axis Wind Turbine. *Appl. Sci.* **2018**, *8*, 262. [CrossRef]
4. Kumar, R.; Raahemifar, K.; Fung, A.S. A critical review of vertical axis wind turbines for urban applications. *Renew. Sustain. Energy Rev.* **2018**, *89*, 281–291. [CrossRef]
5. Li, Q.; Maeda, T.; Kamada, Y.; Ogasawara, T.; Nakai, A.; Kasuya, T. Investigation of power performance and wake on a straight-bladed vertical axis wind turbine with field experiments. *Energy* **2017**, *141*, 1113–1123. [CrossRef]
6. Aslam Bhutta, M.M.; Hayat, N.; Farooq, A.U.; Ali, Z.; Jamil, S.R.; Hussain, Z. Vertical axis wind turbine—A review of various configurations and design techniques. *Renew. Sustain. Energy Rev.* **2012**, *16*, 1926–1939. [CrossRef]
7. Sutherland, H.J.; Berg, D.E.; Ashwill, T.D. *A Retrospective of VAWT Technology*; Technical Report SAND2012-0304; Sandia National Laboratories: Albuquerque, NM, USA, January 2012.
8. Kanyako, F.; Janajreh, I. Numerical Investigation of Four Commonly Used Airfoils for Vertical Axis Wind Turbine. In *ICREGA'14-Renewable Energy: Generation and Application*; Hamdan, M., Hejase, H., Noura, H., Fardoun, A., Eds.; Springer: Cham, Switzerland, 2014; pp. 443–454. [CrossRef]
9. Islam, M. Analysis of Fixed-Pitch Straight-Bladed VAWT with Asymmetric Airfoils. Ph.D. Thesis, University of Windsor, Windsor, ON, Canada, 2008.
10. Kirke, B.K. Evaluation of Self-Starting Vertical Axis Wind Turbines for Stand-Alone Appllications. Ph.D. Thesis, Griffith University, Brisbane, Australia, 1998.
11. Rainbird, J.M.; Bianchini, A.; Balduzzi, F.; Peiró, J.; Graham, J.M.R.; Ferrara, G.; Ferrari, L. On the influence of virtual camber effect on airfoil polars for use in simulations of Darrieus wind turbines. *Energy Convers. Manag.* **2015**, *106*, 373–384. [CrossRef]
12. Mohamed, M.H.; Ali, A.M.; Hafiz, A.A. CFD analysis for H-rotor Darrieus turbine as a low speed wind energy converter. *Eng. Sci. Technol. Int. J.* **2015**, *18*, 1–13. [CrossRef]
13. Carrigan, T.J.; Dennis, B.H.; Han, Z.X.; Wang, B.P. Aerodynamic Shape Optimization of a Vertical-Axis Wind Turbine Using Differential Evolution. *ISRN Renew. Energy* **2012**, *2012*, 1–16. [CrossRef]
14. Asr, M.T.; Nezhad, E.Z.; Mustapha, F.; Wiriadidjaja, S. Study on start-up characteristics of H-Darrieus vertical axis wind turbines comprising NACA 4-digit series blade airfoils. *Energy* **2016**, *112*, 528–537. [CrossRef]

15. Chen, J.; Chen, L.; Xu, H.; Yang, H.; Ye, C.; Liu, D. Performance improvement of a vertical axis wind turbine by comprehensive assessment of an airfoil family. *Energy* **2016**, *114*, 318–331. [CrossRef]

16. Sheldahl, R.E.; Klimas, P.C. *Aerodynamic Characteristics of Seven Sysmmetrical Airfoil Sections Through 180-Degree Angle of Attack for Use in Aerodynamic Analysis of Vertical Axis Wind Turbines*; Technical Report SAND80-2114; Sandia National Laboratories: Albuquerque, NM, USA, March 1981.

17. Klimas, P.C. *Tailored Airfoils for Vertical Axis Wind Turbines*; Technical Report SAND84-1062; Sandia National Laboratories: Albuquerque, NM, USA, November 1984.

18. Islam, M.; Fartaj, A.; Carriveau, R. Design analysis of a smaller-capacity straight-bladed VAWT with an asymmetric airfoil. *Int. J. Sustain. Energy* **2011**, *30*, 179–192. [CrossRef]

19. Islam, M.; Ting, D.S.-K.; Fartaj, A. Desirable Airfoil Features for Smaller-Capacity Straight-Bladed VAWT. *Wind Eng.* **2007**, *31*, 165–196. [CrossRef]

20. Ferreira, C.S.; Geurts, B. Aerofoil optimization for vertical-axis wind turbines. *Wind Energy* **2015**, *18*, 1371–1385. [CrossRef]

21. Ragni, D.; Ferreira, C.S.; Correale, G. Experimental investigation of an optimized airfoil for vertical-axis wind turbines. *Wind Energy* **2015**, *18*, 1629–1643. [CrossRef]

22. Batista, N.C.; Melício, R.; Mendes, V.M.F.; Calderón, M.; Ramiro, A. On a self-start Darrieus wind turbine: Blade design and field tests. *Renew. Sustain. Energy Rev.* **2015**, *52*, 508–522. [CrossRef]

23. Bedon, G.; Betta, S.D.; Benini, E. Performance-optimized airfoil for Darrieus wind turbines. *Renew. Energy* **2016**, *94*, 328–340. [CrossRef]

24. Hand, B.; Kelly, G.; Cashman, A. Numerical simulation of a vertical axis wind turbine airfoil experiencing dynamic stall at high Reynolds numbers. *Comput. Fluids* **2017**, *149*, 12–30. [CrossRef]

25. Sargent, R.W.H.; Ding, M. A New SQP Algorithm for Large-Scale Nonlinear Programming. *SIAM J. Optim.* **2001**, *11*, 716–747. [CrossRef]

26. Andrei, N. Sequential Quadratic Programming Continuous (SQP). In *Nonlinear Optimization for Engineering Applications in GAMS Technology*, 1st ed.; Springer: Cham, Switzerland, 2017; Volume 121, pp. 269–288, ISBN 978-3-319-58355-6.

27. Howell, R.; Qin, N.; Edwards, J.; Durrani, N. Wind tunnel and numerical study of a small vertical axis wind turbine. *Renew. Energy* **2010**, *35*, 412–422. [CrossRef]

28. Airfoil Tools. Available online: http://airfoiltools.com/ (accessed on 1 September 2018).

29. Timmer, W.A.; Van Rooij, R.P.J.O.M. Summary of the Delft University Wind Turbine Dedicated Airfoils. *J. Sol. Energy Eng.* **2003**, *125*, 488–496. [CrossRef]

30. Fuglsang, P.; Bak, C. Development of the Risø wind turbine airfoils. *Wind Energy* **2004**, *7*, 145–162. [CrossRef]

31. Meng, X.; Hu, H.; Yan, X.; Liu, F.; Luo, S. Lift improvements using duty-cycled plasma actuation at low Reynolds numbers. *Aerosp. Sci. Technol.* **2018**, *72*, 123–133. [CrossRef]

32. Anderson, J.D. *Fundamentals of Aerodynamics*, 5th ed.; McGraw-Hill Education: New York, NY, USA, 2011; ISBN 9780073398105.

33. Somers, D.M. *Design and Experimental Results for the S825 Airfoil Period of Performance: 1998–1999 Design and Experimental Results for the S825 Airfoil*; Technical Report NREL/SR-500-36346; National Renewable Energy Laboratory: Golden, CO, USA, 2005.

34. Balduzzi, F.; Bianchini, A.; Maleci, R.; Ferrara, G.; Ferrari, L. Critical issues in the CFD simulation of Darrieus wind turbines. *Renew. Energy* **2016**, *85*, 419–435. [CrossRef]

35. Zhang, X.; Wang, G.; Zhang, M.; Liu, H.; Li, W. Numerical study of the aerodynamic performance of blunt trailing-edge airfoil considering the sensitive roughness height. *Int. J. Hydrogen Energy* **2017**, *42*, 18252–18262. [CrossRef]

 *applied sciences*

Article

Variable Pitch Approach for Performance Improving of Straight-Bladed VAWT at Rated Tip Speed Ratio

Zhenzhou Zhao [1,*], Ruixin Wang [1], Wenzhong Shen [2], Tongguang Wang [3], Bofeng Xu [1], Yuan Zheng [1] and Siyue Qian [1]

[1] College of Energy and Electrical Engineering, Hohai University, Nanjing 210098, Jiangsu, China; wrx123wy@163.com (R.W.); bfxu1985@hhu.edu.cn (B.X.); zhengyuan@hhu.edu.cn (Y.Z.); 15950553095@163.com (S.Q.)

[2] Department of Wind Energy, Technical University of Denmark, 2800 Lyngby, Denmark; wzsh@dtu.dk

[3] Jiangsu Key Laboratory of Hi-Tech Research for Wind Turbine Design, Nanjing University of Aeronautics & Astronautics, Nanjing 210016, Jiangsu, China; tgwang@nuaa.edu.cn

* Correspondence: joephy@163.com

Received: 2 May 2018; Accepted: 5 June 2018; Published: 11 June 2018

Featured Application: This research provides a new pitching approach to increase the maximum power efficiency of vertical axis type wind turbines and water turbines used to absorb marine current energy. This approach did not deny the validity of the traditional methods. On the contrary, the new approach combined with traditional pitching method will improve power efficiency at all tip speed ratios, and well solve the problems of low efficiency and poor self-starting ability of the turbines.

Abstract: This paper presents a new variable pitch (VP) approach to increase the peak power coefficient of the straight-bladed vertical-axis wind turbine (VAWT), by widening the azimuthal angle band of the blade with the highest aerodynamic torque, instead of increasing the highest torque. The new VP-approach provides a curve of pitch angle designed for the blade operating at the rated tip speed ratio (TSR) corresponding to the peak power coefficient of the fixed pitch (FP)-VAWT. The effects of the new approach are exploited by using the double multiple stream tubes (DMST) model and Prandtl's mathematics to evaluate the blade tip loss. The research describes the effects from six aspects, including the lift, drag, angle of attack (AoA), resultant velocity, torque, and power output, through a comparison between VP-VAWTs and FP-VAWTs working at four TSRs: 4, 4.5, 5, and 5.5. Compared with the FP-blade, the VP-blade has a wider azimuthal zone with the maximum AoA, lift, drag, and torque in the upwind half-cycle, and yields the two new larger maximum values in the downwind half-cycle. The power distribution in the swept area of the turbine changes from an arched shape of the FP-VAWT into the rectangular shape of the VP-VAWT. The new VP-approach markedly widens the highest-performance zone of the blade in a revolution, and ultimately achieves an 18.9% growth of the peak power coefficient of the VAWT at the optimum TSR. Besides achieving this growth, the new pitching method will enhance the performance at TSRs that are higher than current optimal values, and an increase of torque is also generated.

Keywords: variable pitch; H-type VAWT; straight blade; DMST model; NACA0012; wind energy; power coefficient; tip speed ratio

---

1. Introduction

Wind turbines absorb wind energy and convert it into mechanical energy, and are classified according to the orientation of their axis of rotation into horizontal-axis wind turbines (HAWTs) and vertical-axis wind turbines (VAWTs). The HAWTs, represented by three-bladed propeller turbines, are the most common wind turbines because of the highest performance and easy manufacturing as a result of the great advance of aerodynamics and material engineering [1]. The VAWTs are further grouped into lift-type and drag-type. The lift-type VAWTs, which employ airfoil section blades to generate lift force as HAWTs, have higher rotation speed and better performance than the drag-type. The lift-type VAWTs, represented by the Φ-type and the straight-bladed type of Darrieus turbine, are also well-known [2]. On the other hand, most of them are used in small-scale applications, although they have several advantages over HAWTs, such as the ability to accept wind from any direction without yawing and the ability to provide direct rotary drive to a fixed load [3]. Some of the disadvantages of the VAWTs include an inability to self-start and relatively lower efficiency [4,5].

Darrieus turbine blades use airfoil sections designed as aircraft wing profiles. The NACA0012, NACA0015, and NACA0018 profiles are commonly used as blade sections [6]. Typically, these are designed to operate at small angles of attack (AoAs), lower than $\pm 10°$. At angles higher than this, the airfoil stalls and the flow separates on the upper surface of the blade, causing a loss of lift and an increase in drag [7]. At the startup of the turbine, a zero-tangential flow speed contributes to a large AoA, which then makes the airfoil stall. In this case, the pressure drag of the blade is so high that the lift cannot overcome the drag to self-start [8], not to mention any additional load from the electricity generator. Previous studies proposed a number of methods to make a self-starting lift-type VAWT. Kirke [9] reviewed those studies, and stated that high solidity turbines have the potential to increase the self-starting torque, but a higher solidity lowers the maximum power coefficient and narrows the operating range. Beri et al. [10] maintained that the Darrieus type VAWTs with asymmetrical blades have better self-starting behavior, but unfortunately, this also causes a reduction in peak efficiency. Zamani et al. [11,12] numerically investigated a 3-kW straight-bladed Darrieus type VAWT with a designed J-shaped profile. Their results showed that the J-shaped profile could suppress the vortices and improve the self-starting of the turbine. The hybrid wind turbine, comprising two vertical co-axial rotors, is normally constructed as a Darrieus with drag-type blades in the middle [13]. The primary purpose of the hybrid design is to promote the lack of torque when self-starting and at low tip speed ratios (TSRs) [4]. They can self-start normally, but after starting, they are less efficient than a turbine with normal lift due to the negative drag created at the middle [14]. It has been found that the above approaches hardly provide a way to promote both the self-starting ability and the peak power coefficient of the VAWTs at same time.

A particular method to address those issues is to develop a variable pitch (VP)-straight-bladed VAWT to control the pitch angles of the blades on each azimuthal angle in order to maximize turbine torque and start up easily [15]. A reduction of stall is the main mechanism whereby VP-technology produces significant improvement of the torque at start up and low TSRs [16]. Besides offering the greatest potential for achieving significantly increased torque at low and intermediate TSRs, VP-technology is also used to promote peak efficiency [17–19]. VP-VAWTs consist of active and passive-type turbines. The active designs are defined as those systems that produce blade pitch change through means other than the direct action of the aerodynamic forces acting on them in passive designs [16]. Staelens et al. [7] introduced three modifications for the increased power output of a straight-bladed VAWT by varying its pitch. The third modification proposed a continuous variation in the local AoA correction during the rotation cycle using a sinusoidal function. Although the power output obtained by using such a modification is less than the other two, it has the inherent advantage of being practically feasible. Also, Kiwata et al. [4] described a micro VP-VAWT that varies the pitch angle according to the azimuthal angle. The performance of the VP-VAWT was measured in an open-circuit wind tunnel, and the results were better than those of the fixed pitch (FP)-VAWT. Erickson et al. [20] tested the effects of cyclic blade pitch actuation on the efficiency and operability of a high-solidity VAWT with a cam and control rod mechanism to prescribe the pitch dynamics in the wind tunnel over a wide range of design and operational variables. The results revealed that a tuned first-order sinusoidal actuation system could achieve a maximum absolute power coefficient of 0.436, which is an increase of 35% over the optimal fixed-blade configuration, with self-starting capabilities and drastically improved performance at a wide range of suboptimal operating conditions. Hwang et al. [21] investigated the performance of a VP-turbine controlled by a cycloidal strategy. Compared with the FP-turbine, the performance at the upwind half-cycle of the VAWT with the optimized blade pitch angle was greatly improved, particularly at the azimuth region from 90° to 180°; the effective region was expanded, and was nearly twice as large as that of the FP-case. Chougule et al. [17] designed a 500-W VAWT to implement the pitch control mechanism using the cyclic VP-technology. Its aerodynamics were predicted using the double multiple stream tube (DMST) model, and their results showed that only a 5° increase of the pitch angle amplitude increased the power coefficient of the VAWT by 12%.

Despite the improvement of the self-starting ability and power coefficient, the above VP-approaches have the same limitation, which is mainly about the performance enhancement of azimuths with high AoAs in the case of the FP. In FP-VAWTs, each blade is constantly subjected to a variation of the AoA. Since the AoA constantly changes, the blade will generate large torque in some azimuthal angles, and low torque in other azimuthal angles. The overall performance is determined by the average torque of the blade in a cycle path line. Evidently, the low torques will produce a lower average value, and then degrade the power efficiency. The azimuthal position with the biggest geometric AoA in FP-VAWTs will achieve the biggest increase in the performance of the blade in VP-VAWTs. Assuming that the biggest geometric AoA is close to and below the stall value at the rated TSR, the magnitude of the maximum increase of the AoA would be much smaller if it still used the above-mentioned VP-approaches; thus, the increase of the AoA in other azimuths and the peak efficiency of the turbine would be much smaller, too.

To change this, we proposed a new VP-approach that was intended to mainly promote the performance of the blade working in the azimuths with smaller AoAs in FP-VAWTs. Ultimately, the aim is to largely enhance the peak efficiency of the VAWT in the rated or larger TSR. In this paper, the effect of the new approach on the performance of the blade or the turbine are investigated using a method implementing the double multiple stream tube (DMST) model.

2. Explanation of Variable Pitch Concept

2.1. Oscillating AoA

The Darrieus-type VAWT is a cross-flow rotor whose axis of rotation meets the flow of the working fluid at right angles. The blades perform the work twice in one cycle: i.e., the upwind half-cycle (azimuthal angle from 0° to 180°, where the 0°, 90°, 180°, and 270° positions are defined in Figure 1) and the downwind half-cycle (ranging from 180° to 360°). The azimuth angle of 0° is defined as a position where the directions of the local wind speed 'V_w' and the tangential flow speed '$U = r\omega$' are the same; the azimuthal angle of 180° is the position where the directions of the two speeds are opposite; as its anticlockwise rotation, the azimuthal angle of 90° is the position moving anticlockwise on a quarter circle from 0°, which is the fixed reference position. The resultant flow, which is the vector sum of the local wind speed 'V_w' and the tangential flow speed '$U = \omega r$', always comes from the upwind side of the blade: i.e., the outer side on the upwind half-cycle and the inner side on the downwind half-cycle [22]. Accordingly, as illustrated in Figure 1, the rotating blade experiences a changing resultant flow, including its magnitude and direction. The changing direction of the resultant velocity and chord line of the airfoil causes the oscillating AoA between the negative and positive values following a sinusoidal function of the azimuthal angle.

As shown in Figure 2, the AoA is determined by:

$$\alpha = \tan^{-1}\left[\frac{a\sin\theta}{\lambda_0 + a\cos\theta}\right] \tag{1}$$

where α is the AoA, a is the axial induction factor defined as $a = V/V_w$ (V is the velocity of the flow in the stream tube, V_w is the velocity of the free wind), θ is the azimuthal angle, and $\lambda_0 = R\omega/V_w$ represents the local TSR, where ω is the rotational speed, R is the rotational radius. The 0° and

180° positions are transformation points of the AoA between the negative and positive values, which determine their zero AoAs. Affected by the oscillating AoA, the blade of the VAWT inevitably produces fluctuating torque, even in steady conditions. For a given airfoil section, the non-dimensional normal and tangential forces are written as:

$$C_N = C_L \cos \alpha + C_D \sin \alpha \tag{2}$$

$$C_T = C_L \sin \alpha - C_D \cos \alpha \tag{3}$$

where C_N and C_T are the normal force and tangential force coefficients, respectively, and C_L and C_D are the lift and drag coefficients, respectively. According to Equations (2) and (3), the blade in the 0° and 180° azimuths will generate a zero-normal force and a negative tangential force, which cause a negative torque, according to Equation (4):

$$M = \frac{1}{2}\rho c R(2H)C_T V_r^2 \tag{4}$$

where M presents the torque of the blade, ρ is the air density, c is the chord length, $2H$ is the height of the blade, V_r is the local resultant velocity, and R is the rotation diameter. Due to the continuity of the blade torque variation, produced by the 0° and 180° azimuths, the torque in a zone around the 0° or 180° azimuth remains negative, and a wider zone may have a positive torque, but is still not big enough. In contrast, the 0° and 180°, and 90° and 270° angles produce the maximum AoAs. This is because the directions of the upwind flow and tangential flow are vertical to each other (Figure 1), and furthermore, the 90° azimuth having the larger upwind flow speed normally shows better aerodynamics than the 270° azimuth.

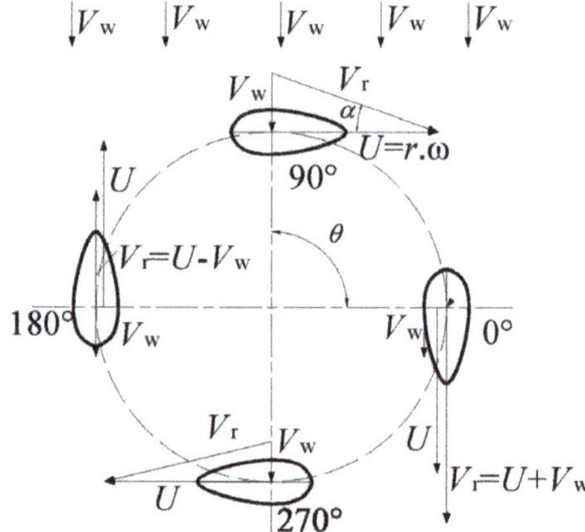

Figure 1. Illustration of Darrieus rotor.

2.2. Aerodynamics in Traditional VP-Technology

The VAWT blade is subject to cyclic variations in wind speed and the AoA, which are efficiency improvements that can result from corresponding variations of the pitch angle of the blade to optimize the AoA at each point in its cycle [20]. Any improvement made to the AoA can be of a cyclic nature, as shown in Figure 3 [23]. Thus, a typical curve of the pitch angle in traditional VP-technology is expressed by Equations (5) and (6):

$$\gamma = A_c \sin(\theta) \tag{5}$$

$$A_c = A_s - \left(A_s \frac{\lambda}{X_0}\right) \sin(\theta) \tag{6}$$

where γ is the variable value of the pitch angle changed with the azimuth, A_c is the maximum pitch angle at a certain TSR, A_s is the maximum pitch angle at all of the TSRs, λ is the TSR, X_0 is the TSR of max C_P at zero A_c, and θ is the azimuthal angle.

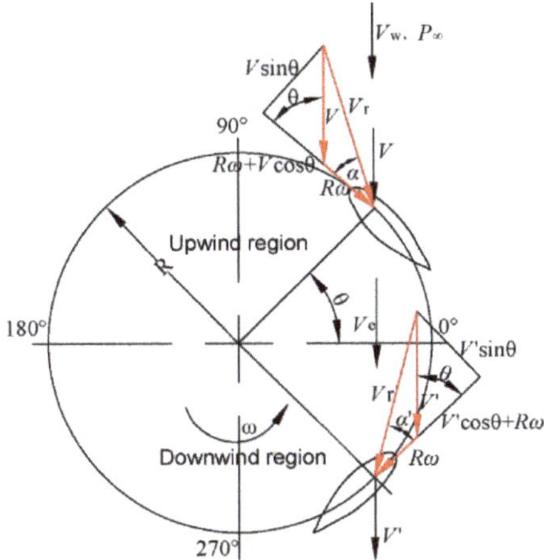

Figure 2. Relative wind speed vectors in the upstream and downstream section.

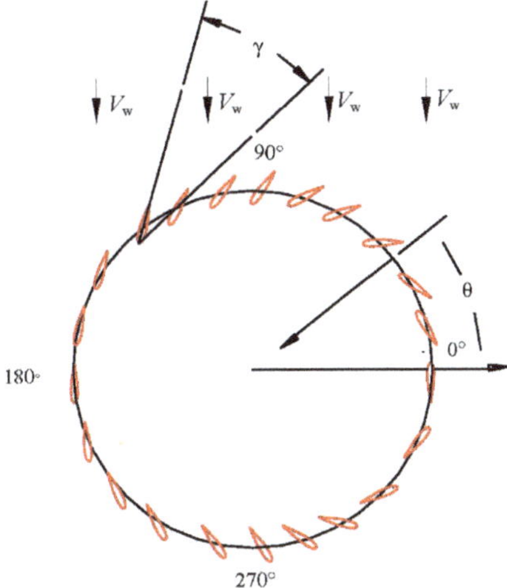

Figure 3. Illustration of cyclical variable pitch.

In another conventional VP-technology of VAWT, the blades have an oscillatory motion around their own axis, which was superimposed on the uniform rotation motion. Considering a cross-section of the turbine, the lines normal to the profile chord for every blade position intersect a given point "*P*" during a complete revolution (Figure 4) [24]. The variation of the pitch angle with respect to the azimuthal angle is expressed as:

$$\gamma = A_c \sin \theta - \arctan\left(\frac{\sin \theta}{\cos \theta + \lambda}\right) \tag{7}$$

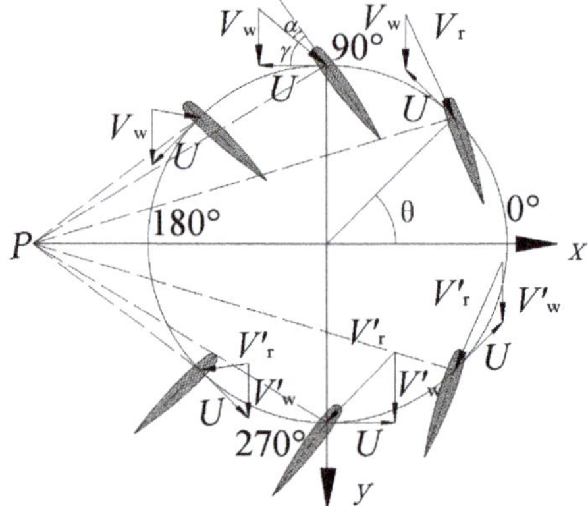

Figure 4. Illustration cycloidal variable pitch.

Two conventional VP-approaches have one point in common: the azimuthal angle with a larger geometric AoA will be given a larger pitch angle. Affected by the performance, as expressed by the lift coefficient ($C_L - \alpha$) of the airfoil, the blade has a stall static AoA of about 10°, as shown in Figure 5a [18]. At the start-up and small TSRs, the maximum AoA is much greater than the stall angle, the lift of the blade will decrease, and the drag will greatly increase after the stall occurs. In such cases, the VP-technology will change the pitch angle in the direction of the decreased AoA to avoid the stall. Two methods were used to achieve this goal. In the first method, the pitch angle changes only in the positions where the local geometric AoA exceeds the stall value, and the final local effective AoA is kept below the static stall angle [7,25], as illustrated in Figure 5b. The advantage of this method is that it makes the blade rotate with the optimum AoA in a much wider azimuthal zone. Its disadvantage is that the change of the pitch angle (represented by a dotted line in Figure 5b) is discontinuous at some points, which may not be practical to implement, and also causes early fatigue due to the abrupt dynamical loads [7]. The other method continuously changes the pitch angle following a sinusoidal function (Figure 5c). Under this control, the maximum amplitude of the sinusoidal function for the pitch angle is set to be equal to the maximum difference between the geometric AoA in the FP-VAWT and the effective AoA in the VP-VAWT. The advantage of this method is that the pitch angle as well as the forces changes smoothly and continuously, and then the method is physically and mechanically feasible to implement; the implementation procedures are presented in Figures 2 and 3. However, the power efficiency, although significantly larger than that of the FP-VAWT, is much lower than that of the other method illustrated in Figure 5b, because the final local effective AoA curve also follows a sinusoidal function in which the blade experiences the maximum performance only at two inflection points. At an optimum or larger TSR, the pitch angle should change toward the direction of the increased AoA for an improved performance in terms of the increased peak power coefficient of the VAWT. Meanwhile, if the variation of the pitch angle follows a sinusoidal function, as shown in Figure 5d, the space for performance enhancement would be exceedingly limited, because the maximum amplitude of the AoA change is much smaller (Figure 5d). A new approach is explored next to solve this problem.

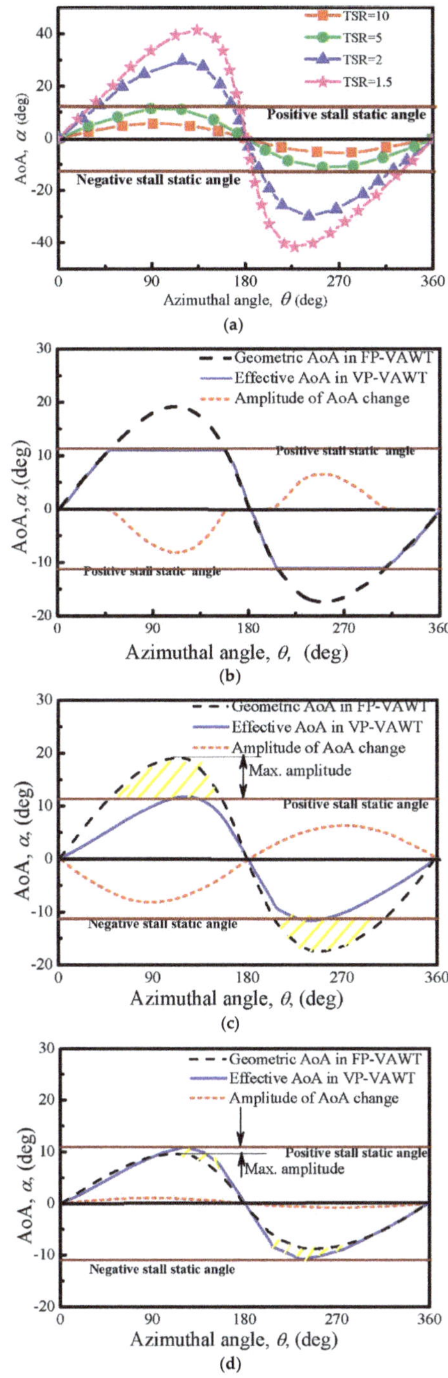

Figure 5. Local angle of attack as a function of the azimuthal angle. (**a**) AoA variation at different TSRs; (**b**) one pitching method at low TSR; (**c**) another pitching method at low TSR; (**d**) pitching method at rated TSR or higher TSR.

At an optimum or larger TSR, the pitch angle should change in the direction of the increased AoA for an improved performance in terms of the increased peak power coefficient of the VAWT. Alternatively, if the variation of the pitch angle follows a sinusoidal function, as shown in Figure 5d, the space for performance enhancement would be exceedingly limited, because the maximum amplitude of the AoA change is much smaller (Figure 5d). A new approach is explored next to solve this problem.

2.3. New VP-Approach

The new VP-strategy proposed in this paper focuses mainly on improving the performance by widening the range of the largest local effective AoA, instead of enlarging the maximum value. In the new VP-approach, the local effective AoA is kept below the stall angle and maintains a constant value in a wide zone during a rotation cycle; the curve is shown as a solid line in Figure 6. During the design of the new approach, four key points are adequately considered.

(1) How to deal with the zero AoA at the azimuthal angles of 0° and 180°: Since the positions of the 0° and 180° angles are the transformation points between the positive and negative AoAs, the design of the new VP-approach cannot avoid the zero value AoAs. The pitch angles in these two positions remain at the zero value.

(2) How to set the value of the constant effective AoA in Figure 6: At the rated TSR, the largest local AoA in a cycle should be designed as the optimum AoA corresponding to the largest lift–drag ratio, rather than the static stall AoA. Thus, if the pitch curve of the new approach is designed for a blade operating at the rated TSR, the constant value in Figure 6 should be equal to the maximum geometric AoA in a FP-VAWT, since both a larger and a smaller value would contribute to a smaller torque. Consequently, the pitch angles are also kept at zero at the two inflection-point locations, namely the azimuthal angles of 90° and 270°.

(3) Is the approach physically and mechanically feasible to implement? The amplitude of the change of the pitch angle is shown in Figure 6 as a dotted curve. In this curve, the pitch angle changes smoothly and continuously, and eliminates the drawback of the design in Figure 5b. The whole variation process of the pitch is illustrated in Figure 7, which is easily accomplished mechanically, such as a specially designed cam. As shown in Figure 7, the pitch angles varies in each 1/4 circle (0°–90°, 90°–180°, 180°–270°, and 270°–360°), and returns to zero at four positions (0°, 90°, 180°, and 270°). Here, we defined the presence of a positive pitch angle when the blade rotates counterclockwise around its own axis, and the leading edge of the blade points toward the inner side of the circular path. In contrast, the angle is negative when the blade rotates and the leading edge points in the opposite direction.

(4) What is the effect of the new VP-approach on the performance of the blade? After the mechanical feasibility of the approach is validated in theory, the following study focuses on the investigation of the effect of the new VP-approach using the DMST model.

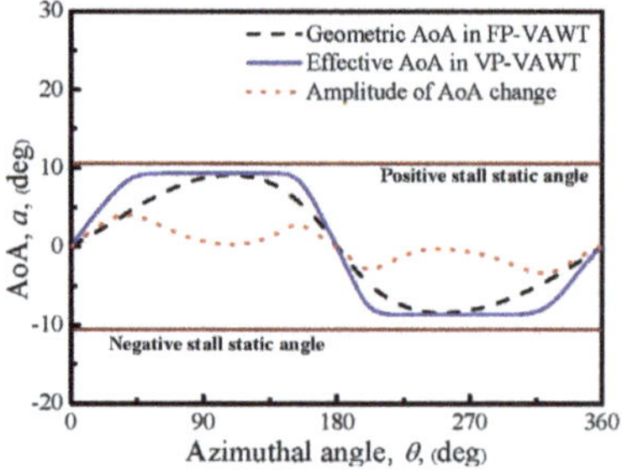

Figure 6. Curve of pitch angle in new approach.

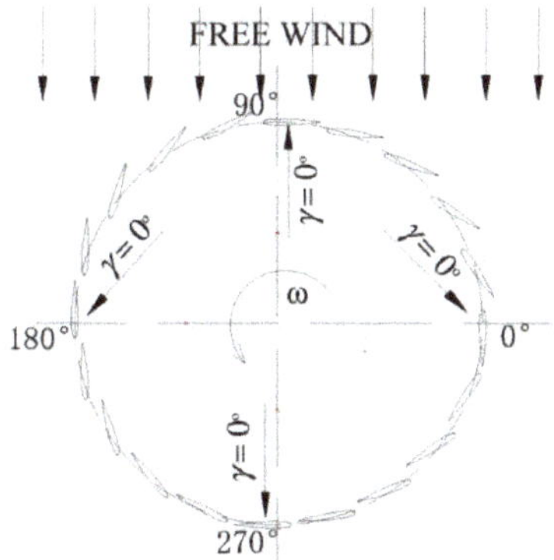

Figure 7. Illustration of new variable pitch.

3. Computational Model

3.1. Geometric Characteristics of VATW

The geometric characteristics of the straight-bladed VAWT studied in this paper are as follows. The NACA0012 blade profile: a height of 2 m (H = 1 m, Figure 8), a rotation radius (R) of 1 m, a chord length of the blade of 0.09 m, with two blades. Two straight-bladed VAWTs are also referred to as H-type VAWTs, since their configurations are much like an "H" shape. The operating conditions of the H-type VAWT aerodynamic parameter chosen for this study are as follows: an air density of 1.225 kg/m^3, a wind speed of 9 m/s, a kinematic viscosity of 1.48×10^{-5} m^2/s. The curve of the pitch angle with respect to the azimuthal angle was designed with the condition of a TSR = 4.5, in which the FP-turbine generates the peak power coefficient. The performance at four TSRs values (i.e., TSR = 4, 4.5, 5.0, and 5.5) was predicted, and their results were compared with each other. In addition, as shown in Figure 8, the effect of the radial arm was not considered, even though it is also a source of drag that reduces the power performance, due to the use of the DMST algorithm in the subsequent research.

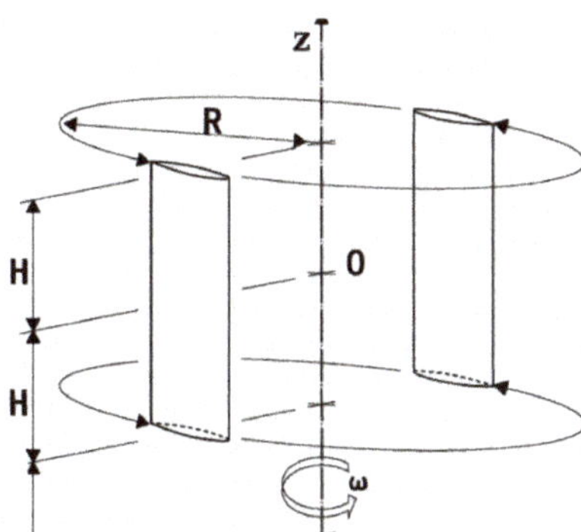

Figure 8. Configuration of an H-type vertical-axis wind turbine (VAWT).

3.2. Aerodynamic Model of Double Multi Stream Tube (DMST) Model

The DMST model makes use of the blade element momentum (BEM) theory, whereby the performance of the rotor is calculated by coupling the momentum equation in the mainstream direction of the wind and the aerodynamics analysis of the interactions between the airfoils in motion and the oncoming flow on the rotor [15,26]. The first approach using a momentum model to analyze the flow field around a VAWT was developed by Templin [27], who considered the rotor as an actuator disk enclosed in a simple stream tube, where the induced velocity through the swept volume of the turbine was assumed to be constant. An extension of this method to the multiple stream tubes model was developed by Strickland [28], who considered the swept volume of the turbine as a series of adjacent stream tubes. Additionally, Paraschivoiu [29] developed an analytical model, the DMST model, which considers a multiple-stream tube system divided into two parts, where the upwind and downwind components of the induced velocities at each level of the rotor are calculated by using the principle of two actuator disks in tandem. Subsequently, many predicted the performance of the Darrieus turbine using the DMST model, and all of the results showed that the DMST model has a high prediction accuracy for predicting the aerodynamics of the lift-type VAWTs [24,30–33]. The Darrieus-type wind turbine at low TSRs normally experiences a large variation within its AoA, and the blades are dominated by dynamic stall. In such conditions, the prediction must consider the effect of the dynamic stall on the accuracy. However, in the paper, the detailed investigation is performed at a high TSR, which is higher than four. Meanwhile, the variation of the AoA is getting much smaller and no stall is occurring; thus it has little effect on the prediction accuracy [22]. Accordingly, the dynamic stall was neglected in the paper. The model relies on pre-existing lift and drag data to calculate the blade aerodynamics. In a cross-flow wind turbine, blade–vortex interactions may occur, but its effects are not clear. There is not a well-developed algorithm for the modification of the lift and drag data due to the impingement of the vortex on the blades [34]. Thus, the interaction was also neglected in our work.

Since this investigation is focused on an H-type VAWT, a simplified version of the DMST model applicable to a straight-bladed rotor is briefly outlined below [35–37]. For an H-type VAWT, the resultant velocity, V_r, during the upwind half-cycle can be determined from the velocity vector triangle shown in Figure 2, which then gives:

$$V_r = V_w \sqrt{(\lambda_0 + a\cos\theta)^2 + (a\sin\theta)^2} \tag{8}$$

Applying the blade element theory and equating the change in momentum to the drag on the rotor for each stream tube, it is found that:

$$fa = \pi\eta(1 - a) \tag{9}$$

where $\eta = r/R$ is the dimensionless parameter of the radius, and f is the unwind function that characterizes the upstream half-cycle of the rotor on the blade element rotating in this zone. The upwind function is given by the following equation:

$$f = \frac{Nc}{8\pi R}\int_0^\pi \left(C_N\frac{\cos\theta}{|\cos\theta|} - C_T\frac{\sin\theta}{|\cos\theta|}\right)\left(\frac{V_r}{V_w}\right)d\theta \tag{10}$$

where N is the number of the blades, and C_N and C_T can be calculated from Equations (2) and (3), respectively. The lift and drag coefficients, C_L and C_D, respectively, are obtained by interpolating the known experimental data using both the local blade section Reynolds number $Re_b = V_r c/V_w$ and the local AoA. The local AoA can be calculated from Equation (1). Also, using Equation (8), Re_b is given as a function of the induced upwind velocity for each blade element in rotation.

3.3. Tip Loss Consideration

For a finite aspect ratio of the blade, tip loss is unavoidable. Thus, the effect of the blade tip loss was considered in the modeling to achieve an accurate performance prediction. The tip loss characterizes the tendency for trailing the vorticity from the blade tip, or from any other point where the circulation distribution changes, thereby reducing the blade's effectiveness. Prandtl's tip loss factor is commonly employed for VAWTs, and Soraghan et al. [30] described the mathematics in detail.

3.4. Computational Procedure

For a given H-Type rotor geometry and rotational speed ω, a value of the local TSR λ is chosen by assuming that the interference factor a is unity. Thus, Re_b and α are evaluated as first approximations, and the airfoil C_L and C_D characteristics are interpolated from the test data reported by Sheldahl and Klimas [38]. Then, using Equation (7), the normal and tangential force coefficients of the blade section are estimated, while Equation (10) allows the upwind function to be evaluated. With the first value of f, Equation (9) is used to calculate another value for the interference factor, and the iterations continue until successive sets of a are reasonably close. Convergence will be achieved, as the error is less than 0.0001. Once the final value of the upwind zone 'a' has been calculated, the local relative velocity V_r is determined from Equation (7) and the local AoA from Equation (1). A similar procedure is repeated for the downwind half-cycle. Additionally, it should be noted that here, the upwind velocity is an equivalent velocity expressed as $V_e = (2a - 1)V_w$.

4. Results Analysis

Due to the consideration of the tip loss of the blade, the results distributed along the spanwise direction of the finite length blade are not constant; their distribution shows symmetric characteristics on both sides of the middle height, namely $H = 0$, as shown in Figure 8. However, the differences between any two heights, except the position that is closest to the tip of the blade, are small for a large aspect ratio. Thus, the data from $H = 0$ are chosen to be analyzed and compared in the following sections.

4.1. AoA Variation

The final variation of the AoA in a full cycle is shown in Figure 9. As required by the original specifications of the design, the new approach increases the AoA in most azimuthal angles, which is much obvious in the upwind region. The downward parabolic AoA curve with respect to the azimuthal angle is more open, and more azimuthal angles have a higher AoA, but the largest value is not changed in the upwind section. In the downwind region, the increase of the AoA is obvious when the TSR = 4, 4.5, and 5, but it decreases as the TSR increases, and even shows a negative tendency when the TSR = 5.5. As the rotation speed increases, more wind energy is absorbed into the stream tubes of the upwind part, and the steam flows though the stream tubes of the downwind part with lower velocity, which causes a lower AoA in the condition of a constant rotation speed. It is also observed in all of the sub-figurations that the real curve of the AoA is not as smooth as the designed one, especially in the downwind region. This is because the AoA strongly depends on the considerably changed axial induction factor.

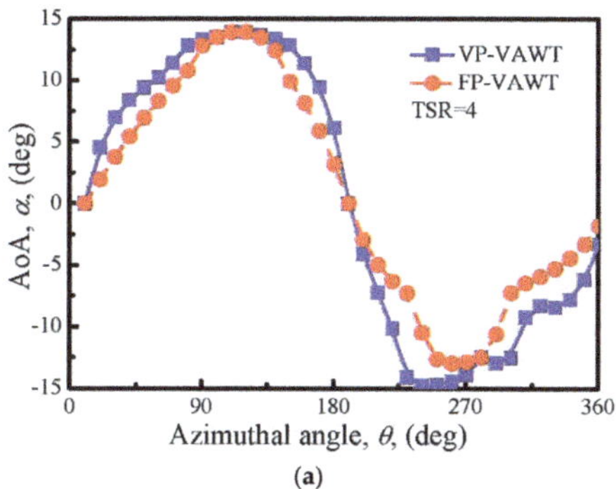

Figure 9. *Cont.*

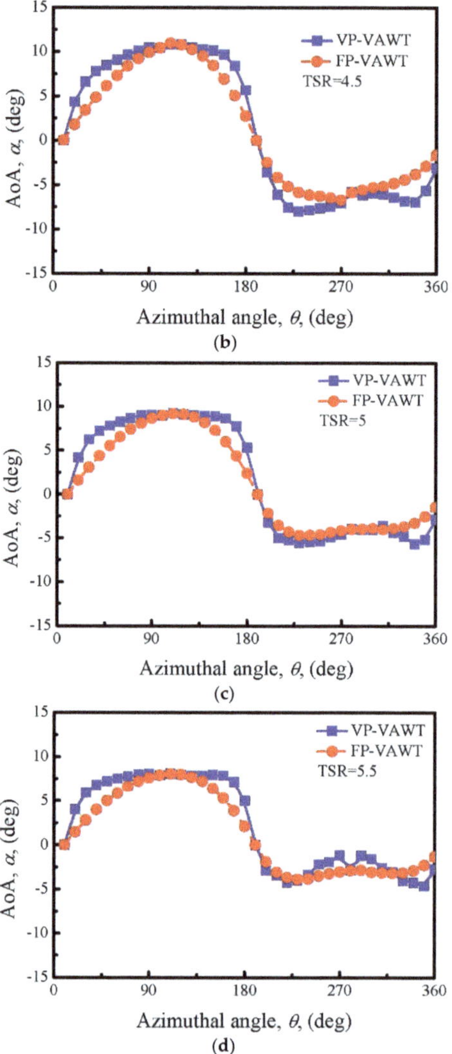

Figure 9. Angle of attack (AoA) as a function of azimuthal angle in variable pitch (VP) and fixed pitch (FP) VAWTs. (**a**) TSR = 4; (**b**) TSR = 4.5; (**c**) TSR = 5; (**d**) TSR = 5.5.

4.2. Lift Force Variation

The lift generated by the blade is a key parameter of a lift-type VAWT, because it determines the magnitude of the tangential force that causes the turbine rotation. One of the effects of the new VP-control changes the lift of the blade. The variation of the local lift coefficient during a complete rotation in four TSRs is illustrated in Figure 10, which reveals that, when the TSR = 4, the lift coefficient fluctuates intensively during a whole range of azimuthal angles, whether in the VP-VAWT or FP-VAWT, and a much lower value is observed at around 90° and 270°. This is because the AoA values in the large region around 90° and 270° are still larger than the stall static angles, which leads to the occurrence of the stall and then a decreased lift coefficient. As can be observed in Figure 10, the VP-blade does not create a larger maximum lift coefficient than the FP-blade, but it does have a wider stall zone.

The reason is that the new VP-approach increases the AoA and creates a wider azimuthal zone having a larger AoA than the stall value, but the maximum effective AoA is not changed. All of these phenomena indicate that the VP-approach designed for the rated TSR is not suitable to be applied to improve the performance of the VAWT at much lower TSRs.

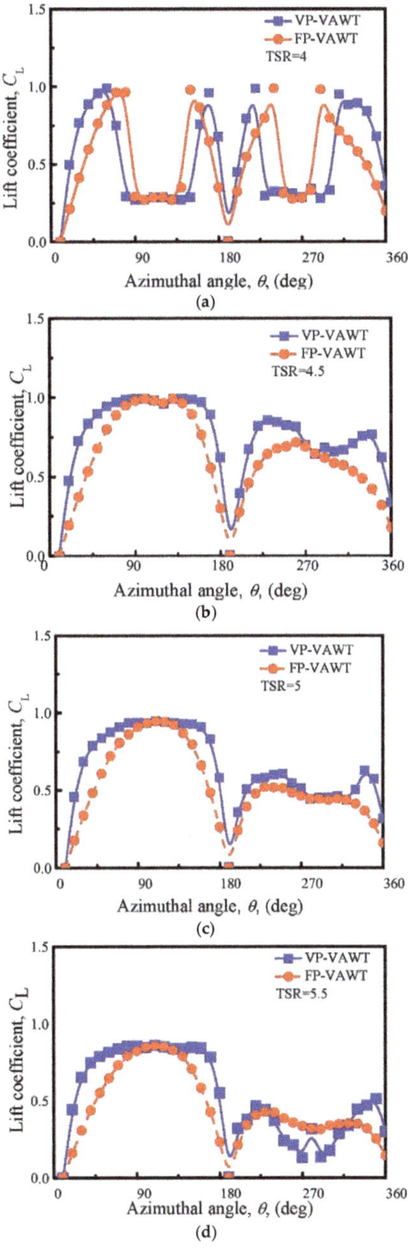

Figure 10. Lift coefficient as a function of azimuthal angle in VP-VAWTs and FP-VAWTs. (**a**) TSR = 4; (**b**) TSR = 4.5; (**c**) TSR = 5; (**d**) TSR = 5.5.

At TSRs larger than 4.5, stall does not occur; the curve of the lift coefficient changes smoothly, and shows an arched shape in the upwind or downwind half-cycles in the FP-VAWT. In the same conditions, the stall does not occur in the VP-VAWT either; larger lift coefficients are obtained almost in the whole rotation zone except four at positions, namely: 0°, 90°, 180°, and 270°. Finally, the novel approach achieves a larger zone with the largest lift coefficients in the upwind half-cycle and two new larger maximum coefficients in the downwind half-cycle. A comparison of the increase among the TSRs reveals that the increase of the lift is in the downstream region as the TSR increases. Even when the TSR = 5.5, the lift is affected by the AoA; as shown in Figure 9, the VP-blade produces lower lift than the FP-blade in an extensive zone of the downwind region.

4.3. Drag Force Variation

The drag coefficients of the blades of the FP-VAWT and VP-VAWT are illustrated in Figure 11. First, when the TSR = 4, the blade that stalls generates a large peak drag in the upwind region, which is almost five times as large as that when the TSR = 4.5, and another larger peak drag in the downwind region of the VP-VAWT. A comparison of the distribution between the VP-VAWTs and FP-VAWTs indicates that the new VP control also increases the drag of the blade because of the increased AoA. The trends of the drag changes between the VP-VAWTs and FP-VAWTs are much like those of the lift changes. However, the increased drag cannot negate the advantage of the new VP-approach. According to the aerodynamics of the airfoil, the AoA that has the largest lift coefficient normally results in a large drag coefficient. The final effect is a promotion of the blade performance or not, which is determined by whether the component of the increase in the tangential direction of the lift is larger or smaller than that of the drag coefficient.

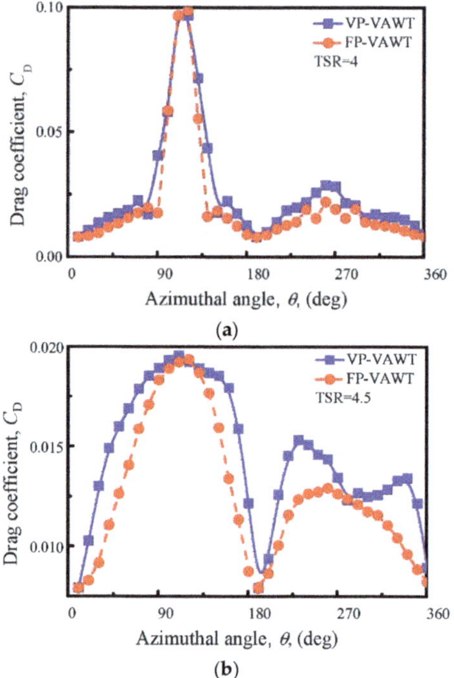

Figure 11. *Cont.*

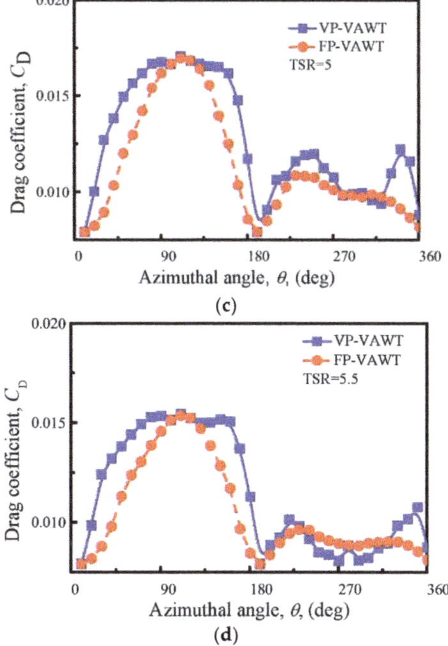

Figure 11. Drag coefficient as a function of the azimuthal angle in VP-VAWTs and FP-VAWTs. (a) TSR = 4; (b) TSR = 4.5; (c) TSR = 5; (d) TSR = 5.5.

4.4. Resultant Velocity Variation

The resultant velocity, which is the vector sum of V_w and U, is depicted as a function of the azimuthal angle in Figure 12. Other than the free wind velocity, the resultant velocity is the effective flow velocity acting on the blade, which is another factor, together with the AoA, that determines the magnitude of all of the forces. The magnitude of the resultant velocity is determined by the tangential flow velocity and flow velocity going through the stream tube according to the local velocity triangle. At the same rotation speed as the VP-VAWT and FP-VAWT, in the upwind half-cycle, the pitch angle does not affect the local tangential flow velocity and wind velocity. This causes slight changes in the resultant velocity in the upwind half-cycle, as shown in Figure 12. The slight changes are also observed when the TSR = 4 and 4.5 in the downstream half-cycle, and large changes occur when the TSR = 5 and 5.5 (Figure 12). The VP-approach decreases the wake velocities of the upwind stream tubes, and changes the velocity of the stream flowing through the stream tubes and the axial induction factor of the downwind half-cycle. The changed stream velocity and axial induction factor determine the different resultant velocities. As depicted in Figure 12, when the TSR = 5.5, the resultant velocity has a tendency to be even in a wide zone of the downwind region.

4.5. Torque of Blade Variation

The torque of a single blade as a function of the azimuthal angle is shown in Figure 13, which reflects well the effects of the new VP-approach on the improvement of the aerodynamics of the blade in every azimuthal position. As expressed by Equations (3) and (4), the torque is extensively affected by lift and drag. For the increase of both forces, the final torque in the VP-VAWT significantly differs from that in the FP-VAWT. At a TSR = 4 and the downwind half-cycle, the torque is increased in some azimuthal angles, but it is also reduced in some zones; thus, it is not easy to directly judge whether the overall torque is increased or decreased (Figure 13). In comparison, when the TSR = 4.5, 5, and 5.5, a different extent of increase is obtained on both sides of the 90° azimuthal angle, where one peak torque is generated in the FP-VAWT, and then the low torque zone gets correspondingly narrower. In the downwind region, the large increase is shown when the TSR = 4.5 and 5; the VP-blade creates two new maximum values on the two sides of the azimuthal angle of 270°, where another peak torque of the FP-blade is generated. It should be noted that the former two new maximum values are larger than the latter ones; thus, the curve of the downwind section exhibits an "M" shape. At a TSR = 5.5, the torque is increased in some positions, and decreases in the zone adjacent to the 270° azimuthal angle.

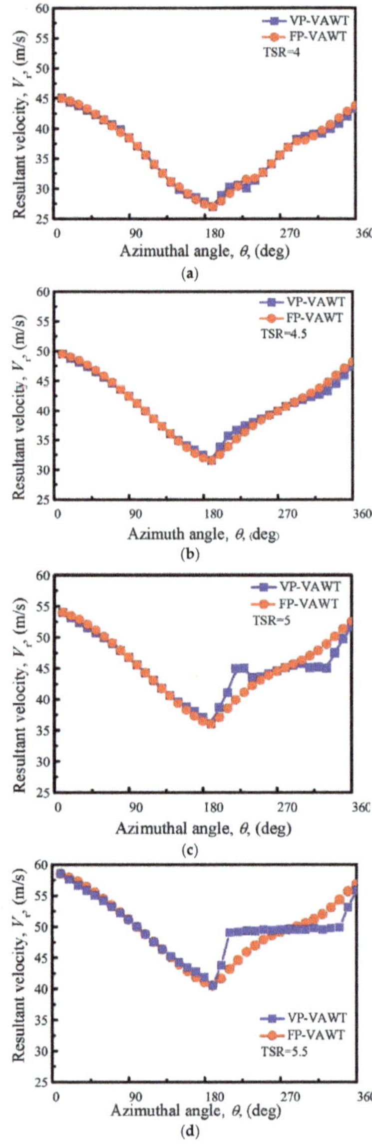

Figure 12. Resultant velocity as a function of the azimuthal angles in VP-VAWTs and FP-VAWTs. (**a**) TSR = 4; (**b**) TSR = 4.5; (**c**) TSR = 5; (**d**) TSR = 5.5.

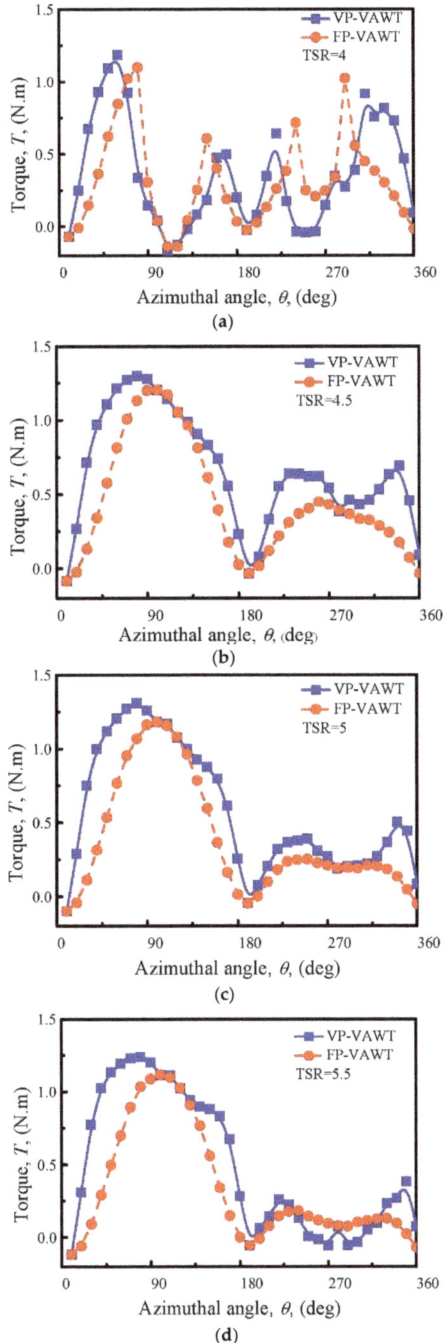

Figure 13. Torque as a function of the azimuthal angles in VP-VAWTs and FP-VAWTs. (**a**) TSR = 4; (**b**) TSR = 4.5; (**c**) TSR = 5; (**d**) TSR = 5.5.

4.6. Power Output

The power output along the radial direction is presented in Figure 14; its value is equal to the sum of the power output of a stream tube in the upwind part and the corresponding downwind part, which can be written as Equation (11). Adding the power outputs of the stream tubes in all of the radial and heights positions, and subsequently dividing the power of the free wind, the total power coefficient is obtained through Equation (12):

$$P_i = P_{up.i} + P_{dw.i} \tag{11}$$

$$C_{P.t} = \sum_{-H}^{H} \sum_{i=-R}^{R} \frac{P_i}{\frac{1}{2} S \rho V_w^3} \tag{12}$$

where C_P denotes the power coefficient; the subscript i is the radial position of the stream tube; up and dw represent the upwind and downwind region, respectively; t means total; and S is the swept area. As shown in Figure 14, the power output of the FP-blade shows the peak value in the radial position of $r = 0$; toward the two sides, the power is gradually reduced, and reaches zero or even a negative power in the $-R$ or R positions. The whole curve of the FP-blade is a downward parabolic line. In comparison, the new VP-approach extends the position generated at the peak power from one point in the FP-VAWT to a wide radial zone. Therefore, the final distribution of the power output is similar to a trapezoid when the TSR = 4.5 and 5. For a TSR = 5.5, although a drop of the torque occurs downstream, as shown in Figure 13, the VP-blade still experiences an increase in the power output in most radial positions, because of a larger torque increase in the upwind region. The results presented in Figure 14 also show a small drop in the position of $r = 0$, but a larger increase occurs in other positions. Thus, the total power output will be absolutely increased.

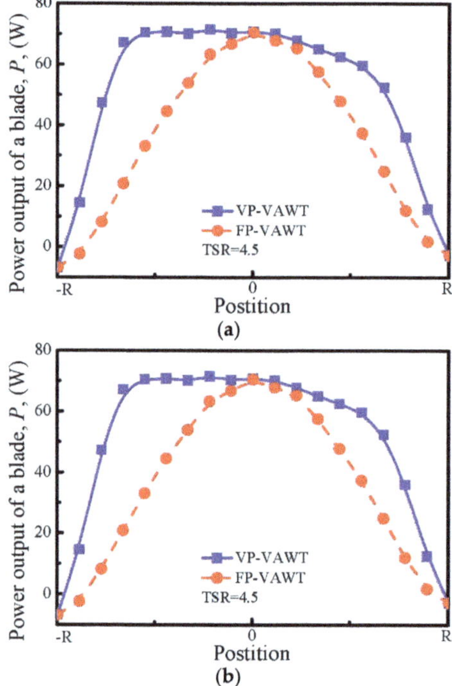

Figure 14. *Cont.*

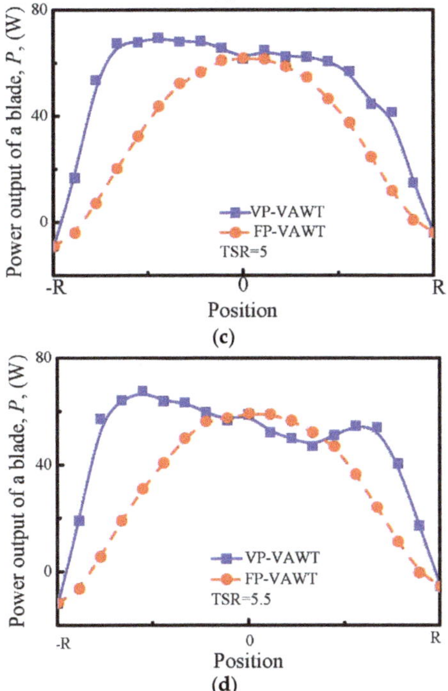

Figure 14. Power output of a blade as a function of position in VP-VAWTs and FP-VAWTs. (**a**) TSR = 4; (**b**) TSR = 4.5; (**c**) TSR = 5; (**d**) TSR = 5.5.

A comparison of the power coefficient of the two types of VAWTs when the TSR = 4 to 5.5 is shown in Figure 15, which reveals the obvious promotion of the power coefficient at a TSR = 4.5, 5, and 5.5. The increase is 16% at a TSR = 4.5, and becomes larger as the TSR increases. Finally, an 18.9% increase of the peak power efficiency of the VP-VAWT is achieved when the TSR = 5. The new curve has a wider TSR zone with a high power coefficient, and the drop in the power coefficient starts when the TSR = 5.5, but at a low rate. The results in Figure 15 clearly indicate that the new approach is highly suitable for improving the peak efficiency of the VAWT at the optimum or larger TSR.

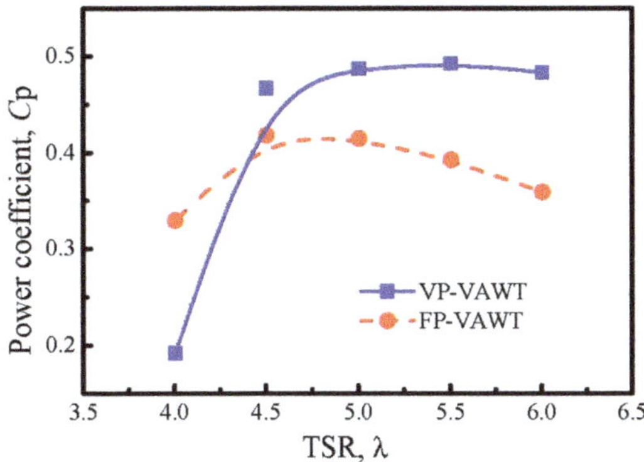

Figure 15. Power coefficient curves of VP-VAWTs and FP-VAWTs.

Since the parasitic drag of the radial arm was not considered in the investigation, the prediction value may be higher than the real one in Figure 15. While the structure of the arm can be well-designed, such as using an airfoil profile and being installed at a zero AoA relative to the free wind, several methods such as this can greatly reduce their parasitic drag. Therefore, the new pitching approach does effectively promote the overall performance of the turbine, even after considering the parasitic drag of the arm.

5. Conclusions

This paper proposed a new VP-approach to increase the peak power coefficient of a straight-blade VAWT. The new approach was designed for the blade at the optimum TSR, and focuses mainly on widening the azimuthal zone of the largest AoA, instead of increasing it. A two-blade H-type VAWT was studied using the DMST model to validate the effect of the new approach, in which the tip loss was evaluated by Prandtl's mathematics. The data of the middle height of the VAWT was adopted to analyze and compare the FP-blade and the VP-blade from six different characteristics, namely the AoA, lift, drag, resultant velocity, torque, and power output. The conclusions drawn from the study are as follows.

(1) Through the application of the new approach, a large increase in the AoA is produced in the upwind half-cycle of the VP-blade. A large increase in the AoA in the downwind half-cycle is also achieved when the TSR = 4, 4.5, and 5, but the increase becomes smaller as the TSR increases, so that a negative growth is shown when the TSR = 5.5.

(2) The new approach greatly enhances the lift of the blade and distributes the maximum lift in a wider zone in the upwind region. In the downwind region, two new and larger maximum lifts are created when the TSR = 4.5 and 5, but a smaller lift is obtained when the TSR = 5.5.

(3) The increased AoA leads to an increase of the drag of the blade in most of the positions. The trend of the distribution of the drag is similar to that of the lift.

(4) The resultant velocity experiences little changes in the upwind region, while in the downwind region, the changes are obvious, and become more obvious as the TSR increases.

(5) Influenced by the lift and drag, the ultimate effect of the torque shows that a large increase is obtained in the upwind region and downwind region at most of the TSRs. The distribution trend is also much like that of the lift.

(6) The new VP-approach also enlarges the azimuthal zone of the blade with the highest power output. Consequently, an 18.9% increase of the peak power efficiency of the VAWT is achieved when the TSR = 4.5. Additionally, the turbines are capable of working with high efficiency in a wider TSR zone. The new VP-approach designed at the rated TSR is suitable to enhance the peak efficiency of VAWTs.

Author Contributions: Z.Z. and T.W. conceived and designed the research; R.W. performed the calculations; R.W. and S.Q. analyzed the data; Y.Z. contributed analysis tools; Z.Z. and B.X. wrote the paper. W.S. revised the paper.

Acknowledgments: This paper was supported by the project of National Natural Science Foundation of China, Project No. 11502070 and 51607058; the Fundamental Research Funds for the Central Universities, Project No. 2018B24914; National Basic Research Program of China ("973" Program), Project No. 2014CB046200.

Conflicts of Interest: The authors declare no conflict of interest.

Nomenclature

Roman Letters

a	induction factor
A_c	amplitude of the pitch angle, [deg]
A_s	Max amplitude of the pitch angle, [deg]
c	length of chord line, [m]

C_D	drag coefficient
C_L	lift coefficient
C_N	normal force coefficient
C_T	tangential force coefficient
C_P	power coefficient
N	number of blade
H	1/2 height of rotor, [m]
R	equatorial radius of rotor, [m]
S	swept area of wind rotor, [m^2]
U	Tangential velocity, [m/s]
V_r	resultant velocity, [m/s]
V_w	free wind velocity, [m/s]
X_0	TSR of max C_P at zero A_c
AoA	angle of attack, [deg]
CFD	computational fluid dynamics
DMST	double multi stream tube (model)
FP	Fixed pitch
HAWT	horizontal axis wind turbine
VAWT	vertical axis wind turbine
TSR	tip speed ratio
VP	variable pitch

Greek Letters

α	angle of attack, [deg]
ρ	density of air, [kg·m^{-3}]
θ	azimuthal angle, [deg]
ω	rotor angular velocity, [rad·s^{-1}]
λ	tip speed ratio (TSR)
γ	pitch angle changed with the azimuth, [deg]
ζ	dimensionless parameter of height

References

1. Kosaku, T.; Sano, M.; Nakatani, K. Optimum pitch control for variable-pitch vertical-axis wind turbines by a single stage model on the momentum theory. In Proceedings of the IEEE Conference on Systems, Man and Cybernetics, Yasmine Hammamet, Tunisia, 6–9 October 2002. [CrossRef]
2. Bhutta, M.M.A.; Hayat, N.; Farooq, A.U.; Ali, Z.; Jamil, S.R.; Hussain, Z. Vertical axis wind turbine–A review of various configurations and design techniques. *Renew. Sustain. Energy Rev.* **2012**, *16*, 1926–1939. [CrossRef]
3. Li, Y.; Zheng, Y.F.; Zhao, S.Y.; Feng, F.; Li, J.Y.; Wang, N.X.; Bai, R.B. A review on aerodynamic characteristics ofstragith-bladed vertical axis wind turbine. *Acta Aerodyn. Sin.* **2017**, *35*, 368–385.
4. Kiwata, T.; Yamada, T.; Kita, T.; Takata, S.; Komatsu, N.; Kimura, S. Performance of a vertical axis wind turbine with variable-pitch straight blades utilizing a linkage mechanism. *J. Environ. Eng.* **2010**, *5*, 213–225. [CrossRef]
5. Aggarwal, A.; Mishra, D.; Chandramaouli, V. Study on optimization of control mechanism in vertical axis wind turbine. *Indian J. Sci. Technol.* **2017**, *10*, 1–10. [CrossRef]
6. Claessens, M.C. The Design and Testing of Airfoils in Small Vertical Axis Wind Turbines. Master's Thesis, Delft University of Technology, Delft, The Netherlands, 2006.
7. Staelens, Y.; Saeed, F.; Paraschivoiu, I. A straight-bladed variable-pitch VAWT concept for improved power generation. AIAA-2003-0524. In Proceedings of the ASME 2003 Wind Energy Symposium, Reno, NV, USA, 6–9 January 2003.
8. Kirke, B.K. Evaluation of Self-Starting Vertical Axis wind Turbines for Stand-Alone Applications. Ph.D. Thesis, Griffith University Gold Coast, Brisbane, Australia, 1998.
9. Cooper, P.; Kennedy, O.C. Development and analysis of a new novel Vertical Axis Wind Turbine. In Proceedings of the 42nd Annual Conference of the Australian and New Zealand Solar Society, Perth, Australia, 30 November–3 December 2004.

10. Beri, H.; Yao, Y. Effect of camber airfoil on self-starting of vertical axis wind turbine. *J. Environ. Sci. Technol.* **2011**, *4*, 302–312. [CrossRef]

11. Zamani, M.; Maghrebi, M.J.; Moshizi, S.A. Numerical study of airfoil thickness effects on the performance of J-shaped straight blade vertical axis wind turbine. *Wind Struct.* **2016**, *22*, 595–616. [CrossRef]

12. Zamani, M.; Maghrebi, M.J.; Varedi, S.R. Starting torque improvement using J-shaped straight-bladed Darrieus vertical axis wind turbine by means of numerical simulation. *Renew. Energy* **2016**, *95*, 109–126. [CrossRef]

13. Sun, X.J.; Chen, Y.J.; Cao, Y.; Wu, G.Q.; Zheng, Z.Q.; Huang, D.G. Research on the aerodynamic characteristics of a lift drag hybrid vertical axis wind turbine. *Adv. Mech. Eng.* **2016**, *8*, 1–11. [CrossRef]

14. Dwiyantoro, B.A.; Suphandani, V. The system design and performance test of hybrid vertical axis wind turbine. *AIP Conf. Proc.* **2017**, *1831*, 020030. [CrossRef]

15. Bianchini, A.; Ferrara, G.; Ferrari, L. Pitch Optimization in small-size Darrieus wind turbines. *Energy* **2015**, *81*, 122–132. [CrossRef]

16. Pawsey, N.C.K. Development and Evaluation of Passive Variable-Pitch Vertical Axis Wind Turbines. Ph.D. Thesis, The University of New South Wales, Sydney, NSW, Australia, 2002.

17. Chougule, P.; Nielsen, S. Overview and Design of self-acting pitch control mechanism for vertical axis wind turbine using multi body simulation approach. *J. Phys.* **2014**, *524*, 012055. [CrossRef]

18. Sagharichi, A.; Maghrebi, M.J.; ArabGolarcheh, A. Variable pitch blades: An approach for improving performance of Darrieuswind turbine. *J. Renew. Sustain. Energy* **2016**, *8*, 053305. [CrossRef]

19. Xisto, C.M.; Páscoa, J.C.; Leger, J.A.; Trancossi, M. Wind energy production using an optimized variable pitch vertical axis rotor. In Proceedings of the Mechanical Engineering Congress and Exposition, Montreal, QC, Canada, 14–20 November 2014.

20. Erickson, D.W.; Wallace, J.J.; Peraire, J. Performance Characterization of Cyclic Blade Pitch Variation on a Vertical Axis Wind Turbine. In Proceedings of the 49th AIAA Aerospace Sciences Meeting including the New Horizons Forum and Aerospace Exposition, Orlando, FL, USA, 4–7 January 2011.

21. Hwang, I.S.; Yun, H.L.; Kim, S.J. Optimization of cycloidal water turbine and the performance improvement by individual blade control. *Appl. Energy* **2009**, *86*, 1532–1540. [CrossRef]

22. Islam, M.; Ting, D.S.K.; Fartaj, A. Aerodynamic models for Darrieus-type straight-bladed vertical axis wind turbines. *Renew. Sustain. Energy Rev.* **2008**, *12*, 1087–1109. [CrossRef]

23. Schonborn, A.; Chantzidakis, M. Development of a hydraulic control mechanism for cyclic pitch marine current turbines. *Renew. Energy* **2007**, *32*, 662–679. [CrossRef]

24. Camporealea, S.M.; Magi, V. Streamtube model for analysis of vertical axis variable pitch turbine for marine currents energy conversion. *Energy Convers. Manag.* **2000**, *41*, 1811–1827. [CrossRef]

25. Zuo, W.; Kang, S. Numerical simulation of aerodynamic performance of H-type wind turbine with pitch angle changing. *J. Eng. Thermophys.* **2015**, *36*, 501–504.

26. Paraschivoiu, I. *Wind Turbine Desigh with Emphasis on Darrieus Concept*; Polytechnic International Press: Montreal, QC, Canada, 2002.

27. Templin, R.J. *Aerodynamic Performance Theory for the NRC Vertical-Axis Wind Turbine*; LTR-LA-160; National Research Council of Canada: Ottawa, ON, Canada, 1974.

28. Strickland, J.H. *Darrieus Turbine: A Performance Prediction Model Using Multiple Streamtubes*; SAND75-0431; Sandia National Laboratories: Livermore, CA, USA, 1975.

29. Paraschivoiu, I.; Delclaux, F. Double multiple streamtube model with recent improvements (for predicting aerodynamic loads and performance of Darrieus vertical axis wind turbines). *J. Energy* **1983**, *7*, 250–255. [CrossRef]

30. Paraschivoiu, I.; Trifu, O.; Saeed, F. H-Darrieus Wind Turbine with Blade Pitch Control. *Int. J. Rotat. Mach.* **2009**, *2009*, 505343. [CrossRef]

31. Soraghan, C.E.; Leithead, W.E.; Yue, H.; Feuchtwang, J. *Double Multiple Streamtube Model for Variable Pitch Vertical Axis Wind Turbines*; American Institute of Aeronautics and Astronautics: Reston, VA, USA, 2013.

32. Saeidi, D.; Sedaghat, A.; Alamdari, P.; Alemrajabi, A.A. Aerodynamic design and economical evaluation of site specific small vertical axis wind turbines. *Appl. Energy* **2013**, *101*, 765–775. [CrossRef]

33. Zhao, Z.Z.; Yan, C.; Wang, T.G.; Xu, B.F.; Zheng, Y. Study on approach of performance improvement of VAWT employing double multiple stream tubes model. *J. Sustain. Renew. Energy* **2017**, *9*, 023305. [CrossRef]

34. Ferrer, E.; Willden, R. Blade-wake interactions in cross-flow turbines. *Int. J. Mar. Energy* **2015**, *11*, 71–83. [CrossRef]
35. Elkhoury, M.; Kiwata, T.; Aoun, E. Experimental and numerical investigation of a three-dimensional vertical-axis wind turbine with variable-pitch. *J. Wind Eng. Ind. Aerodyn.* **2015**, *139*, 111–123. [CrossRef]
36. Firdaus, R.; Kiwata, T.; Kono, T.; Nagao, K. Numerical and experimental studies of a small vertical-axis wind turbine with variable-pitch straight blades. *J. Fluid Sci. Technol.* **2015**, *10*, 14–29. [CrossRef]
37. Sengupta, A.; Biswas, A.; Gupta, R. Studies of some high solidity symmetrical and unsymmetrical blade H-Darrieus rotors with respect to starting characteristics, dynamic performances and flow physics in low wind streams. *Renew. Energy* **2016**, *93*, 536–547. [CrossRef]
38. Sheldahl, R.E.; Klimas, P.C. *Aerodynamic Characteristics of Seven Symmetrical Airfoil Sections through 180-Degree Angle of Attack for Use in Aerodynamic Analysis of Vertical Axis Wind Turbines*; SAND80-2114; Sandia National Laboratories: Livermore, CA, USA, 1981.

Article

A Fully Coupled Computational Fluid Dynamics Method for Analysis of Semi-Submersible Floating Offshore Wind Turbines Under Wind-Wave Excitation Conditions Based on OC5 Data

Yin Zhang and Bumsuk Kim *

Faculty of Wind Energy Engineering Graduate School, Jeju National University, Jeju City 63243, Korea; scarletyuki@jejunu.ac.kr
* Correspondence: bkim@jejunu.ac.kr; Tel.: +82-64-754-4402

Received: 27 September 2018; Accepted: 16 November 2018; Published: 20 November 2018

Abstract: Accurate prediction of the time-dependent system dynamic responses of floating offshore wind turbines (FOWTs) under aero-hydro-coupled conditions is a challenge. This paper presents a numerical modeling tool using commercial computational fluid dynamics software, STAR-CCM+(V12.02.010), to perform a fully coupled dynamic analysis of the DeepCwind semi-submersible floating platform with the National Renewable Engineering Lab (NREL) 5-MW baseline wind turbine model under combined wind–wave excitation environment conditions. Free-decay tests for rigid-body degrees of freedom (DOF) in still water and hydrodynamic tests for a regular wave are performed to validate the numerical model by inputting gross system parameters supported in the Offshore Code Comparison, Collaboration, Continued, with Correlations (OC5) project. A full-configuration FOWT simulation, with the simultaneous motion of the rotating blade due to 6-DOF platform dynamics, was performed. A relatively heavy load on the hub and blade was observed for the FOWT compared with the onshore wind turbine, leading to a 7.8% increase in the thrust curve; a 10% decrease in the power curve was also observed for the floating-type turbines, which could be attributed to the smaller project area and relative wind speed required for the rotor to receive wind power when the platform pitches. Finally, the tower-blade interference effects, blade-tip vortices, turbulent wakes, and shedding vortices in the fluid domain with relatively complex unsteady flow conditions were observed and investigated in detail.

Keywords: computational fluid dynamics; floating offshore wind turbine; dynamic fluid body interaction; semi-submersible platform; OC5 DeepCWind

---

1. Introduction

Energy generation from offshore wind farms has been garnering the attention of researchers, owing to the abundance of resources and low environmental impact. Compared to offshore wind turbines in shallow water, floating offshore wind turbines (FOWTs) have more advantages [1]; i.e., there are several deep-water sites suitable for installing turbines, wind is more abundant in offshore areas, and public concerns on the visual and environmental impacts are minimized with this technology. Some floating wind farms have been established; for instance, the first full-scale 2.3-MW FOWT was installed in Hywind near the coast of Norway, and last year, five 6-MW FOWTs were installed in the North Sea off the coast of Peterhead, Scotland.

However, it is difficult and expensive to operate a real-scale test model and accurately calculate critical loads because the complex multi-physical phenomena are not easy to simulate in reality. In addition, this technology is dependent on extreme weather situations (such as 25 m/s cut-out speeds). Thus, the use of computational methods, involving virtual full-scale modeling, may increase the development of the controllers' reliability (such as structure and loads) of FOWTs, reduce the risks involved, and build confidence in the design stage. Among the codes used, one of the most famous ones is the Fatigue, Aerodynamics, Structures, and Turbulence code (FAST), which was developed by the National Renewable Engineering Lab (NREL) based on the blade element momentum (BEM) theory [2]. However, the BEM theory is seldom applied in FOWT situations owing to its theoretical limitation. In contrast, the fluid structure interaction (FSI) simulations, as a modern computational analysis method, has proven to be an accurate and convincing method for considering aero-hydro-servo-elastic problems; however, complex fluid conditions and blade deformation presents significant computational challenges.

Further, correctly simulating the movement of floaters on free surfaces is also a major challenge; many researchers from different institutions have developed various codes and solvers to simulate the hydrodynamic performance of floaters. Nearly all solvers are based on the following theories: the potential-based panel approach and Morrison equation. The former cannot determine viscous flow details and is usually used together with the damping coefficient obtained from experimental test data. The FAST code HydroDyn module has applied this method. The Morison equation is a semi-empirical equation; this equation mainly describes the inline force in oscillatory flow conditions; this also has theoretical limitations and it cannot adequately describe the time-dependent force. Examples include the wave analysis MIT (WAMIT), TimeFloat, and CHARM3D. However, there are still some physical phenomena that cannot be fully described. Conversely, the unsteady computational fluid dynamics (CFD) approach can simulate with consideration of all physical effects, including flow viscosity, hydrostatic forces, wave diffraction, radiation, wave run-up, and slamming and provide reliable and accurate results regarding the platform movement.

Owing to the reasons mentioned above, the CFD method is widely considered an effective and reliable method to simulate the FOWT problem; to date, several CFD-related investigations have been performed. However, previous studies have used the following methods, ignoring some effects, leading to inaccurate results. First, to investigate the hydrodynamic load and motion response of a platform on an FOWT, some studies just simplified the problem into wind turbine aerodynamic loading or ignored the tower and rotor-nacelle-assembly. Second, some studies focused on aerodynamic loading but restricted the motion of the floating platforms to a prescribed position or did not allow the platform to move with 6 DOF.

Unai Fernandez-Gamiz et al. [3] developed an improved BEM-based solver to verify the NREL 5-MW wind turbine and determined the bending moment and thrust force in the blade root; they also investigated rectangular sub-boundary layer vortex generators using the CFD method [4], which showed the highest vortex generator suitable for separation control. Nematbakhsh et al. [5] developed a CFD spar model and successfully captured strong nonlinear effects, which cannot be captured using the FAST code. Furthermore, their study also observed that when the wave amplitude was large, a discrepancy could exist between CFD and FAST. Vaal et al. [6] used the BEM method to investigate the surge motion of FOWTs. This showed that the BEM method could only provide a reasonable solution under slow surge motion conditions; this is because, in this condition, the wake dynamics can be ignored. Zhao and Wan [7] used a Naoe-FOAM-SJTU simulated OC4 platform to study the effects of the presence of wind turbines. They carried out platform pitch motions at high wind speeds and investigated the wind turbine effect on the floating platform. Tran et al. [8] set the platform to execute a prescribed sinusoidal pitching motion and changed the motion amplitudes and frequencies, instead of modeling a floating platform with 6 DOF using the unsteady BEM theory, generalized dynamic wake (GDW), and CFD; large discrepancies were observed when the pitch amplitude increased to $4°$. Tran et al. [9] analyzed an FOWT system under a prescribed sinusoidal surge motion, and found that thrust and power varied significantly, which is related to the oscillation frequency; the surge motion amplitude also varied significantly. Liu et al. [10] superimposed three DOF platform motions (surge, heave, and pitch) and concluded that the platform motion significantly impacted the thrust and torque of the wind turbine. Ren et al. [11] used FLUENT analysis for a 5-MW tension-leg-platform-type turbine under coupled wave-wind conditions and validated the simulation results against experimental data. They only considered the surge motion and concluded that during the variation in the average/mean surge response of the system, aerodynamic forces played the main role. Quallen et al. [12] performed a CFD simulation involving an OC3 spar-type FOWT model under wind–wave excitation conditions. The mean surge motion predicted using the CFD model was 25% less than that predicted using FAST. Tran and Kim [13] modeled an OC4 semi-submersible FOWT using the dynamic fluid body interaction (DFBI) method and an overset mesh technique under wind–wave excitation conditions. A good overall agreement was found between the CFD results and FAST data. Both codes used the quasistatic method for modeling the mooring lines. S. Gueydon et al. [14] modeled a semi-submersible platform using the aNyPHATAS code to investigate operating rotor effects on drift motions and additional damping. Chen et al. [15] modeled a semi-submersible FOWT with two different blade configurations in a wave basin to further optimize the blade design for FOWTs. A.J. Dunbar et al. [16] developed an open-source CFD/6-DOF solver based on OpenFOAM and compared rotational and translational motions with FAST, demonstrating the accuracy of this tightly coupled solver.

The main purpose of this study was to conduct a virtual test of a real-scale 5-WM semi-submersible FOWT using the advanced CFD method. The hydrodynamic responses were validated using the latest physical test data of the Offshore Code Comparison, Collaboration, Continued, with Correlations (OC5) projects. Full-configuration FOWT simulations, simultaneously considering the rotating blade motion with 6-DOF platform dynamics were performed; a relatively large discrepancy in the predicted power was observed owing to the different properties of the mooring line and rotating inertia moment between the OC4 and OC5 projects. This proves the high infinity result of OC5 project. Further, this study may provide some reference for the validation of the CFD method for use in the OC5 Phase II system and high-fidelity simulation investigations of FOWTs in coupled aero-hydro conditions.

The OC5 DeepCWind semi-submersible floating wind turbine model was used for the investigation, which is briefly described in Section 2. The numerical methods used in the study are introduced in Section 3. The aerodynamic validation studies performed using different modeling tools are briefly presented in Section 4. Section 5 presents the results of the dynamic responses of the floating system under regular wave conditions. Section 6 presents the simulation results of the fully coupled configuration. Section 7 presents the conclusions of the study.

2. Floating Offshore Wind Turbine Model

2.1. Model Description

A semi-submersible FOWT tested in Phase II of the OC5 project was investigated. The design parameters of the full-scale OC5 DeepCWind semi-submersible platform are summarized in Table 1. The NREL 5-MW baseline wind turbine model was set above the tower, 87.6 m from the water surface; we used airfoil data from the DOWEC project, which is also mentioned in Jonkman's work [1] from NREL. The major properties of the NREL 5-MW baseline wind turbine are given in Table 2; the cross-sections of the rotor blade were composed of a series of Delft university of technology (DU) and national advisory committee for aeronautics (NACA) 64 airfoils from the hub to the tip of the outboard section. The computer-aided design (CAD) model of the blade was first developed using Solidworks software (Dassault systems, Velizy-Villacoublay, France), as shown in Figure 1. Except for the cylinder in the blade root and transition section, the control airfoil spread along the blade includes the DU series airfoil from 40% to 21% thickness and a NACA airfoil of 18% thickness. Details concerning the wind blade aerodynamics including the blade twist, chord length, and airfoil designation are presented in Table 3.

Table 1. Full system structural properties.

Parameters	Value
Mass	1.3958×10^7 kg
Draft	20 m
Displacement	1.3917×10^4 m^3
Center of mass (CM) location below seawater level (SWL)	8.07 m
Roll inertia about system CM	1.3947×10^{10} kg/m^2
Pitch inertia about system CM	1.5552×10^{10} kg/m^2
Yaw inertia about system CM	1.3692×10^{10} kg/m^2

Table 2. Blade structural properties.

Parameters	Value
Length (w.r.t. root along axis)	61.5 m
Overall (integrated) mass	2.2333×10^4 kg
Second mass moment of inertia	1.48248×10^7 kg/m^2
First mass moment of inertia	4.5727×10^5 kg/m
CM location	20.475 m

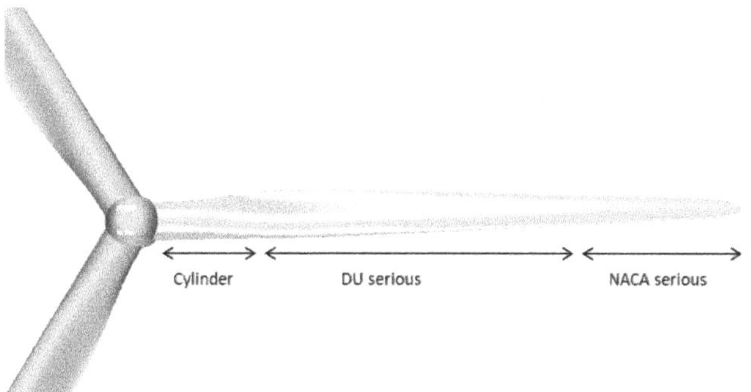

Figure 1. Airfoil construction of National Renewable Engineering Lab (NREL) 5 MW blade.

Table 3. Blade airfoil distribution of National Renewable Engineering Lab (NREL) 5 MW wind turbine.

Node Radius (m)	Twist Angle (deg)	Chord Length (m)	Airfoil Designation
2.867	13.308	3.542	Cylinder
5.600	13.308	3.854	Cylinder
8.333	13.308	4.167	Cylinder
11.750	13.308	4.557	DU 40
15.850	11.480	4.652	DU 35
19.950	10.162	4.458	DU 35
24.050	9.011	4.249	DU 30
28.150	7.795	4.007	DU 25
32.250	6.544	3.748	DU 25
36.350	5.361	3.502	DU 21
40.450	4.188	3.256	DU 21
44.550	3.125	3.010	NACA 64-618
48.650	2.319	2.764	NACA 64-618
52.750	1.526	2.518	NACA 64-618
56.167	0.863	2.313	NACA 64-618
58.900	0.370	2.086	NACA 64-618
61.633	0.106	1.419	NACA 64-618

2.2. OC4 and OC5 Projects

Previous studies mostly used test data from the study by Coulling et al. [17], which was led by the University of Maine at the maritime research institute Netherlands (MARIN) offshore wave basin in 2011; however, the geometrically scaled model did not perform as expected under the low-Reynolds number wind conditions. In addition, only semi-submersible properties (center of mass, inertia force, etc.) were considered in the OC4 project, while the OC5 project considered the properties of the full system, and the mooring line properties could be adjusted. Hence, a new model was built that resulted in better-scaled thrust and torque loads. However, this turbine was retested in 2013, and this turbine was examined in Phase II of the OC5 project [18]. The different physical properties of the semi-submersible platforms in the two projects are summarized in Table 4.

Table 4. Comparison of OC4 and OC5 project.

Semisubmersible Platform	OC5 Phase II	OC4 Phase II
Mass	12,919,000 kg	13,444,000 kg
Draft	20 m	20 m
CM below SWL	14.09 m	14.4 m
Roll inertia	7.5534×10^9 kg/m^2	8.011×10^9 kg/m^2
Pitch inertia	8.2236×10^9 kg/m^2	8.011×10^9 kg/m^2
Yaw inertia	1.3612×10^{10} kg/m^2	1.391×10^{10} kg/m^2
Buoyancy center below SWL	13.15 m	-
Mooringline anchors from center	837.6 m	837.6 m
Mooringline fairlead from center	40.868 m	40.868 m
Unstretched mooringline length	835.5 m	835.5 m
Mooringline mass density	Line 1: 125.6 kg/m Line 2: 125.8 kg/m Line 3: 125.4 kg/m	113.35 kg/m
Mooringline extensional stiffness	Line 1: 7.520 E8 N Line 2: 7.461 E8 N Line 3: 7.478 E8 N	7.536 E8 N
6 degrees of freedom (DOF) Natural Periods	Surge: 107 s Sway: 112 s Heave: 17.5 s Roll: 32.8 s Pitch: 32.5 s Yaw: 80.8 s	Surge: 107 s Sway: 113 s Heave: 17.5 s Roll: 26.9 s Pitch: 26.8 s Yaw: 82.3 s

The turbine is a 1/50th scale horizontal-axis model of the NREL 5-MW reference wind turbine with a flexible tower affixed atop a semi-submersible platform. The DeepCwind semisubmersible platform is composed of the main column and three offset columns linked to the main column via several pontoons and braces, as mentioned in the OC5 report [18]; the 1/50th scale and full-scale models are shown in Figure 2. A 5-MW baseline wind turbine is vertically mounted on the main column so that the hub height from the sea surface is 90 m. In addition, the platform is moored with three catenary mooring lines, with fairleads located at the base columns. The anchors are located 200 m below the sea surface, on the seabed. One mooring line is aligned in the wave direction, which is also the platform surge direction; the other two mooring lines are distributed around the platform uniformly and the attachment angle between each mooring line is 120°.

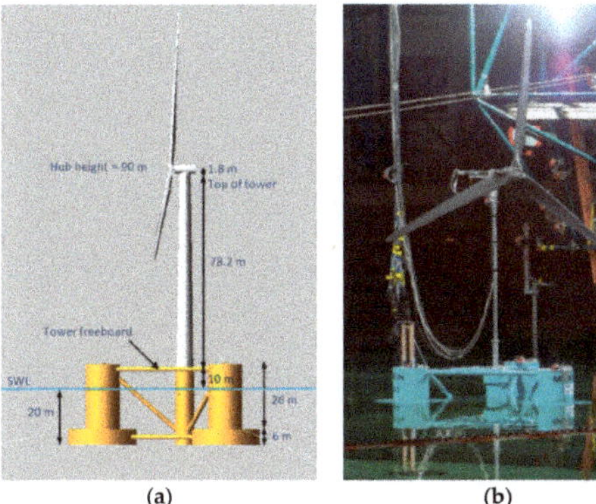

(a)   (b)

Figure 2. Semi-submersible model: (**a**) Full-scale model; (**b**) The 1/50th scale model in maritime research institute Netherlands (MARIN) wave basin.

3. Simulation Method

3.1. Numerical Setting and Governing Equations

This paper presents a numerical modeling tool using commercial CFD software, STAR-CCM+(V12.02.010) (Siemens, Munich, Germany), to perform a fully coupled dynamic analysis of the DeepCwind semi-submersible floating platform with the NREL 5-MW baseline wind turbine model under combined wind-wave excitation conditions.

This investigation used the unsteady incompressible Navier–Stokes equations, according to the first principles of the conservation of mass and momentum. To solve the pressure–velocity coupling, a semi-implicit method was used, which involved a predictor–corrector approach. Second-order upwind and central difference schemes were used for the convection terms and temporal time discretization, respectively. Additionally, the shear stress transport (SST) k-ω turbulence model (Menter's Shear Stress Transport) is a robust two-equation, eddy-viscosity turbulence model used for many aerodynamic applications to resolve turbulent behavior in the fluid domain and was first introduced in 1995 by F.R. Menter [19]. The model combines the k-ω and k-e turbulence models; therefore, the k-ω turbulence model can be used in the inner region of the boundary, and the k-e turbulence model can be used in free shear flow. Menter's SST turbulence model can be expressed as follows:

$$\frac{\partial(\rho k)}{\partial t} + \frac{\partial(\rho u_j k)}{\partial x_j} = P - \beta^* \rho \omega k + \frac{\partial}{\partial x_j}\left[(\mu + \sigma_k \mu_t)\frac{\partial k}{\partial x_j}\right] \tag{1}$$

$$\frac{\partial(\rho \omega)}{\partial t} + \frac{\partial(\rho u_j \omega)}{\partial x_j} = \frac{\gamma}{v_t} P - \beta \rho \omega^2 + \frac{\partial}{\partial x_j}\left[(\mu + \sigma_\omega \mu_t)\frac{\partial \omega}{\partial x_j}\right] + 2(1 - F_1)\frac{\rho \sigma_{\omega 2}}{\omega}\frac{\partial k}{\partial x_j}\frac{\partial \omega}{\partial x_j} \tag{2}$$

To obtain details of the free surface between air and water, the unsteady CFD method with the volume of fraction (VOF) approach coupled with the 6 DOF solver was used for the hydrodynamic analysis considering the surge, sway, heave, roll, pitch, and yaw motions of the platform.

3.2. Dynamic Fluid Body Interaction (DFBI) Method

The DFBI method was applied to simulate the motion of the rigid FOWT body in response to pressure and shear forces in the fluid domain and consider the recovery force from the mooring lines. STAR-CCM+ was used to calculate the resultant force and moment acting on the body due to various influences and also to solve the governing equations of rigid body motion to determine the new position of the rigid body. A flow chart illustrating the DFBI method is shown in Figure 3.

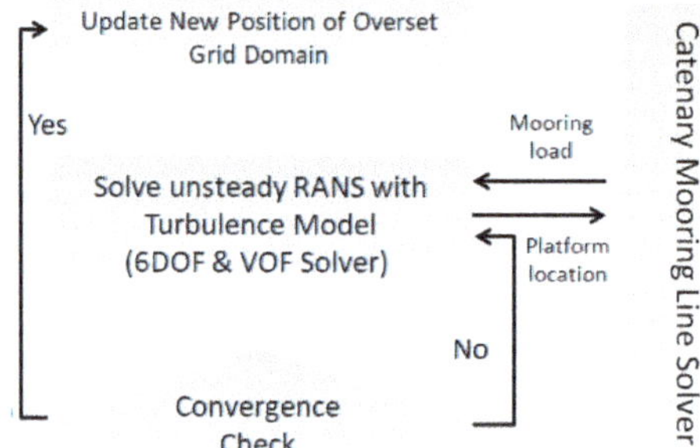

Figure 3. Flowchart of Dynamic Fluid Body Interaction (DFBI) method.

3.3. Overset Mesh Technology

The overset mesh technique, also called overlapping or chimera grids, was applied. A new internal interface node was created within the overset mesh region. This volume-type interface enables the coupling of solutions on the domains using automatically generated sets of acceptor cells in one mesh and donor cells in another mesh. Varying values of the donor cells affect the values of the acceptor cells based on interpolation. This method can handle complex geometries and body motions in dynamic simulations.

3.4. Mooring Line Modeling and Damping

The catenary coupling model was used to model an elastic, quasi-stationary catenary, hanging between two endpoints and subject to its own weight in the field of gravity. In a local Cartesian coordinate system, the shape of the catenary is given by the following set of parametric equations:

$$x = au + bsin(u) + \alpha \tag{3}$$

$$y = acosh(u) + \frac{b}{2}sinh^2(u) + \beta \tag{4}$$

$$for\ u_1 \leq u \leq u_2 \tag{5}$$

In addition, a wave-damping area was applied, considering the wave reflection near the outlet boundary; this treatment includes a wave-damping zone. The wave-damping area was designed to minimize the effects of wave reflections on the far downstream outlet boundary. As a result, the VOF wave could be damped in the pressure outlet boundary to reduce wave oscillations. This damping introduces vertical resistance to the vertical motion of the wave.

4. Aerodynamic Validation of Wind Turbine

4.1. Numerical Setting and Mesh Convergence Test

An aerodynamic simulation of the rotor part was performed to validate the accuracy of the 3D model and the numerical model using the CFD method. The hexahedral computation domain size and boundary type was 8D(x) × 5D(y) × 3D(z) and extended up to 2.5D and 5.5D in the upstream and downstream x-directions from the wind turbine, respectively.

Both poly grids and trim grids were tried in the mesh test procedure, and the trim mesh was finally selected owing to its robust features and relatively low computational cost. Nineteen layers of prism grids were generated to determine the blade-attached flow, and the near-wall first boundary layer thickness was 3.82×10^{-6} m. A mesh convergence test was also performed, and the results are given in Table 5; five sets of grids were generated with different grid densities while all the other parameters remained unchanged. A total of 22 million grids were selected with both a short calculation time and acceptable power loss. Details of the trim mesh around the blade tip are presented in Figure 4; at the same time, denser grids around the leading and trailing edges of the airfoils were used to detect fluid details. Y plus is a dimensionless value used to measure the mesh quality; a value below 1 is produced when the k-ω SST turbulence model is applied. In this study, the Y plus value was much lower than 1 in all five cases.

Table 5. Mesh convergence test.

Mesh	Thrust (kN)	Power (kW)	Power Variation
14 millions	726	4980	−5.0%
20 millions	728	4990	−4.8%
22 millions	731	5080	−3.0%
30 millions	730	5120	−2.3%
38 millions	734	5240	0%

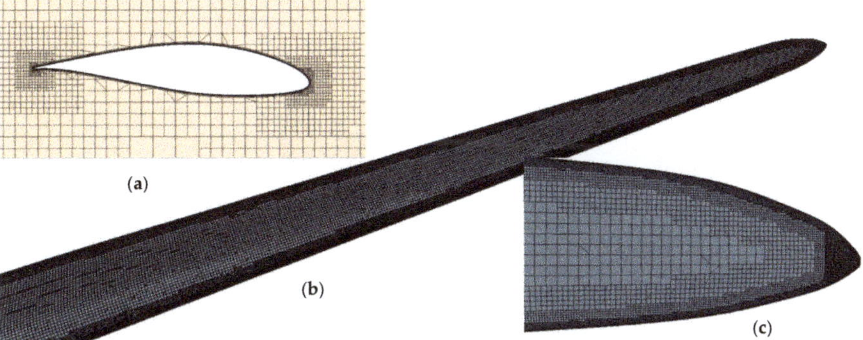

Figure 4. Trim cell detail: (**a**) Mesh section near 45 m spinwise airfoil; (**b**) Trim mesh around blade tip; (**c**) Blade surface mesh.

4.2. Validation of Rotor Aerodynamic Performance

Currently, there are three main methods used for simulating aerodynamic performance: the BEM method, generalized dynamic wake (GDW) model, and CFD model. The results of each model are shown in Figure 5. The CFD results are in good agreement with those of the other codes with regard to both power and thrust prediction, which were collected by Gyeongsang national university (GNU) using FAST code. The BEM method was observed to overestimate results at a relatively high wind speed, which is also noted in other studies [20]. Sivalingam et al. also compared results between the CFD and BEM methods; there was a good agreement in terms of the thrust and torque results below the rated speed. However, because of tip loss factors at relatively high wind speeds, a deviation of axial induction factors was shown by the BEM method, while the CFD method captured wake rotation accurately [21].

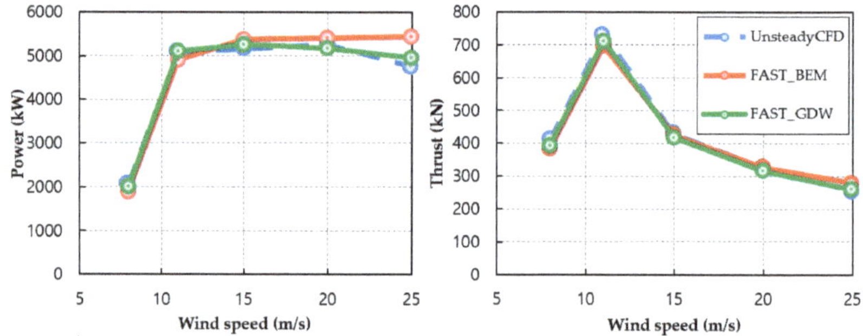

Figure 5. Power and thrust in 8, 11, 15, 20, and 25 m/s uniform wind speeds.

4.3. Study of Wind Profile and Tower Dam Effect Under Onshore Wind Turbine Generator Conditions

The aerodynamics of the NREL 5-MW fixed wind turbine was studied on a full-scale without the floating platform. The results will later be compared with the data for a floating wind turbine. All the numerical settings used in the CFD-rigid body motion (RBM) approach in the previous simulations were applied to the rotor part, except the inlet uniform wind was replaced with the wind profile.

The wind profile shows variations in the horizontal wind speed with height, which may result in increased fatigue loading and reduced power output, usually characterized by the power law, as follows:

$$v = v = v_{hub}(H/H_{hub})^{\alpha} \tag{6}$$

The wind shear exponent (alpha) was 0.12 for flat onshore conditions. Figure 6 illustrates the computational mesh domain; the upstream boundary of the inlet is defined as velocity inlet and the pressure outlet is defined as the downstream boundary. Symmetric boundary conditions were applied in the far field region and a no-slip wall condition was imposed on the surface. Tran et al. [20] had carried out a convergence test to determine the fluid domain size; herein, we used a hexahedral computational domain size of 1000 m × 600 m × 275 m (length, width, height), the same as that used by Tran et al, in the x-direction; the fluid domain extended 313 m upstream and 687 m downstream to help analyze the fluid domain and consider the impact of the vortex after the tower, considering the time-dependent motion of the rotating blades; we used the RBM method in this study.

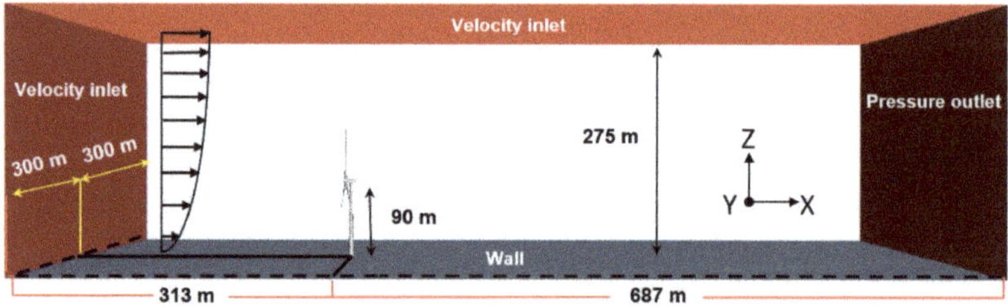

Figure 6. Onshore 5 MW wind turbine fluid domain and boundary conditions.

Aerodynamic simulations of onshore wind turbines were conducted using the RBM method under unsteady conditions; the obtained torque output under 11 m/s wind conditions was compared with the results of the FSI method obtained at the University of California. As shown in Figure 7, there was good agreement between both, but the predictions from our simulation were slightly higher, which may be because blade deformation was ignored in the RBM method [22,23].

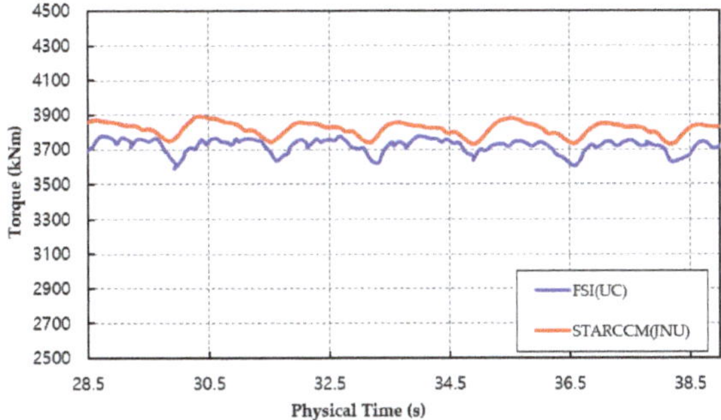

Figure 7. Torque curve between fluid structure interaction (FSI) method and unsteady method in 11 m/s wind speed.

5. Hydrodynamic Response of Floating Platform

5.1. Free-Decay Test

Free-decay tests are methods generally used in a wave tank to determine the natural period of a floating system. The OC4 project was based on a 1/50th-scale semi-submersible platform in the MARIN wave basin in 2011. This model was revised two years later, to provide more precise test data for the OC5 project [18]. Owing to the different properties of the mooring line (line stiffness) and rotating inertia forces in the two tests, the OC4 and OC5 data reveal good agreement in terms of translation motions (surge, sway, and heave) but a relatively large discrepancy in terms of rotating motions (roll, pitch, and yaw), as observed in Table 6. Six-DOF free decay tests were conducted to determine the hydrodynamic damping characteristics of the OC5 semi-submersible platform.

Table 6. Natural period in OC4 and OC5 project.

DOF	OC5 Natural Period	OC4 Natural Period
Surge	107 s	107 s
Sway	112 s	112 s
Heave	17.5 s	17.5 s
Roll	32.8 s	26.9 s
Pitch	32.5 s	26.8 s
Yaw	80.8 s	82.3 s

The wave mode was set as still water, and the air density was zero. Only the platform was considered to simplify the simulation; however, the gross mass should also be considered. The platform was given a prescribed displacement and released to move freely from the initial position. This test considered only three rigid-body DOFs, that is, the surge, pitch, and heave motions.

The results are presented in Figure 8, along with the simulation results from GNU and the wave basin test results from phase II of the OC5 project. The heave and surge time-domain responses for the platforms are in good agreement, as similar results were obtained in the heave and surge periods in the OC4 and OC5 projects. However, in the case of pitch, a relatively large discrepancy was observed in the time-domain response for the GNU simulation. As mentioned above, this effect may be owing to the different properties of the mooring line and rotating inertia forces of the platforms used in the two projects. Based on the natural period of the pitch, the pitch results were in good agreement with the OC5 test data.

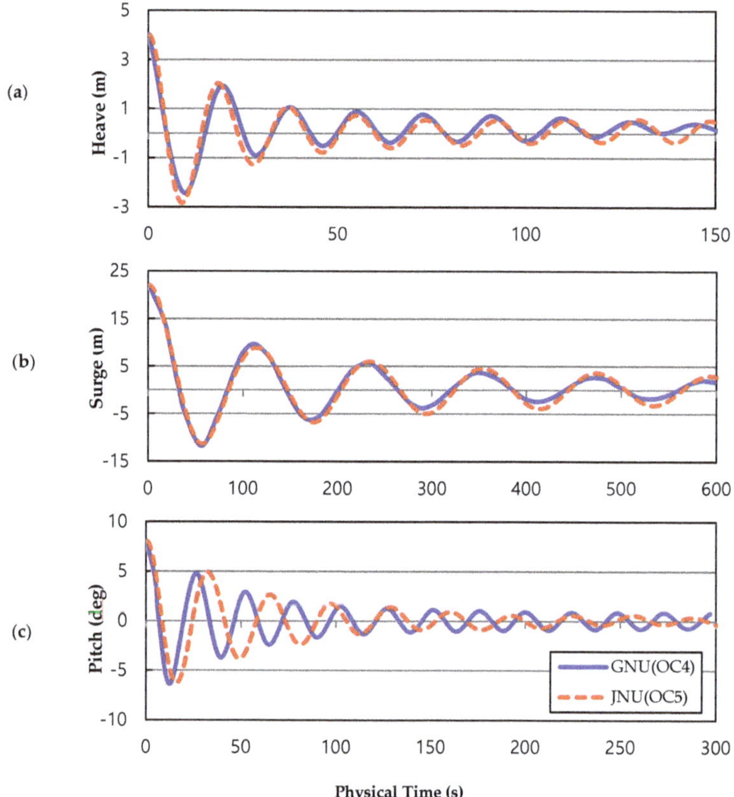

Figure 8. 3-DOF movement of platform by time domain: (**a**) Heave movement; (**b**) Surge movement; (**c**) Pitch movement.

5.2. Hydrodynamic Response Under Regular Waves

The characteristics of the DeepCWind platform under regular wave conditions were investigated by calculating the response amplitude operators (RAOs). An RAO is the normalized value of the amplitude of the periodic response of a field variable divided by the amplitude of a regular wave [24]. The platform was initialized at a static position, and a regular wave was introduced. The regular wave had an amplitude of 3.79 m and a period of 14.3 s. A fifth-order wave was applied in the regular wave test; here, the fifth-order wave was modeled with a fifth order approximation to the Stokes theory of waves [25]. This wave more closely resembled a real wave than one that was generated by the first-order method. The transient start-up period should not be considered in the results. After simulation runs for 400 s, the platform movement achieved a periodic quasi-steady state. The surge, heave, and pitch motion amplitudes were calculated by averaging the amplitudes over the last eight wave periods [24]. These values were then normalized using the amplitude of the regular wave to obtain the RAO. The RAOs of Phase II of the OC5 project were higher for the surge, heave, and pitch DOFs, similar to our simulation results obtained using the unsteady CFD method, as shown in Figure 9.

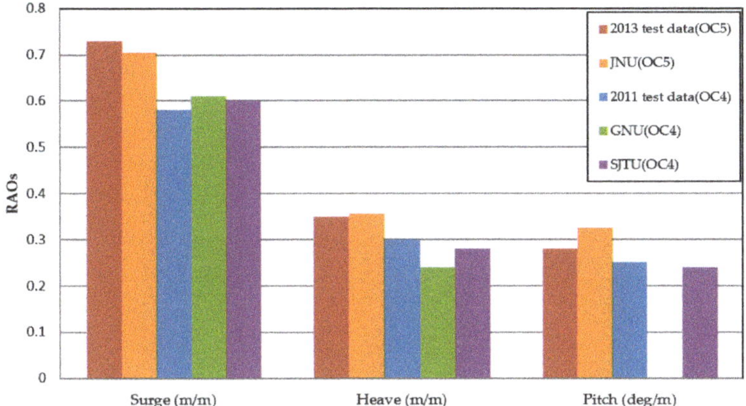

Figure 9. Comparison of response amplitude operators (RAOs) for surge, heave, and pitch.

6. Fully Coupled Wind–Wave Simulation

The full-scale DeepCwind OC5 model in a coupled wind–wave excitation condition was finally conducted using the DFBI method mentioned above. Figure 10 shows the fluid domain, together with an xz-plane section of the mesh distribution in the whole fluid domain. To obtain the fluid details near the free surface, as well as those near the turbine blade tip and vortex regions after the tower, mesh refinement was performed around the blades and platform, as shown in Figure 11. Nearly 27 million cells were generated using the built-in trim mesh feature in STAR-CCM+. The wind speed, V, was assumed to be the rated wind speed (11.4 m/s); the wave height and wave period were assumed to be 7.58 m and 12.1 s, respectively, similar to the MARIN wave basin.

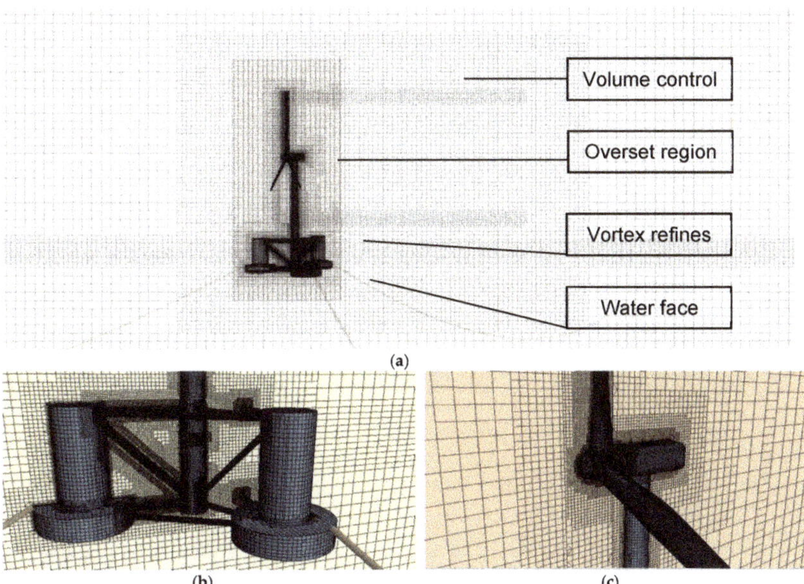

Figure 10. Full-coupled floating offshore wind turbines (FOWT) domain: (**a**) Mesh distribution in xz-plane; (**b**) Close-up view of mesh around platform; (**c**) Close-up view of mesh around wind turbine.

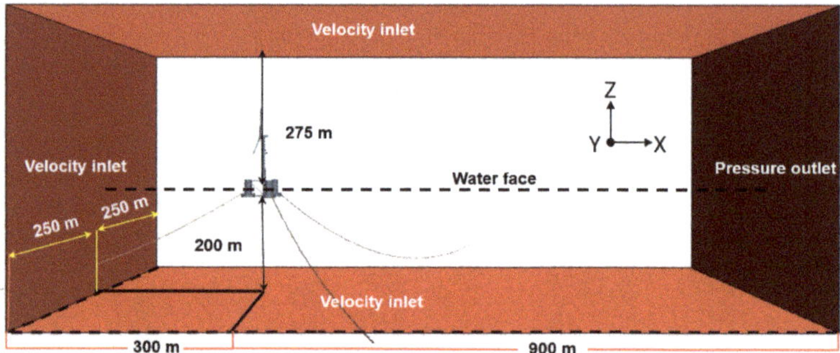

Figure 11. Fully coupled floating offshore wind turbines (FOWT) domain fluid, domain size, and boundary conditions.

After a 10 s start-up time, the aerodynamic output for the wind turbine was stabilized and the platform was released to move. A computational flow chart for the FOWT is presented in Figure 12.

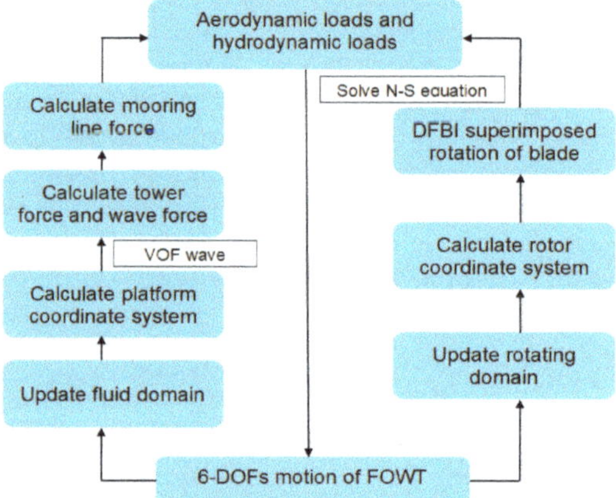

Figure 12. Flow chart of fully coupled simulation in the wind–wave condition.

The time for one revolution of the blades with a rotation speed of 12.1 rpm is 4.96 s. The time-step size (dt) of 0.07009 s utilized here corresponds to a 5° increment in the azimuth angle of the blade for each time-step. The wave heading angle is 0° and the wave is parallel to the direction of mooring line 2, which is also parallel to the platform surge direction.

All computations of the FOWT considering the wind–wave coupling were performed using a 4U multi D500 server. The elapsed real central processing unit (CPU) time for parallel processing per time-step with 15 sub-iterations was 6 min when using 66 CPUs. The total number of iterations for a simulation runtime of 300 s was approximately 30,000. The total simulation time taken to obtain the results using 66 CPUs was 20 days.

Figure 13 demonstrates the vortex contours with 0.5 Q-criteria, colored according to the velocity magnitude component, where the free surface is colored according to the surface elevation. After post-processing, as easily observed in the figure, strong vortices appear near the blade tips and roots. The presence of the tower caused a complex flow wake because of the interaction between the tower and flow. Such a detailed flow map is useful to identify the means for improving the wind turbine power output in the design stage.

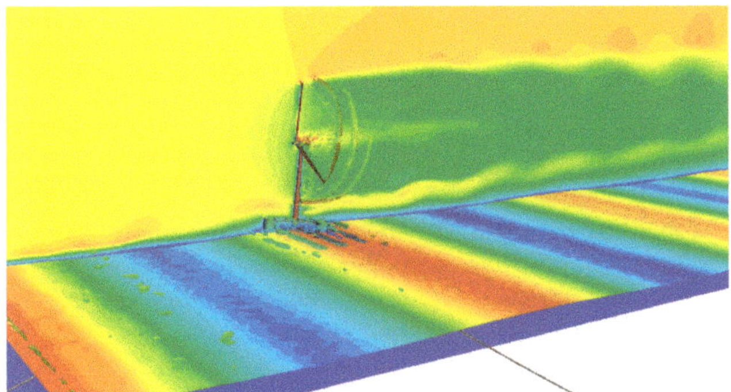

Figure 13. Velocity scene and water elevation.

This is an advantage of the CFD method, which is not present in other codes such as FAST. The size of the vortex tubes gradually decreases with time, and the patterns can be described by an iso-vorticity value. Herein, one wave period is separated into eight steps, i.e., T1 to T8; the duration from T1 to T8 represents one period of the platform surge motion. When the wind turbine moves backward, the number of vortex tubes increases, and the gaps between the blade tip vortices tend to continuously decrease at the same time. Figure 14 shows the fluid field and turbulence wakes between the fluid and tower and nacelle during the platform surge motion at different times. It shows that the generated vortexes near the tower and nacelle configurations diffuse outward as the platform moves backward, and vice versa.

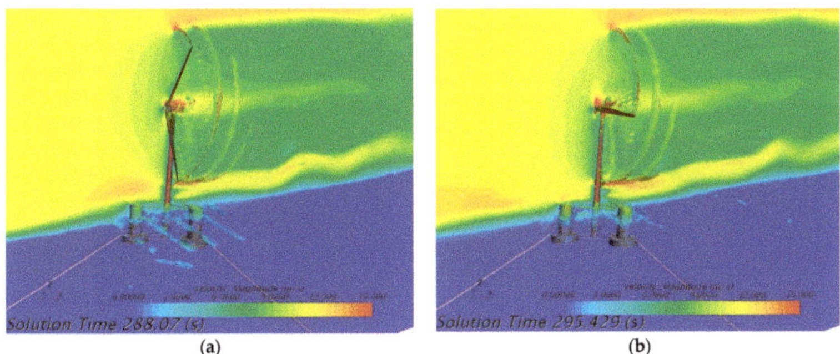

(a)   (b)

Figure 14. Instantaneous iso-velocity contours within one period: (**a**) upwind direction movement; (**b**) downwind direction movement.

The RAOs of surge, heave, and pitch in the fully coupled configuration under wind–wave excitation conditions are given in Table 7. Compared to the result of the regular wave test, where only regular waves exist, under a no-wind condition some discrepancies can be observed. The motion RAOs and the time-average values over the last four wave periods for 3-DOF were compared. Due to the unavailability of MARIN test data for the wind and wave conditions simulated herein, the comparison was only performed with the result from a previous study. All the 3-DOFs, i.e., heave, surge, and pitch showed small amplitudes compared to the results of the regular wave test without wind conditions. The incoming wind from the x-direction obviously has a significant effect on the restoring force of the mooring line; hence, the FOWT system cannot be restored to the equilibrium position as in the regular wave test. The incoming wind increased the aerodynamic thrust towards the floating system and pushed the platform further away in the backward direction, also leading to an increase in the mean surge. Nevertheless, the close agreement between the results for the 3-DOFs demonstrates the capability of this method.

Table 7. Motion response amplitude operators (RAOs) in different environment conditions.

Motion	Wave Only	Wind-Wave Coupled
Surge (m/m)	0.70	0.12
Heave (m/m)	0.36	0.13
Pitch (deg/m)	0.33	0.07

Thrust and power are two critical aerodynamic performance factors for evaluating a wind turbine. Thrust is defined as the integrated force component normal to the rotor plane. The power output and thrust force time-histories for the coupled simulation are presented in Figure 15 along with the dynamic responses of the pitch motion rate of the platform.

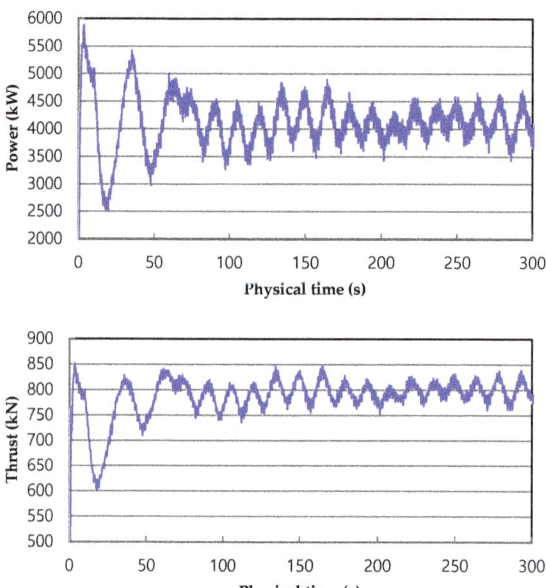

Figure 15. Fully coupled FOWT aerodynamic performance.

The response curves of power and thrust act at the same frequency as the incident wave. Due to the tower shade effect, the curves of power and thrust force exhibit periodical fluctuations with a period of 120° for the blade rotation. However, the effect of tower dam effects on the power output of the wind turbine is less than 5%. The variation of pitch motion also acts at the same frequency as the inlet wave. When the platform moves in the upwind direction, the power output and thrust force both increase, while the aerodynamic load decreases as the sign of the pitch motion changes. This is because the upwind pitch motion of the FOWT increases the relative velocity between the wind turbine and the inlet wind, and the angle of attack for each blade section increases correspondingly.

The dynamic responses of the wind turbine performance and typical platform motions after 300 s of the simulation are presented in Table 8. The power output varies from 3446 kW to 4698 kW at a rated wind speed. The variation in power is larger than that in the thrust force, that is, the power output is more sensitive than the thrust force to platform motion. Then, aerodynamic performances of the onshore fixed wind turbine and offshore floating wind turbine were compared and the average thrust value was calculated over the last four periods. In the case of thrust, a 7.8% increase was observed in a floating offshore turbine, the floating offshore wind turbine had an average thrust around 796 kN and the onshore wind turbine had an average thrust of 738 kN, which indicated a relatively small load on the hub and blades. This is because of the thrust force acting on the top of the tower, meaning the platform always moves in the upwind direction to offset the capsizing moment induced by the thrust force. In the case of the power curve, the average power value was calculated over the last four periods. A 10% decrease was observed in the floating offshore wind turbine, which is likely due to the smaller project area and relative income wind speed when the platform pitches, as shown in Figure 16. Accordingly, the platform surges from 7.9 to 9.8 m, which is also due to the thrust force that must be offset by the mooring line tension.

Table 8. Dynamic response of floating offshore wind turbines (FOWT) in the wind–wave condition.

	Parameters	Value
Power output	Range (kW)	3446–4689
	Mean value (kW)	4181
Thrust force	Range (kN)	759–838
	Mean value (kN)	801
Pitch angle	Range (deg)	4.5–5.3
	Mean value (deg)	4.9
Surge motion	Range (m)	7.9–9.8
	Mean value (m)	9.1
Heave motion	Range (m)	−0.7–0.7
	Mean value (m)	0.1

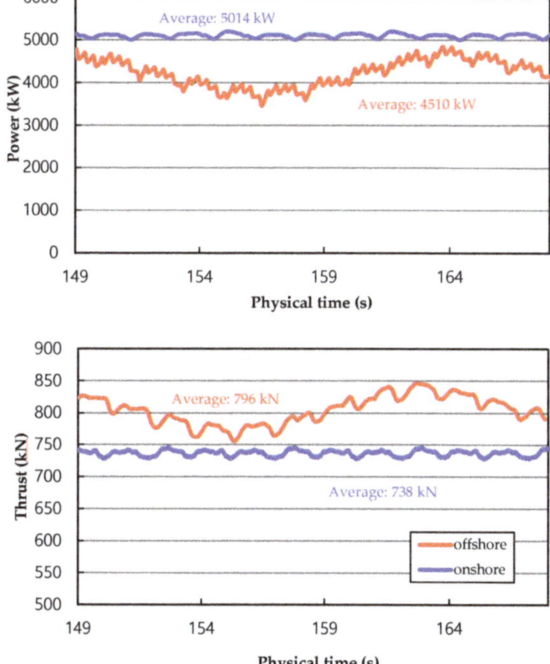

Figure 16. Comparison between onshore and offshore.

7. Conclusions

This investigation performed a CFD numerical analysis for a semi-submersible-type FOWT used in Phase II of the OC5 project, by advanced DFEI method and overlap mesh techno ledge. The full-configuration FOWT in a wind–wave excitation condition has been successfully performed, with simultaneous consideration of the wind turbine movement due to 6-DOF platform dynamics. The RAOs of the surge, heave, and pitch were compared to MARIN test data and data reported in previous studies when only the wave condition was considered. A slight discrepancy was observed between the CFD studies with regard to the pitch, possibly because of the different physical properties of the platform and mooring lines in the OC4 and OC5 projects. There was a relatively large discrepancy in the hydrodynamic response, which can induce large deviations in the prediction procedure. Particularly, pitch natural period showed a 21% discrepancy between OC4 and OC5 projects, as indicated by the results of the free decay test of the pitch and the numerical discrepancy in the RAOs in the regular wave test.

Besides, unsteady blade-tip vortices and strong flow interactions with the turbulent wakes of the tower due to the surge motion of the platform were successfully simulated and visualized using the advanced DFBI and VOF methods. The power and thrust force of the FOWT increased when the floating platform moved in the upwind direction, while the aerodynamic loads decreased as the pitch motion reversed direction. This can be explained by the variation in the angle of attack for each blade section when the FOWT system experiences pitch motion. All the 3-DOFs, including heave, surge, and pitch had smaller amplitudes compared with the results in the regular wave test without wind conditions. Incoming wind from the x-direction obviously has a large effect on the restoring force in the mooring line, and, as a result, the whole FOWT system cannot be restored back to the equilibrium position as in the regular wave test.

In addition, a relatively heavy load on the hub and blades was observed for the FOWT compared with the onshore wind turbine. This is because of the thrust force acting on the top of the tower, due to which the platform moves in the upwind direction to offset the capsizing moment induced by the thrust force. With regard to the power curve, a 10% decrease was observed for the floating offshore wind turbine, which is likely due to the smaller project area and relative income wind speed when the platform experiences pitch motion. Overall, there is a greater variation in the power than in the thrust force, that is, the power output is more sensitive than the thrust force to platform motions.

Until now, all published papers based on an OC4 project which was carried out in 2013 (this was code-to-code comparison project 5 years ago), which found a large discrepancy from the experimental test data of the OC5 project. This study could be a good insight for future studies, as there has not been any specific CFD research based on OC5 test data until now. Examination of the OC5 Phase II project, with a computational fluid dynamics code (which has a higher-fidelity model of the underlying physics), could help determine if there are some deficiencies in the hydrodynamic models being employed by participants in an OC5 code-to-test project [18].

Author Contributions: This research was supervised by B.K. All laboratory work was down by Y.Z.

Funding: This works was supported by "Development of multi-class large capacity wind power generator system specialized in Korea wind site" and "Human Resources in Energy Technology" of KETEP (Grant Nos 20173010024930 and 20184030202200) granted financial resource from the Ministry of Trade, Industry & Energy (MOTIE), Korea.

Conflicts of Interest: The authors declare no conflict of interest.

References

1. Jonkman, J.M. Dynamics Modeling and Loads Analysis of an Offshore Floating Wind Turbine. Ph.D. Thesis, University of Colorado, Denver, CO, USA, 2007.
2. Sebastian, T.; Lackner, M.A. Development of a free vortex wake method code for offshore floating wind turbines. *Renew. Energy* **2012**, *46*, 269–275. [CrossRef]
3. Fernandez-Gamiz, U.; Zulueta, E.; Boyano, A.; Ansoategui, I.; Uriarte, I. Five Megawatt Wind Turbine Power Output Improvements by Passive Flow Control Devices. *Energies* **2017**, *10*, 742. [CrossRef]
4. Fernandez-Gamiz, U.; Errasti, I.; Gutierrez-Amo, R.; Boyano, A.; Barambones, O. Computational Modelling of Rectangular Sub-Boundary Layer Vortex Generators. *Appl. Sci.* **2018**, *8*, 138. [CrossRef]

5.   Nematbakhsh, A.; Bachynski, E.E.; Gao, Z.; Moan, T. Comparison of wave load effects on a TLP wind turbine by using computational fluid dynamics and potential flow theory approaches. *Appl. Ocean Res.* **2015**, *53*, 142–154. [CrossRef]

6.   De Vaal, J.B.; Hansen, M.O.L.; Moan, T. Effect of wind turbine surge motion on rotor thrust and induced velocity. *Wind Energy* **2014**, *17*, 105–121. [CrossRef]

7.   Zhao, W.; Wan, D. Numerical study of interactions between phase II of OC4 wind turbine and its semi-submersible floating support system. *J. Ocean Wind Energy* **2015**, *2*, 45–53.

8.   Tran, T.T.; Kim, D.H.; Song, J. Computational fluid dynamic analysis of a floating offshore wind turbine experiencing platform pitching motion. *Energies* **2014**, *7*, 5011–5026. [CrossRef]

9.   Tran, T.T.; Kim, D.H. The coupled dynamic response computation for a semi-submersible platform of floating offshore wind turbine. *J. Wind Eng. Ind. Aerodyn.* **2015**, *147*, 104–119. [CrossRef]

10.   Liu, Y.; Xiao, Q.; Incecik, A.; Wan, D.C. Investigation of the effects of platform motion on the aerodynamics of a floating offshore wind turbine. *J. Hydrodyn. Ser. B* **2016**, *28*, 95–101. [CrossRef]

11.   Ren, N.; Li, Y.; Ou, J. Coupled wind-wave time domain analysis of floating offshore wind turbine based on Computational Fluid Dynamics method. *J. Renew. Sustain. Energy* **2014**, *6*, 1–13. [CrossRef]

12.   Quallen, S.; Xing, T.; Carrica, P.; Li, Y.; Xu, J. CFD simulation of a floating offshore wind turbine system using a quasi-static crowfoot mooring-line model. In Proceedings of the Twenty-third International Offshore and Polar Engineering Conference, Anchorage, AK, USA, 30 June–5 July 2013; International Society of Offshore and Polar Engineers: Mountain View, CA, USA, 2013.

13.   Tran, T.T.; Kim, D.H. Fully coupled aero-hydrodynamic analysis of a semisubmersible FOWT using a dynamic fluid body interaction approach. *Renew. Energy* **2016**, *92*, 244–261. [CrossRef]

14.   Gueydon, S. Aerodynamic damping on a semisubmersible floating foundation for wind turbines. *Energy Procedia* **2016**, *94*, 367–378. [CrossRef]

15.   Chen, J.; Hu, Z.; Wan, D.; Xiao, Q. Comparisons of the dynamical charateristics of a semi-submersible floating offshore wind turbine based on two different blade concepts. *Ocean Eng.* **2018**, *153*, 305–318. [CrossRef]

16.   Dunbar, A.J.; Craven, B.A.; Paterson, E.G. Development and validation of a tightly coupled CFD/6-DOF solver for simulating floating offshore wind turbine platforms. *Ocean Eng.* **2015**, *110*, 98–105. [CrossRef]

17.   Coulling, A.J.; Goupee, A.J.; Robertson, A.N.; Jonkman, J.M.; Dagher, D.J. Validation of a FAST semi-submersible floating wind turbine numerical model with DeepCwind test data. *J. Renew. Sustain. Energy* **2013**, *5*, 16–23. [CrossRef]

18.   Robertson, A. OC5 Project Phase II: Validation of Global Loads of the DeepCwind Floating Semisubmersible Wind Turbine. *Energy Procedia* **2017**, *137*, 38–57. [CrossRef]

19.   Menter, F.R. Two-Equation Eddy-Viscosity Turbulence Models for Engineering Applications. *AIAA J.* **1994**, *32*, 1598–1605. [CrossRef]

20.   Tran, T.T.; Kim, D.H. A CFD study into the influence of unsteady aerodynamic interference on wind turbine surge motion. *Renew. Energy* **2016**, *90*, 204–228. [CrossRef]

21.   Sivalingam, K.; Narasimalu, S. Floating Offshore Wind Turbine Rotor Operating State-Modified Tip Loss Factor in BEM and Comparison with CFD. *Int. J. Tech. Res. Appl.* **2015**, *3*, 179–189.

22.   Hsu, M.C.; Bazilevs, Y. Fluid-Structure Interaction Modeling of Wind Turbines: Simulating the Full Machine. Ph.D. Thesis, Iowa State University, Ames, IA, USA, 2012.

23.   Siddiqui, M.S.; Rasheed, A.; Kvamsdal, T. Quasi-Static & Dynamic Numerical Modeling of Full Scale NREL 5MW Wind Turbine. *Energy Procedia* **2017**, *137*, 460–467.

24.   Liu, Y.; Xiao, Q.; Incecik, A.; Wan, D. Establishing a fully coupled CFD analysis tool for floating offshore wind turbines. *Renew. Energy* **2017**, *112*, 280–301. [CrossRef]

25.   Tezdogan, T.; Demirel, Y.K.; Kellett, P.; Khorasanchi, M.; Incecik, A.; Turan, O. Full-scale unsteady RANS CFD simulations of ship behavior and performance in head seas due to slow steaming. *Ocean Eng.* **2015**, *97*, 186–206. [CrossRef]

Article

Aerodynamic Force and Comprehensive Mechanical Performance of a Large Wind Turbine during a Typhoon Based on WRF/CFD Nesting

Shitang Ke [1,2,*], Wenlin Yu [1], Jiufa Cao [2] and Tongguang Wang [2]

[1] Department of Civil Engineering, Nanjing University of Aeronautics and Astronautics, 29 Yudao Road, Nanjing 210016, China; nuaayuwenlin@163.com

[2] Jiangsu Key Laboratory of Hi-Tech Research for Wind Turbine Design, Nanjing University of Aeronautics and Astronautics, Nanjing 210016, China; caojiufa98@163.com (J.C.); tgwang@nuaa.edu.cn (T.W.)

* Correspondence: keshitang@163.com

Received: 22 August 2018; Accepted: 15 October 2018; Published: 19 October 2018

Featured Application: The main research conclusions of this paper could provide references for load prediction of similar wind turbine systems during the occurrence of typhoon, and deepen the understanding on meso/microscale wind field nesting mechanism.

Abstract: Compared with normal wind, typhoons may change the flow field surrounding wind turbines, thus influencing their wind-induced responses and stability. The existing typhoon theoretical model in the civil engineering field is too simplified. To address this problem, the WRF (Weather Research Forecasting) model was introduced for high-resolution simulation of the Typhoon "Nuri" firstly. Secondly, the typhoon field was analyzed, and the wind speed profile of the boundary layer was fitted. Meanwhile, the normal wind speed profile with the same wind speed of the typhoon speed profile at the gradient height of class B landform in the code was set. These two wind speed profiles were integrated into the UDF (User Defined Function). On this basis, a five-MW wind turbine in Shenzhen was chosen as the research object. The action mechanism of speed was streamlined and turbulence energy surrounding the wind turbine was disclosed by microscale CFD (Computational Fluid Dynamics) simulation. The influencing laws of a typhoon and normal wind on wind pressure distribution were compared. Finally, key attention was paid to analyzing the structural response, buckling stability, and ultimate bearing capacity of the wind turbine system. The research results demonstrated that typhoons increased the aerodynamic force and structural responses, and decreased the overall buckling stability and ultimate bearing capacity of the wind turbine.

Keywords: typhoon; wind turbine; meso/microscale; aerodynamic force; mechanical performance

---

1. Introduction

A large wind turbine structure is more flexible [1] and has a more prominent wind-induced dynamic effect [2]. It is a typical wind-sensitive structure. Wind load is the control load of its structural design [3,4]. In particular, the wind-induced failure of large wind turbine structures occur frequently during strong typhoons [5,6]. For example, the Typhoon "Cuckoo" in 2003 affected the normal operation of wind power plants in south China, which halted 13 wind turbines. The Typhoon "Saomai" in 2006 landed on Zhejiang, and caused different degrees of failure of 20 units in the Cangnan Wind Power Plant. In 2014, the Typhoon "Rammasun" landed at Hainan Province, which caused the collapse of—and serious damage to—the 1.5-MW wind turbine with a 77-m blade diameter in the wind power plant. Compared with normal wind, typhoons form a more complicated near-ground wind field. The wind field characteristics of high turbulence, changeable direction, and great changes of shearing wind speed will intensify the fluctuation of the wind turbine structure. Under this circumstance, airflow movement on the near-wall surface is disorganized, and the surface pressure is changed significantly, which further caused significant changes in the wind-induced responses and the stability of the wind turbine system [7,8]. For these reasons, a contrastive analysis of aerodynamic performances and comprehensive performances of the large wind turbine system during typhoons and normal wind was of important theoretical significance and engineering values.

For studies concerning wind turbine performance during typhoon, Takeshi et al. [9] studied influences of blade, wind direction, and the relative position of a doorway on the collapse of wind turbines according to wind turbine collapse data and measured typhoon speed in Japan. Tang et al. [10] and Chou and Tu [11] analyzed stresses on the wind turbine structure under typhoon load by the simplified static analysis method. Li et al. [12] analyzed typhoon characteristics and the damages of different parts of wind turbines. They pointed out that the design of different parts of a wind turbine has to strengthen the structural strength and pay attention to the yaw of the wind turbine under typhoon load. Wang et al. [6] and Luo et al. [13] carried out a systematic analysis on wind-induced static/dynamic responses of the tower in the wind turbine system during a typhoon by combining a theoretical deduction and fluctuating wind spectrum. Some suggestions regarding the design of the typhoon resistance of the tower body in the wind turbine system were given. Lian et al. [14] and Zhou et al. [15] carried out a systematic analysis on the aerodynamic load on blades of the wind turbine system under typhoon loads by combining a theoretical analysis and measured data. They established a typhoon-resistance aerodynamic blade design optimization model for wind turbines. Cao [16] and An [17] analyzed the flow field and pressure field and structural stresses of the wind turbine under typhoon loads by theoretical studies and CFD numerical simulation. A new typhoon-resistance wind turbine design philosophy and structural design was proposed. Utsunomiya et al. [18] and Ma et al. [19] carried out a statistical analysis on the dynamic response and mooring linear tension response of a submarine platform of an offshore wind turbine under typhoon loads. They attached high attentions to the aerodynamic shape design and typhoon resistance characteristics of blades.

Most of the above studies concerning the typhoon resistance of a wind turbine are based on theoretical analysis and measured data. The theoretical systems in these studies were too simplified to reflect the specific wind profile and landing attenuation effect of a mesoscale typhoon field. Nowadays, mesoscale models have been well applied in case analyses of typhoons, early warnings, and disaster forecasts. Common models include Weather Research Forecasting (WRF), MM5 (Mesoscale Model Version 5), and ETA. Among these models, MM5 enjoyed widespread applications in earlier studies, because it was studied and used earlier. However, it has been never adopted by the NCEP (National Centers for Environmental Prediction), on the grounds that its dynamical framework is obsolete, and its procedures are not fairly standard. Utilized by the NCEP for operational forecast, ETA can hardly promptly assimilate the outstanding research findings of scientific research institutes and universities, although its simulation results are reliable. As a result, its promotion is restricted. Comparatively, the WRF model based on fluid dynamics and thermodynamics can not only effectively simulate wind speed, pressure, and temperature in a typhoon field, it can also comprehensively consider the evolution process, strong specificity, and attenuation characteristics of typhoons. In spite of its parametric schemes and the more complicated setting of some of its parameters, it takes more complete and accurate environmental parameters into account. Therefore, WRF is used most widely among all of the existing models. However, due to the large sphere of scope (hundreds of kilometers) of typhoons, grid resolution of the typhoon field is generally at the magnitude of kilometers. However, the overall dimension of a large wind turbine system is only at the magnitude of hectometers. To predict the aerodynamic load at blade edges accurately, it has to deepen into the near-ground boundary layer. The minimum grid size of the near-wall surface is generally lower than 10^{-2} m. The mesoscale model in WRF fails completely at this magnitude. Hence, it will be helpful for observing the airflow motion in a typhoon field by a more refined simulation of microscale typhoon fields based on the mesoscale WRF. WRF and Computational Fluid Dynamics (CFD), which are new-generation mesoscale and microscale forecast models respectively, are employed most widely as technologies for high-resolution and refined wind field forecasts. In the literature [20,21], refined numerical simulations were conducted on the bridge sites of mountainous areas by multi-scale coupling. The results suggested that the results of WRF could be more adaptable to inlet boundaries during numerical simulations of these wind fields by CFD. Thus, the existing problem about the mean wind speed at the inlet during the numerical simulations was successfully solved. In the literature [22–24], the complicated wind fields of lakes on the underlying surface were simulated meticulously by coupling WRF and CFD. According to the simulation results, WRF–CFD coupling better eliminates the mean deviation and the root mean squared error than WRF. Besides, it reflects the atmospheric characteristics of actual microscale topographical changes more accurately. Therefore, the meso/microscale nesting WRF/CFD has to be used for the high-precision simulation of a multi-scale flow field of large wind turbines under typhoon loads in order to solve various problems, such as the high-precision transition of parameters and flowing structure, multilayer and multi-scale grid nesting, multi-time scale control, and multi-scale mutation [25].

To solve these problems, the mesoscale WRF model was introduced for high temporal spatial resolution simulation of Typhoon "Nuri". After the typhoon field information was analyzed, the wind speed profile in the simulation region was acquired based on the nonlinear least square fitting. On this basis, the normal wind speed profile with the same wind speed of the typhoon speed profile at the gradient height of the class B landform was set. Later, the aerodynamic distribution on the surface of a five-MW horizontal axis wind turbine in a wind power plant in Shenzhen (China) under typhoon loads and normal wind was studied by microscale CFD numerical simulation. Finally, the mechanical performance and stability performance of a large wind turbine system under typhoon loads and normal wind were compared by combining them with the finite element method.

2. Mesoscale Typhoon Field Simulation and Analysis

2.1. WRF Model

WRF is the mesoscale weather research and forecasting system that was developed by the America Environmental Forecasting Center, National Center of Atmospheric Research of the United States and Rainstorm Analysis and Forecasting Center of University of Central Oklahoma [25]. The Euler equation model based on a non-static force equilibrium was used as the WRF–ARW (Advanced Research WRF) model of the dynamic framework. The horizontal computational domain adopted the Arakawa-C meshing scheme [26] (Figure 1a). The Arakawa-C grid can express scalar and vector parameters simultaneously, which is conducive to increase the accuracy of a high-resolution simulation. Besides, the Arakawa-C grid has good dispersion properties and conservativeness. The vertical computational domain applied the Euler mass coordinate η along the terrain (Figure 1b):

$$\eta = (p_h - p_{ht})/\mu, \quad \mu = p_{hs} - p_{ht} \tag{1}$$

where p_h is the air pressure at the target point p_{ht} is the air pressure of the top layer, which is a constant, and p_{hs} is the near-ground air pressure.

The WRF model is transferable when it is run on the Linux system. With considerations to the influences of physical processes such as water vapor, long-wave radiations, short-wave radiations, cumulus cloud, and the underlying surface, the WRF model can simulate airflow, air pressure, and wind field characteristics in a large region. The results can be used as the input boundary conditions for CFD numerical simulation.

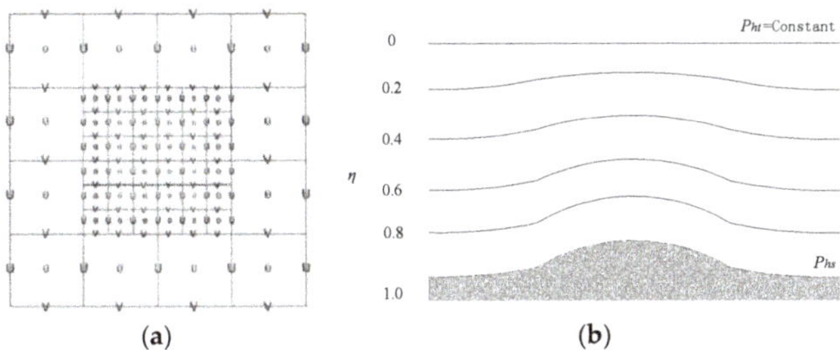

Figure 1. Meshing of the Weather Research Forecasting (WRF) model. (**a**) Horizontal grids (space between coarse and thin grids is 3:1); (**b**) Vertical grids.

2.2. Selection of Physical Scheme and Parameter Settings

The #12 Typhoon "Nuri" in 2008 was chosen as the simulation object. Based on the information provided by the Typhoon Website of China Meteorological Administration, it landed on coastal regions of Saigon, Hong Kong on 16:55 on 22 August and the maximum wind near the center was 12-level (33 m/s). On the same day, it landed again at Nanlang Town, Zhongshan City, Guangdong Province at 22:10, and the maximum wind near the center was eight-level (25 m/s). For effective numerical simulation of typhoon path as well the air pressure field and wind speed field, the simulation covers the whole process from the first landing at Saigon, Hong Kong to the second landing at Zhongshan City, Guangdong Province. The initial atmospheric conditions and time-dependent boundary conditions, which were applied in the WRF model, were all based on the FNL (Final Global Data Assimilation System) [27–30]. It covered 27 atmospheric layers. The sampling interval was set at 6 h, and the spatial resolution was $1° \times 1°$. The WRF simulation region had 37 vertical layers, and the air pressure at the top layer was set 5000 Pa.

The simulation center chose a district in north Shenzhen. Typhoon "Nuri" ran through this region at 20:00 on 22 August. With comprehensive considerations to data demands and computational conditions, the three-dimensional (3D) nesting program was used to calculate typhoon field in the WRF model. The horizontal mesh resolution was 13.5 km, 4.5 km, and 1.5 km, respectively. The longitude and latitude of center of the simulation region was (114.1° E, 22.5° N). The coarse grid space D01 in the external surface was 13.5 km, and the number of grids was 211×211. The grid space D02 in the second layer was 4.5 km, and the number of grids was 217×217. The grid space D03 in the inner layer was 1.5 km, and the number of grids was 241×241. The map projection applied the Lambert program. A total of 37 vertical layers were set, forming upper-sparse and lower-dense stratification patterns. There were 19 layers below 1000 m. The simulation domain is shown in Figure 2. The altitudes of different vertical layers are shown in Table 1.

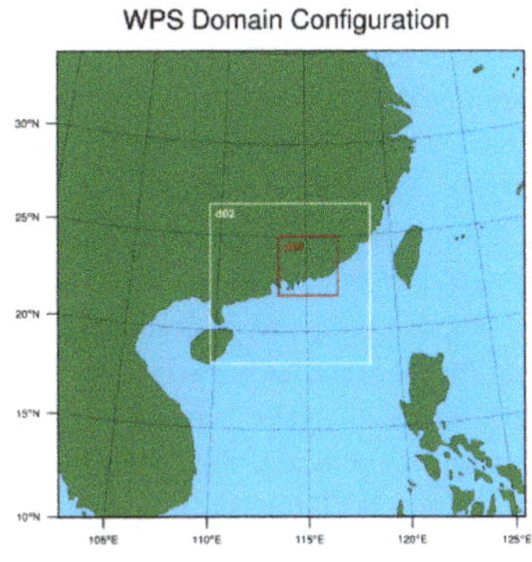

Figure 2. Simulation domain of the Typhoon "Nuri".

Table 1. Altitudes of different vertical layers.

Layers	Altitude/m	Layers	Altitude/m
0	0	19	997.85
1	5.41	20	1441.03
2	21.64	21	1864.44
3	43.28	22	2383.35
4	64.96	23	2929.90
5	86.67	24	3508.54
6	108.42	25	4124.07
7	130.20	26	4780.99
8	152.01	27	5484.65
9	179.34	28	6243.02
10	217.70	29	7067.05
11	272.71	30	8489.99
12	350.13	31	10,070.22
13	439.15	32	11,332.60
14	528.81	33	12,805.37
15	619.11	34	14,592.36
16	710.08	35	17,014.78
17	824.83	36	19,617.33
18	987.29	37	——

The hub height of the wind turbine is generally lower than 150 m, which is below the thickness of the boundary layer. The hub height is significantly influenced by airflow mass, humidity, and heat transmission in the boundary layer [31–35]. Therefore, the settings of the physical parameterization scheme can affect the simulation accuracy of boundary layer of the typhoon field directly. In this paper, the physical parameters of WRF model are listed in Table 2.

Table 2. Physical parameter settings of the WRF model.

WRF Parameters	Main Zone (D01)	Nesting Zone (D02)	Nesting Zone (D03)
Horizontal resolution	13.5 km	4.5 km	1.5 km
Integral time step		180 s	
Microphysical process scheme		Lin	
Long-wave radiation		RRTM	
Short-wave radiation		Dudhia	
Near-ground layer scheme		Monin-Obukhov	
Land surface process scheme		Noah	
Planet boundary layer scheme		MYJ	
Cumulus convective parameterization scheme		Kain-Fritsch	

2.3. Simulation Results and Discussions

To verify the validity of the proposed simulation model of Typhoon "Nuri", the typhoon path and minimum sea level pressure throughout the simulation are shown in Figure 3. The simulated path was drawn according to the minimum pressure in the cyclonic center of the WRF model. The measured path was the best path provided by the Typhoon Website of the China Meteorological Administration, and it could also refer to some literature cites [36,37]. The simulated path and the measured path were stable, and both paths moved toward the northwest. In early simulations, the simulated path was close to the measured path. However, the typhoon path deflected gradually as time went on, which might be caused by the intensifying initial error with the increase of simulation time. The evolution of typhoon intensity was mainly represented by the minimum sea level pressure. The simulated value and measured value of minimum sea level pressure in the simulation period differed slightly, but the distribution trend was basically consistent. Both the simulated value and the measured value increased slowly with time, and increased sharply in the late simulation period. At this moment, the typhoon became weak and disappeared gradually.

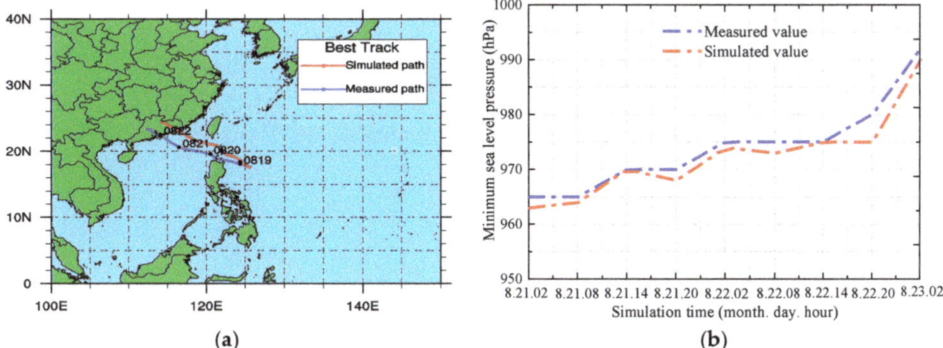

Figure 3. Typhoon path and minimum sea level pressure throughout the simulation. (**a**) Typhoon path; (**b**) Minimum sea level pressure.

Simulation results and measured results in the simulation regions at typical height for the landing of the typhoon at 20:00 on 22 August are shown in Figures 4 and 5. These nephograms are conducive to analyzing the wind field information when the Typhoon "Nuri" ran through the center of the simulation region. It can be concluded from the analysis that:

(1) The typhoon had a low-pressure center. The land underlying the surface was rougher than the ocean after landing, thus cutting the heat source. Hence, the upflow was weakened, while the surrounding airflows continued radiation toward the typhoon center, which gradually further increased the air pressure. After comparison, it was found that the simulated pressure distribution agreed well with the measured results.

(2) Wind speed increased due to the invasion of the peripheral nephsystem of the typhoon. The wind speed increased continuously as the typhoon approached. The wind speed closer to the center was the higher.

(3) Influenced by the downward transmission of atmospheric heats, the MYJ boundary layer scheme contributed a high computational accuracy of turbulent fluctuation fluxes, such as heats and momentum in the boundary layer of the region. This implied that the low atmosphere heats were mainly carried into the boundary layer by the upper atmosphere through high temperature, which caused the hybrid heating of the low atmosphere.

(4) Cloud precipitation close to the typhoon belongs to the long-term convective precipitation. The large rainfall volume in this region is related to the typhoon strength and the central distance. The Noah pavement scheme can simulate the surface flux and airflow convergence field well.

(5) There is a minor difference between the simulation results and the measured results of the typhoon, but the distribution trend is basically consistent. In the central area of the typhoon simulation, the ranges of wind pressure, wind speed, temperature, and rainfall were 860–920 Pa, 22–26 m/s, 294–297 K, and 40–48 mm/h, respectively. These data are close to the measured data on the website of the National Meteorological Center. The maximum error is lower than 15%.

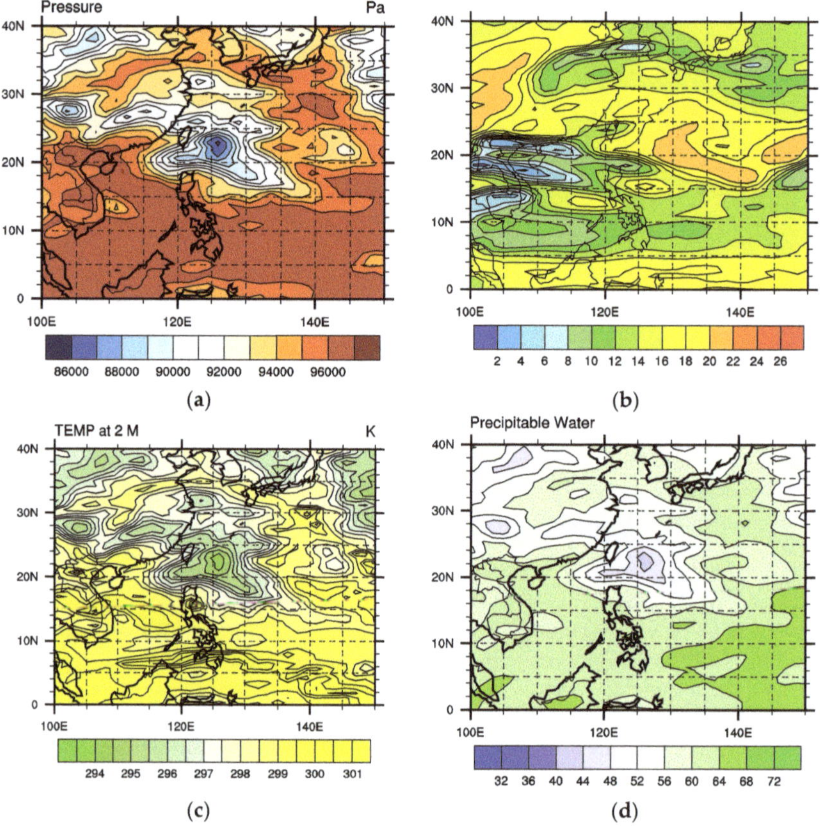

Figure 4. Simulation results for landing of typhoon. (**a**) Air pressure nephogram; (**b**) Wind speed nephogram; (**c**) Temperature nephogram; (**d**) Rainfall nephogram.

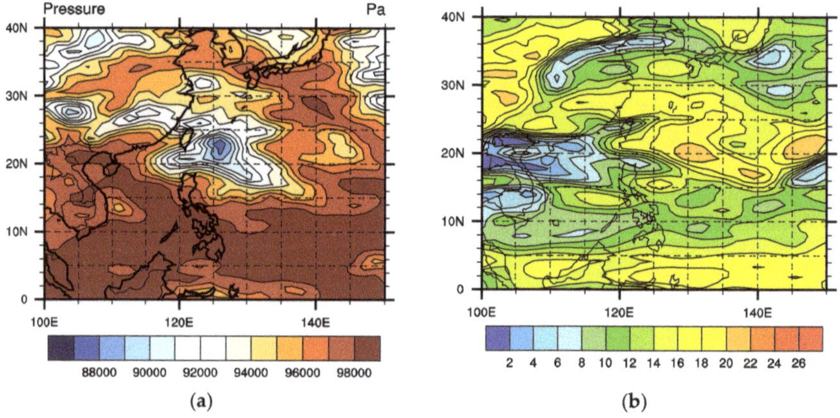

Figure 5. *Cont.*

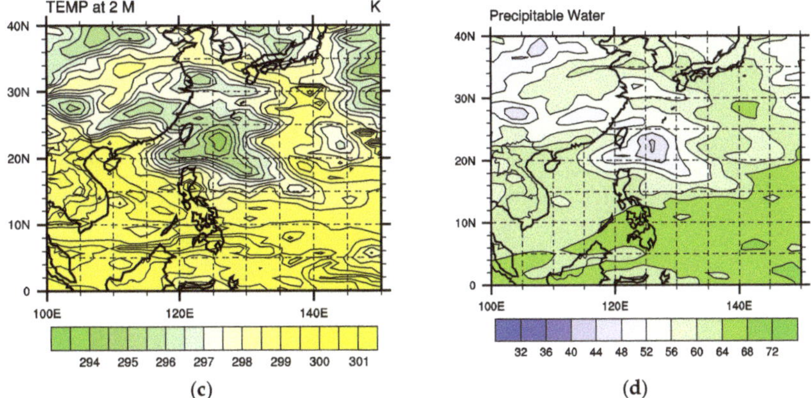

Figure 5. Measured results for landing of typhoon. (**a**) Air pressure nephogram; (**b**) Wind speed nephogram; (**c**) Temperature nephogram; (**d**) Rainfall nephogram.

The wind speed streamlines before, at, and after the landing of the typhoon are shown in Figure 6. No strong convection was observed before the landing of the typhoon. However, it turned to a southeaster gradually at the landing of the typhoon. After the landing, surrounding airflows continued to radiate toward the typhoon center. Wind speed streamlines are disorganized, without a regular development trend.

The wind speed profiles close to the typhoon center at different moments are shown in Figure 7a. According to analysis, wind speed distributes uniformly in the near-ground region at different moments, but wind speed distributes disorderly at high altitude. The typhoon speed at 20:00 on 22 August was significantly lower than those at other moments. This might be because the typhoon had landed at Shenzhen at this moment, but it was on the sea at other moments. The near-ground wind speed in the center of the simulation region and the corresponding nonlinear least square fitting curves are shown in Figure 7b. The fitting value of the wind speed profile index was 0.118, which was smaller than the surface roughness index (0.15) of class B landform under normal wind, indicating the good fitting effect in the near-ground typhoon field (simulation accuracy reached 95.97%). The typhoon speed at 10 m of height was high, and increased slowly with the increase of height. Moreover, it defined that the typhoon field at 350 m altitude of class B landform in the code [38] was equal to the wind speed in the normal wind field for the purpose of qualitative and quantitative analyses of the difference between the normal wind field and the typhoon field. The near-ground typhoon profile and the normal wind profile were integrated into the UDF and used as the initial boundary conditions in the follow-up microscale CFD numerical simulation.

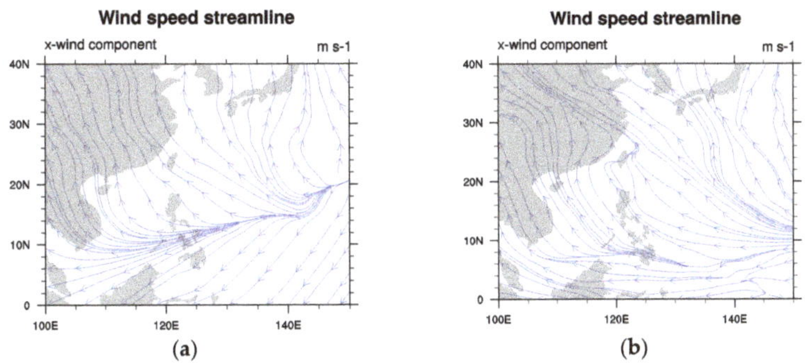

Figure 6. *Cont.*

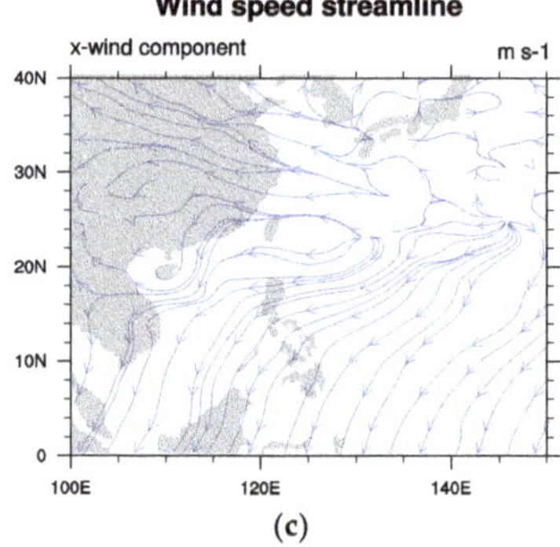

(c)

Figure 6. Wind speed streamlines before, at, and after landing of the typhoon. (a) Before landing (14:00 on 22 August); (b) At landing (20:00 on 22 August); (c) After landing (02:00 on 23 August).

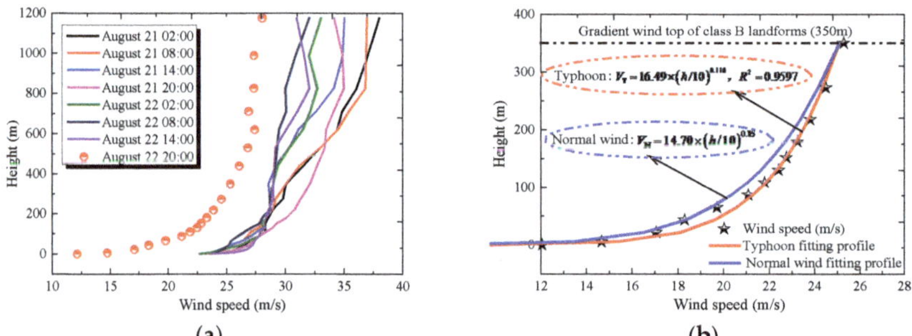

Figure 7. Wind speed profile close to the typhoon center and near-ground wind speed and fitting curves in the center of the simulation region at different moments. (a) Wind speed profile close to the typhoon center at different moments; (b) Near-ground wind speed and fitting curves in the center of the simulation region.

3. Microscale CFD Numerical Simulation

3.1. Brief Introduction to the Project

The main structural design parameters and model of a five-MW windward horizontal axis three-blade wind turbine in a wind power plant in Shenzhen, China are shown in Table 3. This five-MW wind turbine was developed in the Jiangsu Key Laboratory of Hi-Tech Research for Wind Turbine Design of Nanjing University of Aeronautics and Astronautics [39]. The tower body was a long thickness-variable structure. The upper wall thickness was 40 mm, and the bottom wall thickness was 90 mm. The nacelle size was 18 m (Length) × 6 m (Width) × 6 m (Height). The dip angle of the blade was 5°, and the cut-out speed was 25 m/s. The blades were distributed uniformly along the circumferential direction at an angle of 120°, and the blade length was 60 m. The parameters of the different blade sections along the span are listed in Table 4. The airfoil types of the blade were NH02_40, NH02_35, NH02_30, NH02_25, NH02_21, NH02_18, and NH02_15. Given the coordinates, chord lengths, and installation angles of the discrete points on the different blade profile sections in the wind turbine blades, the three-dimensional (3D) coordinates of different blade profile sections were gained through coordinate transformation. Finally, the 3D model of the blade was constructed.

Specific steps of coordinate transformation included [40]:

(1)   The coordinates of discrete points on different blade profile sections were calculated from relevant software. The normalized coordinates (x_0, y_0) of different blade elements were transformed to the coordinates (x_1, y_1) of discrete point under the coordinate system 1:

$$(x_1, y_1) = c \times (x_0, y_0) \tag{2}$$

(2)   The coordinates were further transformed to the coordinates (x_2, y_2) under the coordinate system 2:

$$(x_2, y_2) = (x_1, y_1) - (X, Y) \tag{3}$$

where (X, Y) are the coordinates of the pressure center.

(3)   The coordinates were further rotated to the actual spatial coordinates (x, y, z):

$$x = \sqrt{x_2^2 + y_2^2} \times \cos\left(\arctan\frac{y_2}{x_2} + \theta\right) \tag{4}$$

$$y = \sqrt{x_2^2 + y_2^2} \times \sin\left(\arctan\frac{y_2}{x_2} + \theta\right) \tag{5}$$

$$z = r \tag{6}$$

where θ is the installation angle, and r is the turning radius of the blade elements.

The inflow angle is the included angle between the plane of rotation and inflow speed in the traditional theory of blade momentum. However, the calculation formula of the inflow angle during the blade design stage of the wind turbine according to the Baez limit is:

$$\cot \varphi = \frac{3}{2}\lambda \tag{7}$$

where λ is the tip velocity ratio.

According to the relationship between the tip velocity ratio and the power coefficient, the optimal tip velocity ratio can be determined, and the inflow angle of each blade element is:

$$\cot \varphi = \frac{3}{2}\lambda_r \tag{8}$$

where λ_r is the optimal tip velocity ratio.

The tower body, blade, and nacelle models were constructed successively. Finally, the 3D physical model of a large wind turbine was formed by Boolean operation.

Table 3. Main structural design parameters and model of the five-MW wind turbine.

Parameters	Numerical Value	3D Blade Model	3D Wind Turbine Model
Tower height	124 m		
Radius at tower top	3.0 m		
Radius at tower bottom	3.5 m		
Thickness at tower top	0.04 m		
Thickness at tower bottom	0.09 m		
Blade length	60 m		
Nacelle size	18 m × 6 m × 6 m		
Cut-out speed	25 m/s		
Yaw rotation speed	0.5°/s		

Table 4. Blade parameters of wind turbine.

Position/%	Blade Span/m	Chord Length/m	Installation Angle/°	Blade Pitch Angle/°	Position/%	Blade Span/m	Chord Length/m	Installation Angle/°	Blade Pitch Angle/°
5	3	2.9	0.823	37.14	55	33	1.95	0.169	−0.293
10	6	3.66	0.64	26.672	60	36	1.75	0.156	−1.072
15	9	4.41	0.507	19.069	65	39	1.58	0.144	−1.736
20	12	4.56	0.414	13.692	70	42	1.42	0.134	−2.31
25	15	4.25	0.346	9.83	75	45	1.27	0.125	−2.81
30	18	3.91	0.296	6.976	80	48	1.12	0.118	−3.25
35	21	3.59	0.258	4.802	85	51	0.98	0.111	−3.64
40	24	3.05	0.229	3.103	90	54	0.83	0.105	−3.987
45	27	2.63	0.205	1.742	95	57	0.69	0.099	−4.299
50	30	2.29	0.186	0.63	100	60	0.54	0.095	−4.58

3.2. Computational Domain and Meshing

To assure the full development of the wake flow of the wind turbine [40,41], the computational domain was determined as 12D × 5D × 5D (flow direction X × extension direction Y × vertical direction Z, and D is the diameter of the wind wheel). The wind turbine was placed 3D to the entrance of the computational domain. To achieve satisfying computational efficiency and accuracy [42,43], due to the complicated blade appearance, the hybrid grid discrete form was applied to divide the whole computational domain into the internal and external parts. The local encrypted region covered the wind turbine model, which used the non-structured meshing. Non-structured meshing produces a tetra/mixed tetrahedral mesh in ICEM CFD automatically by Robust (Octree). The peripheral region had a regular shape that used the high-quality structured meshing. The grid number, grid quality, and windward pressure coefficient of the tower body under different meshing schemes are shown in Table 5. Grid quality includes the minimum orthogonal quality of the grids and grid skewness. The minimum orthogonal quality of the grids shall be larger than 0.1, and it is best to be higher than 0.2. No negative volume is allowed. The grid skewness shall be lower than 0.95, and it is suggested to be lower than 0.9 [44]. It can be seen from Table 5 that with the increase of the total number of grids, the grid quality and windward pressure coefficient converged gradually. No significant differences were observed in grid quality and computational results under a 9.3 million meshing program and a 28.4 million meshing program. Considering computational accuracy and efficiency, this paper chose the 9.3 million meshing program. When processing flow in the near-wall region, the standard wall function was applied to connect the physical variables on the wall and the unknown variables in the turbulence core region directly. In this paper, the cooperation of the standard wall function and standard k-ε model (which will be introduced in Section 3.3) can increase the simulation accuracy more significantly. Essentially, the flowing conditions in the turbulence core region were solved by the k-ε model. The flow in the wall surface was not solved. Instead, the physical parameters (e.g., speed) of this region were solved by a semi-empirical formula directly. Since the near-wall flow was processed by the standard wall function in this paper, no encryption of the wall surface was needed during meshing, and it only has to place the first internal node in the logarithmic law region. The calculated results demonstrated that the $y+$ of the model wall could assure bottom grids in the logarithmic law, which met common engineering requirements. The meshing schemes of the overall computational domain and model are shown in Figure 8. Both the number of grids and grid quality met the requirements of calculation.

Table 5. Grid number, grid quality, and windward pressure coefficient on the tower body under different meshing schemes.

Meshing Schemes	1	2	3	4	5
Total number of grids	1.1 million	4.5 million	7.4 million	9.3 million	28.4 million
Minimum orthogonal quality of grids	0.13	0.36	0.53	0.60	0.64
Grid skewness	0.95	0.87	0.82	0.74	0.71
Windward pressure coefficient	0.92	0.88	0.85	0.80	0.79

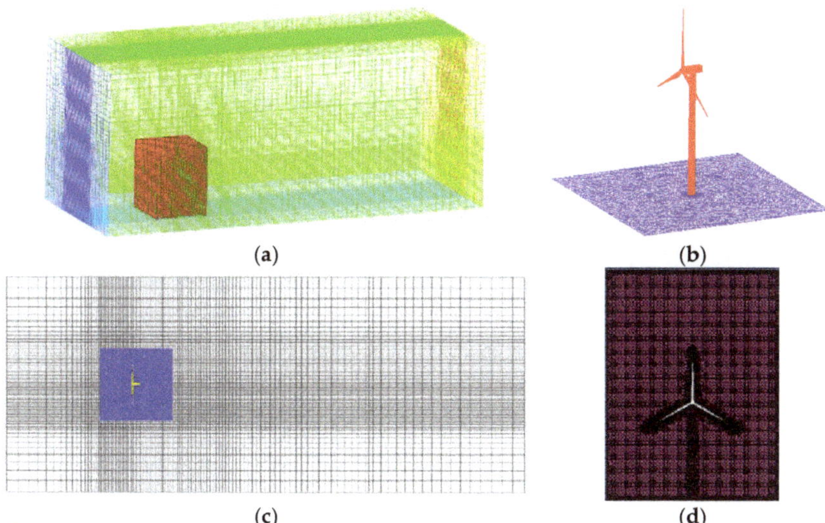

Figure 8. Computational domain and meshing schemes. (**a**) Meshing of the overall computational domain; (**b**) Meshing of local computational domain; (**c**) Meshing of the *x-y* plane; (**d**) Meshing of the encrypted *y-z* plane.

3.3. Selection of Turbulence Model

The 3D steady implicit algorithm was used in this paper. The air wind field chose the incompressible flow field and the turbulence model chose the standard *k-ε* model [45]. The speed vector and pressure coupling problem in the momentum equation was solved by the SIMPLEC algorithm. The turbulent kinetic energy, turbulent dissipation term, and momentum equation in the calculation all adopted the second-order upwind discretion, and the calculation residual error of the governing equation (Navier-Stokes(N-S) equation) was set $1 \times 10-6$. During the simulation, the enhanced wall surface function model was started to assure that the logarithmic distribution of the bottom grids was true.

The continuity equation and momentum equation of the viscous incompressible N-S equation based on Reynolds average are described by Equations (9)–(13).

$$\frac{\partial \mu_i}{\partial x_i} = 0 \tag{9}$$

$$\frac{\partial (\rho \overline{\mu_i})}{\partial t} + \frac{\partial (\rho \overline{\mu_i \mu_j})}{\partial x_j} = -\frac{\partial (\overline{p})}{\partial x_i} + \frac{\partial}{\partial x_j}(\mu \frac{\partial \overline{\mu_i}}{\partial x_j} - \rho \overline{\mu_i' \mu_j'}) \tag{10}$$

where $i, j = 1, 2$; ρ is the air density, which is generally 1.225 kg/m³L and μ is the dynamic coefficient of viscosity, which is generally 1.7894×10^{-5} kg/(m·s)

$$\frac{\partial (\rho_m k)}{\partial t} + \frac{\partial (\rho_m k \mu_j)}{\partial x_j} = P - \rho_m \varepsilon + \frac{\partial}{\partial x_j}[(\mu + \frac{\mu_t}{\sigma_k})\frac{\partial k}{\partial x_j}] \tag{11}$$

$$\frac{\partial (\rho_m \varepsilon)}{\partial t} + \frac{\partial (\rho_m \zeta \mu_j)}{\partial x_j} = C\zeta_1 \frac{\zeta}{k} P_t - C\zeta_2 \rho_m \frac{\zeta^2}{k} + \frac{\partial}{\partial x_j}[(\mu + \frac{\mu_t}{\sigma_k})\frac{\partial \zeta}{\partial x_j}] \tag{12}$$

$$\mu_t = \frac{C_\mu \rho_m k^2}{\zeta} \tag{13}$$

where k and ε are the turbulence energy and the turbulence dissipation rate, P_t is the generating term of turbulence energy, and μ_t is the viscosity coefficient of turbulence. The model constants are $C_{\zeta 1} = 1.44$, $C_{\zeta 2} = 1.92$, $\sigma_\zeta = 1.3$, $\sigma_k = 1.0$, and $C_\mu = 0.09$.

Under the premise of meeting the above mass and momentum conservation equations, the basic flow characteristics of turbulence are described by Equations (14)–(16).

$$\frac{\partial H}{\partial t} + \frac{\partial(\mu H)}{\partial x} + \frac{\partial(v H)}{\partial y} + \frac{\partial(w H)}{\partial z} = 0 \tag{14}$$

$$\frac{\partial(\mu H)}{\partial t} + \frac{\partial}{\partial x}\left| u^2 H + \frac{1}{2}gH^2 \right| + \frac{\partial(\mu v H)}{\partial y} = Hfv - gH\frac{\partial Z_b}{\partial x} - \frac{H}{\rho}\frac{\partial p}{\partial x} + \frac{\tau_{ax} - \tau_{bx}}{\rho} \tag{15}$$

$$\frac{\partial(v H)}{\partial t} + \frac{\partial(\mu v H)}{\partial x} + \frac{\partial}{\partial y}\left| v^2 H + \frac{1}{2}gH^2 \right| = -Hfu - gH\frac{\partial Z_b}{\partial y} - \frac{H}{\rho}\frac{\partial p}{\partial y} + \frac{\tau_{ay} - \tau_{by}}{\rho} \tag{16}$$

where u, v, and w are components of the speed vector $u(z)$ on the x, y, and z directions; H is the reference height; g is the gravitational acceleration; w is the angular velocity of ground rotation; φ is the geological latitude of the calculation point; Z_b is the elevation of ground; τ_{ax} and τ_{ay} are the wind stress components along the x and y directions; τ_{bx} and τ_{by} are the ground frictional force components along the x and y directions; p is the pressure difference; and ρ is the air density. f can be expressed as:

$$f = 2w\sin\varphi \tag{17}$$

The k-ε model is a turbulence model that is applicable to the calculation of high Reynolds numbers. It can predict the fully developed turbulence flow well and it is suitable to the simulation of the surrounding flow fields of similar megawatt wind turbines [46].

3.4. Settings of Boundary Conditions

The inlet of the computational domain was used as the speed inlet boundary, and the outlet was the pressure outlet boundary. Two side walls and the top surface used the symmetric boundaries. The wind turbine and ground set the wall boundary. The overlapping surface of the local and peripheral computational domain was set as the interface. The wind field computational domain and its boundary conditions are shown in Figure 9. Wind velocity profiles were set as the indexes of the atmospheric boundary layer in the computational domain, which are based on the profiles of typhoon and normal wind that are shown in Figure 7b of the revised manuscript, respectively. The velocity profiles of the typhoon and normal wind were determined by formulas (18) and (19) as follows:

$$V_{\mathrm{T}} = 16.49 \times (h/10)^{0.118} \tag{18}$$

$$V_{\mathrm{N}} = 14.70 \times (h/10)^{0.15} \tag{19}$$

where, V_{T} and V_{N} were velocity of typhoon and normal wind at the height of h. They equaled to 16.49 and 14.70 respectively at the height of 10 m. 0.118 was the index on the velocity profiles of typhoon fit by nonlinear least squares, while 0.15 was the roughness index of surface on Type B landforms as specified by Chinese Code [38].

In this study, turbulence of wind fields near the ground was determined by Formulas (20) and (21) under typhoon and normal wind respectively.

$$I_{\mathrm{T}} = 0.20 \times (h/10)^{-0.118} \tag{20}$$

$$I_{\mathrm{N}} = 0.14 \times (h/10)^{-0.15} \tag{21}$$

where 0.20 was the nominal turbulence of the typhoon "Nuri" at the height of 10 m, which was determined based on the synchronous monitoring results of the typhoon [37], and 0.14 was the nominal turbulence on Type B landforms at the height of 10 m, as specified by Chinese code [38].

In typhoon field simulation, fluid parameters (e.g., mean wind profile, scale of turbulence, turbulence energy, turbulence integral scale, and specific dissipation rate) in the fluctuating wind field were integrated into Fluent by UDF based on the above k-ε model and fitting typhoon profile (Figure 10). This was to accomplish the simulation of the typhoon field in a meso/microscale nesting large wind turbine.

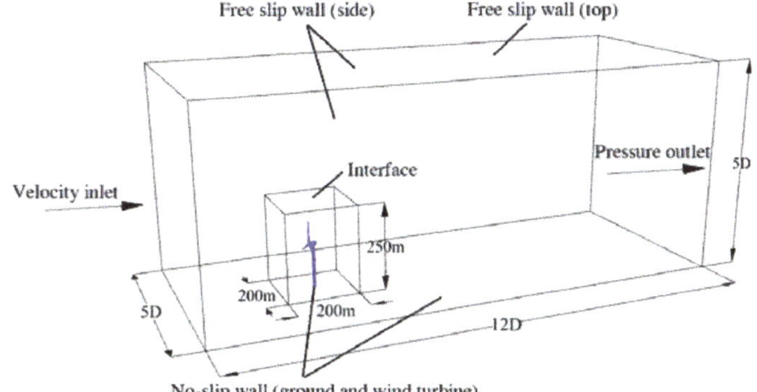

Figure 9. Boundary conditions of the computational domain.

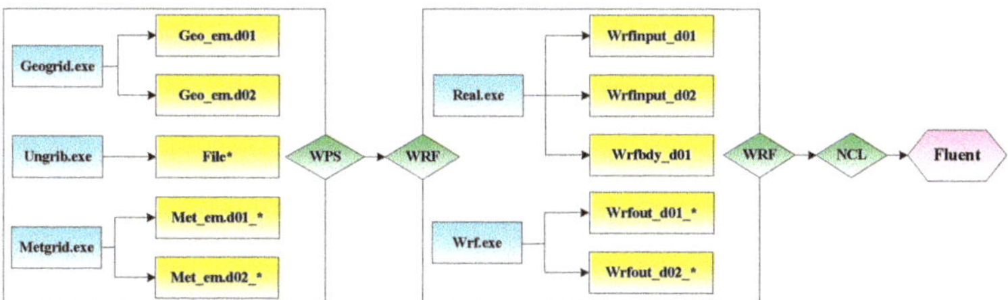

Figure 10. Procedures of WRF-CFD (Computational Fluid Dynamics) computation and nesting.

3.5. Analysis of Aerodynamic Results

The surface pressure coefficient on the wind turbine tower is a main index for evaluating the aerodynamic performances of a wind turbine system. Figure 11 shows the typical cross-section circumferential curve of the pressure coefficient for a wind turbine tower under normal wind conditions, and compares the coefficient with the code value [38]. The analysis suggested that the numerical simulation results of the wind turbine tower under normal wind conditions were generally in accord with the circumferential distribution laws and numerical values specified by Chinese codes. They were only lower than the value specified by the codes out of wind, but the maximum error was below 10%. This demonstrated that the numerical simulation results of CFD were effective.

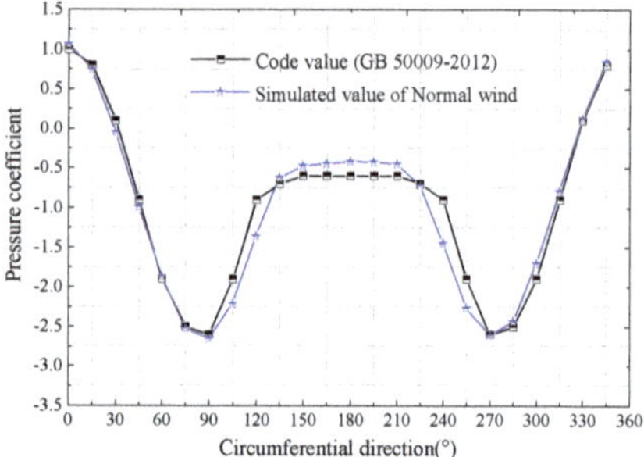

Figure 11. Numerical simulation results under normal wind conditions versus value specified by Chinese code.

The wind speed streamlines and turbulence energy distribution on typical sections of the tower body under normal wind and typhoon loads are shown in Figures 12 and 13. H was the height of the tower body. The wind speed streamlines presented relatively consistent distribution under normal wind and typhoon loads. The wind speed surrounding the tower body and blade under the typhoon loads was higher than that under normal wind. The inflow split at $0°$ of the windward side of the tower body, thus generates reflux and different sizes of eddies on the leeside of the tower body. Additionally, the inflow moved along the blade surface and ran through the front and rear edges of the blades, which formed small-sized eddies on the blade leeside. At the same time, the turbulence energy surrounding the tower body and blades under typhoon loads was significantly stronger than that under normal wind. Wakes on the blade and tower body developed the large-scaled eddy increment region. Besides, the interaction between the blade and the tower body was significant on the middle and upper sections of the tower body.

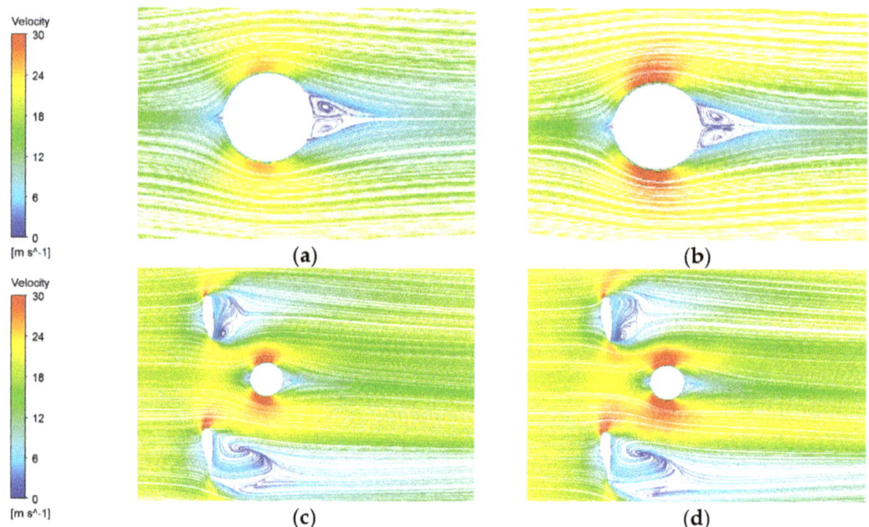

Figure 12. Wind speed streamlines at a typical section of the tower body under normal wind and typhoon loads. (**a**) Normal wind (0.3H); (**b**) Typhoon (0.3H); (**c**) Normal wind (0.8H); (**d**) Typhoon (0.8H).

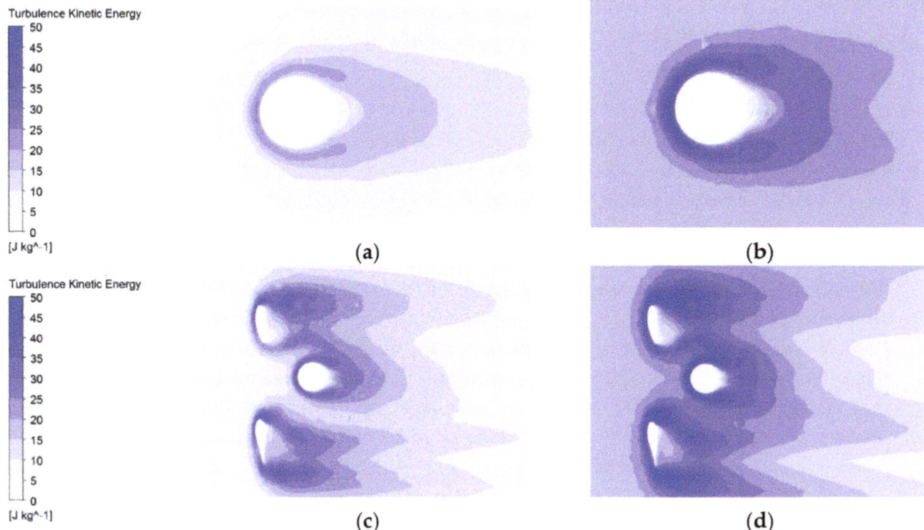

Figure 13. Turbulence energy distribution at a typical section of the tower body under normal wind and typhoon loads. (**a**) Normal wind (0.3H); (**b**) Typhoon (0.3H); (**c**) Normal wind (0.8H); (**d**) Typhoon (0.8H).

The distribution curves of the wind pressure coefficient at the typical section of the tower body under normal wind and typhoon loads are shown in Figure 14. Under normal wind and typhoon loads, the circumferential wind pressure coefficient on the tower body was symmetric along the windward side. With the increase of circumferential angle, it decreased firstly and then increased until it stabilized close to the leeside. However, extremum positive pressure on the windward side and extremum negative pressure on the lateral surface under the typhoon were significantly higher compared with those under normal wind. The maximum increase of the wind pressure coefficient on the windward side was 61.1%, and the maximum increase of the extreme negative pressure on the lateral surface was 44.9%. Both were achieved at bottom of the tower body.

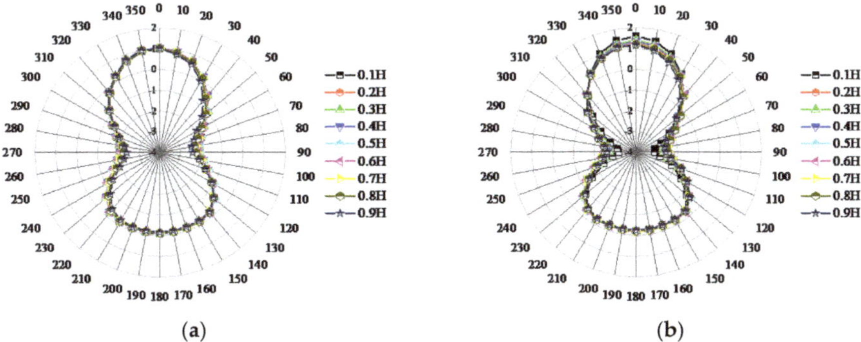

Figure 14. Distribution curves of wind pressure coefficient on tower body under normal wind and typhoon loads. (**a**) Normal wind; (**b**) Typhoon.

The straight-up blade was defined Blade A. The remaining two clockwise rotating blades were Blade B and Blade C. Wind pressure coefficient nephograms on the windward side and leeside of blade as well as mean along the blade span under normal wind and typhoon loads are shown in Figure 15.

The distribution curves of the overall wind pressure coefficient of the blades under normal wind and typhoon loads are shown in Figure 16. It can be seen that:

(1)   The wind pressure on the windward side of all of the blades was positive under normal wind and typhoon loads. Negative wind pressure only occurred at the front blade edges and rear edges of the blade root. However, the wind pressure on the leeside of blades was basically negative.

(2)   The wind pressure on the windward side of Blade A generally increased along the blade span. However, the positive pressure on the windward sides of Blade B and Blade C changed slightly along the span, without irregular changes. Negative pressure on the leeside of the blades decreased firstly, and then increased along the span direction. If these two effects were superposed, the overall wind pressure coefficient of the different blades decreased firstly, and then increased along the span direction.

(3)   Positive wind pressure on the windward side of the blades under typhoon loads was higher than that under normal winds. The negative wind pressure on the leeside was slightly different from that under normal winds. As a result, the overall wind pressure coefficient on different blades under typhoon loads was higher than that under normal winds. The maximum increase was 23.6%, which occurred at the tip of Blade C.

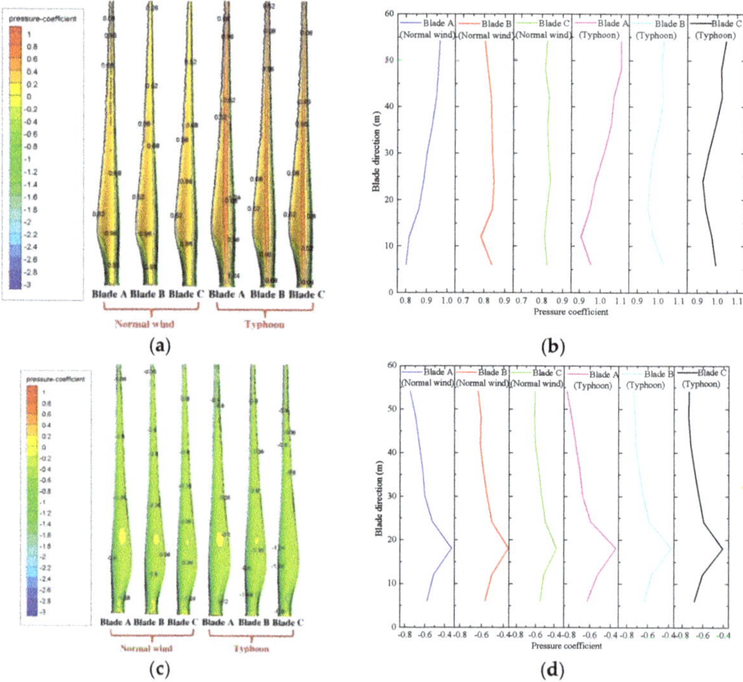

Figure 15. Wind pressure coefficient nephograms on the windward and leeside of blades as well as the mean along the blade under normal wind and typhoon loads. (**a**) Wind pressure coefficient nephogram on the windward side; (**b**) Wind pressure coefficient curves on the windward side; (**c**) Wind pressure coefficient nephogram on the leeside; (**d**) Wind pressure coefficient curves on the leeside.

The distribution curves of the lift coefficient and drag coefficient on the different height sections of the tower body under normal wind and typhoon loads are shown in Figure 17. The lift–drag coefficient ratio is shown in Table 6. The calculation formulas of the across-wind lift coefficient (C_L) and downwind drag coefficient (C_D) are shown as follows [4]:

$$C_L = \frac{\sum_{i=1}^{n} C_{Pi} A_i \sin \theta_i}{A_T} \tag{22}$$

$$C_D = \frac{\sum_{i=1}^{n} C_{Pi} A_i \cos \theta_i}{A_T} \tag{23}$$

where Cpi is the mean wind pressure coefficient at the measuring point i on the tower body, A_i is the pressure coverage area at the measuring point i, θ_i is the included angle between the pressure at measuring point i and the wind axis, and A_T is the projection area of the overall structure along the wind axis.

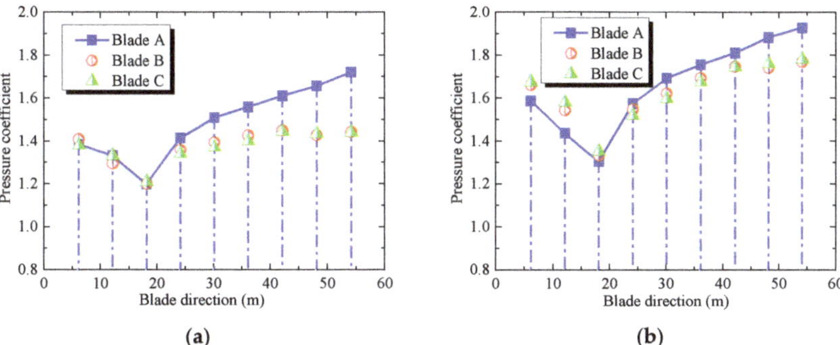

Figure 16. Distribution curves of overall wind pressure coefficient on blades under normal wind and typhoon loads. (**a**) Normal wind; (**b**) Typhoon.

The following can be seen from Figure 17 and Table 6. (1) The lift coefficient was negative under normal wind, but it was alternatively positive and negative under typhoon loads. However, the distribution law of the lift coefficient was consistent under normal wind and typhoon loads, and the numerical value was small; (2) The drag coefficient was mainly in the range of 0.35~0.4 under normal winds, and it was in the range of 0.45~0.65 under typhoon loads. The drag coefficient at different heights under the typhoon loads increased dramatically than that under the normal wind. The maximum was about 0.18, which was at the tower top. This was mainly caused by the large disturbance of the tower top to blades.

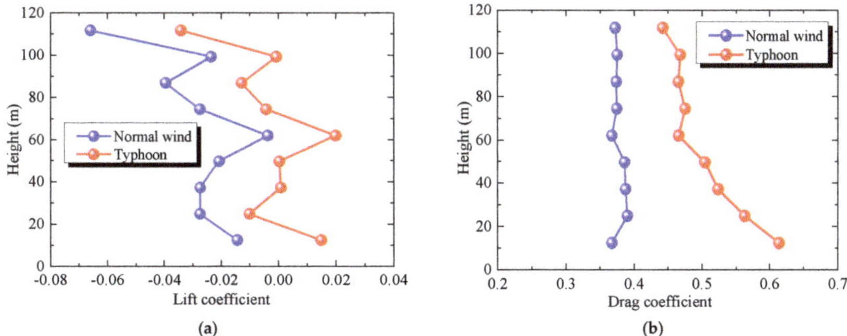

Figure 17. Distribution curves of lift coefficient and drag coefficient of the tower body under normal wind and typhoon loads. (**a**) Lift coefficient. (**b**) Drag coefficient.

Table 6. Lift–drag coefficient ratio of the tower body under normal wind and typhoon loads.

C_L/C_D	Height of the Tower Body								
	12.4 m	**24.8 m**	**37.2 m**	**49.6 m**	**62 m**	**74.4 m**	**86.8 m**	**99.2 m**	**111.6 m**
Normal wind	−0.040	−0.071	−0.072	−0.055	−0.011	−0.074	−0.107	−0.064	−0.179
Typhoon loads	0.024	−0.019	0.001	0.001	0.042	−0.010	−0.028	−0.002	−0.078

4. Response Analysis of the Wind Turbine Coupling Structure

4.1. Finite Element Modeling and Dynamic Characteristics

The tower–blade integrated finite element coupling model was constructed by the large universal finite element analysis software ANSYS. The tower body and blade were simulated by Shell63 unit, whereas the nacelle and internal structure were simulated as a whole by using Beam 189 units. The round raft basis unit was Solid65. The interaction between the foundation and the basis was simulated by a Combin14 unit. After finite element models of different wind turbine parts were constructed, a coupling connection may cause the relative slippage of models under loads, which was attributed to the different unit types of different parts. Therefore, the multipoint restriction unit coupling command was used to connect different parts and construct the tower–blade integrated simulation model. According to the principle of efficiency and the accuracy balance, the model was divided into 4122 units.

The natural frequency of vibration and the mode of vibration of the wind turbine were solved by a block Lanczos algorithm after modeling [47,48]. The finite element model of the wind turbine, the first 100 orders of the frequency distribution, and the mode of vibration at typical orders are shown in Figures 18 and 19. The tower–blade integrated model had a low natural frequency of vibration. The fundamental frequency was 0.137 Hz, and the first 10 orders of frequency were lower than 1 Hz. At the same time, the natural frequency of the vibration of the structure was low and distributed densely. The structural frequency achieved approximately linear growth with the number of vibration orders, but the growth rate decreased after 90 orders. In addition, three blades in the tower–blade integrated model made front-to-back movement at the first, fifth, and 10th order. The blades made complicated front-to-back and left-to-right swings at the 30th order and 50th order, which were accompanied by the bending deformation of the tower body. According to the multi-order mode of vibration, the low-order mode of vibration in the tower–blade integrated model was mainly manifested by the front-to-back and left-to-right swing of the blades. Under the high order of vibration, the tower body and blades developed deformations and buckling failure.

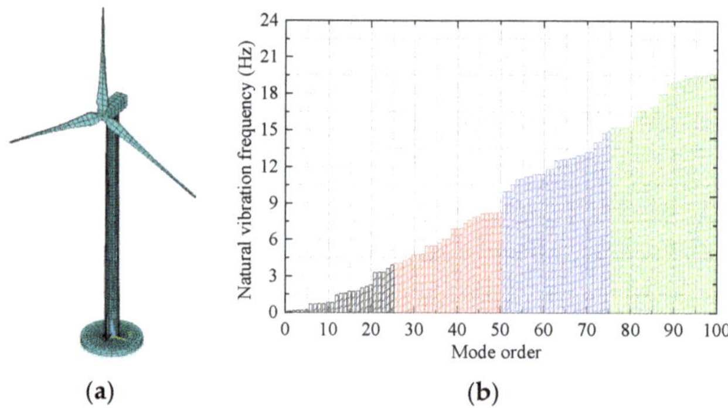

(a)   (b)

Figure 18. Finite element model of wind turbine and the first 100 orders of inherent frequency. (a) Finite element model; (b) First 100 orders of inherent frequency.

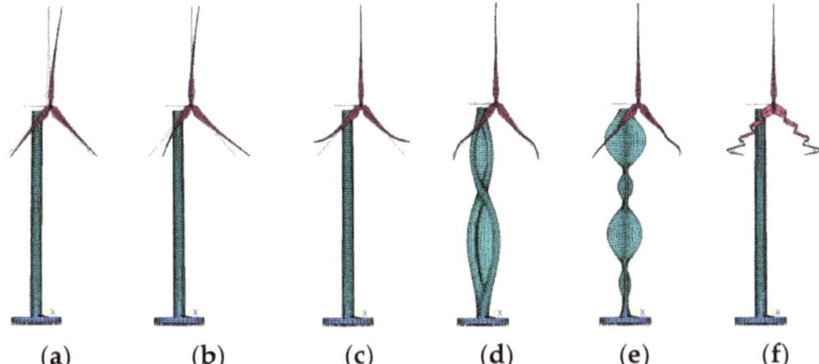

Figure 19. Mode of vibration of the tower–blade coupling model at different orders. (**a**) First order; (**b**) Fifth order; (**c**) 10th order; (**d**) 30th order; (**e**) 50th order; (**f**) 100th order.

4.2. Stress Analysis on the Tower Body

The 3D distributions of the radial displacement of the tower body and distributions of internal stress responses at the tower bottom under normal wind and typhoon loads are shown in Figures 20 and 21. It can be seen that: (1) the radial distribution of the tower body presented relatively similar distribution patterns under normal wind and typhoon loads. The radial displacement increased gradually going up of the tower body, forming two extreme regions on the windward side and leeside. The maximum positive and negative displacements were at 180° and 0°; (2) Different wind field mainly affected the radial displacement at the middle and upper parts of the tower body. Under the typhoon loads, the maximum radial displacement of the lower tower body was 0.117 m, which was increased by 28.6% compared with that under normal wind; (3) Internal stresses at the tower bottom were symmetric along the windward side under normal wind and typhoon loads. Despite the slight difference in the meridian axial force at the tower bottom, other internal stresses presented basically consistent circumferential distribution laws under normal wind and typhoon loads. All of the internal stresses achieved the maximum value close to the circumferential 330° of the tower body; (4) Extreme internal stresses at the tower bottom under typhoon loads were stronger than those under normal wind loads. The maximum meridian axial force, maximum shearing force, maximum circumferential bending moment, and maximum meridian bending moment were increased by 36.1%, 34.8%, 34.2%, and 46.5%, respectively. All occurred close to the circumferential 330°.

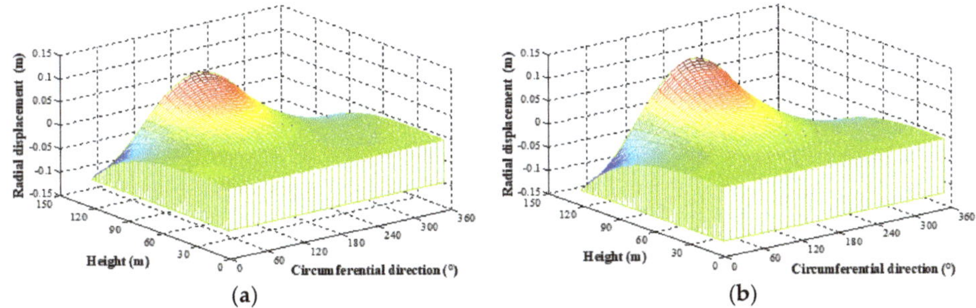

Figure 20. 3D radial displacement of tower body under normal wind and typhoon loads. (**a**) Normal wind; (**b**) Typhoon.

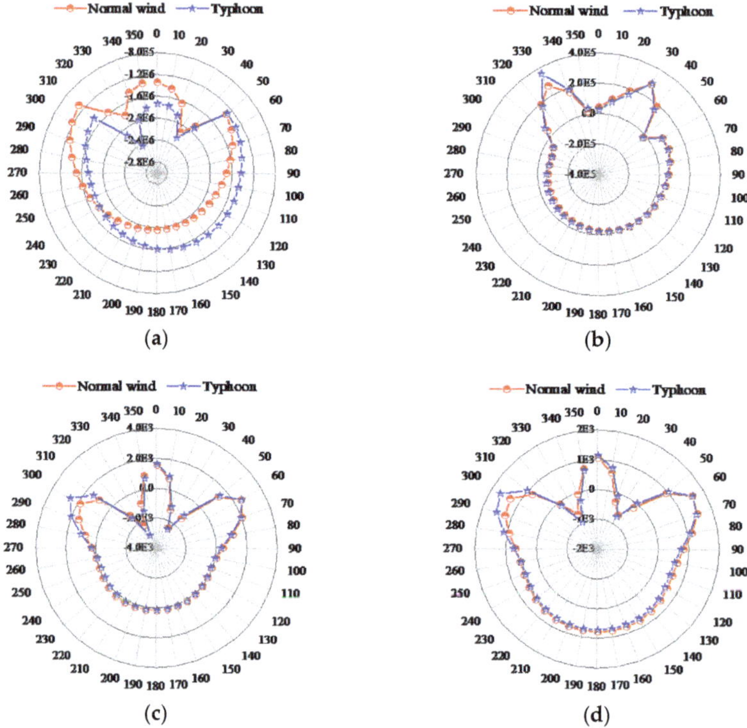

Figure 21. Internal stress responses at the tower bottom of the wind turbine system under normal wind and typhoon loads. (**a**) Meridian axial force Ty/(N); (**b**) Shearing force Txy/(N); (**c**) Circumferential bending moment Mx/(N·m); (**d**) Meridian bending moment My/(N·m).

4.3. Stress Analysis of Blades

The internal stress responses at blade roots and downwind displacement distribution along the blade span under normal wind and typhoon loads are shown in Figures 22 and 23. Downwind displacements at the blade tips are shown in Table 7. It was found that: (1) under normal wind and typhoon loads, the internal stresses at the root of Blade A were significantly higher than those of Blade B and Blade C. The numerical values of Blade B and Blade C were similar. Moreover, the internal stresses at the blade roots under typhoon loads were significantly higher than those under normal wind loads. The maximum increase of the shearing force was 41.3%. The maximum increase of the circumferential bending moment was 39.6%. The maximum increase of the meridian bending moment was 39.5%. They were all achieved by Blade C; (2) The downwind displacements of Blade B and Blade C along the span were basically same under normal wind and typhoon loads. However, they were significantly different from those of Blade A. With the extension along the blade span, the downwind displacement of Blade A was higher than those of Blade B and Blade C, and then became smaller. Next, the numerical values that were close to the middle of the blades were the same. However, the numerical values that were close to the middle of Blade A exceeded those of Blade B and Blade C. The numerical values of the three blades were equal at about 35 m of the blade span under normal wind, and at about 30 m of the blade span under typhoon loads; (3) The downwind displacements of the blades were larger under typhoon loads than those under normal wind loads. The maximum increase of downwind displacement was 49.3%, which was achieved at the tip of Blade A.

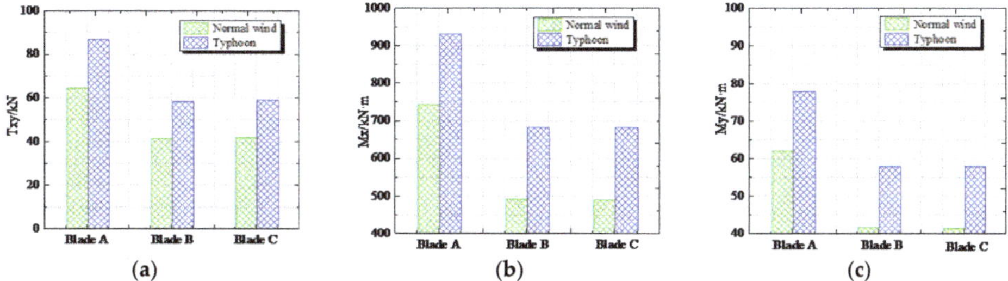

Figure 22. Internal stress responses at the blade roots of the wind turbine under normal wind and typhoon loads. (**a**) Shearing force (Txy); (**b**) Circumferential bending moment (Mx); (**c**) Meridian bending moment (My).

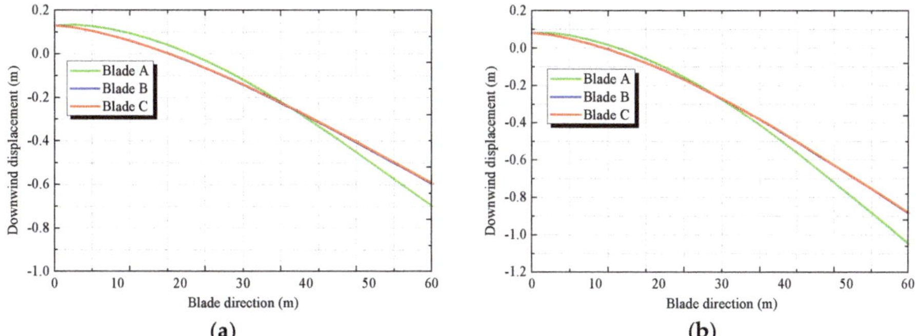

Figure 23. Distribution curves of downwind displacements of three blades under normal wind and typhoon loads. (**a**) Normal wind; (**b**) Typhoon.

Table 7. Downwind displacements at blade tips under normal wind and typhoon loads.

Blade A		Blade B		Blade C	
Normal Wind	Typhoon	Normal Wind	Typhoon	Normal Wind	Typhoon
-0.701 m	-1.047 m	-0.599 m	-0.885 m	-0.596 m	-0.884 m

5. Wind-Induced Stability Analysis of Wind Turbine System

5.1. Buckling Stability Analysis

The buckling mode and eigenvalues of the tower-blade integrated model under normal wind and typhoon loads are shown in Table 8. Based on comparison, the critical wind speed was higher than the designed wind speed under normal wind and typhoon loads, meeting the design requirements. The buckling position of the wind turbine was at the tower top. The buckling displacement and critical wind speed of the large wind turbine system were significantly influenced by different wind fields. The buckling coefficient, critical wind speed, and maximum displacement under typhoon loads were increased by 1.3%, 30.9%, and 28% compared with those under normal wind loads.

Table 8. Comparison of buckling mode and eigenvalues of wind turbine system under normal wind and typhoon loads.

Buckling Eigenvalue under Normal Wind		Buckling Mode under Normal Wind	Buckling Mode under Typhoon Loads	Buckling Eigenvalue under Typhoon Loads	
Buckling coefficient	2.955			Buckling coefficient	2.992
Critical wind speed/(m/s)	24.12			Critical wind speed/(m/s)	31.57
Maximum displacement/m	0.025			Maximum displacement/m	0.032

5.2. Ultimate Bearing Capacity

Changes in the buckling displacement of the wind turbine system with wind speed under normal wind and typhoon loads are shown in Figure 24. It was found that with the leveling loading of the wind speed, the buckling displacement of the wind turbine system under normal wind and typhoon loads increased firstly, and then decreased, and finally increased slightly. It reached the peak at about 20 m/s of the reference wind speed, and reached the minimum at about 40 m/s of the reference wind speed. The above trend was due to the "reverse effect" [46,49]. Moreover, the typhoon effect would significantly decrease the ultimate bearing capacity of the wind turbine system. The buckling displacement was increased by 38.5% at 40 m/s of the reference wind speed.

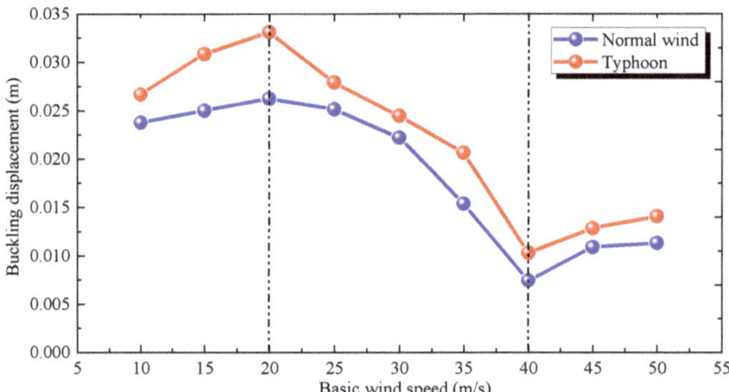

Figure 24. Changes of the buckling displacement of the wind turbine with wind speed under normal wind and typhoon loads.

6. Summary and Conclusions

High-resolution simulation of the Typhoon "Nuri" was accomplished based on the mesoscale WRF model. Later, differences regarding the aerodynamic force, wind-induced responses, buckling stability, and ultimate bearing capacity of the wind turbine system under normal wind and typhoon loads were analyzed by microscale CFD numerical simulation and the finite element method. Some major conclusions could be drawn:

- A high-resolution simulation of Typhoon "Nuri" was carried out through the mesoscale WRF model. The typhoon profile index that was based on the least square fitting was 0.118, which was about 20% smaller than the wind profile index under normal wind loads. In this paper, the meso/microscaled nested downscaling method can effectively simulate a three-dimensional typhoon field of similar structures and provide effective input loads for follow-up comprehensive stress analysis.

- The flow field distribution surrounding the wind turbine is basically similar under normal wind and typhoon effects. However, the airflow under typhoon conditions showed higher wind speed and turbulence compared with those under normal wind. Positive pressure on windward surface of the tower body, negative pressure on the crosswind surface of the tower body, the overall pressure coefficient of the blade, the resistance coefficient of the tower body, and the coupling structural response all increased significantly. The buckling stability and ultimate bearing capacity of the coupling system decreased significantly.

- The pressure coefficient on the windward surface of the tower body under typhoon conditions increased by 61.1% to the maximum extent. The maximum increment of the negative pressure extremum on the crosswind surface was 44.9%. Both were observed at the bottom of the tower body. The typhoon influenced the pressure coefficient on the windward surface of the blades significantly, but it only affected the leeside surface slightly. The maximum increment of the overall pressure coefficient of the blade was 23.6%, which occurred at the tip of Blade C. Under typhoon conditions, the resistance coefficient of the tower body increased significantly, and the maximum increment, 67.5%, occurred at the bottom of the tower body.

- Under typhoon conditions, the radial displacement at the tower top of the wind turbine increased by 28.6%. The increments of the maximum meridian axial force, maximum shearing force, maximum circumferential bending moment, and maximum meridian bending moment were 36.1%, 34.8%, 34.2%, and 46.5%, respectively. All were observed close to the circumferential 330°. The maximum increments of the shearing force, circumferential bending moment, and meridian bending moment at the blade root under typhoon conditions were 41.3%, 39.6%, and 39.5%, which were observed on Blade C. The maximum increment of downwind displacement at the blade tip occurred on Blade A, which reached 49.3%.

- Under typhoon conditions, the buckling coefficient, critical wind speed, and maximum buckling displacement of the wind turbine system increased by 1.3%, 30.9%, and 28%, respectively. The wind turbine system achieved the maximum buckling displacement when the basic wind speed was about 20 m/s, and the minimum buckling displacement at about 40 m/s of the basic wind speed. When the basic wind speed was about 40 m/s, the buckling displacement under typhoon conditions increased by 38.5%.

Author Contributions: Conceptualization, S.K.; Data curation, S.K.; Formal analysis, W.Y.; Funding acquisition, S.K.; Investigation, J.C.; Methodology, T.W.; Software, W.Y.; Writing—original draft, S.K.; Writing—Review & Editing, S.K.

Funding: This work was jointly funded by the National Basic Research Program of China ("973" Program) under Grant (2014CB046200), Natural Science Foundation of China (51878351, 51761165022, U1733129), open fund for Jiangsu Key Laboratory of Hi-Tech Research for Wind Turbine Design (ZAA20160013), Jiangsu Outstanding Youth Foundation (BK20160083), and China Postdoctoral Science Foundation (2015T80551).

Acknowledgments: The authors would like to thank Yaojun Ge and Yukio Tamura for valuable discussions on this article. The authors are sincerely grateful to the reviewers for their valuable review comments, which substantially improved the paper.

Conflicts of Interest: The authors declare no conflict of interest.

References

1. Murtagh, P.J.; Basu, B.; Broderick, B.M. Along-wind response of a wind turbine tower with blade coupling subjected to rotationally sampled wind loading. *Eng. Struct.* **2005**, *27*, 1209–1219. [CrossRef]
2. Ke, S.T.; Ge, Y.J.; Wang, T.G.; Cao, J.F.; Tamura, Y. Wind field simulation and wind-induced responses of large wind turbine tower-blade coupled structure. *Struct. Des. Tall Spec. Build.* **2015**, *24*, 571–590. [CrossRef]
3. Hansen, M.O.L.; Sørensen, J.N.; Voutsinas, S.; Sørensen, N.; Madsen, H.A. State of the art in wind turbine aerodynamics and aeroelasticity. *Prog. Aerosp. Sci.* **2006**, *42*, 285–330. [CrossRef]
4. Ke, S.T.; Yu, W.; Wang, T.G.; Zhao, L.; Ge, Y.J. Wind loads and load-effects of large scale wind turbine tower with different halt positions of blade. *Wind Struct. Int. J.* **2016**, *23*, 559–575. [CrossRef]

5. Utsunomiya, T.; Sato, I.; Yoshida, S.; Ookubo, H.; Ishida, S. Dynamic response analysis of a floating offshore wind turbine during severe typhoon event. In Proceedings of the ASME 2013 32nd International Conference on Ocean, Offshore and Arctic Engineering, Nantes, France, 9–14 June 2013; p. V008T09A032.

6. Wang, Z.Y.; Zhang, B.; Zhao, Y.; Liu, G. H.; Jiang, J.Q. Study on the vibration response of wind turbine tower under typhoons. *Acta Energ. Sol. Sin.* **2013**, *34*, 1434–1442. (In Chinese)

7. Chen, X.; Xu, J.Z. Structural failure analysis of wind turbines impacted by super typhoon Usagi. *Eng. Fail. Anal.* **2016**, *60*, 391–404. [CrossRef]

8. Ishihara, T.; Yamaguchi, A.; Takahara, K.; Mekaru, T.; Matsuura, S. An analysis of damaged wind turbines by typhoon Maemi in 2003. In Proceedings of the 6th Asia-Pacific Conference on Wind Engineering, Seoul, Korea, 12–14 September 2005; pp. 1413–1428.

9. Takeshi, I.; Van, P.P.; Fujino, Y.; Takahara, K.; Mekaru, T. A field test and full dynamic simulation on a stall regulated wind turbine. In Proceedings of the Sixth Asia-Pacific Conference on Wind Engineering (APCWE-VI), Seoul, Korea, 12–14 September 2005; Volume 48, pp. 599–612.

10. Tang, W.L.; Yuan, Q.; Han, Z.H. Withstanding typhoon design of wind turbine tower. *Acta Energ. Sol. Sin.* **2008**, *29*, 422–427. (In Chinese)

11. Chou, J.S.; Tu, W.T. Failure analysis and risk management of a collapsed large wind turbine tower. *Eng. Fail. Anal.* **2011**, *18*, 295–313. [CrossRef]

12. Li, Z.Q.; Chen, S.J.; Ma, H.; Feng, T. Design defect of wind turbine operating in typhoon activity zone. *Eng. Fail. Anal.* **2013**, *27*, 165–172. [CrossRef]

13. Luo, Y.; Chen, Q.; Ren, D.; Li, X.; Liao, Y. The light-weight design of tower for under typhoon wind turbine based on the typhoon control strategy. In Proceedings of the International Conference on Sensors, Mechatronics and Automation, Zhuhai, China, 12–13 November 2016.

14. Lian, J.; Jia, Y.; Wang, H.; Liu, F. Numerical study of the aerodynamic loads on offshore wind turbines under typhoon with full wind direction. *Energies* **2016**, *9*, 613. [CrossRef]

15. Zhou, P.Z.; Long, S.H.; Chen, R. Optimal design of aerodynamic shape of wind turbine blade against typhoon based on genetic algorithm. *J. Changsha Univ. Technol. (Nat. Sci.)* **2015**, *12*, 89–94. (In Chinese)

16. Cao, Z. Design of a vertical axis wind turbine to resist typhoon. *Appl. Energy Technol.* **2015**, *35*, 50–52. (In Chinese)

17. An, Y. *Study on the Performance of Large Double Runner Wind Turbine against Typhoon*; South China University of Technology: Guangzhou, Guangdong, 2011. (In Chinese)

18. Utsunomiya, T.; Sato, I.; Yoshida, S.; Ookubo, H.; Ishida, S. At-sea experiment of a spar-type floating wind turbine (behavior during typhoon attack). *Wind Energy* **2015**, *39*, 86–97.

19. Ma, Z.; Li, W.; Ren, N.X.; Ou, J. The typhoon effect on the aerodynamic performance of a floating offshore wind turbine. *J. Ocean Eng. Sci.* **2017**, *35*, 87–99. [CrossRef]

20. Lundquist, J.K.; Mirocha, J.D.; Chow, F.K.; Kosovic, B.; Lundquist, K.A. Simulating atmosphere flow for wind energy applications with WRF-LES. *Univ. North Texas* **2008**, *7*, 281–295.

21. Shen, L.; Han, Y.; Dong, G.C.; Cai, C.S.; Zheng, J.R. Numerical simulation of wind field on bridge site located in mountainous area and gorge based on WRF. *China J. Highw. Transp.* **2017**, *30*, 104–113. (In Chinese)

22. Papanastasiou, D.K.; Melas, D.; Lissaridis, I. Study of wind field under sea breeze conditions; an application of WRF model. *Atmos. Res.* **2010**, *98*, 102–117. [CrossRef]

23. Dimitrov, N.; Natarajan, A.; Mann, J. Effects of normal and extreme turbulence spectral parameters on wind turbine loads. *Renew. Energy* **2017**, *10*, 1180–1193. [CrossRef]

24. Li, J.; Song, X.P.; Cheng, X.L.; Hu, F.; Zhu, R. Dynamical simulation of wind flow from synoptic scale to turbine scale. *Acta Energ. Sol. Sin.* **2015**, *36*, 806–811. (In Chinese)

25. Maruyama, T.; Tomokiyo, E.; Maeda, J. Simulation of strong wind field by non-hydrostatic mesoscale model and its applicability for wind hazard assessment of buildings and houses. *Hydrol. Res. Lett.* **2010**, *16*, 40–44. [CrossRef]

26. Skamarock, W.C. A description of the advanced research WRF version 3. *NCAR Tech.* **2005**, *113*, 7–25.

27. Cha, D.H.; Wang, Y.A. Dynamical Initialization Scheme for Real-time Forecasts of Tropical Cyclones Using the WRF Model. *Mon. Weather Rev.* **2013**, *141*, 964–986. [CrossRef]

28. Gan, C.M.; Hogrefe, C.; Mathur, R.; Pleim, J.; Xing, J.; Wong, D.; Gilliam, R.; Pouliot, G.; Wei, C. Assessment of the effects of horizontal grid resolution on long-term air quality trends using coupled WRF-CMAQ simulations. *Atmos. Environ.* **2016**, *132*, 207–216. [CrossRef]

29. Zou, Z.C.; Deng, Y.C.; Deng, Z.P. Analysis on the effect of different initial fields on wind field simulation of WRF model. *Renew. Energy* **2015**, *33*, 1048–1053. (In Chinese)

30. Pan, X.D.; Li, X. Research on the effect of horizontal resolution on WRF model—Taking Heihe river basin WRF simulation as an example. *Sci. Res. Inf. Technol. Appl.* **2011**, *2*, 126–137. (In Chinese)

31. Yang, J. *Numerical Simulation Study of Local Typhoon Wind Fields Nested by WRF and CFD*; Harbin Institute of Technology: Harbin, Heilongjiang, 2015. (In Chinese)

32. Miao, Y.; Liu, S.; Chen, B.; Zhang, B.; Wang, S.; Li, S. Simulating urban flow and dispersion in Beijing by coupling a CFD model with the WRF model. *Adv. Atmos. Sci.* **2013**, *30*, 1663–1678. [CrossRef]

33. Liu, Y.S.; Miao, S.G.; Zhang, C.L.; Cui, G.X.; Zheng, Z.S. Study on micro-atmospheric environment by coupling large eddy simulation with mesoscale model. *J. Wind Eng. Ind. Aerodyn.* **2012**, *s107–s108*, 106–117. [CrossRef]

34. Davis, C.; Wang, W.; Chen, S.S.; Chen, Y.; Corbosiero, K.; DeMaria, M.; Dudhia, J.; Holland, G.; Klemp, J.; Michalakes, J.; et al. Prediction of Landfalling Hurricanes with the Advanced Hurricane WRF Model. *Month. Weather Rev.* **2008**, *136*, 1990–2005. [CrossRef]

35. Lou, X.F.; Hu, Z.J.; Wang, P.Y.; Zhou, X.J. Introduction of mesoscale model cloud precipitation physics scheme. *J. Appl. Meteorol.* **2003**, *14*, 49–59. (In Chinese)

36. Yi, J.K. Diagnosis and analysis of the movement, landing site and intensity change of typhoon No. 0812 'Nuri'. In Proceedings of the 26th Annual Meeting of the Chinese Meteorological Society, Hangzhou, Zhejiang, 14 October 2009. (In Chinese)

37. Song, L.L.; Pang, J.B.; Jiang, C.L.; Huang, H.H.; Qin, P. Measurement and analysis of turbulence characteristics of 'Nuri' typhoon in Macao friendship bridge. *Chin. Sci. Technol. Sci.* **2010**, *40*, 1409–1419. (In Chinese)

38. *Load Code for the Design of Building Structures*; GB 50009-2012; The Ministry of Structure of the People's Republic of China: Beijing, China, 2012. (In Chinese)

39. Wang, X.H.; Ke, S.T. Study on aerodynamic performance of large wind turbine tower considering blade yawing and interference effect. *Chin. J. Electr. Eng.* **2018**, *38*, 4546–4554, 4655. (In Chinese)

40. Zhang, L.D.; Ren, L.C.; Lian, L.J.; Chen, R.S. Study on shape design and three-dimensional solid modeling of wind turbine blades. *Acta Energ. Sol. Sin.* **2008**, *29*, 1177–1180.

41. Luo, K.; Zhang, S.; Gao, Z.; Wang, J.; Zhang, L.; Yuan, R.; Fan, J.; Cen, K. Large-eddy simulation and wind-tunnel measurement of aerodynamics and aeroacoustics of a horizontal-axis wind turbine. *Renew. Energy* **2015**, *77*, 351–362. [CrossRef]

42. Kuo, J.Y.J.; Romero, D.A.; Beck, J.C.; Amon, C.H. Wind farm layout optimization on complex terrains—Integrating a CFD wake model with mixed-integer programming. *Appl. Energy* **2016**, *178*, 404–414. [CrossRef]

43. Chattot, J.J. Effects of blade tip modifications on wind turbine performance using vortex model. *Comput. Fluids* **2009**, *38*, 1405–1410. [CrossRef]

44. Jiang, F. *Advanced Applications and Case Study of Fluent*; Tsinghua University Press: Beijing, China, 2008. (In Chinese)

45. Genç, M.S. Numerical simulation of flow over a thin aerofoil at a high reynolds number using a transition model. *J. Mech. Eng. Sci.* **2010**, *1*, 1–10. [CrossRef]

46. Ke, S.T.; Xu, L.; Ge, Y.J. The aerostatic response and stability performance of a wind turbine tower-blade coupled system considering blade shutdown position. *Wind Struct. Int. J.* **2017**, *25*, 507–535.

47. Ke, S.T.; Wang, T.G.; Ge, Y.J.; Tamura, Y. Wind-induced responses and equivalent static wind loads of tower-blade coupled large wind turbine system. *Struct. Eng. Mech. Int. J.* **2014**, *52*, 485–505. [CrossRef]

48. Cárdenas, D.; Escárpita, A.A.; Elizalde, H.; Aguirre, J.J.; Ahuett, H.; Marzocca, P.; Probst, O. Numerical validation of a finite element thin-walled beam model of a composite wind turbine blade. *Wind Energy* **2012**, *15*, 203–223. [CrossRef]

49. Ke, S.T.; Wang, X.H. Wind vibration response and stability analysis of large-scale wind turbine systems considering the yaw and interference effects of blades. *J. Hunan Univ. (Nat. Sci. Ed.)* **2018**, *45*, 61–70. (In Chinese)

 *applied sciences*

Article

Aerodynamics and Motion Performance of the H-Type Floating Vertical Axis Wind Turbine

Ying Guo, Liqin Liu, Xifeng Gao * and Wanhai Xu

State Key Laboratory of Hydraulic Engineering Simulation and Safety, Tianjin University, Tianjin 300072, China; yynocry@tju.edu.cn (Y.G.); liuliqin@tju.edu.cn (L.Q.L.); xuwanhai@tju.edu.cn (W.H.X.)
* Correspondence: gaoxifeng@tju.edu.cn

Received: 15 December 2017; Accepted: 7 February 2018; Published: 9 February 2018

Abstract: Aerodynamics and motion performance of the floating vertical wind turbine (VAWT) were studied in this paper, where the wind turbine was H-type and the floating foundation was truss spar type. Based on the double-multiple-stream-tube theory, the formulae were deduced to calculate the aerodynamic loads acting on the wind turbine considering the motions of the floating foundation. The surge-heave-pitch nonlinear coupling equations of the H-type floating VAWT were established. Aerodynamics and motion performance of a 5 MW H-type floating VAWT was studied, and the effect of the floating foundation motions on the aerodynamic loads was analyzed. It is shown that the motions of the floating foundation on the aerodynamics cannot be ignored. The motion of the H-type floating VAWT was also compared with that of the Φ-type floating VAWT: they have the same floating foundation, rated output power, mooring system and total displacement. The results show that the H-type floating VAWT has better motion performance, and the mean values of surge, heave and pitch of the H-type floating VAWT are much smaller comparing with the Φ-type floating VAWT.

Keywords: H-type floating VAWT; truss Spar floating foundation; coupling of aerodynamics and hydrodynamics

---

1. Introduction

With the increase of unit capacity, size of the offshore wind turbine is getting bigger and bigger. The instability of floating horizontal axis wind turbine (HAWT) may become obvious as the huge drive system is placed on the top of the high tower [1]. The drive system of Vertical Axis Wind Turbine (VAWT) is installed on the bottom, which has little effect on the dynamic behavior of the tower. This makes floating VAWT more advantageous in the development of large-scale offshore wind power [2].

Darrieus wind turbine is the earliest lift type VAWT, and it has the higher power coefficient comparing with other VAWT [1]. According to the form of the blade, the Darrieus wind turbine includes curved blade and straight blade, withib which the Φ-type and the H-type are the most common. With the development of the offshore wind farm, the feasibility of the Darrieus wind turbine is assessed to be used in deep water and supported by the floating foundation.

The research on the floating VAWT has been carried out, including the aerodynamic load calculations of the floating VAWT, the motion comparison of the VAWT supported by different floating foundations, the influences between aerodynamic and hydrodynamic loads, and so on. Some methods and codes were developed to calculate the floating VAWT, including the FloVAWT code [3], the enhanced Hawc2 code [4], Simo-Riflex-DMS (AC) code [5,6], and a rigid-flexible coupling code [7]. The motion performances of the floating VAWT were investigated by these codes. Vita studied the feasibility of a 5 MW Φ-type wind turbine supported by a rotating spar type foundation by using the enhanced Hawc2 code [4]. Borg and Collu calculated the responses of vertical wind turbines supported by different floating foundations by using the FloVAWT code [8]. Cheng et al. implemented the AC method in the Simo-Reflex to conduct the fully coupled aero-hydro-servo-elastic simulation of the floating VAWT [6,9]. Cahay et al. put forward the concept of a three-blade H-type wind turbine installed on a semi-submersible floating foundation [10].

In the above research, the form of the floating foundations mainly includes Spar type, Semi-submersible type, and TLP type. Liu et al. studied the motion performance of a 5 MW VAWT supported by the truss Spar platform [11], which had shorter length than the existent spar type foundation [12–14] and can be used in the moderate water depth with good motion performance, and had lighter weight than the existing semi-submersible type foundation [15]. The above codes cannot be used to calculate the truss Spar type VAWT, as the nonlinear coupling between the heave and pitch of truss Spar foundation is important, similar to the deep see Spar platform [16,17], and this is not considered in the above methods.

Compared with Φ-type, H-type is easier to manufacture and has higher power output [18]. In this paper, we studied the motion performances of the H-type VAWT supported by a truss Spar type floating foundation. The aerodynamic loads were calculated considering the dynamic stall and motions of the floating foundation. The surge–heave–pitch coupling nonlinear motion equations of the floating VAWT were established, and the motions of the floating VAWT were assessed.

2. The Aerodynamic Loads of H-Type Floating VAWT

Based on the double multiple stream tube (DMST) theory, the aerodynamic loads acting on the blades were calculated considering dynamic stall, tip losses, aspect ratio, tower shadow effect and the influence of the surge–heave–pitch motions of the floater. The whole floating VAWT was dealt with as a rigid body.

The stream tubes were divided into upwind and downwind sectors, respectively. Coordinate system is shown in Figure 1, where the coordinate origin o is in the center of gravity of the floating VAWT, and ξ_1, ξ_3, and ξ_5 are the displacements of surge, heave and pitch of the floating VAWT, respectively. The positive directions of the surge and heave motions are along the positive directions of the x axis and z axis, respectively; the positive direction of the pitch motion is clockwise. The vector relations of the wind turbine are shown in Figure 2.

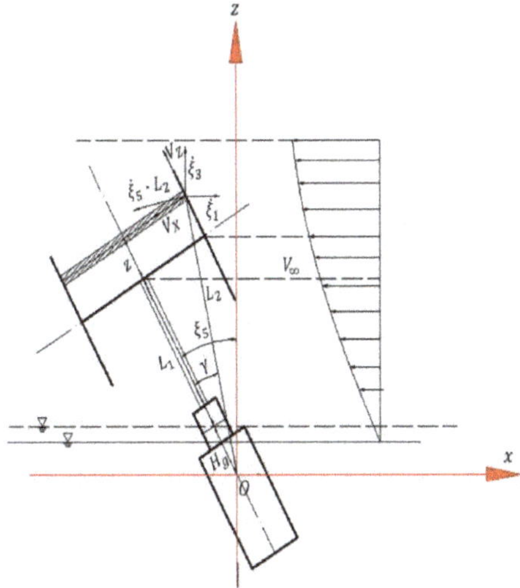

Figure 1. Coordinate system.

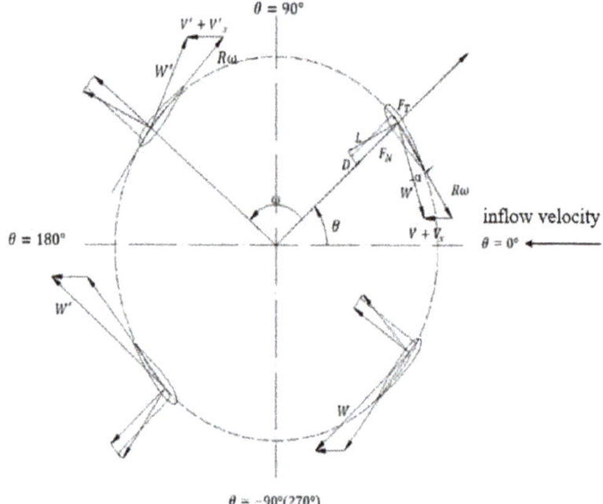

Figure 2. Equatorial velocity vector.

The steady wind was considered here, it was a function of the height above the water level, and for a local height z_i, the wind speed can be written as follow,

$$V_{\infty i}(z_i) = V_\infty (z_i/z_{ref})^a,\tag{1}$$

where V_∞ is the average wind velocity at a reference height z_{ref}, the vertical distance from the center of the blades to the water level is taken as the reference height in this paper, and with power $a = 0.14$ [19].

Assuming the velocity parallel to the direction of the steam tube affects the induced velocities and the velocity perpendicular to the direction of the steam tube only affects the local relative velocities, the coordinate and the velocity of each stream tube were deduced considering the motions of the floating foundation in the following parts.

Induced velocities at the upwind and downwind sectors V and V', and the induced equilibrium velocity V_e are given by [20],

$$V = u V_{\infty i} \cos \xi_5,\tag{2}$$

$$V_e = (2u - 1) V_{\infty i} \cos \xi_5,\tag{3}$$

$$V' = u' V_e = u'(2u - 1) V_{\infty i} \cos \xi_5.\tag{4}$$

where u is the interference factor, and u' is the second interference factor.

For the upwind sector, the distance from local blades to the gravity center of floating wind turbine system is expressed as,

$$L_2 = \sqrt{(R \cos \theta)^2 + (z + L_1 + H_g)^2},\tag{5}$$

where R is the radius of rotor, θ is the azimuth of rotor, z is the distance from each stream tube to equator (below equator is negative direction), L_1 is the distance from equator to still water level, and H_g is the distance from still water level to the gravity center of floating wind system when the system is in a state of stillness. The angle between the local position of the blade and the tower is written as

$$\gamma = \arctan\left(\frac{R \cos \theta}{z + L_1 + H_g}\right),\tag{6}$$

Then, the local wind velocity of the blade is given by,

$$V_{\infty i} = V_\infty \cdot \left(\frac{L_2 \cos(\xi_5 - \gamma) - (H_g - \xi_3)}{H + L_1}\right)^a,\tag{7}$$

where H is the half-height of rotor. The velocities of local blades can be derived from the surge, heave and pitch velocities of floater by using transformation matrix. Then, the resultant velocities, which are caused by the motion of the floating foundation, parallel to the direction of the steam tube for the local position can be written as,

$$V_x = -\dot{\zeta}_1 \cos \zeta_5 + \dot{\zeta}_3 \sin \zeta_5 - \dot{\zeta}_5 \cdot L_2 \cdot \cos \gamma. \tag{8}$$

Using a similar deducing progress, we get the resultant velocities parallel to and vertical to the direction of the steam tube for the downwind sector, as follows,

$$V_x' = -\dot{\zeta}_1 \cos \zeta_5 + \dot{\zeta}_3 \sin \zeta_5 - \dot{\zeta}_5 \cdot L_2 \cdot \cos \gamma. \tag{9}$$

For the rotor of upwind sector, namely, the range of azimuth θ is $[-\pi/2, \pi/2]$, and the local relative velocity of the rotor can be written as,

$$W = \sqrt{(V + V_x)^2 \cos \theta^2 + (R\omega - V \sin \theta)^2}, \tag{10}$$

where ω is the rotation velocity of the blade.

The local attack of angle is derived as,

$$\alpha = \arcsin\left[\frac{(V + V_x) \cos \theta}{W}\right]. \tag{11}$$

The azimuthal angle of upwind sector of the rotor is divided into several angular tubes, and the angle of each tube is $\Delta\theta$. Assuming the induced velocity for each angular tube is a constant, the interference factor of each angular tube is a constant, it can be presented as,

$$u(\theta) = \frac{KK_0}{KK_0 + \int_{\theta-\Delta\theta/2}^{\theta+\Delta\theta/2} f_{up}(\theta) d\theta}, \tag{12}$$

where K, K_0 and $f(\theta)$ can be given by following expressions,

$$K = 8\pi R/(nc), \tag{13}$$

$$K_0 = \sin(\theta + \Delta\theta/2) - \sin(\theta - \Delta\theta/2), \tag{14}$$

$$f_{up}(\theta) = (W/V)^2 [C_N \cos \theta + C_T \sin \theta], \tag{15}$$

where n is the number of blade, c is the chord length, and C_N and C_T are the normal force and tangential force coefficients, respectively, they can be given by,

$$C_N = C_L \cos \alpha + C_D \sin \alpha, \tag{16}$$

$$C_T = C_L \sin \alpha - C_D \cos \alpha, \tag{17}$$

where C_L and C_D are the lift coefficient and the drag coefficient, respectively. 2D static coefficient C_L^{2D} and C_D^{2D} can be obtained by interpolation of the experimental data [21] considering the local Reynolds number and the local attack angle α, and 3D static coefficient C_L^{3D} and C_D^{3D} can be derived by taking into account the influence of tip losses and finite aspect ratio [22]. Finally, the coefficients can be corrected by the Gormont–Berg method about the dynamic stall [23].

The local Reynolds number is presented as,

$$\mathrm{Re}_t = \frac{cW}{v_\infty}, \tag{18}$$

where $v_\infty = 1.47 \times 10^{-5} \, (m^2/s)$ is the coefficient of kinematic viscosity.

For the given the geometric lines of rotor, rotational speed and wind speed of each stream tube, the final induced factor $u(\theta)$ can be derived by iteration of Equations (2)–(18). Furthermore, the normal force and tangential force coefficients can be obtained.

The normal force F_N and tangential force F_T of the blade at different θ are further calculated by the following formulae,

$$F_N = 0.5\rho c H \int_{-1}^{1} C_N W^2 d\zeta, \tag{19}$$

$$F_T = 0.5\rho c H \int_{-1}^{1} C_T W^2 d\zeta, \tag{20}$$

where $\zeta = H_0/H$ is local height of the stream tube to equator plane of the rotor.

The thrust force F_X and side force F_Y of the blade are calculated as following,

$$F_X = F_N \cos\theta + F_T \sin\theta \tag{21}$$

$$F_Y = F_N \sin\theta - F_T \cos\theta \tag{22}$$

The power coefficient is presented as following,

$$C_{P1} = (R\omega/V_\infty) \times \frac{ncH}{2\pi S} \int_{-\frac{\pi}{2}}^{\frac{\pi}{2}} \int_{-1}^{1} C_T (W/V_\infty)^2 d\zeta d\theta. \tag{23}$$

Similarly, the normal force F_N' and the tangential force F_T' can be calculated for the downwind sector, and obtain the power coefficient C_{p2}. The power coefficient for one cycle can be written as,

$$C_P = C_{P1} + C_{P2}. \tag{24}$$

In this paper, the steady wind speed was used to calculate the aerodynamics loads. The lift and thrust forces acting on the heave and surge of the floating VAWT can be obtained by decomposing the normal and tangential forces. Calculating the moment of the thrust force about the center of gravity of the floating VAWT, the pitch moment of the aerodynamics acting of the pitch motions of the floating VAWT can be obtained.

Taking a fixed H-type wind turbine VAWT260 [24] as an example, the rotor power was calculated (the coupling of aerodynamics and hydrodynamics was not considered in this part). The VAWT260 H-rotor has two blades with chord length of 1.02 cm and airfoil of NACA0018. The lift and drag coefficients used in this paper are from static test, and iteration residual standard is set to 10^{-4}. The results of rotor power obtained by the numerical simulation were compared with those of model experiment [24], as shown in Figure 3.

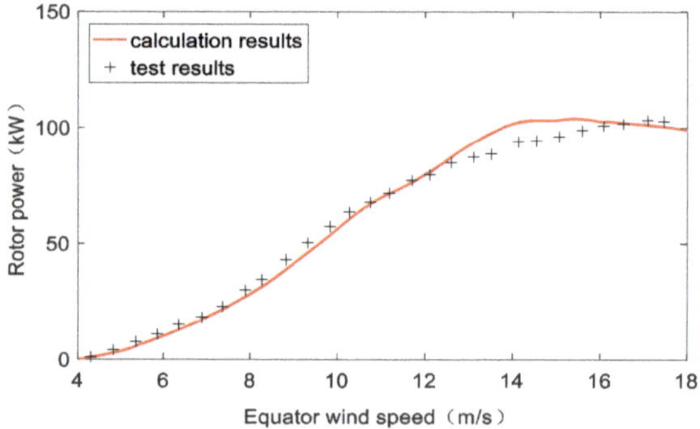

Figure 3. Validity checking of the aerodynamic loads.

It can be seen in Figure 3 that the rotor power obtained by the numerical simulation agrees well with the results of the experiment.

To verify the aerodynamic loads, comparisons of tangential and normal force coefficients with the calculation results of Paraschivoiu [25] are presented in Figure 4. A NACA0015 blade profile, a chord of 0.2 m and 6 m high H-type blade is used here. The rotational speed is 125 rpm, and the tip speed ratio is 5.2. The double multiple stream tube method with considering finite aspect ratio is used in [25].

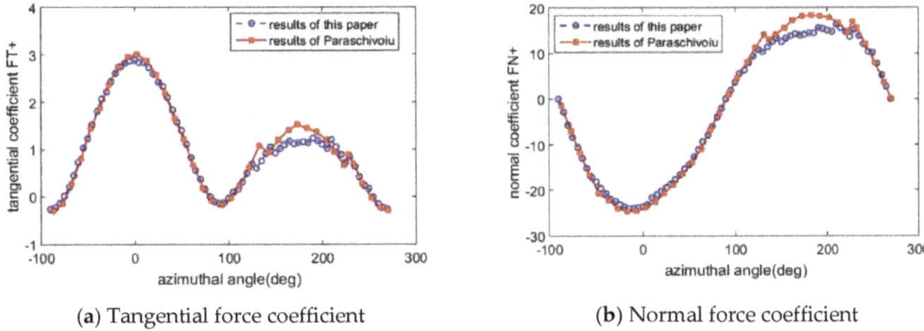

(a) Tangential force coefficient   (b) Normal force coefficient

Figure 4. Validity checking of the aerodynamic loads.

Figure 4 shows that the aerodynamic load coefficients mainly agree well with the results of Paraschivoiu. The reason for the relatively obvious differences appearing downwind is that the tower effect is ignored in Paraschivoiu's calculation. The thrust force and side force used in the dynamical equation in Section 3 can be checked by Equations (21) and (22).

3. The Dynamical Equations of the Floating VAWT

3.1. The Nonlinear Coupled Motion Equations

Liu et al. [16] established the nonlinear coupled motion equations of the Spar platform, and verified the equations by comparing with test modal. In this paper, the method is used to analyze the dynamic motion of the floating VAWT. Regarding the whole floating VAWT as a rigid body, considering the inertia force (moment), damping force (moment), restoring force (moment), mooring force, wind and wave loads, the surge–heave–pitch coupled equations of the floating VAWT can be presented as follows,

$$(M + A_\infty)\ddot{X}(t) + \int_0^t h(t - \tau)\dot{X}(t)d\tau + K(X(t))X(t) = F(t, X, \dot{X}), \tag{25}$$

where M is the mass matrix of the floating VAWT, A_∞ is the mass matrix at infinite frequency, and $h(t - \tau)$ is the retardation function accounting for the fluid memory effect X, $\dot{X}$ and $\ddot{X}$ represents the displacement, velocity and acceleration vectors of the floating VAWT at its center of gravity, respectively, and $X = [\xi_1 \, \xi_3 \, \xi_5]^T$. K is the restoring matrix, including the hydrostatic restoring, restoring of the mooring system, and the nonlinear hydrostatic restoring. $F(t, X, \dot{X})$ is the exciting force vector, which includes the Froud–Krylov force $F_{fk}(t)$, diffraction force $F_d(t)$, wind loads $F_w(X, \dot{X}, t)$ (including the aerodynamic loads $F_w{}^a(X, \dot{X}, t)$ and the wind loads $F_w{}^p$ calculated by the wind pressure theory), and hydrodynamic viscous force $F_{v1} = D_1\dot{X} + D_2\dot{X}|\dot{X}|$, where D_1 is the matrix of viscous damping coefficient,

$$F(t, X, \dot{X}) = F_{fk}(t) + F_d(t) + F_w(X, \dot{X}, t) + F_{v1}, \tag{26}$$

For the Spar type floating foundation, the nonlinear coupling between the heave and pitch motions is significant. Referring to the nonlinear heave–pitch coupling motion equations of the deepsea truss Spar platform [16,17], the restoring matrix can be written as follows,

$$K = \begin{bmatrix} f_1(\xi_1) & 0 & 0 \\ 0 & f_3(\xi_3) + \rho g A_w & -\frac{H_g}{2}\rho g A_w \xi_5 \\ 0 & \frac{1}{2}\rho g (\nabla + 2A_w\overline{GM})\xi_3 & \rho \nabla g \overline{GM} \end{bmatrix}, \tag{27}$$

where $f_i(\xi_i)$ $(i = 1, 3)$ represents restoring stiffness of the mooring system in surge and heave of the floating VAWT. Aw is water plane area of the floating foundation, $\overline{GM}$ is the initial stability height of the floating VAWT, ∇ is the displacement, ρ is the density of seawater, and g is gravity acceleration. The hydrostatic restoring loads are calculated on the static equilibrium position of the floating VAWT.

3.1. Wind Pressure

The wind pressure and wind heeling moment acting on the tower and the blades (in the parked state) were calculated in accordance with the rules of the API [26], as follows,

$$F = 0.613 \times Cs \times Ch \times \overline{S} \times V_\infty{}^2, \tag{28}$$

$$M = F \times \overline{H}, \tag{29}$$

where Cs is the shape coefficient of the wind-bearing component; Ch is the height coefficient of the wind-bearing component; $\overline{S}$ is the frontal projected area of the wind-bearing component in the wind direction; $\overline{H}$ is the distance from the wind force acting points to the rotation center of the floating VAWT; and V_∞ is shown as Equation (1).

3.2. Hydrodynamic Loads

Hydrodynamic loads were calculated by using the potential theory and the Morison's equation [27]. The potential theory was used to calculate the wave loads, added mass and radiation damping of the large scale structure (including the upper buoyancy tank, upper mechanical tank, heaving plates, and bottom ballast tank). The first-order wave loads and second-order mean drift forces were considered in this paper.

The Morison equation was used to calculate the inertial load and viscous drag load of the slender structure (truss structures, $D/\lambda < 0.2$, where D is the diameter of the structure and λ is the wave length). The transverse hydrodynamic force per unit length can be written as follows,

$$dF = \rho\pi\frac{D^2}{4}\dot{u}_w + \rho\pi C_a\frac{D^2}{4}(\dot{u}_w - \dot{u}_b) + \frac{1}{2}\rho C_d D(u_w - u_b)|u_w - u_b|, \tag{30}$$

where C_a and C_d are the added mass and quadratic drag coefficients, respectively; u_w is the transverse wave particle velocity; and u_b is the local transverse body velocity.

The Wadam [28] was used to calculate the hydrodynamics in this paper. The wind turbine was in the parked state and the wind loads and mooring system were not considered in the hydrodynamic calculation, while the mass and shape of the blade were considered.

3.3. Damping and Mooring System

The heave plates were used to reduce the heave motions of the floating VAWT, and the viscous damping of floating foundation was important. In this paper, the viscous damping was obtained by the experiment of free decay model [29].

The mooring forces acting on the floating foundation in the directions of surge and heave were calculated by quasi-static catenary method, considering catenary mooring line and the fairlead set near the gravity center of the floating VAWT.

3.4. The Motions Calculation of the Floating VAWT

The motions calculation of the floating VAWT was carried out using the method of Runge–Kutta, and the computational procedures were as follows: give initial displacements and velocities of the floating VAWT to calculate the aerodynamic loads of the blades according to Equations (2)–(22); solve Equation (23) considering the wind and wave loads, and mooring loads; and calculate the new aerodynamic loads according to the new displacements and velocities of the floating VAWT. The coupling between aerodynamics and hydrodynamics was obtained by repeating the process above.

4. Results and Analysis

4.1. The Parameters of a 5 MW Floating VAWT

The H-type floating VAWT studied as an example is shown Figure 5. The upper wind turbine refers to the VAWT proposed by Collu et al. [15], and the floating foundation is Truss Spar type, which consists of buoyancy tank, the upper machinery tank, truss structure, heave plates and ballast tank. The specific parameters of the floating VAWT are shown in Table 1. The floating VAWT was located by four groups (one line for each group), the angle between each group of mooring lines was 90°. The lines were designed in the form of chain–cable–chain, and were connected with the floating foundation by the fairleads. The specific parameters of the mooring line are indicated in Table 2.

Table 1. Parameters of the floating VAWT. VAWT: Vertical Axis Wind Turbine.

Items	Values
Rated power	5.00 (MW)
Number of blade	2.00
Radius of rotor	37.00 (m)
Height of blade	78.75 (m)
Chord of blade	3.50 (m)
Hub height	72.00 (m)
Airfoil profile	NACA0018
Rated wind speed	15.00 (m/s)
Rated rotor speed	12.00 (rpm)
Mass of blades	76.75 (ton)
Draft of the foundation	65.00 (m)
Total water displacement	7475.10 (t)

Table 2. Parameters of the mooring line. EA: XXX.

Items	Chain	Cable
Length (m)	55	135
Diameter (mm)	147	84
Wet weight (Kg/m)	375.7	25.6
Dry weight (Kg/m)	432.2	29.5
Axial rigidity EA (N)	1.6×10^9	3.9×10^9

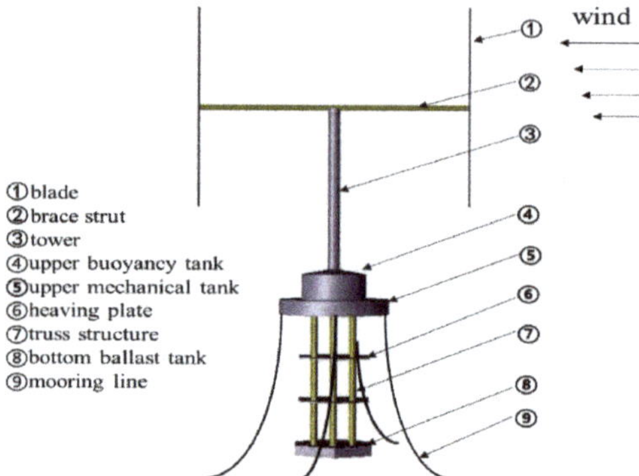

Figure 5. The H-Type floating VAWT system. VAWT: Vertical Axis Wind Turbine.

4.2. Control Strategy of Rotational Speed

The control strategy of rotor speed of the floating VAWT is as follows: for wind speeds less than 12 m/s, the rotational speed is designed to obtain the maximum power capture; the rotational speed is set to the rated rotational speed when wind speed is in the range of 12–15 m/s; and for wind speeds more than the rated wind speed, the rotational speed should be set to maintain the rated output power. Output power for different rotor speed and wind speed are shown in Figure 6, and rotor speed and output power for the different wind speed are shown in Figure 7.

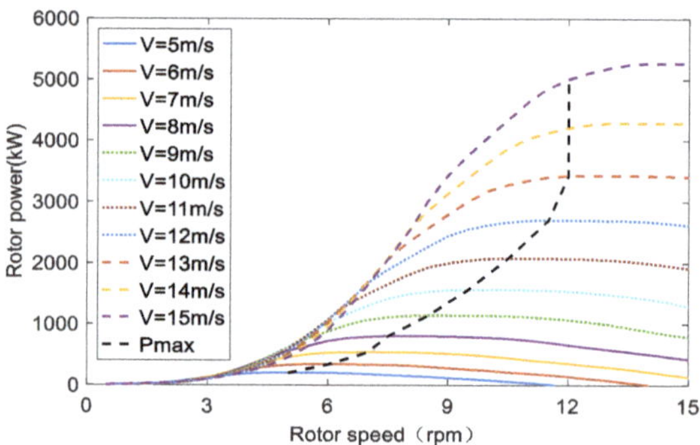

Figure 6. Power for different rotational speed and wind speed.

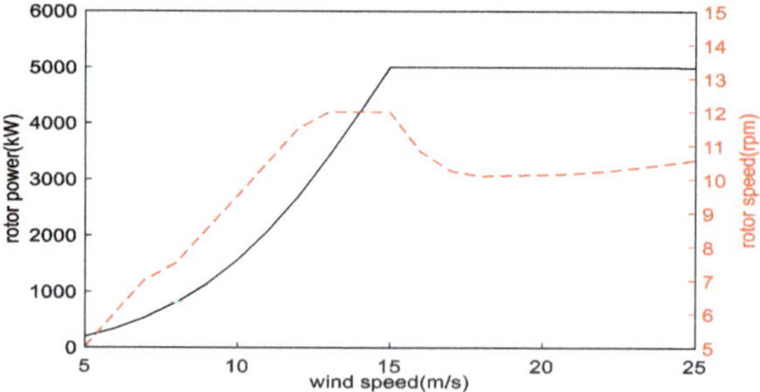

Figure 7. Power and rotational speed for different wind speed.

4.3. Results and Analysis

4.3.1. Free Decay Motions of the Floating VAWT

The free decay motions were carried out in this section. The wind turbine was in the parked state and wind and wave loads were not considered. The natural periods of surge, heave and pitch motions of the system are obtained as shown in Table 3.

Table 3. Natural periods of the floating VAWT.

DOFs	Natural Period (s)
surge	51.2
heave	20.5
pitch	29.3

4.3.2. Motions of the Floating VAWT Considering the Wind and Regular Wave Loads

Motions of the floating VAWT were calculated considering the wind loads and regular wave loads in this section. The parameters of wave and wind are shown in Table 4.

Table 4. Regular wave and wind cases (LC3).

Items	Wave Height (m)	Wave Period (s)	V_∞ (m/s)	Total Time (s)
LC3.1	2.10	9.74	5	3600
LC3.2	3.62	10.29	15	3600
LC3.3	6.02	11.38	25	3600

The time histories and the corresponding spectra (obtained by the Fast Fourier Transform) of the floating VAWT motions are shown in Figures 8–10, and the mean displacements and the corresponding standard deviations (STD) are shown in Table 5.

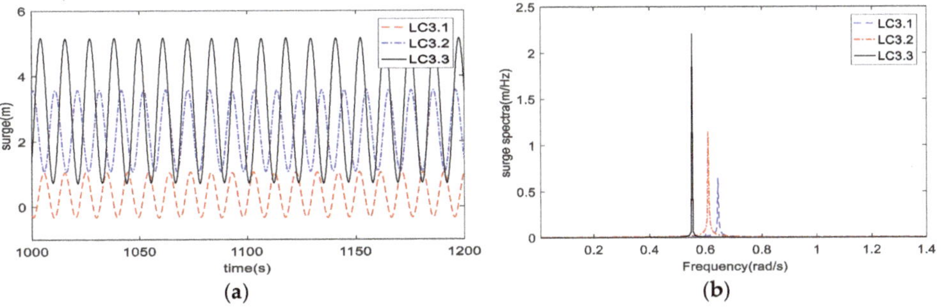

Figure 8. Surge responses of the floating VAWT (LC3): (**a**) time histories; and (**b**) response spectra.

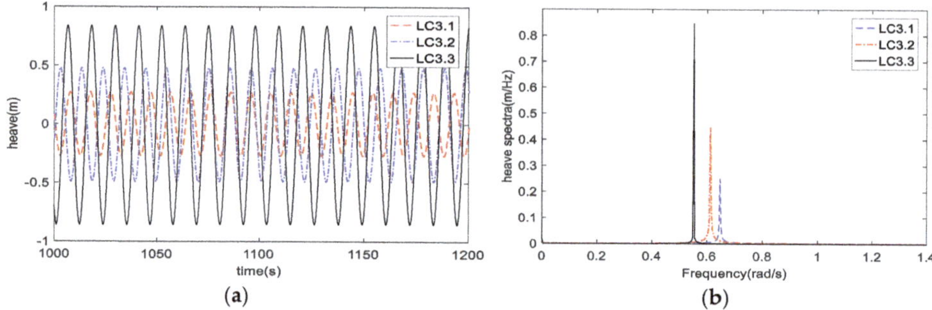

Figure 9. Heave responses of the floating VAWT (LC3): (**a**) time histories; and (**b**) response spectra.

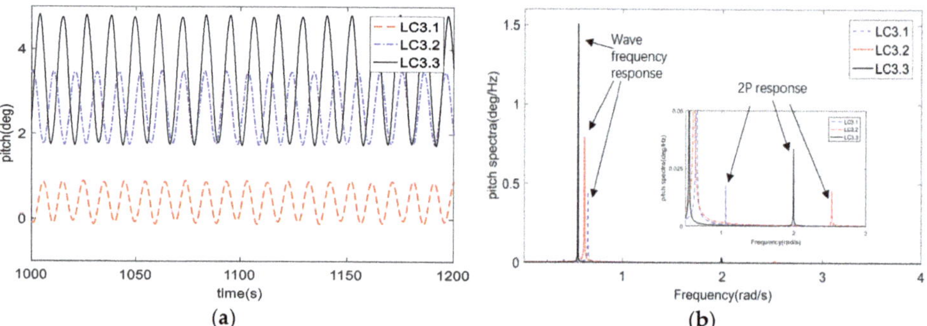

Figure 10. Pitch responses of the floating VAWT (LC3): (**a**) time histories; and (**b**) response spectra.

Table 5. The standard deviations and mean values (LC3). STD: standard deviations.

Case	STD			Mean Value		
	Surge (m)	Heave (m)	Pitch (deg)	Surge (m)	Heave (m)	Pitch (deg)
LC3.1	0.50	0.20	0.84	0.36	−0.00	0.40
LC3.2	0.89	0.35	1.49	2.32	0.006	2.62
LC3.3	1.55	0.60	2.64	2.95	0.011	3.27

Table 5 shows that, for the three cases, the max STDs of the surge, heave and pitch motions of the floating VAWT are 1.55 m and 0.60 m and 2.64 degrees, respectively; the mean values of heave are all small, and the max mean values of the surge and the pitch motions of the three cases are 2.95 m and 3.27 degrees, respectively. Therefore, the floating VAWT has good motion performance. Compared with LC3.2, the pitch STD of the floating VAWT in LC3.3 increases by approximately 77%, and the corresponding mean value increases by approximately 25%.

In Figures 8–10, we can find that the motion frequencies of the floating VAWT are dominated by the wave frequencies for the three cases. For the pitch frequencies of the floating VAWT, there are components of the rotor rotational frequency (2P response). However, they are not significant compared with the wave–frequency responses.

4.3.3. The Aerodynamics of the Floating VAWT

To analyze the effect of floating foundation's motion on the aerodynamics, two conditions are considered for LC3.2: (a) the foundation is floating, and the couple motions of floating foundation are considered in aerodynamics calculation; and (b) the wind turbine is considered to be set on a fixed foundation. The normal and tangential forces acting on the equator of one blade are calculated, and the results of time histories and spectra are shown in Figures 11 and 12.

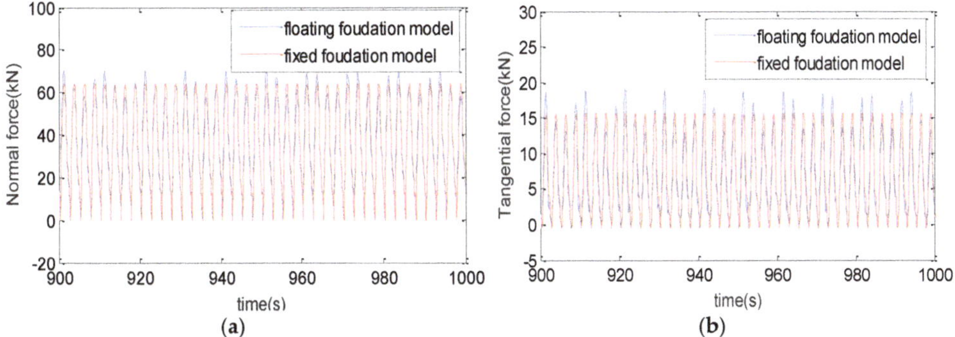

Figure 11. Time histories of aerodynamic force on the blade: (**a**) Normal force; and (**b**) Tangential force.

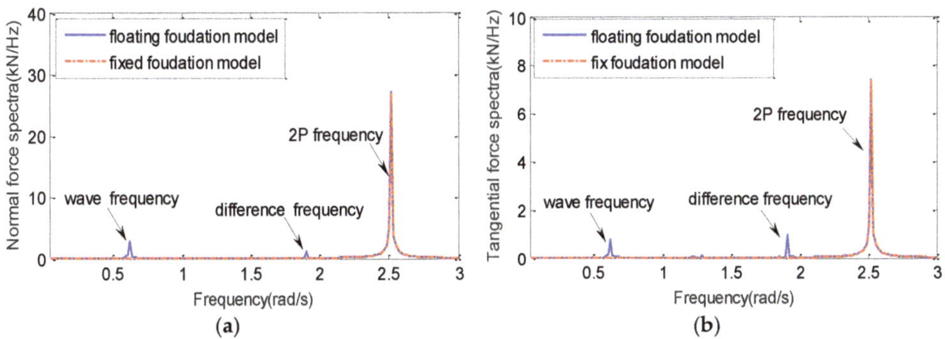

Figure 12. Spectra of blade aerodynamics: (**a**) Normal force spectra; and (**b**) Tangential force spectra.

Figures 11 and 12 show that the normal force is much bigger than tangential force. Compared with the fixed foundation case, the amplitudes of normal force and tangential force of the floating foundation cases increase up to 9.4% and 16.3%, respectively. As shown in Figure 12, the main frequency of aerodynamic force is 2P frequency. In addition, for the floating foundation model, there are frequency components of wave frequency and the difference frequency between 2P and wave frequency.

4.3.4. Motions of the Floating VAWT Considering the Wind and Irregular Wave Loads

Motions of the floating VAWT were calculated considering the wind loads and irregular wave (JONSWAP sea spectrum) loads in this section. The calculation parameters are shown in Table 6.

Table 6. Irregular wave and wind cases (LC4).

Items	Significant Wave Height (m)	Spectrum Peak Period (s)	V_∞(m/s)	Total Time (s)
LC4.1	2.1	9.74	5	3600
LC4.2	3.62	10.29	15	3600
LC4.3	6.02	11.38	25	3600

The time histories and spectra of pitch motions of LC4.2 and LC4.3 are presented in Figure 13. The max values, mean values and STDs of surge, heave and pitch responses of the floating VAWT are calculated, and the results are shown in Table 7.

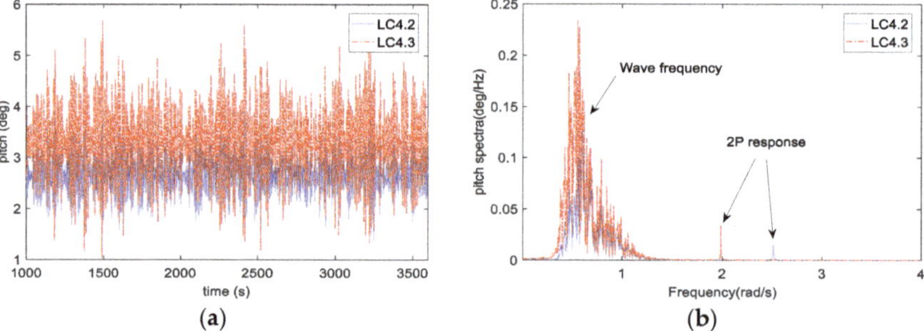

Figure 13. Pitch responses of the floating VAWT: (**a**) time histories; and (**b**) response spectra.

Table 7. Response statistics of the floating VAWT (cases LC4).

Items	Surge (m)			Heave(m)			Pitch(deg)		
	STD	Mean	Max	STD	Mean	Max	STD	Mean	Max
LC4.1	0.36	0.32	1.46	0.27	0.17	0.52	0.19	0.42	1.14
LC4.2	0.58	2.33	4.28	0.34	0.17	0.72	0.39	2.61	3.97
LC4.3	1.04	2.95	6.45	0.49	0.17	1.15	0.72	3.28	5.67

Figure 12 shows that the pitch motions of the floating VAWT increase with the increase of wave height and wind speed, and they are still dominated by the wave–frequency motions. Comparing Figure 12b with Figure 9b, the 2P frequency caused by the aerodynamic loads is more significant in the case LC4. Because of the effect of heave plates, the heave responses are small and the max values of heave is 1.15 m for LC4. In Table 7, we can find that the STDs of surge and pitch motions of the floating VAWT are small, which shows the floating VAWT has good motion performance in the wave and wind conditions.

4.3.5. Comparison with Φ-Type Floating VAWT

Liu et al. studied a 5 MW Φ-Darrieus type wind turbine supported by the same truss Spar type floating foundation and with the same mooring system, and its parameters are presented in Reference [11]. Compared with the Φ-type wind turbine (blade height is 129.56 cm, mass of two blades is 308 tons), the blade height and blades mass of this paper decreased by 39% and 75%, respectively. The two floating VAWTs have the same total mass.

Motions of the two floating VAWTs were compared considering the wind loads and regular wave loads, and the calculation parameters are shown in Table 8. The mean value and STDs of the surge, heave, and pitch motions of the two floating VAWTs are presented in Figures 14–16. The mean generator powers of the two floating VAWTs are shown in Figure 17.

Table 8. Comparison cases (LC5).

Case	Wind Turbine Type	Wave Height (m)	Wave Period (s)	V_∞ (m/s)
LC5.1	H-type Φ-type	2.10	9.74	5 5
LC5.2	H-type Φ-type	3.62	10.29	15 (rated wind speed) 14 (rated wind speed)
LC5.3	H-type Φ-type	6.02	11.38	25 25

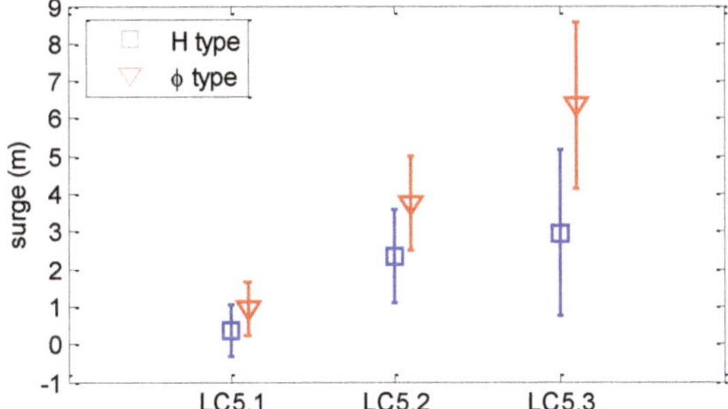

Figure 14. Surge comparison of two floating VAWTs (cases LC5).

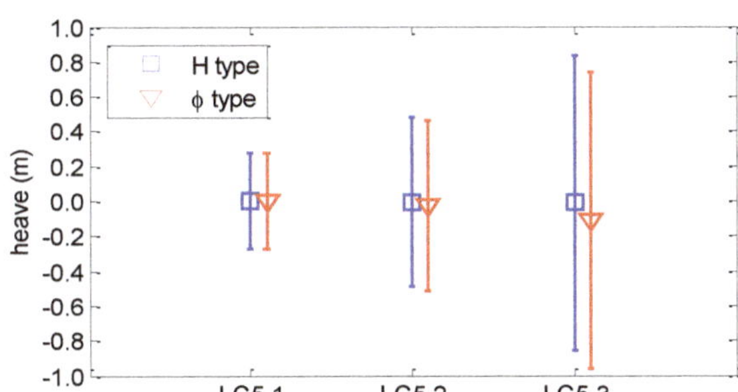

Figure 15. Heave comparison of two floating VAWTs (cases LC5).

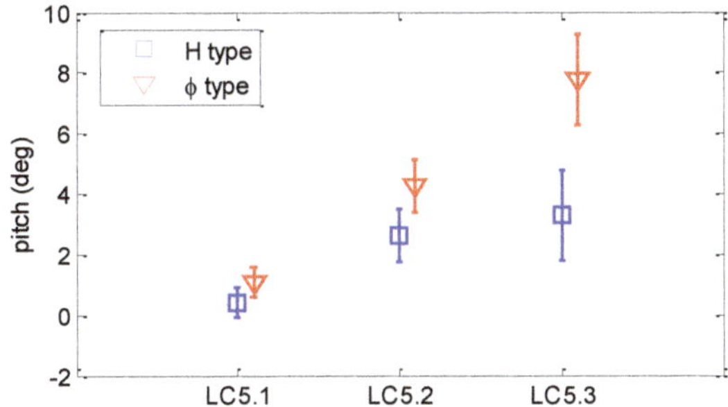

Figure 16. Pitch comparison of two floating VAWTs (cases LC5).

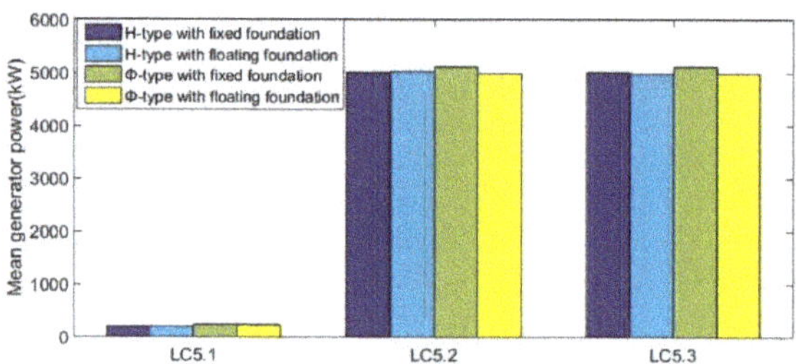

Figure 17. Mean generator power of two floating VAWTs (cases LC5).

Figures 14–16 show that compared with the Φ-type floating VAWT, the mean values of the H-type floating VAWT's motions are much smaller. Taking the rated wind speed case (LC5.2) as an example, the mean values of surge, heave and pitch of H-type floating VAWT decrease 38%, 74.2% and 38.6%, respectively. Therefore, H-type VAWT withstand smaller wind loads compared with the Φ-type VAWT. The STDs of motions of the two floating VAWTs are almost same, as they are supported by the same floating foundation, and the STD of the floating VAWT motions mainly depends on the wave loads. Figure 17 shows that the motion of floating VAWTs has insignificant effect on the mean generator power for both type wind turbines. It is a satisfactory result for the prospective development of a floating VAWT.

5. Conclusions

The surge–heave–pitch coupling nonlinear motion equations of the H-type floating VAWT were established considering the coupling between aerodynamic and hydrodynamic loads, and the motions of a 5 MW H-type floating VAWT were studied. The motion of the H-type floating VAWT was also compared with that of the Φ-type floating VAWT; they have the same floating foundation and total displacement. The main conclusions are as follows:

1. The natural periods of surge, heave, and pitch motions of the H-type floating VAWT are 51.2 s, 20.5 s, and 29.3 s, respectively. The small STDs of the LC3 and LC4 show the good motion performance of the H-type floating VAWT in the wave and wind conditions.
2. The effects of wind and wave loads on the motions of the H-type floating VAWT are assessed. For LC3, the motion frequencies of the floating VAWT are dominated by the wave frequencies, and there are small components of 2P frequency for the pitch motions. The 2P frequency is more significant in irregular wave conditions (LC4).
3. The effects of the floating foundation motions on the aerodynamics are also assessed. The results show that the normal force is much bigger than the tangential force. Compared with the fixed foundation case, the amplitudes of normal and tangential forces increase up to 9.4% and 16.3%, respectively, for the floating foundation case. Besides the 2P frequency, the frequencies of aerodynamics also include the wave frequency and the difference frequency between 2P and wave frequency for the floating foundation case. The effect of the floating foundation motion on the aerodynamics cannot be ignored.
4. The H-type VAWT withstand smaller wind loads compared with the Φ-type VAWT, and mean values of the H-type floating VAWT's motions are much smaller compared with the Φ-type floating VAWT. The STDs of motions of the two floating VAWTs are almost same, as they mainly depend on the wave loads.

Acknowledgments: The project was supported by the Natural Science Foundation of China (No. 51579176), the Natural Science Foundation of Tianjin (No. 16JCYBJC21200), and the Fund of the State Key Laboratory of Ocean Engineering, Shanghai Jiao Tong University (No. 1501).

Author Contributions: Ying Guo and Liqin Liu deduced the formulae and carried out numerical calculation; Xifeng Gao analyzed the data; and Wanhai Xu wrote the paper.

Conflicts of Interest: The authors declare no conflict of interest.

References

1. Paraschivoiu, I. *Wind Turbine Design with Emphasis on Darrieus Concept*; Polytechnic International Press: Montreal, QC, Canada, 2002.
2. Tjiu, W.; Marnoto, T.; Mat, S.; Ruslan, M.H.; Sopian, K. Darrieus vertical axis wind turbine for power generation II: Challenges in HAWT and the opportunity of multi-megawatt Darrieus VAWT development. *Renew. Energy* **2015**, *75*, 560–571. [CrossRef]
3. Collu, M.; Borg, M.; Shires, A.; Brennan, F.P. Progress on the development of a coupled model of dynamics for floating offshore vertical axis wind turbines. In Proceedings of the ASME 32nd International Conference on Ocean, Offshore and Arctic Engineering, Nantes, France, 9–14 June 2013. OMAE2013-54337.
4. Vita, L. Offshore Floating Vertical Axis Wind Turbines with Rotating Platform. Ph.D. Thesis, Danish Technical University, Copenhagen, Denmark, 2011.
5. Wang, K.; Hansen, M.O.L.; Moan, T. Dynamic analysis of a floating vertical axis wind turbine under emergency shutdown using hydrodynamic brake. *Energy Procedia* **2014**, *53*, 53–69. [CrossRef]
6. Cheng, Z.S.; Madsen, H.A.; Gao, Z.; Moan, T. A fully coupled method for numerical modeling and dynamic analysis of floating vertical axis wind turbines. *Renew. Energy* **2017**, *107*, 604–619. [CrossRef]
7. Owens, B.C.; Hurtado, J.E.; Paquette, J.A. Aero-elastic modelling of large offshore vertical-axis wind turbines: development of the offshore wind energy simulation toolkit. In Proceedings of the 54th AIAA/ASME/ASCE/A-HS/ASC Structures, Structural Dynamics, and Materials Conference, Boston, MA, USA, 8–11 April 2013.
8. Borg, M.; Collu, M. A comparison on the dynamics of a floating vertical axis wind turbine on three different floating support structures. *Energy Procedia* **2014**, *53*, 268–279. [CrossRef]
9. Cheng, Z.S.; Madesn, H.A.; Gao, Z.; Moan, T. Numerical study on aerodynamic damping of floating vertical axis wind turbines. *J. Phys. Conf. Ser.* **2016**, *753*, 102001. [CrossRef]
10. Cahay, M.; Luquiau, E.; Smadja, C.; Silvert, F. Use of a vertical wind turbine in an offshore floating wind farm. *J. Pet. Technol.* **2011**, *63*, 105–107.
11. Liu, L.Q.; Guo, Y.; Jin, W.C.; Yuan, R. Motion performances of a 5 MW VAWT supported by spar floating foundation with heave plates. In Proceedings of the ASME 36th International Conference on Ocean, Offshore and Arctic Engineering, Trondheim, Norway, 25–30 June 2017. OMAE2017-62625.
12. Paulsen, U.S.; Borg, M.; Madsen, H.A. Outcomes of the DeepWind conceptual design. *Energy Procedia* **2015**, *80*, 329–341. [CrossRef]
13. Bedon, G.; Paulsen, U.S.; Madsen, H.A.; Belloni, F.; Castelli, M.; Benini, E. Aerodynamic Benchmarking of the Deepwind Design. *Energy Procedia* **2015**, *75*, 677–682. [CrossRef]
14. Battisti, L.; Benini, E.; Brighenti, A.; Castelli, M.R.; Dell, A.S. Wind tunnel testing of the DeepWind demonstrator in design and tilted operating conditions. *Energy Procedia* **2016**, *111*, 484–497. [CrossRef]
15. Collu, M.; Borg, M.; Manuel, L. On the relative importance of loads acting on a floating vertical axis wind turbine system when evaluating the global system response. In Proceedings of the ASME 35th International Conference on Ocean, Offshore and Arctic Engineering, Busan, Korea, 19–24 June 2016. OMAE2016-54628.
16. Liu, L.Q.; Zhou, B.; Tang, Y.G. Study on the nonlinear dynamical behavior of deep sea spar platform by numerical simulation and model experiment. *J. Vib. Control* **2014**, *20*, 1528–1553. [CrossRef]
17. Keyvan, S.; Atilla, I.; Martin, J.D. Response analysis of a truss spar in the frequency domain. *J. Mar. Sci. Technol.* **2004**, *8*, 126–137.
18. Sandra, E. Evaluation of different turbine concepts for wind power. *Renew. Sustain. Energy Rev.* **2008**, *12*, 1419–1434.
19. International Electrochemical Commission. *Wind Turbines, Part 3: Design Requirements for Off-Shore Wind Turbines*; IEC61400-3; International Electrochemical Commission: Geneva, Switzerland, 2009.
20. Paraschivoiu, I. Aerodynamic loads and performance of the Darrieus rotor. *J. Energy* **1982**, *6*, 406–412. [CrossRef]

21. Sheldahl, R.E.; Klimas, P.C. *Aerodynamic Characteristics of Seven Symmetrical Airfoil Sections through 180-Degree Angle of Attack for Use in Aerodynamic Analysis of Vertical Axis Wind Turbines*; Sandia National Laboratories: Albuquerque, NM, USA, 1981.

22. Andrew, S. Development and Evaluation of an Aerodynamic Model for a Noval Vertical Axis Wind Turbine Concept. *Energies* **2013**, *6*, 2501–2520.

23. Berg, D.E. An improved double-multiple stream tube model for the Darrieus type vertical-axis wind turbine. In Proceedings of the Sixth Biennial Wind Energy Conference and Workshop, Minneapolis, NM, USA, 1–3 June 1983; pp. 231–233.

24. Morgan, C.A.; Gardner, P.; May, I.D.; Anderson, M.B. The Demonstration of a Stall Regulated 100 Kw Vertical Axis Wind Turbine. In Proceedings of the European Wind Energy Conference, Glasgow, UK, 10–13 July 1989.

25. Paraschivoiu, I.; Desy, P.; Massont, C. Blade Tip, Finite Aspect Ratio, and Dynamic Stall Effects on the Darrieus Rotor. *J. Propuls. Power* **1988**, *4*, 73–80. [CrossRef]

26. American Petroleum Institute. *Design and Analysis of Station keeping Systems for Floating Structures*; API RP-2SK; American Petroleum Institute: Washington, DC, USA, 2005.

27. Faltinsen, O.M. *Sea Loads on Ships and Offshore Structures*; Cambridge University Press: Cambridge, UK, 1990.

28. Pan, Z.Y.; Vada, T.; Finne, S. Benchmark study of numerical approaches for wave-current interaction problem of offshore floaters. In Proceedings of the ASME 35th International Conference on Ocean, Offshore and Arctic Engineering, Busan, Korea, 19–24 June 2016. OMAE2016-54411.

29. Liu, L.Q.; Guo, Y.; Zhao, H.X.; Tang, Y.G. Dynamic modeling, simulation and model tests research on the floating VAWT. *Chin. J. Theor. Appl. Mech.* **2017**, *49*, 299–307. (In Chinese)

MDPI

St. Alban-Anlage 66

4052 Basel

Switzerland

Tel. +41 61 683 77 34

Fax +41 61 302 89 18

www.mdpi.com

Applied Sciences Editorial Office

E-mail: applsci@mdpi.com

www.mdpi.com/journal/applsci